Interactions between Soil Particles and Microorganisms

IUPAC SERIES ON ANALYTICAL AND PHYSICAL CHEMISTRY
OF ENVIRONMENTAL SYSTEMS

Series Editors

Jacques Buffle, *University of Geneva, Geneva, Switzerland*
Herman P. van Leeuwen, *Wageningen University, Wageningen,*
The Netherlands

Series published within the framework of the activities of the IUPAC Commission on
Fundamental Environmental Chemistry, Division of Chemistry and the Environment.

INTERNATIONAL UNION OF PURE AND APPLIED CHEMISTRY (IUPAC)
Secretariat, PO Box 13757, 104 T. W. Alexander Drive, Building 19,
Research Triangle Park, NC 27709-3757, USA

Previously published volumes (Lewis Publishers):
Environmental Particles Vol. 1 (1992) ISBN 0-87371-589-6
Edited by Jacques Buffle and Herman P. van Leeuwen
Environmental Particles Vol. 2 (1993) ISBN 0-87371-895-X
Edited by Jacques Buffle and Herman P. van Leeuwen

Previously published volumes (John Wiley & Sons, Ltd):
Metal Speciation and Bioavailability in Aquatic Systems Vol. 3 (1995)
 ISBN 0-471-95830-1
Edited by André Tessier and David R. Turner
Structure and Surface Reactions of Soil Particles Vol. 4 (1998)
 ISBN 0-471-95936-7
Edited by P. M. Huang, N. Senesi and J. Buffle
Atmospheric Particles Vol. 5 (1998)
 ISBN 0-471-95935-9
Edited by Roy M. Harrison and René E. van Grieken
In Situ Monitoring of Aquatic Systems Vol. 6 (2000)
 ISBN 0-471-48979-4
Edited by Jacques Buffle and George Horvai
The Biogeochemistry of Iron in Seawater Vol. 7 (2001)
 ISBN 0-471-49068-7
Edited by David R. Turner and Keith A. Hunter

Contents

P. M. Huang received his Ph.D. degree in soil science at the University of Wisconsin, Madison, in 1966. He is currently Professor of Soil Science at the University of Saskatchewan, Saskatoon, Canada. His research work has significantly advanced the frontiers of knowledge on the nature and surface reactivity of mineral colloids and organomineral complexes of soils and sediments and their role in the dynamics, transformations, and fate of nutrients, toxic metals, and xenobiotics in terrestrial and aquatic environments. His research findings, embodied in over 270 refereed scientific publications, including research papers, book chapters, and 10 books, are fundamental to the development of sound strategies for managing land and water resources.

He has developed and taught courses in soil physical chemistry and mineralogy, soil analytical chemistry, and ecological toxicology. He has successfully trained and inspired M.Sc. and Ph.D. students and postdoctoral fellows, and received visiting scientists world-wide. He has served on numerous national and international scientific and academic committees. He also has served as a member of many editorial boards such as the *Soil Science Society of America Journal, Geoderma, Chemosphere,* and *Advances in Environmental Science.* He is currently a titular member of the Commission of Fundamental Environmental Chemistry of the International Union of Pure and Applied Chemistry and is the founding and current Chairman of the Working Group MO 'Interactions of Soil Minerals with Organic Components and Microorganisms' of the International Union of Soil Sciences. He received the distinguished Researcher Award from the University of Saskatchewan and the Soil Science Research Award from the Soil Science Society of America. He is a fellow of the Canadian Society of Soil Science, the Soil Science Society of America, the American Society of Agronomy, and the American Association for the Advancement of Science.

Jean-Marc Bollag is Professor of Soil Biochemistry and Director of the Center for Bioremediation and Detoxification, the Environmental Resources Research Institute at The Pennsylvania State University. His major research goals are area of bioremediation: to employ enzymes, microorganisms or plants for the transformation, degradation contamination of pollutants in aquatic and environments. Dr Bollag has advised many co-workers—graduate students, post-doctoral scholars and visiting scientists. He has published 220 research articles and book chapters, and presently he is editor of the book series *Soil Biochemistry.* has served as associate editor of several journals and he is still on the editorial boards of four international journals.

Dr Bollag is a recipient of the Julius Baer Fellowship, the Gamma Sigma Delta Research Award, and the Badge of Merit from the Polish Ministry of Agriculture. He is a Fellow of the American Academy of Microbiology, the Soil Science Society of America and the American Society of Agronomy, and he has received the Environmental Quality Research Award from the American Society of Agronomy. He is also a member of the American Society for Microbiology, the International Union of Soil Sciences, and the International Humic Substances Society. Dr Bollag received the Ph.D. degree from the University of Basel (Switzerland), and conducted postdoctoral work at the Weizmann Institute of Science, Rehovoth, Israel, and at Cornell University, Ithaca, New York. He was also a Visiting Scientist at Ciba-Geigy, Basel, Switzerland.

Nicola Senesi is Professor of Soil Chemistry at the Faculty of Agriculture, Department of Agroforestal and Environmental Biology and Chemistry at the University of Bari, Italy, where he has been actively involved in research and teaching since 1969. He has served as the President of the International Humic Substances Society and the Chairman of Commission II—Soil Chemistry of the International Society of Soil Science. He is currently titular member of the IUPAC Commission of Fundamental Environmental Chemistry and member of the European Environmental Research Organization. He spent a sabbatical year (1975–76) at the Soil Research Institute, Agriculture Canada, Ottawa, with the support of a NATO Science—CNR Fellowship and various sabbatical leaves (1982–92) for a total of one and half years, at the University of California, Department of Soil and Environmental Sciences in Riverside, and Department of Soil Science in Berkeley. He has been a visiting professor for various periods (1978–83) for a total of 2 years, at the Faculty of Agriculture of the University of Mogadishu, Somalia. He has also presented short courses at universities in Indonesia, Switzerland, Argentina, Venezuela and Colombia. He is currently an Associate Editor of *Geoderma, Soil Science, European Journal of Soil Science, Journal of Environmental Sciences and Health, Part B* and *Humic Substances in the Environment*. Dr Senesi earned a university degree in chemistry in 1966 and an academic degree in soil chemistry in 1970 at the University of Bari. He has been conferred a Doctorate Honoris causa by the Institut National Polytechnique de Toulouse (INPT) in 2000. He has authored about 190 technical papers and 40 book chapters and invited reviews, and co-edited five books. His research focuses on fundamental and applied aspects of the chemistry of organic matter from soils and other systems and materials, and its interactions with soil-applied organic chemicals and trace metals, by the use of advanced physicochemical techniques. He is particularly interested in chemical aspects related to abiotic interactions occurring between herbicides and humic substances, complexation mechanisms occurring between trace metals of environmental importance and natural and artificial humic materials, and soil fertility and environmental implications deriving from recycling of organic wastes to soils. He is pioneering the application of fractal geometry to the study of molecular conformation and aggregation processes of natural soil organic colloids.

List of Contributors

T. A. Anderson
TIEHH, Texas Tech. University, 1207 Gilbert Drive, Lubbock, TX 79416, USA

H. Awata
The Institute of Environmental and Human Health, Texas Tech. University, Lubbock, TX, USA

J. A. Baldock
CSIRO, Division of Land and water, PMB No. 2, Glenn Osmond, SA 5064, Australia

T. J. Beveridge
Department of Microbiology, University of Guelph, Guelph, ON N1G 2W1, Canada

L. Boddy
School of Biosciences, University of Wales, P.O. Box 915, Cardiff CF1 3TL, Wales, UK

J.-M. Bollag
Environmental Resources Research Institute, 129 Land and Water Building, the Pennsylvania State University, University Park, PA 16802, USA

C. Chenu
INRA, United de Science du sol, Centre de Recherche de Versallies, Route de Saint-Cyr, 78026 Versailles Cedex, France

J. Dec
Environmental Resources Research Institute, 129 Land and Water building, The Pennsylvania State University, University Park, PA 16802, USA

H. L. Ehrlich
Department of Biology, Rensselaer Polytechnic Institute, Troy, NY 12180, USA

M. Elimelech
Department of Chemical Engineering, Environmental Engineering Program, Yale University, 9 Hillhouse Avenue, New Haven, CT 06520–8286, USA

J. J. Germida
Department of Soil Science, University of Saskatchewan, 51 Campus Drive, Saskatoon, SK S7N 5A8, Canada

G. Guggenberger
University of Bayreuth, Institute of Soil Science and Soil Geography, 95440 Bayreuth, Germany

K. M. Haider
Kastanienallee 4, 82041 Deisenhofen, Germany

P. M. Huang
Department of Soil Science, University of Saskatchewan, 51 Campus Drive, Saskatoon, SK S7N 5A8, Canada

G. S. R. Krishnamurti
CSIRO Land and Water, PMB 2, Glen Osmond, SA 5064, Australia

E. Kurek
Laboratory of Environmental Microbiology, University of Maria Curie-Sklodowska, 19 Akademicka, 20–033 Lublin, Poland

J.-U. Lee
Department of Microbiology, University of Guelph, Guelph, ON N1G 2W1, Canada

J. S. McLean
Department of Microbiology, University of Guelph, Guelph, ON N1G 2W1, Canada

J. N. Ryan
Department of Civil and Environmental Engineering, University of Colorado, Boulder, CO 80309, USA

N. Senesi
Dipartimento di Biologia e Chimica Agroforestale e Ambieutale, Università de Bari, Via G. Amendola – 165/A, 70126 Bari, Italy

D. P. Shupack
Bradburne, Briller & Johnson LLC, USA

G. Stotzky
Department of Biology, New York University, New York, NY, USA

A. Violante
Dipartimento di Scienze Chemico-Agrarie, Universita degli studi di Napoli Federico II, 80055 Portici (Na), Via Università, Italy

Series Preface

The main purpose of the IUPAC Series on Analytical and Physical Chemistry of Environmental Systems is to make chemists and other scientists aware of the most important bio-physicochemical conditions and processes that define the behavior of environmental systems. Thus the various volumes of the series emphasize the fundamental theoretical concepts of environmental and bio-environmental processes, taking into account their specific aspects such as physical and chemical heterogeneity. Another major goal of the series is to discuss the analytical tools which exist or should be developed to study these processes. Indeed, there is a great need for methodology developed specifically for the field of analytical/physical chemistry of the environment, in close connection with the corresponding process studies.

The present volume of the series focuses on major elements of the analytical and physical chemistry of soil particles in their interaction with microorganisms and organic components in the medium. It critically discusses a variety of aspects ranging from fractal structures and aggregate formation to microbial mobilization processes and bioremediation. The volume was realized within the frame of activities of the IUPAC Commission on Fundamental Environmental Chemistry/Division of Chemistry and the Environment. We thank the responsible IUPAC officers, especially the executive director, Dr John Jost, for their support and assistance. We also thank the International Council of Scientific Unions (ICSU) for financial support of the work of the Commission. This enabled us to materialize the contributors' discussion meeting (Salt Lake City, 1999) that formed such an essential step in the preparation and harmonization of the various chapters of this book.

The series is growing prosperously. Volume 9, on the 'Physicochemical Kinetics and Transport at Chemical-Biological Interphases', is at an advanced stage of preparation. It will provide us with an integral treatment of another part of the interface between physical chemistry and (micro)biology of environmental systems. As with the earlier books in the series, the volume collects critical reviews that characterize the current state of the art and provide recommendations for directing future research.

Jacques Buffle and Herman P. van Leeuwen
Series Editors

Preface

Soil is an integral compartment of the environment. Minerals, organic components, and microorganisms are three major solid components of the soil. They profoundly affect the physicochemical and biological properties of terrestrial systems. To date, scientific accomplishments in individual disciplines of the chemistry of soil minerals, the chemistry of soil organic matter and soil microbiology are commendable. However, minerals, organic matter, and microorganisms should not be considered as separate entities but rather as a united system constantly in close association and interactions with each other in the terrestrial environment. Interactions of these components have enormous impact on terrestrial processes critical to environmental quality and ecosystem health.

As a recognition of the significance of the interactions of soil minerals with organic matter and microorganisms, we initiated the eighth volume of the IUPAC book series on analytical and physical chemistry of environmental systems. The volume consists of 12 chapters that address a variety of issues on fundamentals of soil mineral–organic component–microorganism interactions at the molecular and microscopic levels and the impact on the terrestrial ecosystem. This book covers an overview on interactions of soil particles and microorganisms, a fractal approach for studying interactions between soil particles and microorganisms, microbial mobilization of metals from minerals, fine-grained mineral development and bioremediation, and the impact of the various interactions on formation of metal oxides, development of aggregates, ion cycling and organic pollutant transformation, rhizosphere chemistry and biology, and anaerobic and transport processes in the terrestrial environment.

The book is expected to provide the scientific community with a critical evaluation of the state-of-the-art on the interactions of soil minerals with organic components and microorganisms. It is hoped that this book would advance the knowledge on reactions and processes occurring at the interface between chemistry and biology of the soil and related environments. This book is a definitive guide to chemists and biologists working on soil systems. It would also provide a basis for stimulating further research to uncover the dynamics and mechanisms of environmental processes in nature. Fundamental understanding of these interactions at the molecular and microscopic levels is essential for restoring ecosystem health on a global scale. This book should be useful for scientists, professors, students and consultants working in environmental systems.

This volume was completed under the auspices of the IUPAC Commission of Fundamental Environmental Chemistry. It was co-sponsored by the International Union of Soil Sciences (IUSS). We wish to sincerely thank all of the external referees for their critical review comments. We would also like to express our appreciation to Professors Jacques Buffle, Herman P. van Leeuwen, and David Turner for their insightful criticism.

P. M. Huang, J.-M. Bollag and N. Senesi

Part I Fundamentals of Soil Particle–Microorganism Interactions

1 Interactions between Microorganisms and Soil Particles: An Overview

C. CHENU
Unité de Science du Sol, Versailles, France

G. STOTZKY
New York University, USA

Interactions between Soil Particles and Microorganisms
Edited by P. M. Huang, J.-M. Bollag and N. Senesi. © 2002 John Wiley & Sons, Ltd

1 INTRODUCTION

Understanding how soil characteristics control the activity of microorganisms in soil is essential to predict the occurrence and rate of microbially mediated functions that are of agronomic and environmental importance, such as nitrogen mineralization, denitrification, biological nitrogen fixation, turnover of C and N, stability of soil structure, biodegradation of organic pollutants, soil-borne pathogenicity, etc. Such knowledge is also needed to optimize the success of microorganisms purposely introduced to soil and to improve the bioremediation of pollutants.

Microorganisms in soil live in an ecosystem that is dominated by solid particles, some of which have large surface area (Table 1). Soils have specific surface areas that are variable and depend on the texture and mineralogy of the soils (Table 2). The colloidal fraction of these particles can exhibit permanent

Table 1. Surface area and cation-exchange capacities of soil particles

Soil constituent	Specific surface area*		Cation-exchange capacity
	Total $(m^2\,g^{-1})$	External $(m^2\,g^{-1})$	$(cmol\,kg^{-1})$
Kaolinite	11–26[†]	11–26[†]	2–5[‡]
Illite	24–93[#]	101[‡]	27[‡]
Montmorillonite	800[†]	14–33[#]	76–120[‡]
Goethite	17–81[†]	17–81[†]	–
Allophane	638–897[†]	292–582[†]	–
Ferrihydrite	225–340[†]	–	–
Humic acids	121[¥]	4.9[¥]	300–600[°]
Particulate organic matter > 50 μm[Δ]	–	0.60, 7	20, 10
Coarse dense fraction > 50 μm[Δ]	–	0.90, 8.3	0.7, 1.8
Fraction 0.2–2 μm[Δ]	–	24, 42	22, 15
Fraction < 0.2 μm[Δ]	–	48, 73	33, 19

* Unless otherwise specified, total surface areas were measured with ethylene glycol monoether (EGME) adsorption and external surface areas with adsorption of N_2. The applicability of these methods to swelling clays and to organic matter is still the subject of debate [200–202].
† From [200].
‡ From [203].
# From [204].
¥ From [202]. Total surface areas are measured by CO_2 adsorption.
° From [205].
Δ From [206]. Values in plain style are from a sandy Ferric Lixisol and those in italics from a clayey Ferric Acrisol. Particle size fractions were separated after dispersion of the soil. Particulate organic matter is separated as the light fraction > 50 μm. The coarse dense fraction is composed of primary minerals > 50 μm. Clay–sized fractions are composed of kaolinite, oxides, and organic matter.

Table 2. Selected characteristics of soils with different textures

Characteristic	Dominant texture		
	Sand*	Silt†	Clay‡
External surface area (m^2 g^{-1})[#]	3	17	60
Total surface area (m^2 g^{-1})[¥]	3	55	208
Cation-exchange capacity (cmol kg^{-1})	3.6	13.3	34.5
Porosity (cm^3 g^{-1})	0.21	0.40	0.51
Accessible porosity filled with water at -0.01 MPa (cm^3 g^{-1})	0.07	0.17	0.17
Surfaces developed by soil bacteria[°]			
In % external surface area of soil	2.26%	0.40%	0.11%
In % total surface area of soil	2.26%	0.12%	0.03%
Volume of soil bacteria (cm^3 g^{-1} soil)[°]			
In % soil porosity	0.79%	0.41%	0.32%
In % accessible pore space at -0.01 MPa	2.42%	0.97%	0.95%

* Sandy soil (sand = 79%, silt = 15.3%, clay = 5.7%) from [207].
† Orthic luvisol (sand = 33.3%, silt = 50.3%, clay = 16.5%) (T. Halilat, 1995, personal communication).
‡ Calcisol (sand = 3%, silt = 28%, clay = 69%) from [208].
[#] Measured by adsorption of N_2.
[¥] Measured by adsorption of ethylene glycol monoether (EGME).
[°] A bacterial population of 10^{10} bacteria g^{-1} soil is assumed; with cells being 1 μm long and 0.5 μm in diameter. The total surface area of the bacterial population is then 0.0157 m^2 g^{-1} soil, and the volume of the population is 0.0016 cm^3 g^{-1} soil.

(e.g., most clay minerals) or variable charges (e.g., oxyhydroxides, organic matter) (Table 1). These colloids are considered to be *surface-active particles*. The surface area attributable to a bacterial population of 10^{10} cells g^{-1} of soil is small compared with the total specific surface area or the external surface area of soils with different textures (Table 2). If the bacteria were spread as a monolayer on the surface area of soil particles, they would probably cover only a small fraction of it. Moreover, the ratio of the surface area of the solid particles to the volume of the liquid phase is high in soils. Hence, the surfaces of soil particles, especially those of surface-active particles, are likely to act as sinks for microbial metabolites.

The spatial arrangement of the solid particles results in a complex and discontinuous pattern of pore spaces of various sizes and shapes that are more or less filled with water or air (Table 2 and Figure 1). This is the habitat of soil microorganisms. Soil may be best defined as the juxtaposition of a multitude of microenvironments or microhabitats, characterized by a variety of physical and chemical conditions. Some of these microhabitats are represented in Figure 2. Bacteria and unicellular algae are 'aquatic' organisms in the sense that they rely on diffusion of organic and inorganic compounds in water

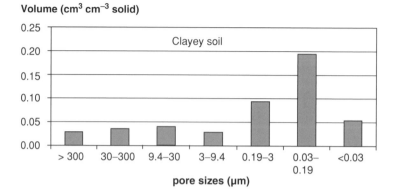

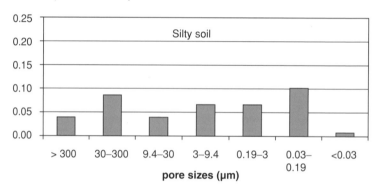

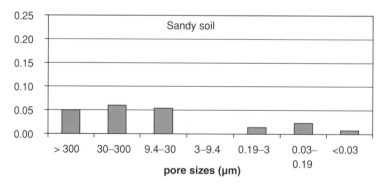

Figure 1. Pore size distribution of soils with different textures (see Table 1 for soil textures). As soil dries and water potential decreases, the coarser pores empty, e.g., at −0.01 MPa or pF 2, all pores larger than 30 μm are filled with air. Calculated from [207,208] and T. Halilat (1995, personal communication)

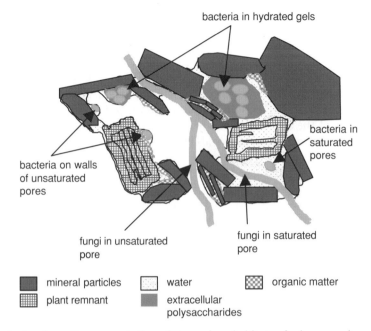

bacteria in hydrated gels

bacteria in saturated pores

bacteria on walls of unsaturated pores

fungi in unsaturated pore

fungi in saturated pore

■ mineral particles ▨ water ▨ organic matter

▦ plant remnant ▨ extracellular polysaccharides

Figure 2. A schematic representation of the various habitats of microorganisms in soils

for their nutrition. Bacteria, including actinomycetes, and unicellular algae can be free or attached to surfaces of particles, located in water-filled pores, or surrounded by a film of water on the walls of air-filled pores (Figure 2). When attached to surfaces, bacteria may occur as scattered individual cells, micro-colonies, or as biofilms. Fungi can occupy the same locations; in addition, fungal hyphae can extend through unsaturated pores (Figure 2). Soil micro-habitats are discrete, not always accessible to microorganisms, and not always interconnected. However, the accessible pore space, i.e., the pore space that is filled with water and in which microorganisms may enter (pore necks larger than the dimensions of microorganisms), is unlikely to be limiting in soils, because microorganisms occupy only a small volume of soil (Table 2).

The survival of microorganisms and the rate of their biological functions have been shown to vary widely with soil type and soil management. However, despite the many studies that have been performed and the statistical relations that have been established between soil characteristics and biological functions, they do not enable sufficient prediction and control of most functions (e.g., [1,2]). The precise nature of the interactions between soil constituents and micro-organisms must be known to identify the soil characteristics that actually control survival and activity of microorganisms. This should enable relevant descriptive parameters (e.g., pore size distribution, abundance of montmorillonite) to be defined and, then, to allow realistic predictions of the overall phenomena, especially in simulation models.

The interactions between microorganisms and soil particles can be broadly classified into biological and abiological. *Biological interactions* involve the growth and multiplication of cells and the secretion of organic substances, such as enzymes and other biopolymers. *Abiotic interactions* involve physical and physicochemical interactions. Physical interactions relate to the geometry and cohesion of the soil matrix. They include, for example, pore size distribution, water retention, aggregate stability, and mechanical properties of soil. Physical interactions are highly dependent on the size, shape, and spatial arrangement of soil particles, as well as on their surface properties. Physicochemical interactions include processes at interfaces or in the soil solution, e.g., sorption, dissolution, hydrolysis, oxidation, and parameters such as pH. Characteristics of the surface of particles, i.e., surface area, electrostatic charge, surface free energy, and functional groups, are important here. Interactions between microorganisms and the soil environment often simultaneously involve biological, physical, and physicochemical processes.

Interactions between soil microorganisms and soil particles are bidirectional. Soil particles influence the survival and biological activity of microorganisms, partly by controlling the geometry of pores in which microorganisms live and the local physicochemical conditions. Microorganisms, despite being minor soil constituents, affect soil particles by modifying their arrangement or aggregating

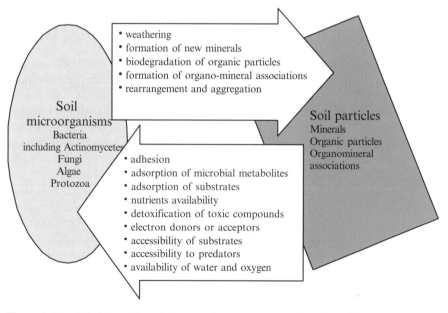

Figure 3. Possible interactions between microorganisms and soil particles

them, weathering mineral particles or contributing to the precipitation and formation of new mineral particles, and degrading organic particles.

This chapter provides an overview of the interactions between soil microorganisms and mineral and organic particles in soil. The overview is not exhaustive, and its main purpose is to emphasize recent developments as well as important gaps in knowledge. Possible interactions between microorganisms and soil particles are summarized in Figure 3, of which only some will be discussed. The reader is referred to other reviews for further information, e.g., [3–7]. Surface interactions will be discussed before interactions that are related to the spatial arrangement of soil particles.

2 SURFACE INTERACTIONS BETWEEN MICROORGANISMS AND SOIL PARTICLES

Surface interactions between microorganisms and soil particles can be *indirect*, as the surface properties of soil particles can affect the composition of the soil solution (e.g., the cation-exchange capacity (CEC) of soil particles controls, to a great extent, the cation concentration of cations in the soil solution). *Adsorption*, which is the concentration of *adsorbates* (microbial cells, organic molecules) at the interface between the soil solution or atmosphere and a solid particle, the *adsorbent*, is a *direct* surface interaction between microorganisms and soil particles. Such concentration at the interface is the result of physical or chemical forces between the adsorbate and the adsorbent. The term *adhesion* is generally used when the adsorbate is also a solid particle, e.g., a bacterium. The term *binding* will be used herein to describe nonreversible adsorption [4]. The term *sorption* is often used in a more general sense to describe situations where a molecule or biological entity is absent from the soil solution, as the result of adsorption, entrapment in soil aggregates, dissolution in the hydrophobic portions of soil organic matter, or even absorption by living organisms.

2.1 ADHESION OF MICROORGANISMS TO SURFACES

Surface interactions of microorganisms with soil particles involve several steps: (i) transport to the surface, (ii) contact and initial adhesion, (iii) firmer attachment, and then (iv) growth, to form adhering microcolonies or biofilms. The initial adhesion is rapid (seconds, minutes) and can be reversible or nonreversible. It is a physicochemical process, which is described and reasonably well predicted by theories of colloid chemistry, such as the Derjaguin–Landau–Verwey–Overbeek (DLVO) theory for electrostatic interactions (e.g., [8]) and Lewis-acid/base hydrophobic interactions [9,10]. Most data on microbial adhesion have been obtained with bacteria and have shown that adhesion depends on the surface properties of the cells and on their physiological state [9,11,12].

Bacteria and fungi may, in a second step with a time scale of hours or days, synthesize extracellular polymers, such as polysaccharides or proteins, which have been shown to anchor the cells more firmly [13–16].

Data exist on the extent and mechanisms of adhesion of microbial cells to pure solid surfaces, such as those of polymers, anion exchange resins, quartz, or mica, from studies with model systems. However, polymeric materials differ from soil particles in their composition and properties. Quartz and mica particles are present in soil, but they have surface electrostatic charges and reactivity very different from those of clay minerals (Table 1). Few studies have focused on the adhesion of bacteria and fungi to clay minerals [4,17–19], because of experimental difficulties. Measuring adhesion between particles of different nature and of similar small size is particularly difficult because these particles cannot easily be distinguished and separated. The adhesion of bacteria to clay particles also poses a conceptual problem as both constituents are usually net negatively charged. According to the diffuse double layer theory, bacterial cells have to overcome a potential energy barrier to come into contact with clay surfaces. Possible ways to overcome this barrier are cation bridging, water bridging, polymer bridging through extracellular polymers, or high ionic strength of the soil solution [4,20]. Furthermore, many of the surfaces of soil particles are not clean but are coated with mineral or organic compounds that may change the overall surface properties of the particles.

In soils, microorganisms are observed to adhere to particles larger than their cells, such as sand grains (Figure 4a) or plant residues (Figure 4b and c). Adhesion of bacteria to smaller-sized particles results in mineral coatings of the cell envelope, often described as 'bacterial microaggregates' (Figure 4d). Biofilms are also observed adhering to the surface of soil aggregates (Figure 4e). Other than such morphological evidence, scant direct quantification exists of adhesion of microorganisms to mineral surfaces in soils [4]. It is often suggested that the limited extent of leaching of bacteria from soils and the difficulty in separating microorganisms from soils [21,22] are the result of adhesion. However, in all experiments where the retention of microorganisms by soils was quantified, no clear distinction was made between adhesion to surfaces and entrapment in soil aggregates or narrow pores (e.g., [23]).

There is relative agreement in the literature that solid surfaces can influence microbial activities, but experimental observations are often contradictory, and most data are restricted to model systems and to particles other than clays. Positive and negative effects of adhesion on substrate utilization, nitrification, respiration, and growth have been demonstrated and extensively reviewed by Stotzky [4]. These effects were, or could be, interpreted by an indirect action of surfaces, as most observed effects were the result of a modification of the composition of the soil solution through the adsorption or desorption of protons, substrates, nutrients, or enzymes on the mineral surfaces. For example, montmorillonite affected bacterial metabolism in part by buffering the pH of

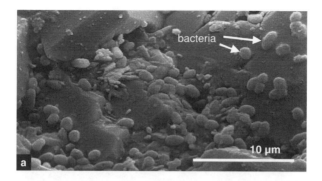

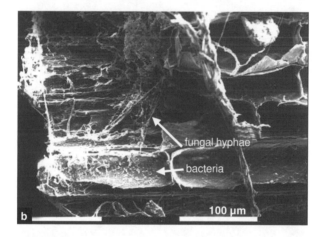

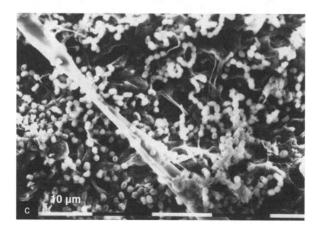

Figure 4.

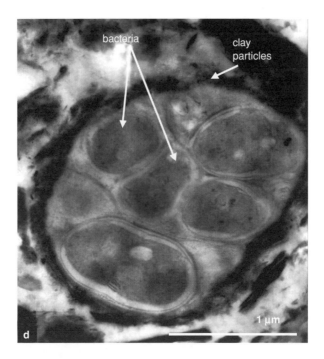

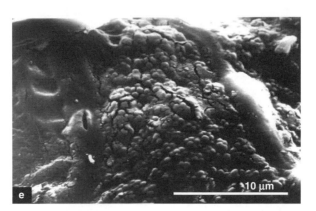

Figure 4. Scanning and transmission electron microscopy of microorganisms attached to soil particle surfaces. (a) *Pseudomonas* cells adhering to sand grains. (b) Bacteria and fungi attached to wheat straw surfaces. (c) Bacteria attached to wheat straw surfaces (detail of previous photograph). (d) Bacterial microaggregate. (e) Bacterial biofilm on the surface of a soil aggregate. (a) Reprinted with permission from [209]. Copyright (1993), with permission from Elsevier Science. (b) and (c) from C. Chenu and S. Recous, 1996, unpublished. (d) Reprinted from [3] by courtesy of Marcel Dekker, Inc. (e) From [211]

the suspension and maintaining it within the optimal range for growth [4]. The decreased rate of growth of microcolonies of *Pseudomonas fluorescens* when adhering to glass was explained by limitations in diffusion of the substrates in the immediate vicinity of the cell [24]. Similarly, Harms and Zehnder [25] found that the degradation of polychlorinated dibenzofuran by *Sphingomonas* cells was lower for cells attached to glass beads than for free ones. Specific cell activities were unchanged, but substrate transport was limited for adhering cells because less liquid medium surrounded attached than free cells. In addition, the diffusion zones of adjacent adhering cells overlapped. There has been much debate on the idea that bacteria at surfaces are at an advantage because of the concentration of mineral nutrients and organic substrates at the interface. Experimental studies on substrate utilization by free and attached bacteria have shown variable results, confirming this hypothesis in a few cases (e.g. [26,27]) but not in general [4,8].

In soils or sediments, adhesion of cells to solid surfaces may result in the creation of a new microenvironment for microbes, consisting of extracellular slime, clay coatings, or both. In these microenvironments, diffusion of substrates or O_2 is likely to be limited (see next section). In suspensions, the presence of clay limited respiration in pure culture of many fungal species [28,29] and decreased the uptake of soluble substrates by bacteria [30–32]. The physical barrier created by the adsorbed clay platelets presumably hindered the uptake of O_2 or organic substrates by the cells. Clay coatings that masked the surface of cells were also invoked to explain the protection of bacteria from predation in clayey soils (e.g. [33]).

According to van Loosdrecht *et al.* [8], there is no definitive evidence of any direct influence of adhesion on microbial metabolism. This is partly the result of the difficulty of controlling and measuring accurately all the possible indirect effects of surfaces on the surroundings of adhering cells. The use of reporter-gene methodology offers a new perspective in this field [34]. For example, Davies and Geesey [35] showed that *Pseudomonas* cells adhering to a glass surface for at least 15 min exhibited expression of the gene *algC* that controls alginate biosynthesis. Synthesis of this extracellular polysaccharide increased the attachment of cells to the glass surface. Knowledge of how adhesion modifies cell metabolism is necessary, as predictions of the functioning of microorganisms in soils are still based on the assumption that physiological traits measured in pure culture can be extrapolated to soil.

2.2 THE ADSORPTION OF MICROBIAL METABOLITES ON SOIL PARTICLES AND ITS CONSEQUENCES

2.2.1 Adsorption of Extracellular Polysaccharides

Many soil bacteria and fungi have the ability to produce extracellular polysaccharides (EPS) in pure culture. It is assumed that they do so in soils, provided that availability of carbon is sufficient [36]. Indeed, layers of EPS around bacteria can be observed in thin sections of soils and rhizosphere [36–42] (Figure 4d). In laboratory experiments, microbial polysaccharides of net zero, negative, or positive charge have been shown to adsorb rapidly and irreversibly to clay minerals [43–45]. The adsorption involves mainly weak bonds, such as hydrogen bonding and London–van der Waals forces. Electrostatic interactions are also involved in the case of charged EPS, and di- or trivalent cations enhance the adsorption of negatively charged EPS to negatively charged clay minerals [19].

Extracellular polysaccharides constitute the outermost surface of those microorganisms that produce them. The adsorption of EPS to mineral surfaces causes or strengthens the attachment of bacteria to soil particles. Exudation of EPS results in the establishment of a porous hydrated continuum between the cell and its mineral surroundings (see Figures 2 and 4d) [36]. The rates of water loss and the rates of decrease in water potential of pure clay–polysaccharide or sand–polysaccharide complexes with desiccation are less with than without EPS [46]. Furthermore, the diffusion of substrates is slow through EPS [47]. Consequently, bacterial microaggregates, in which the bacteria are surrounded by EPS and clay, are presumed to be microenvironments that are well buffered against fluctuations in physical and chemical conditions in soils but in which the rate of movement of nutrients and substrates would be reduced. The adhesion of ligninolytic and cellulolytic fungi to plant residues involves EPS, which serve as supporting structures for the extracellular enzymes of these microorganisms involved in plant decay [48–50]. Excreted enzymes are thereby produced in close vicinity to their targets, with minimal loss to the soil solution.

The major unanswered question about EPS is the quantitative importance of their production in soils. There is no quantification of either the proportion of bacteria that produce EPS in soils or of the amounts produced, because there are no chemical criteria that enable the distinction of EPS from other soil polysaccharides (see ref. [36] for a review). Recent promising developments indicate that proton-spin relaxation editing (PSRE) nuclear magnetic resonance (NMR) might enable the indirect quantification of the adsorption of microbial extracellular metabolites, detected as o-alkyl carbon, to mineral surfaces, at least in model systems [51].

2.2.2 Adsorption of Biologically Active Macromolecules

A number of macromolecules having specific roles and targets, such as extracellular enzymes, toxins, or DNA, are released by microorganisms in the soil environment and are susceptible to adsorption by soil particles, especially by clay minerals and humic substances.

Soil microorganisms release extracellular enzymes to ensure the cleavage of polymers into smaller molecules that can be taken up by cells. Extracellular enzymes adsorb readily to clay minerals through H-bonding and electrostatic and hydrophobic interactions [52]. Adsorption generally leads to a decrease in enzyme activity and to an apparent shift of the optimal pH towards alkaline values [53–58]. At pH values below the isoelectric point of an enzyme, strong electrostatic interactions occur between the net negatively charged clay surface and the net positively charged enzyme. These interactions cause the protein to change conformation, spreading out on the clay surface, as shown by Fourier–transform infrared analysis (FITR) and NMR [52,59], thereby partly loosing its catalytic activity.

Soils have a catalytic activity that results from enzymes that are not attached to the cells that produced them but that are associated with soil colloids and presumably adsorbed. The kinetic behavior of so-called 'immobilized' enzymes is poorly known, but it is different from that of free enzymes. This is probably the result of (i) changes in conformation of the enzyme, (ii) indirect influences of surfaces on the physicochemical characteristics of the environment of the enzyme, and (iii) possible restrictions in the diffusion of substrates and products to and from the catalytic sites [60]. Immobilized enzymes have more stability against thermal and proteolytic inactivation than free enzymes [61]. Extracellular enzymes in soil are intriguing in terms of ecology, as soil microorganisms produce these enzymes with a high expenditure of C, N, and energy, and high proportion of these may become rapidly adsorbed and inactive after release. However, Quiquampoix [7] reported on phytases from soil fungi having high conformational stability and, thus, being rather unaffected by adsorption. Hence, soil microorganisms may have adapted to their environment in producing extracellular enzymes able to remain active when adsorbed to clays. Although not a macromolecule, the growth of *Histoplasma capsulatum*, a fungal pathogen of human beings, was inhibited by clay minerals, the result, in part, of the binding by clays of the siderophores necessary for the iron nutrition of the fungus [62].

Attention has also been directed to other proteins. Subspecies of the bacterium *Bacillus thuringiensis* produce parasporal protein crystals (protoxins) that, when ingested by insect larvae, are cleaved by species-specific proteases in their alkaline mid-gut to toxic insecticidal proteins that are named toxins. Genes that code for the synthesis of the toxins have been inserted in the genome of genetically modified bacteria. Thus, proteins with insecticidal activities are being released to the soil environment in root exudates [63,64], in pollen [65], or when crop residues are incorporated in soil. The toxins active against Lepidoptera (from *B. thuringiensis* subsp. *kurstaki*, Btk) and against Coleoptera (from *B. thuringiensis* subsp. *tenebrionis*, Btt) adsorbed rapidly, in large amounts, and almost irreversibly to soil, clay minerals, and humic acids [5,66–68]. The bound toxins retained insecticidal activity [66,69]. The binding of the toxins to clays and humic acids reduced their availability to microorganisms, which is probably responsible for the observed persistence of the toxins in soil [70]. Hence, the interaction of these toxins with soil particles may affect non-target organisms.

DNA is also released in soils from lysed or by growing bacteria. Extracellular DNA can be taken up by competent bacteria and its genetic information incorporated in their genome. It is generally considered that this process, called 'natural transformation', is unlikely to occur in soils, because DNA is suscep-tible to biodegradation [71]. Several studies have shown that chromosomal as well as different forms of plasmidic DNA are adsorbed to soil constituents, such as clay minerals [72–76] and humic acids [77]. DNA bound to clay minerals and humic acids was capable of transforming cells of *Bacillus subtilis* [77–79]. Bound DNA was protected against degradation by nucleases [72] but was susceptible to amplification by the polymerase chain reaction (PCR) [78,79]. DNA is bound to the edges of clay particles primarily by only one end [80], so that the other end would be free to interact with a competent cell. The binding of DNA to clays presumably alters its conformation, so that nucleases are not able to recognize or interact with it and, therefore, would not hydrolyze it. In addition to transformation, clays are also important in gene transfer by conjugation and transduction among bacteria in soils [81,82] .

These examples of interactions between biologically active macromolecules and soil particles, which have been discussed in greater detail elsewhere [5] demonstrate that surface interactions allow such molecules to persist in the soil environment, even though their bioactivity may often be diminished. From an environmental perspective, the ability of soils to store and preserve enzy-matic, toxic, and genetic activity needs further assessment.

2.3 EFFECTS OF ADSORPTION OF ORGANIC SUBSTRATES ON THEIR BIODEGRADATION

Adsorption to soil particles decreases the rates of biodegradation of a variety of compounds, including biogenic molecules and organic pollutants. However, the rates differ between small molecules and larger polymeric ones.

Small organic molecules are degraded in cells after diffusion or transport into cells by permeases. In some cases, they may be degraded outside of the cell by extracellular enzymes. The rate of biodegradation of small organic molecules was slower in the presence of particles, such as clay minerals [30–32,57,83–85], aluminum hydroxides [32,86], hydroxyapatite [87], or soil (e.g. [26,88]), than in their absence. For example, the mineralization of citrate by soil bacteria was reduced in the presence of clay minerals and iron hydroxide, and the decrease depended on the extent of adsorption of citrate to these minerals (Figure 5). For several substrates, the kinetics of biodegradation can be best described by assuming that the adsorbed substrate was unavailable to microorganisms and that its utilization depended on its rate of desorption from the surface [26,30,53,85,86]. For example, Ogram *et al.* [26] studied the biodegradation of the herbicide, 2,4-dichlorophenoxyacetate (2,4-D), in suspensions of soil and of clay that were inoculated with a pure culture of a bacterium able to degrade the herbicide. The patterns of disappearance of 2,4-D from the different systems was best explained by a model that assumed that adsorbed 2,4-D was

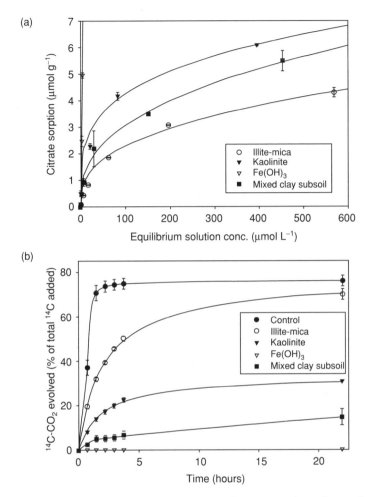

Figure 5. Inhibition of the biodegradation of organic compounds as the result of their adsorption to the surface of solid particles. (a) Adsorption of citrate to illite-mica, kaolinite, Fe hydroxide and a mixed clay subsoil. (b) Cumulative ^{14}C-CO_2 production from suspensions of soil bacteria after the addition of ^{14}C-citrate, either in the absence (control) or in the presence of different soil particles. Reprinted from [32]. Copyright (1998), with permission from Elsevier Science

not utilized by the bacterium and that both free and adhering cells of the bacterium utilized 2,4-D in solution at the same rate. In a few cases where the substrates were not desorbed (i.e., the molecules were tightly bound), unavailability of the bound substrate to bacteria was directly shown [84]. However, unambiguous demonstration that adsorbed compounds are not utilized is difficult, because very slow rates of desorption can occur, and they are difficult to measure. Furthermore, the presence of bacteria may facilitate desorption, e.g., by production of surfactants. The availability of an adsorbed substrate to a microorganism can depend on the relative affinity of the substrate for the solid surface and for permeases of the cell. For example, a strain of *Pseudomonas putida* utilized naphthalene sorbed on soil or smectite, whereas a strain of *Alcaligenes* sp. could not utilize it [83,88,89]. *Agrobacterium radiobacter* was able to utilize arginine, but not cysteine or aspartic acid, adsorbed on montmorillonite [84]. The adsorption of polymeric substances was found in a number of studies to decrease their rate of biodegradation, e.g., polysaccharides [90,91], proteins [5,53,66,70], DNA [72,74,77]. In contrast to substrates of small molecular mass, the adsorption of macromolecules is generally nonreversible. The adsorption of polymers is often associated with conformational changes that render the molecules unavailable to the action of extracellular enzymes (e.g. [80]).

To predict the fate of organic pollutants in the environment, it is necessary to take into account the impact of adsorption and binding of the pollutant on surface-active particles on biodegradation. Several simulation models of different complexities have been proposed for the degradation of organic pollutants in soil (see review by Scow and Johnson [92]). The simplest category of models assumes that the adsorbed substrate is not metabolized and predicts that biodegradation follows first-order rate equations, in which the concentration of substrate available to microorganisms is the concentration of soluble substrate. The distribution of the substrate between adsorbed and free states is calculated using, for example, linear or Langmuir sorption isotherms [26,85]. In two-compartment models, the substrate exists in two phases: in one, the adsorbed substrate is unavailable to microorganisms or is metabolized at a very slow rate, and in the other compartment, the substrate is readily available to microorganisms. This compartment is rapidly depleted, and the subsequent overall rate of biodegradation of the chemical is governed by the rate of replenishment of the 'available' compartment from the 'resistant' one, i.e., by the rate of desorption [93]. Although such models may describe adequately the biodegradation of organic molecules in soils [92], there is the difficulty of separating the effects of adsorption from those of diffusion in controlling the concentration of the substrate in solution in experiments with soil [83,88,94,95]. Few studies have explicitly addressed the two processes independently [92].

3 INTERACTIONS BETWEEN SOIL PARTICLES AND MICROORGANISMS AT THE MICROSTRUCTURE SCALE

3.1 ACCESSIBILITY OF SUBSTRATES TO MICROORGANISMS

Adsorption is not the only process by which soil particles limit the extent of biodegradation of organic molecules in soils. There is ample evidence of limited accessibility of substrates to microorganisms in soils, which results from the spatial arrangement of soil particles. Organic compounds that should be readily biodegradable and are not strongly sorbed by soil, such as ethylene dibromide, can persist in soils for decades, even though the environmental conditions are favorable for microbial growth and biodegradation [96,97]. Dispersion of the structural units of soil has been found to stimulate the mineralization of soil organic matter as well as that of organic pollutants. Dispersing treatments included soil tillage [98], soil crushing (e.g., [99–102]), sonication [103], alternate wetting and drying (e.g. [102]), or freezing and thawing [104]. The more pronounced the dispersion of soil, the greater the mineralization [98]. Dispersion treatments are assumed to enable contact between previously sequestered organic substrates and microorganisms. Experiments with model systems have confirmed this by showing that mineralization of organic substrates was retarded when they were placed in aggregates having pores in the micrometer or nanometer range, and from which bacteria and fungi were excluded because of their sizes [94,105–107].

Biodegradation requires contact between the substrate and the microbial cell, in the case of small substrates, or between polymeric substrates and extracellular enzymes. However, such contact is infrequent in soil matrices, as both the substrate and microorganisms occupy a very small proportion of the soil volume, and they have a heterogeneous spatial distribution. Visual observations of the soil matrix with electron microscopy [38,39,108,109] and studies based on physical fractionation of soil attest to the unevenness of the distribution of soil microorganisms at scales from a centimeter down to a micrometer. Furthermore, the soil matrix is compartmentalized, and transfers of substrates, enzymes, and microbial cells in soils are probably limited. Organic substrates can be located in pores to which microorganisms do not have access, because the pore necks are too small or, in the case of bacteria, yeasts and protozoa, the water pathways are discontinuous [110,111]. From Figure 1, it can be deduced that 15% of the soil porosity in a sandy soil to 52% in a clayey soil is inaccessible to microorganisms, because the diameters of pore necks are smaller than 0.2 μm. At a water tension of -0.1 MPa, an additional 53% of the soil porosity in the sandy soil and 14% in the clayey soil is filled with air, and thus the diffusion of nutrients and substrates and the movement of bacteria will be limited. Diffusion pathways in soils are tortuous and often discontinuous. The lower the water potential, the lower the rate of diffusion of substrates. Physical inaccessibility applies also to the pores of macromolecules in soils, i.e., those of humic substances, plant debris, and extracellular polymers, which may trap substrates and nutrients. Nonpolar substrates, such as hydrophobic organic molecules, will tend to remain partitioned in hydrophobic regions of humic substances and not diffuse in the aqueous soil solutions where microorganisms and their enzymes are located [105,106,112–114]. For example, Nam and Alexander [106] showed that phenanthrene was rapidly mineralized by bacteria in the presence of porous silica beads or nonporous polystyrene beads, but there was only limited biodegradation in the presence of porous polystyrene beads having nanometer-size pores in which the phenanthrene sorbed and persisted.

Simulation models have been developed to account for the impact of physical inaccessibility on the biodegradation rates of organic compounds in soils. The major difficulty of such modeling is the description and the amount of simplification that can be made of soil microstructure and of the spatial distribution of microorganisms. Most models have two compartments, and microorganisms are excluded from one of them. For example, Scow and Alexander [107] analyzed the biodegradation of phenol by a *Pseudomonas* strain in the presence of kaolinite aggregates in a well-mixed suspension. The substrates had access to the intra-aggregate and inter-aggregate porosity, whereas bacteria were restricted to the inter-aggregate solution. The biodegradation rate of phenol in this simple system was adequately described using a two-compartment model, in which phenol diffused from the internal porosity of aggregates to the inter-aggregate pores, where it was available to the bacteria. Radial diffusion–biodegradation models have also been successfully developed to describe the biodegradation of naphthalene in suspensions of 0.5–30 μm soil aggregates, assuming that the aggregates were spherical and that bacteria were located only on the outside of the aggregates [115] (Figure 6). Priesack and Kisser-Priesack [116] predicted the mineralization of glucose in remolded spherical soil aggregates of 18 mm diameter with a radial diffusion–biodegradation model. Initially, microorganisms were located only inside the aggregates and glucose diffused from the surface of the aggregates inwards. Rijnaarts *et al.* [117] described the biodegradation of α-hexachlorocyclohexane in a soil slurry with a similar model, assuming that, at the onset of biodegradation, bacteria and the α-hexachlorocyclohexane were located within aggregates. However, the actual distribution of microorganisms in their system was unclear.

The assumed distributions of microorganisms in these models are simplistic when applied to soil. Actual knowledge of the spatial distribution of bacteria, including actinomycetes, and fungi in soil is insufficient to implement models with realistic, but simple, assumptions. Several studies have reported contrasting microbial distributions among aggregate classes [41,118–126] and between operationally defined outer and inner compartments of aggregates [127–129]. However, no clear trend has emerged, as the aggregate classes or fractions that have been separated varied among the studies because of the fractionation methods, and, thus, the degree of dispersion of soil differed widely. The factors that govern the spatial distribution of soil microorganisms, e.g., location of food sources, local physicochemical conditions, and predation, have not been sufficiently analyzed. Furthermore, the distribution of microorganisms in the soil matrix most probably changes with time, as substrates become available by diffusion and are degraded. For example, Priesack and Kisser-Priesack [116] showed that microbial populations, indirectly assessed by measurement of ATP,

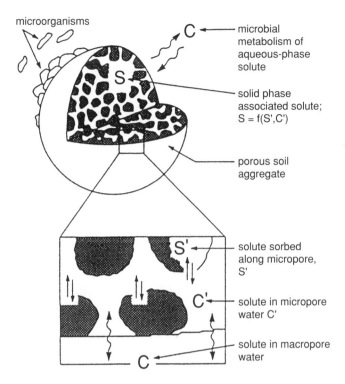

microorganisms

C ← microbial metabolism of aqueous-phase solute

S

solid phase associated solute; S = f(S',C')

porous soil aggregate

S' — solute sorbed along micropore, S'

C' — solute in micropore water C'

C — solute in macropore water

Figure 6. Schematic representation of a diffusion–biodegradation model. Reprinted with permission from [115]. Copyright (1991) American Chemical Society

increased on the surface of artificial soil aggregates after glucose application to the surface. Ronn *et al.* [130] and Gaillard *et al.* [131] showed changes in the spatial distribution of soil microorganisms at the millimeter scale, as the result of the diffusion of soluble compounds from decomposing plant material. These studies indicate hot spots of microbial activity in soil that are temporary and whose contribution to the overall mineralization of C and N is as yet unknown.

3.2 GEOMETRIC HINDRANCE AND PREDATION

One factor that regulates bacterial populations in soils is their predation by protozoa, which accounts for most of the decline in cell numbers of introduced bacteria [33]. For example, Heijnen *et al.* [132] found that amending sandy loams with montmorillonite or kaolinite increased the survival of introduced *Rhizobium leguminosarum*. Clay minerals, in particular montmorillonite, provided protective pores that were < 6 μm in diameter, thus shielding the bacteria from larger protozoa [133,134]. In model systems, the accessibility of pores to protozoa, which was monitored either with artificial aggregates [135] or by changes in water potential [136], determined the predation of bacteria. In grassland soils, total protective capacity (i.e., number of pores $< 6\,\mu m$ in diameter) correlated well with the biomass of bacteria [111]. Fine soil texture is thought to reduce the extent of predation because of the exclusion of predators by small pore size and by imposing larger distances for protozoa to travel for feeding [135,137]. Görres et al. [138] suggested that with drying, nematodes and bacteria became trapped in large water-filled pore space with no connection to other pores in which predation was favored.

Pore size distribution and the proportions of pores filled with water determine the possibilities of contact between soil organisms and molecules and are, thus, likely to influence interactions between microorganisms. For example, the competition between inoculated bacteria was affected by the respective location of the introduced strains and of carbon substrates [139].

3.3 SOIL STRUCTURE DEFINES LOCAL PHYSICOCHEMICAL CONDITIONS

The packing and spatial arrangement of soil particles define the pores that are the habitat of soil microorganisms. An adequate supply of water is probably the most important factor affecting microbial life in soils. Water is retained by adsorption to the surfaces of hydrophilic soil particles, especially clays, biopolymers, and humic substances, and held within pores by capillary forces. Some of the water adsorbed to surfaces of soil particles is retained with energies far too high for microorganisms to utilize the water, e.g., the water on interlayer surfaces in Ca-saturated smectites [3,140]. According to capillary laws, only pores having diameters smaller than a critical value are filled with water at a given water potential (Table 3). The pore-size distribution of a soil is a critical property that determines the relative abundance of habitats with different sizes and water regimes. The smaller the pore size, the lower is the water potential value of the air-entry point, and thus, the less often the pores will be filled with

Table 3. Relation between soil water potential Ψ, relative humidity and maximum diameter of pores filled with air. From [210]

Ψ (pF)	Ψ (MPa)	Ψ (bars)	Ψ (kPa)	Relative humidity (%)	Maximum diameter of water-saturated pores (μm)
1	−0.001	−0.01	−1	99.999	300
2	−0.01	−0.1	−10	99.99	30
3	−0.1	−1	−100	99.93	3
4	−1	−10	−1000	99.27	0.3
5	−10	−100	−10000	92.70	0.03
6	−100	−1000	−100000	50	0.03

air. Pore size distributions in soil are determined by Hg porosimetry and with water content–water potential curves (e.g., Table 3). However, their interpretation is limited by the fact that most soils do not fulfill the conditions of Poiseuille's law: i.e., rigidity of the pore system and well-interconnected cylindrical pores. Hence, the pore-size distribution of clay soils varies with water potential [141], and pore necks (i.e., the diameters of pore openings to adjacent pores) are measured rather than pore diameters. The pore-size distribution of soils, in turn, depends on the size and shape of soil particles (e.g., Table 2) as well as on their degree of association with organic matter and oxyhydroxides in aggregates [142,143]. Hence, few models enable the prediction of the pore-size distribution of soils only from their texture and mineralogy.

Clay minerals provide microbial habitats with specific characteristics. Clay minerals in soil occur as complex associations of clay layers into particles called tactoids or domains, which have internal pores ranging from 1 nm to 0.1 μm. Bacteria are excluded from these, but they can partially utilize the water contained therein (Table 3). The spatial arrangement of these particles define pores from 0.1 to a few microns [140]. Such pores are buffered in terms of water regime, because they remain filled with water in a wide range of water potentials and because fluctuations in water potentials with desiccation and rehydration are slow. However, these pores undergo shrinkage with drying [140]. These characteristics of clay minerals in soils may explain partly why the survival of soil bacteria during desiccation depends on the abundance and mineralogy of clays [144–149]. Microorganisms located in pores with small sizes defined by the arrangement of clay particles may experience limitations in O_2 [4] and compaction with drying [150], but they should benefit from an environment that is buffered in water potential and water content. Despite decades of studies, precise knowledge and the ability to predict the physical and physicochemical conditions that are experienced by microorganisms in soil are still in their infancy. The main difficulty arises from the spatial heterogeneity in soil at scales as small as a micron. Consequently, properties measured on bulk soil can seldom be related to those experienced by microorganisms and to fluxes caused by these organisms. One example is the availability of O_2, which controls the activity of denitrifiers. The occurrence of anaerobic sites in unsaturated soils depends locally on the distance to large pores and fissures filled with air and on the presence of organic residues that locally increase the consumption of O_2 [151]. Detailed studies of the spatial heterogeneity under natural soil conditions are needed to determine the actual range of variation of various physicochemical parameters at the spatial scale of microorganisms and to identify the appropriate spatial scales for study. Miniature sensors for measuring local concentrations of O_2 [151–153] and pH [154,155] have great potential in this respect. These studies should be supplemented by studies on the spatial distribution of microorganisms, to determine the conditions that they are likely to experience.

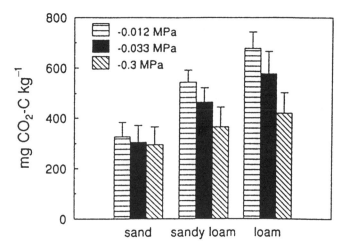

Figure 7. Mineralization of native soil organic matter as a function of the soil texture and water potential. From [156]. Reproduced by permission of Soil Science Society of America

An additional difficulty derives from the multiplicity of factors and their interactions, so that the actual factors affecting a variable are difficult to identify. Water potential is the classic example of a parameter that affects the availability of water, as well as the availability of O_2 and the diffusion of nutrients. For example, Scott *et al.* [156] found that water potential influenced the decomposition of native soil organic matter over short-term incubation and that its impact depended on soil texture (Figure 7). The water-filled pore space, which depends on both texture and water potential, explained the variability in C mineralization better than water potential or soil texture alone. However, it was unclear whether water-filled pore space accounted for changes in the soil microenvironment in terms of availability of O_2 or water to microorganisms or whether it accounted for changes in the diffusion of substrates. Soroker [157] showed that the inhibition of microbial activity at low water potentials was the result of a direct effect on microbial cells and of a limitation in substrate diffusion to cells, and she quantified these effects.

Studies with model systems are very useful, because they enable a reduction in spatial heterogeneity and the manipulation of a limited number of parameters. Fruitful model systems include cores made of pure clay minerals [150], porous glass beads [135], remolded soil aggregates [158,159], and cores made from repacked soil aggregates [131,160,161].

3.4 ARRANGEMENT AND AGGREGATION OF SOIL PARTICLES

The basic mechanisms by which soil microorganisms can modify the arrangement of soil particles and the physical stability of the resulting aggregates

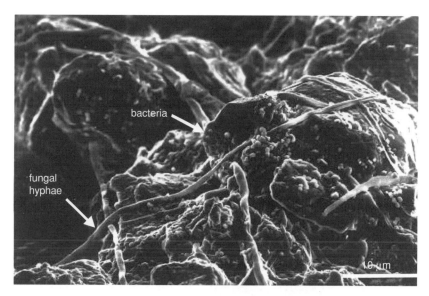

Figure 8. Binding of soil particles by fungal hyphae. Reprinted with permission from [162]. Copyright CRC Press, Boca Raton, Florida

have been reasonably well described. At a large scale ($> 20\,\mu$m), the physical entanglement of soil particles by fungi is easily observed [162–165] (Figure 8). Binding of soil particles by hyphae [131,162–165] is invoked when aggregate stability is positively correlated with hyphal length [164,166,167], which is not always the case [168]. Despite their relatively small size, fungal hyphae can locally modify the arrangement of clay particles, because of the pores that hyphae leave behind (Cabidoche, Y. M., 1998, personal communication). Local compaction or desiccation by microbial cells tend to orient clay particles parallel to the cell wall over a few microns, especially when EPS link the clay particles to the surface of the cells [165]. The secretion of EPS by bacteria, unicellular algae, or fungi, and the adsorption of EPS to soil particles (see section 2.2.1) tend, at a small scale ($< 20\,\mu$m), to aggregate the soil particles [18,37,39,40,165,169,170]. These constitute the nuclei of stable microaggregates, i.e., zones where interparticle cohesion is enhanced by polymer bridges [171–173] (Figure 9).

Increasing interparticle cohesion is not the only mechanism by which micro-organisms can increase the stability of aggregates against the disruptive action of water. The low wettability of aggregates, especially aged aggregates containing hydrophobic organic materials, slows their rate of wetting and, thus, limits the extent of slaking. Microbial growth decreased the wettability of aggregates in soil [174]. In model systems, hydrophobic molecules adsorbed to clay minerals

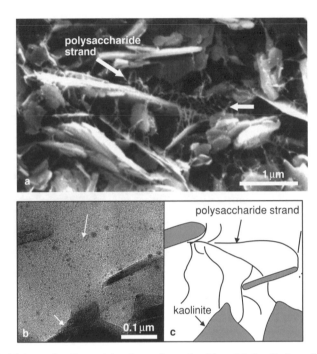

Figure 9. Bridging of soil particles by polysaccharides. (a) kaolinite–microbial polysaccharide (scleroglucan) association as observed with low temperature scanning electron microscopy. (b) kaolinite–microbial polysaccharide (scleroglucan) association as observed with transmission electron microscopy after staining the polysaccharide with silver. (c) Schematic representation of (b), showing the polymer bridges. From [46]

decreased the contact angles of the clays with water [175–177]. The content of lipids in soils, presumably of microbial origin, correlated well with aggregate stability [178]. The presence of a group of glycoproteins, named *glomalins*, which are exuded in copious amounts by vesicular arbuscular mycorrhizal fungi [179], correlated well with aggregate stability [180], presumably because glomalins render the clay surfaces more hydrophobic.

However, there is little, if any, quantification of microbial EPS and microbial hydrophobic substances that may be involved in aggregate stabilization in soils *in situ*. Several studies have shown that fungi have a more dominant role in soil aggregation than bacteria, because of the physical enmeshment of soil particles by fungal hyphae [181–184]. Quantification of the relative importance of these mechanisms is insufficient, even though numerous studies have searched for relevant indicators of microbially mediated aggregation, such as microbial biomass, hyphal length (e.g., [164,166–168,181,183]), and carbohydrates soluble in hot water (e.g., [183,185–187]).

The factors that govern the importance of microbially mediated aggregation are also poorly known. It seems unlikely that EPS diffuse any distance from the producer cells. Furthermore, the spatial distribution of bacteria, fungi, and algae probably influences the fraction of the soil volume in which microorganisms aggregate soil particles. Factors that stimulate production of EPS are expected to enhance stable aggregation in soils. The abundance of polysaccharides derived from microbes and bound to soil minerals was affected by the fertilization and salinity of soil [188,189]. The effects of wetting and drying events in soils on microbially mediated aggregation are insufficiently described. For example, do extracellular substances with hydrophobic moieties reorient their functional groups upon drying, as suggested by Ma'shum and Farmer [190]? Does drying of EPS increase their bonding to clay particles [168]? These mechanisms could account for the observed increased stability of aggregates after drying and rewetting [191,192]. The observed effects of microbial growth on the cracking patterns of soils [193] may be a result of local changes in the shrink and swell properties of soil resulting from binding of soil particles by EPS and fungal hyphae.

Knowledge of the mechanisms by which soil microorganisms aggregate soil particles and of the factors involved in such aggregation is needed, if soil microorganisms are to be manipulated to improve soil tilth. Such manipulation of the soil microbiota involves the *in situ* stimulation of specific microorganisms (e.g., [188,194]) and the proposed inoculation of EPS-producing microorganisms into soils [169,195–198].

4 FUTURE PROSPECTS

4.1 EXTEND STUDIES FROM MODEL SYSTEMS TO NATURAL SYSTEMS

Many studies on interactions between microorganisms and particles have been done with pure clay minerals from geological deposits. Future studies should focus on clay minerals from soils, which are more complex in their constitution, often smaller in size, and coated with humic substances, biopolymers, and metal oxides and hydroxides. Na-saturated montmorillonites have been used in many *in vitro* studies; however, clays with expanded interlayers are a curiosity in most soils [199]. In soils, humic acids occur often as coatings on clay minerals or as small, humified debris. Experimental studies should consider these rather than humic acids extracted from soil.

Studies performed in batch systems often adequately address the mechanisms of interactions between the surface of soil particles and microorganisms. However, the importance of these interactions is likely to be different in soils, where particles are aggregated. The surface areas accessible to microorganisms will be much less than in dilute suspension, and the ratio of the surface area of particles to the volume of solution will increase. More studies in natural aggregated soils are obviously needed. Given the complexity of the microstructure of soils, designing meaningful microcosms for such studies is a true challenge.

4.2 DEFINE THE SPATIAL DISTRIBUTION OF MICROORGANISMS IN SOILS

As emphasized in this brief overview, the location of microorganisms in soil determines, to a large extent, their access to food sources, their exposure to predation and drying, and the physical and chemical characteristics of their immediate environment. There have been numerous studies in this area, but there is often a poor definition of the structural units of soil that are analyzed. This hampers the elucidation of the factors that govern the spatial distribution of microorganisms in soil structure. Studies are needed to relate, beyond simple descriptions, the location of microbes to specific microbial functions and interactions.

5   SUMMARY

Soil microorganisms live in an environment that is dominated by solid particles of various natures, sizes, and shapes and having complex spatial arrangements. Microorganisms are influenced by the nature, properties, and arrangement of soil particles, and they also modify these particles and their arrangements.

Direct surface interactions between soil particles and microorganisms involve adsorption processes. Bacteria and fungi are generally considered to adsorb to soil particles, although there is little quantification of adhesion in soils. Data from *in vitro* systems indicate that adhesion can modify the activity of microorganisms. In soils, microbial life is also indirectly influenced by surface interactions, as molecules involved in microbial activity can adsorb to soil particles, e.g., C and N substrates, extracellular polysaccharides, extracellular enzymes, DNA, insecticidal proteins. Adsorption generally reduces the rate of biodegradation of molecules and their activity (e.g., enzymatic activity, genetic transformation, toxicity).

Soil particles also influence microbial life through their spatial arrangements that define voids of different size, shape, and connectivity and which are the habitats of soil microorganisms. Soil microstructure, thereby, controls the local physicochemical conditions that are experienced by soil microorganisms, in particular the availability of water and O_2. The arrangement of soil particles is an important regulator of biogeochemical cycling in soil because it reduces the accessibility of substrates and of microorganisms to their predators. In turn, bacteria, algae, and in particular, fungi modify the arrangement of soil particles and the stability of soil aggregates through binding actions or through the secretion of extracellular substances.

Interactions between soil microorganisms and soil particles are bidirectional and numerous, but knowledge of these interactions is scant. Studies are needed to identify and describe such interactions, to quantify their effects on microbial life and soil properties, and to analyze the factors that influence these interactions. Future studies should, in particular, focus on natural soil systems rather than on model systems and address specifically the spatial distribution of soil microorganisms in the structure of soil. Studies are needed to assess the relative importance of interactions at the scale of both surface and microstructure, to define and integrate relevant descriptive parameters and to obtain a comprehensive view of the functioning of soil *in situ*. To attain this, it is necessary to define and quantify the effects of biophysicochemical interactions between soil microorganisms and soil particles.

REFERENCES

1. Hénault, C. and Germon, J. C. (1995). Quantification de la dénitrification et des émissions de protoxyde (N_2O) par les sols, *Agronomie*, **15**, 321.
2. Ladd, J. N., Foster, R. C. and Skjemstad, J. O. (1993). Soil structure: carbon and nitrogen metabolism, *Geoderma*, **56**, 401.
3. Robert, M. and Chenu, C. (1992). Interactions between microorganisms and soil minerals. In *Soil Biochemistry*, Vol. 7, ed. Stotzky, G. and Bollag, J. M., Marcel Dekker, New York, p. 307.
4. Stotzky, G. (1986). Influence of soil mineral colloids on metabolic processes, growth, adhesion, and ecology of microbes and viruses. In *Interactions of Soil Minerals with Natural Organics and Microbes*, Vol. 17, ed. Huang, P. M. and Schnitzer, M., Soil Science Society of America, Madison, WI, p. 305.
5. Stotzky, G. (2000). Persistence and biological activity in soil of insecticidal proteins from *Bacillus thuringiensis* and of bacterial DNA bound on clays and humic acids, *J. Environ. Qual.*, **29**, 691.
6. Violante, A. and Gianfreda, L. (2000). Role of biomolecules in the formation and reactivity towards nutrients and organics of variable charge minerals and organo-mineral complexes in soil environment. In *Soil Biochemistry*, Vol. 10, ed. Bollag, J. M. and Stotzky, G., Marcel Dekker, New York, p. 207.
7. Quiquampoix, H. (2000). Mechanisms of protein adsorption on surfaces and consequences for soil extracellular enzyme activity. In *Soil Biochemistry*, Vol. 10, ed. Bollag, J. M. and Stotzky, G., Marcel Dekker, New York, p. 171.
8. van Loosdrecht, M. C., Lyknema, J., Norde, W. and Zehnder, A. B. (1990). Influence of interfaces on microbial activity, *Microbiol. Rev.*, **54**, 75.
9. Grasso, D., Smets, B. F., Strevett, K. A., Machinist, B. D., van Oss, C. J., Giese, R. F. and Wu, W. (1996). Impact of physiological state on surface thermodynamics and adhesion of *Pseudomonas aeruginosa*, *Environ. Sci. Technol.*, **30**, 3604.
10. Bellon-Fontaine, M. N., Mozes, N., Mei, H. C., van der Sjollema, J., Cerf, O., Rouxhet, P. G. and Busscher, H. J. (1990). A comparison of thermodynamic approaches to predict the adhesion of dairy microorganisms to solid substrata, *Cell Biophys.*, **17**, 93.

11. Dufrêne, Y. F. and Rouxhet, P. G. (1996). Surface composition, surface properties, and adhesiveness of *Azospirillum brasilense*—variation during growth, *Can. J. Microbiol.*, **42**, 548.
12. Huysman, F. and Verstraete, W. (1993). Effect of cell surface characteristics on the adhesion of bacteria to soil particles, *Biol. Fertil. Soils*, **16**, 21.
13. Dufrêne, Y., Boonaert, C. J. P. and.Rouxhet, P. G. (1999). Role of proteins in the adhesion of *Azospirillum brasilense* to model substrata. In *Effect of Mineral–Organic–Microorganisms Interactions on Soil and Freshwater Environments*, ed. Berthelin, J. Huang, P. M., Bollag, J. M. and Andreux, F., Kluwer–Plenum, New York, p. 261.
14. Hermesse, M. P., Dereppe, C., Bartholome, Y. and Rouxhet, P. G. (1998). Immobilisation of *Acetobacter aceti* by adhesion, *Can. J. Microbiol.*, **34**, 638.
15. Robbins, R. J., Hall, D. O., Turner, R. J. and Rhodes, M. J. C. (1986). Mucilage acts to adhere cyanobacteria and cultured plant cells to biological and inert surfaces, *FEMS Microbiol. Lett.*, **34**, 155.
16. Bashan, Y. and Levanony, H. (1988). Active attachement of *Azospirillum brasillense* to quartz sand and to a light textured soil by polymer bridging, *J. Gen. Microbiol.*, **134**, 2269.
17. Hattori, T. (1970). Adhesion between cells of *E. coli* and clay particles, *J. Gen. Appl. Microbiol.*, **16**, 351.
18. Kiremidjian, L. and Stotzky, G. (1973). Effects of natural microbial preparations on the electrokinetic potential of bacterial cells and clay minerals, *Appl. Microbiol.*, **25**, 964.
19. Santoro, G. and Stotzky, G. (1967). Effect of electrolyte composition and pH on the particle size distribution of microorganisms and clay minerals as determined by the electrical sensing zone method, *Arch. Biochem. Biophys.*, **122**, 664.
20. Theng, B. K. G. and Orchard, V. A. (1995). Interactions of clays with microorganisms and bacterial survival in soil : a physicochemical perspective. In *Environmental Impact of Soil Component Interactions. Metals, Other Inorganics and Microbial Activities*, Vol. 2, ed. Huang, P. M., Berthelin, J., Bollag, J. M., Mc Gill, W. B. and Page, A. L., CRC Lewis, Boca Raton, FL, p. 123.
21. Ramsay, A. (1984). Extraction of bacteria from soil: efficiency of shaking and ultrasonication as indicated by direct counts and autoradiography, *Soil Biol. Biochem.*, **16**, 475.
22. Richaume, A., Steinberg, C., Jocteur-Monrozier, L. and Faurie, G. (1993). Differences between direct and indirect enumeration of soil bacteria: the influence of soil structure and cell location, *Soil Biol. Biochem.*, **25**, 641.
23. Ozawa, T. and Yamaguchi, M. (1986). Fractionation and estimation of particle-attached and unattached *Bradyrhizobium japonicum* strains in soil, *Appl. Environ. Microbiol.*, **52**, 911.
24. Caldwell, D. E. and Lawrence, J. R. (1986). Growth kinetics of *Pseudomonas fluorescens* microcolonies within the hydrodynamic boundary layers of surface microenvironments, *Microbiol. Ecol.*, **12**, 299.
25. Harms, H. and Zehnder, A. J. B. (1994). Influence of substrate diffusion on degradation of dibenzofuran and 3–chlorodibenzofuran by attached and suspended bacteria, *Appl. Environ. Microbiol.*, **60**, 2736.
26. Ogram, A., Jessup, R. E., Ou, L. T. and Rao, P. S. C. (1985). Effects of sorption on biological degradation rates of (2,4–dichlorophenoxy) acetic acid in soils, *Appl. Environ. Microbiol.*, **49**, 582.
27. Griffith, P. C. and Fletcher, M. (1991). Hydrolysis of protein and model di-peptide substrates by attached and nonattached marine *Pseudomonas* sp. strain NCIMB 2021, *Appl. Environ. Microbiol.*, **57**, 2186.

28. Lavie, S. and Stotzky, G. (1986). Adhesion of the clay minerals, montmorillonite, kaolinite, and attapulgite, reduces respiration of *Histoplasma capsulatum*. *Appl. Environ. Microbiol.*, **51**, 65.

29. Stotzky, G. and Rem, L. T. (1967). The influence of clay minerals on microorganisms. IV. Montmorillonite and kaolinite on fungi, *Can. J. Microbiol.*, **13**, 1535.

30. Smith, S. C., Ainsworth, C. C., Trainer, S. J. and Hicks, R. J. (1992). Effects of sorption on biodegradation of quinoline, *Soil Sci. Soc. Am. J.*, **56**, 737.

31. Magdaliniuk, S., Block, J. C., Leyval, C., Bottero, J. Y., Villemin, G. and Babut, M. (1995). Biodegradation of naphthalene in montmorillonite/polyacrylamide suspensions, *Wat. Sci. Technol.*, **31**, 85.

32. Jones, D. L. and Edwards, A. C. (1998). Influence of sorption on the biological utilization of two simple carbon substrates, *Soil Biol. Biochem.*, **30**, 1895.

33. England, L. S., Lee, H. and Trevors, J. T. (1993). Bacterial survival in soil: effect of clays and protozoa, *Soil Biol. Biochem.*, **25**, 525.

34. Dagostino, L., Goodman, A. E. and Marshall, K. C. (1991). Physiological responses induced in bacteria adhering to surfaces, *Biofouling*, **4**, 113.

35. Davies, D. G. and Geesey, G. G. (1995). Regulation of the alginate biosynthesis gene algC in *Pseudomonas aeruginosa* during biofilm development in continuous culture, *Appl. Environ. Microbiol.*, **61**, 860.

36. Chenu, C. (1995). Extracellular polysaccharides: An interface between microorganisms and soil constituents. In *Environmental Impact of Soil Component Interactions*, Vol. I. *Natural and Anthropogenic Organics*, ed. Huang, P. M., Berthelin, J., Bollag, J. M., McGill, W. B. and Page, A. L., Lewis, Boca Raton, FL, p. 217.

37. Dormaar, J. F. and Foster, R. C. (1991). Nascent aggregates in the rhizosphere of perenial ryegrass (*Lolium perenne L.*). *Can. J. Soil Sci.*, **71**, 465.

38. Tessier, D. and Chenu, C. (1997). Importance of clay fabrics in soils, an approach by electron microscopy, *Geol. Carpathica*, **6**, 55.

39. Foster, R. C. (1988). Microenvironments of soil microorganisms, *Biol. Fertil. Soils*, **6**, 189.

40. Guckert, A., Breisch, H. and Reisinger, O. (1975). Interface sol-racine. I-Etude au microscope électronique des relations mucigel–argiles–microorganismes, *Soil Biol. Biochem.*, **7**, 241.

41. Chotte, J. L., Jocteur Monrozier, L., Villemin, G. and Albrecht, A. (1993). Soil microhabitat and the importance of the fractionation method. In *Soil Organic Matter Dynamics and Sustainability of Tropical Agriculture*, ed. Mulongoy K. and Merckx, R.. John Wiley & Sons, Leuwen, p. 39.

42. Foster, R. C. and Martin, J. K. (1981). In situ analysis of soil components of biological origin. In *Soil Biochemistry*, Vol. 5., ed. Paul, E. A. and Ladd, J. N., Marcel Dekker, NewYork, p. 75.

43. Chenu, C., Pons, C. H. and Robert, M. (1987). Interaction of kaolinite and montmorillonite with neutral polysaccharides. In *Proceedings International Clay Conference, 1985 Denver*, ed. Schultz, L. G., van Olphen, H., and Mumpton, F. A., The Clay Minerals Society, Bloomington, MN, p. 375.

44. Khandall, R. K., Chenu, C., Lamy, I. and Tercé, M. (1992). Adsorption of different polymers on kaolinite and their effect on flumequine adsorption, *Appl. Clay Sci.*, **6**, 343.

45. Malik, M. and Letey, J. (1991). Adsorption of polyacrylamide and polysaccharide polymers on soil materials, *Soil Sci. Soc. Am. J.*, **55**, 380.

46. Chenu, C. (1993). Clay– or sand–polysaccharides associations as models for the interface between microorganisms and soil: water-related properties and microstructure, *Geoderma*, **56**, 143.

47. Chenu, C. and Roberson, E. B. (1996). Diffusion of glucose in microbial extracellular polysaccharide as affected by water potential, *Soil Biol. Biochem.*, **28**, 877.
48. Barrasa, J. M., Gutierrez, A., Escaso, V., Guillen, F., Martinez, M. J. and Martinez, A. T. (1998). Electron and fluorescence microscopy of extracellular glucan and arylalcohol oxidase during wheat-straw degradation by *Pleurotus eryngii*, *Appl. Environ. Microbiol.*, **64**, 325.
49. Green, I., F., Clausen, C. A., Larsen, M. J. and Highley, T. J. (1992). Immunoscanning electron microscopic localization of extracellular wood-degrading enzymes within the fibrillar sheath of the brown-rot fungus *Postia placenta*, *Can. J. Microbiol.*, **38**, 898.
50. Ruel, K. and Joseleau, J. P. (1991). Involvement of an extracellular glucan sheath during degradation of populus wood by *Phanerochaete chrysosporium*. *Appl. Environ. Microbiol.*, **57**, 374.
51. Golchin, A., Clarke, P. and Oades, J. M. (1996). The heterogeneous nature of microbial products as shown by solid-state C-13 CP/MAS NMR spectroscopy, *Biogeochem.*, **34**, 71.
52. Quiquampoix, H., Abadie, J., Baron, M. H., Leprince, F., Maromoto-Pintro, P., Ratcliffe, R. G. and Staunton, S. (1995). Mechanisms and consequences of protein adsorption on soil mineral surfaces, *ACS Symp. Ser.*, **602**, 321.
53. Calamai, L., Lozzi, L., Stotzky, G., Fusi, P. and Ristori, G. (2000). Interaction of catalase on montmorillonite homonionic to cations with different hydrophobicity: effect on enzyme activity and microbial utilization, *Soil Biol. Biochem.*, **32**, 815.
54. Claus, H. and Filip, Z. (1990). Effects of clays and other solids on the activity of phenoloxidases produced by some fungi and actinomycetes, *Soil Biol. Biochem.*, **22**, 483.
55. Gianfreda, L., Rao, M. A. and Violante, A. (1991). Adsorption, activity and kinetic properties of urease on montmorillonite, aluminium hydroxide and $Al(OH)_x$–montmorillonite complexes, *Soil Biol. Biochem.*, **23**, 581.
56. Leprince, F. and Quiquampoix, H. (1996). Extracellular enzyme activity in soil: effect of pH and ionic strength on the interaction with montmorillonite of two acid phosphatases secreted by the ectomycorrhizal fungus *Hebeloma cylindrosporum*. *Eur. J. Soil Sci.*, **47**, 511.
57. Masaphy, S., Fahima, T., Levanon, D., Henis, Y. and Mingelgrin, U. (1996). Parathion degradation by *Xanthomonas* and its crude enzyme extract in clay suspensions, *J. Environ. Qual.*, **25**, 1248.
58. Quiquampoix, H., Staunton, S., Baron, M. H. and Ratcliffe, R. G. (1993). Interpretation of the pH dependence of protein adsorption on clay mineral surfaces and its relevance to the understanding of extracellular enzyme activity in soil. *Colloids Surf. A: Physicochem. Eng. Aspects*, **75**, 85.
59. Quiquampoix, H. and Ratcliffe, R. G. (1992). A ^{31}PNMR study of the adsorption of bovine serum albumin on montmorillonite using phosphate and the paramagnetic cation Mn^{2+}: modification of conformation with pH. *J. Colloid Interface Sci.*, **148**, 343.
60. Burns, R. G. (1990). Microorganisms, enzymes and soil colloid surfaces. In *Soil Colloids and their Associations in Aggregates*, ed. de Boodt, M., Hayes, M. and Herbillon, A., Plenum, New York, p. 337.
61. Ruggiero, P., Dec, J. and Bollag, J. M. (1996). Soil as a catalytic system. In *Soil Biochemistry*, Vol. 9, ed. Stotzky, G. and Bollag, J. M., Marcel Dekker, New York, p. 79
62. Lavie, S. and Stotzky, G. (1986). Interaction between clay minerals and siderophores affect the respiration of *Histoplasma capsulatum*. *Appl. Environ. Microbiol.*, **51**, 74.

63. Saxena, D., Flores, S. and Stotzky, G. (1999). Insecticidal toxin in root exudates of *Bt* corn, *Nature*, **402**, 480.
64. Saxena, D. and Stotzky, G. (2000). Insecticidal toxin from *Bacillus thuringiensis* is released from roots of transgenic *Bt* corn *in vitro* and *in situ*. *FEMS Microbiol. Ecol.*, **33**, 35.
65. Losey, J. E., Raynor, L. S. and Cater, M. E. (1999). Transgenic pollen harms monarch larvae, *Nature* (London), **399**, 214.
66. Crecchio, C. and Stotzky, G. (1998). Insecticidal activity and biodegradation of the toxin from *Bacillus thuringiensis* subsp *kurstaki* bound to humic acids from soils, *Soil Biol. Biochem.*, **30**, 463.
67. Tapp, H., Calamai, L. and Stotzky, G. (1994). Adsorption and binding of the insecticidal proteins from *Bacillus thuringensis* subsp. *kurstaki* and subsp *tenebrionis* on clay minerals, *Soil Biol. Biochem.*, **26**, 663.
68. Venkateswerlu, G. and Stotzky, G. (1992). Binding of the protoxin and toxin of *Bacillus thuringiensis* subsp. *kurstaki* on clay minerals, *Curr. Microbiol.*, **25**, 1.
69. Tapp, H. and Stotzky, G. (1995). Insecticidal activity of the toxins from *Bacillus thuringiensis* subspecies *kurstaki* and *tenebrionis* adsorbed and bound on pure and soil clays, *Appl. Environ. Microbiol.*, **61**, 1786.
70. Koskella, J. and Stotzky, G. (1997). Microbial utilization of free and clay-bound insecticidal toxins from *Bacillus thuringiensis* and their retention of insecticidal activity after incubation with microbes, *Appl. Environ. Microbiol.*, **63**, 3561.
71. Lorenz, M. G. and Wackernagel, W. (1994). Bacterial gene transfer by natural genetic transformation in the environment, *Microbiol. Rev.*, **58**, 563–602.
72. Khanna, M. and Stotzky, G. (1992). Transformation of *Bacillus subtilis* by DNA bound on montmorillonite and effect of DNase on the transforming ability of bound DNA, *Appl. Environ. Microbiol.*, **58**, 1930.
73. Ogram, A., Sayler, G. S., Gustin, D. and Lewis, R. J. (1988). DNA adsorption to soils and sediments, *Environ. Sci. Technol.*, **22**, 982.
74. Paget, E., Jocteur Monrozier, L. and Simonet, P. (1992). Adsorption of DNA on clay minerals: protection against DNase I and influence on gene transfer, *FEMS Microbiol. Lett.*, **97**, 31.
75. Poly, F., Chenu, C., Simonet, P., Rouiller, J. and Jocteur Monrozier, L. (2000). Differences between linear chromosomal and supercoiled plasmid DNA in their mechanisms and extent of adsorption on clay minerals, *Langmuir*, **16**, 1233.
76. Gallori, E., Bazzicallupo, M., Dal Canto, L., Fani, R., Nannipieri, P., Vettori, C. and Stotzky, G. (1994). Transformation of *Bacillus subtilis* by DNA bound on clay in nonsterile soil, *FEMS Microbiol. Ecol.*, **15**, 119.
77. Crecchio, C. and Stotzky, G. (1998). Binding of DNA on humic acids: effect on transformation of *Bacillus subtilis* and resistance to DNase, *Soil Biol. Biochem.*, **30**, 1061.
78. Alvarez, A. J., Khanna, M., Toranzos, G. A. and Stotzky, G. (1998). Amplification of DNA bound on clay minerals, *Mol. Ecol.*, **7**, 775.
79. Vettori, C., Paffetti, D., Pietramellara, G., Stotzky, G. and Gallori, E. (1996). Amplification of bacterial DNA bound on clay minerals by the random amplified polymorphic DNA (RAPD) technique, *FEMS Microbiol. Ecol.*, **20**, 251.
80. Khanna, M., Yoder, M., Calamai, L. and Stotzky, G. (1998). X-ray diffractometry and electron microscopy of DNA bound on clay minerals, *Sci. Soils, http://link-springer.de/link/service/journals*, **3**, 1.
81. Stotzky, G. (1989). Gene transfer among bacteria in soil. In *Gene Transfer in the Environment*, ed. Levy, S. B. and Miller, R. V., McGraw-Hill, New York, p. 165.

82. Yin, X. and Stotzky, G. (1997). Gene transfer among bacteria in natural environments, *Adv. Appl. Microbiol.*, **45**, 153.
83. Crocker, F. H., Guerin, W. F. and Boyd, S. A. (1995). Bioavailability of naphthalene sorbed to cationic surfactant- modified smectite clay, *Environ. Sci. Tech.*, **29**, 2953.
84. Dashman, T. and Stotzky, G. (1986). Microbial utilization of amino acids and a peptide bound on homoionic montmorillonite and kaolinite, *Soil Biol. Biochem.*, **18**, 5.
85. Miller, M. E. and Alexander, M. (1991). Kinetics of bacterial degradation of benzylamine in a montmorillonite suspension, *Environ. Sci. Technol.*, **25**, 240.
86. Bolton, H., Jr and Girvin, D. C. (1996). Effect of adsorption on the biodegradation of nitrilotriacetate by *Chelatobacter heintzii*. *Environ. Sci. Tech.*, **30**, 2057.
87. Gordon, A. S. and Millero, F. J. (1985). Adsorption mediated decrease in the biodegradation rate of organic compounds, *Microb. Ecol.*, **11**, 289.
88. Guerin, W. F. and Boyd, S. A. (1997). Bioavailability of naphthalene associated with natural and synthetic sorbents, *Wat. Res.*, **31**, 1504.
89. Guerin, W. F. and Boyd, S. A. (1992). Differential bioavailability of soil sorbed naphthalene to two bacterial species, *Appl. Environ. Microbiol.*, **58**, 1142.
90. Cortez, J. (1977). Adsorption sur les argiles de deux lipopolysaccharides rhizosphériques, *Soil Biol. Biochem.*, **9**, 25.
91. Olness, A. and Clapp, C. E. (1972). Microbial degradation of a montmorillonite-dextran complex, *Soil Sci. Soc. Am. Proc.*, **36**, 179.
92. Scow, K. M. and Johnson, C. R. (1997). Effect of sorption on biodegradation of soil pollutants, *Adv. Agron.*, **58**, 1.
93. Scow, K. M., Simkins, S. and Alexander, M. (1986). Kinetics of mineralization of organic compounds at low concentrations in soils, *Appl. Environ. Microbiol.*, **51**, 1028.
94. Barlett, J. R. and Doner, H. E. (1988). Decomposition of lysine and leucine in soil aggregates: adsorption and compartmentalization, *Soil Biol. Biochem.*, **20**, 755.
95. Fu, M. H., Mayton, H. and Alexander, M. (1994). Desorption and biodegradation of sorbed styrene in soil and aquifer solids, *Environ. Toxicol. Chem.*, **13**, 749.
96. Steinberg, S. M., Pignatello, J. J. and Sawhney, B. L. (1987). Persistence of 1,2-dibromoethane in soils: entrapment in intraparticle micropores, *Environ. Sci. Technol.*, **21**, 1201.
97. Pignatello, J. J., Sawney, B. L. and Frink, C. R. (1987). EDB : Persistence in soil, *Science*, **237**, 898.
98. Balesdent, J., Chenu, C. and Balabane, M. (2000). Relationship of soil organic matter dynamics to physical protection and tillage, *Soil Till. Res.*, **53**, 215.
99. Beare, M. H., Cabrera, M. L., Hendrix, P. F. and Coleman, D. C. (1994). Aggregate-protected and unprotected organic matter pools in conventional-tillage and no-tillage soils, *Soil Sci. Soc. Am. J.*, **58**, 787.
100. Hassink, J. (1992). Effects of soil texture and structure on carbon and nitrogen mineralization in grassland soils, *Biol. Fertil. Soils*, **14**, 126.
101. Powlson, D. S. (1980). The effect of grinding on microbial and non-microbial organic matter in soil, *J. Soil Sci.*, **31**, 77.
102. White, J. C., Quinones-Rivera, A. and Alexander, M. (1998). Effect of wetting and drying on the bioavailability of organic compounds sequestered in soil, *Environ. Toxicol. Chem.*, **17**, 2378.
103. Gregorich, E. G., Kachanovski, R. G. and Voroney, R. P. (1989). Carbon mineralization in soil size fractions after various amounts of aggregate disruption, *J. Soil Sci.*, **28**, 417.

104. Breland, T. A. (1994). Enhanced mineralization and denitrification as a result of heterogeneous distribution of clover residues in soil, *Plant Soil*, **166**, 1.
105. Hatzinger, P. B. and Alexander, M. (1997). Biodegradation of organic compounds sequestered in organic solids or in nanopores within silica particles, *Environ. Toxicol. Chem.*, **16**, 2215.
106. Nam, K. and Alexander, M. (1998). Role of nanoporosity and hydrophobicity in sequestration and bioavailability: Tests with model solids, *Environ. Sci. Technol.*, **32**, 71.
107. Scow, K. M. and Alexander, M. (1992). Effect of diffusion on the kinetics of biodegradation: Experimental results with synthetic aggregates, *Soil Sci. Soc. Am. J.*, **56**, 128.
108. Chotte, J. L., Villemin, G., Guilloré, P. and Jocteur Monrozier, L. (1994). Morphological aspects of microorganism habitats in a vertisol. In *Soil Micromorphology. Proc. IX Int. Working Meeting on Soil Micromorphology*, Vol. 22, ed. Ringrose-Voase, A. J. and Humpreys, G. S., Elsevier, Amsterdam, p. 395.
109. Kilbertus, G. (1980). Etude des microhabitats contenus dans les agrégats de sol, leur relation avec la biomasse bactérienne et la taille des procaryotes présents, *Rev. Ecol. Biol. Sols*, **17**, 205.
110. Hassink, J., Matus, F., Chenu, C. and Dalenberg, J. W. (1997). Interactions between soil biota, soil organic matter and soil structure. In *Soil Ecology in Sustainable Agricultural Systems*, ed. Brussard, L. and Ferrara Ceccato, R., CRC Lewis, Boca Raton, FL, p. 15.
111. Hassink, J., Bouwman, J. and Brussaard, L. B. (1993). Relationships between habitable pore space, soil biota and mineralization rates in grassland soils, *Soil Biol. Biochem.*, **25**, 47.
112. Manilal, V. B. and Alexander, M. (1991). Factors affecting the microbial degradation of phenanthrene in soil, *Appl. Microbiol. Biotechnol.*, **35**, 401.
113. Nam, K., Chung, N. and Alexander, M. (1998). Relationship between organic matter content of soil and the sequestration of phenanthrene, *Environ. Sci. Technol.*, **32**, 3785.
114. NamHyun., C. and Alexander, M. (1998). Differences in sequestration and bioavailability of organic compounds aged in dissimilar soils, *Environ. Sci. Technol.*, **32**, 855.
115. Mihelcic, J. R. and Luthy, R. G. (1991). Sorption and microbial degradation of naphthalene in soil-water suspensions under denitrification conditions, *Environ. Sci. Technol.*, **25**, 169.
116. Priesack, E. and Kisser-Priesack, G. M. (1993). Modelling diffusion and microbial uptake of 13C-glucose in soil aggregates, *Geoderma*, **56**, 561.
117. Rijnaarts, H. H., Bachmann, A., Jumemer, J. C. and Zehnder, A. J. B. (1990). Effect of desorption and intraparticle mass transfer on the aerobic biomineralization of α-hexachlorocyclohexane, *Environ. Sci. Technol.*, **24**, 1349.
118. Gupta, V. V. S. R. and Germida, J. J. (1988). Distribution of microbial biomass and its activity in different soil aggregate size classes as affected by cultivation, *Soil Biol. Biochem.*, **20**, 777.
119. Jocteur Monrozier, L., Ladd, J. N., Fitzpatrick, R. W., Foster, R. C. and Raupach, M. (1991). Physical properties, mineral and organic components and microbial biomass content of size fraction in soils of contrasting aggregation, *Geoderma*, **49**, 37.
120. Kabir, M., Chotte, J. L., Rahman, M., Bally, R. and Jocteur Monrozier, L. (1994). Distribution of soil fractions and location of soil bacteria in vertisol under cultivation and perennial grass, *Plant Soil*, **163**, 243.

121. Lensi, R., Clays-Josserand, A. and Jocteur Monrozier, L. (1995). Denitrifiers and denitrifying activity in size fractions of a Mollisol under permanent pasture and continuous cultivation, *Soil Biol. Biochem.*, **27**, 61.

122. Mendes, I. C. and Bottomley, P. J. (1998). Distribution of population of *Rhizobium leguminosarum* bv *trifolii* among different size classes of soil aggregates, *Appl. Environ. Microbiol.*, **64**, 970.

123. Recorbet, G., Richaume, A. and Jocteur-Monrozier, L. (1995). Distribution of a genetically-engineered *Escherichia coli* population introduced into soil, *Lett. Appl. Microbiol.*, **21**, 38.

124. Scheu, S., Maraun, M., Bonkowski, M. and Alphei, S. (1996). Microbial biomass and respiratory activity in soil aggregates of different sizes from three beechwood sites on a basalt hill, *Biol. Fertil. Soils*, **21**, 69.

125. Singh, S. and Singh, J. S. (1995). Microbial biomass associated with water-stable aggregates in forest, savanna and cropland soils of a seasonally dry tropical region, India, *Soil Biol. Biochem.*, **27**, 1027.

126. Van Gestel, M., Merckx, R. and Vlassak, K. (1996). Spatial distribution of microbial biomass in microaggregates of a silty-loam soil and their relation with the resistance of microorganisms to soil drying, *Soil Biol. Biochem.*, **28**, 503.

127. Drazzkiewicz, M. (1994). Distribution of microorganisms in soil aggregates: effect of aggregate size, *Folia Microbiol.*, **39**, 276.

128. Hattori, T. (1988). Soil aggregates as microhabitats for microorganisms, *Rep. Inst. Agr. Res. Tohoku Univ.*, **37**, 23.

129. Ranjard, L., Richaume, A., Jocteur Monrozier, L. and Nazaret, S. (1997). Response of soil bacteria to Hg(II) in relation to soil characteristics and cell location, *FEMS Microbiol. Ecol.*, **24**, 321.

130. Ronn, R., Griffiths, B., Ekelund, F. and Christensen, S. (1996). Spatial distribution and successional pattern of microbial activity and micro-faunal populations on decomposing barley roots, *J. Appl. Ecol.*, **33**, 662.

131. Gaillard, V., Chenu, C., Recous, S. and Richard, G. (1999). C, N and microbial gradients induced by plant residues decomposing in soil, *Eur. J. Soil Sci.*, **50**, 567.

132. Heijnen, C. E., van Elsas, J., Kuikman, P. J. and van Veen, J. J. (1988). Dynamics of *Rhizobium leguminosarum* biovar *trifollii* introduced to soil; the effect of bentonite clay on predation by protozoa, *Soil Biol. Biochem.*, **20**, 483.

133. Heijnen, C. E., Chenu, C. and Robert, M. (1993). Micromorphological studies on clay amended and unamended loamy sand, relating survival of introduced bacteria and soil structure, *Geoderma*, **56**, 195.

134. Heijnen, C. E. and van Veen, J. A. (1991). A determination of protective microhabitats for bacteria introduced into soil, *FEMS Microbiol. Ecol.*, **85**, 73.

135. Young, I. M., Roberts, A., Griffiths, B. S. and Caul, S. (1994). Growth of a ciliate protozoan in model ballotini systems of different particle sizes, *Soil Biol. Biochem.*, **26**, 1173.

136. Vargas, R. and Hattori, T. (1986). Protozoan predation of bacterial cells in soil aggregates, *FEMS Microbiol. Ecol.*, **38**, 233.

137. Rutherford, P. M. and Juma, N. G. (1991). Influence of the texture on habitable pore space and bacterial-protozoa populations in soil, *Biol. Fertil. Soils*, **12**, 221.

138. Görres, H., Savin, M. C., Naher, D. A., T. R., W. and Amador, J. A. (1999). Grazing in a porous environment: 1. The effect of pore structure on C and N mineralization, *Plant Soil*, **212**, 75.

139. Duquenne, P., Chenu, C., Richard, G. and Catroux, G. (1999). Effect of carbon source supply and its location on competition between inoculated and established bacterial strains in sterile soil microcosm, *FEMS Microbiol. Ecol.*, **29**, 331.

140. Tessier, D. (1990). Behaviour and microstructure of clay minerals. In *Soil Colloids and their Associations in Aggregates*, Nato ASI Series, Vol. 215, ed. De Boot, M. F., Hayes, M. H. B. and Herbillon, A., Plenum, New York, p. 387.

141. Bruand, A. and Prost, R. (1987). Effect of water content on the fabric of a soil material: an experimental approach, *J. Soil Sci.*, **38**, 461.

142. Bruand, A., Tessier, D. and Baize, D. (1988). Clayey soils water retention: significance of the clay fabrics, *C.R. Acad. Sci. Paris*, **307(II)**, 1937.

143. Tessier, D., Lajudie, A. and Petit, J. C. (1992). Relation between macroscopic behaviour of clays and their microstructural properties, *Appl. Geochem.*, Suppl. Issue No **1**, 151.

144. Biederbeck, V. O. and Geissler, H. J. (1993). Effect of storage temperatures on *Rhizobium meliloti* survival in peat- and clay-based inoculants, *Can. J. Plant Sci.*, **73**, 101.

145. Bushby, H. V. A. and Marshall, K. C. (1977). Some factors affecting the survival of root-nodule bacteria on desiccation, *Soil Biol. Biochem.*, **9**, 143.

146. Chao, W. L. and Alexander, M. (1982). Influence of soil characteristics on the survival of *Rhizobium* in soils undergoing drying, *Soil Sci. Soc. Am. J.*, **46**, 949.

147. Hartel, P. G. and Alexander, M. (1986). Role of extracellular polysaccharide production and clays in the desiccation tolerance of cowpea *Bradyrhizobia*, *Soil Sci. Soc. Am. J.*, **50**, 1193.

148. Osa-Afiana, L. O. and Alexander, M. (1982). Clays and the survival of *Rhizobium* in soil during desiccation, *Soil Sci. Soc. Am. J.*, **46**, 285.

149. Morra, M. J., Chaverra, M. H., Dandurand, L. M. and Orser, C. S. (1998). Survival of *Pseudomonas fluorescens* 2–79RN10 in clay powders undergoing drying, *Soil Sci. Soc. Am. J.*, **62**, 663.

150. Schmit, J., Prior, P., Quiquampoix, H. and Robert, M. (1990). Studies on survival and localization of *Pseudomonas solanacearum* in clays extracted from Vertisols. In *Plant Pathogenic Bacteria*, ed. Klement, Z., Akademia Kiado, Budapest, p. 1001.

151. Sierra, G., Renault, P. and Vallès, V. (1995). Anaerobiosis in saturated soil aggregates: modelling and experiment, *Eur. J. Soil Sci.*, **46**, 519.

152. Holberg, O. and Sorensen, J. (1993). Microgradients of microbial oxygen consumption in a barley rhizosphere model system, *Appl. Environ. Microbiol.*, **59**, 431.

153. Sextone, A. J., Revsbech, N. P., Parkin, T. B. and Tiedje, J. M. (1985). Direct measurement of oxygen profiles and denitrification rates in soil aggregates, *Soil Sci. Soc. Am. J.*, **49**, 645.

154. Jaillard, B., Ruiz, L. and Arvieu, J. C. (1996). pH mapping in transparent gel using color indicator videodensitometry, *Plant Soil*, **183**, 85.

155. Revsbech, N. P., Pedersen , O., Reichardt, W. and Briones, A. (1999). Microsensor analysis of oxygen and pH in the rice rhizosphere under field and laboratory conditions, *Biol. Fertil. Soils*, **29**, 379.

156. Scott, N. A., Cole, C. V., Elliott, E. T. and Huffman, F. A. (1996). Soil textural control on decomposition and soil organic matter dynamics, *Soil Sci. Soc. Am. J.*, **60**, 1102.

157. Soroker, E. (1990). *Low water content and low water potential as determinants of microbial fate in soils*, thesis, Univ. California Berkeley.

158. Philippot, L., Renault, P., Sierra, J., Hénault, C., Clays-Josserand, A., Chenu, C., Chaussod, R. and Lensi, R. (1996). Dissimilatory nitrite-reductase provides a competitive advantage to *Pseudomonas* sp. RTC01 to colonise the center of soil aggregates, *FEMS Microbiol. Ecol.*, **21**, 175.

159. Sierra, J. and Renault, P. (1995). Oxygen consumption by soil microorganisms as affected by oxygen and carbon dioxide levels, *Appl. Soil Ecol.*, **2**, 175.
160. Fazzolari, E., Guérif, J., Nicolardot, B. and Germon, J. C. (1998). A method for the preparation of repacked soil cores with homogeneous aggregates for studying microbial nitrogen transformations under highly controlled physical conditions, *Eur. J. Soil Biol.*, **34**, 39.
161. Fruit, L., Recous, S. and Richard, G. (1999). Plant residue decomposition: effect of soil porosity and particle size. In *Effect of Mineral–Organic–Microorganisms Interactions on Soil and Freshwater Environments*, ed. Berthelin, J., Huang, P. M., Bollag, J. M. and Andreux, F., Kluwer/Plenum, New York, p. 189.
162. Angers, D. A. and Chenu, C. (1997). Dynamics of soil aggregation and C sequestration. In *Soil Processes and the Carbon Cycle*, ed Lal, R., Kimble, J., Follet, R. F. and Stewart, B. A., CRC Press, Boca Raton, FL, p. 199.
163. Chenu, C. and Tessier, D. (1995). Low temperature scanning electron microscopy of clay and organic constituents and their relevance to soil microstructures, *Scanning Microsc.*, **9**, 989.
164. Degens, B. P., Sparling, G. P. and Abbott, L. K. (1996). Increasing the length of hyphae in a sandy soil increases the amount of water-stable aggregates, *Appl. Soil Ecol.*, **3**, 149.
165. Dorioz, J. M., Robert, M. and Chenu, C. (1993). The role of roots, fungi and bacteria on clay particle organization. An experimental approach, *Geoderma*, **56**, 179.
166. Tisdall, J. M. (1991). Fungal hyphae and structural stability of soil, *Aust. J. Soil Res.*, **29**, 729.
167. Miller, R. M. and Jastrow, J. D. (1990). Hierarchy of root and mycorrhizal fungal interactions with soil aggregation, *Soil Biol. Biochem.*, **22**, 579.
168. Degens, B. P., Sparling, G. P. and Abbott, L. K. (1994). The contribution from hyphae, roots and organic carbon constituents to the aggregation of a sandy loam under long term clover-based and grass pastures, *Eur. J. Soil Sci.*, **45**, 459.
169. Falchini, L., Sparvoli, E. and Tomaselli, L. (1997). Effect of *Nostoc* (Cyanobacteria) inoculation on the structure and stability of clay soils, *Biol. Fertil. Soils*, **23**, 346.
170. Amellal, N., Bartoli, F., Villemin, G., Talouizte, A. and Heulin, T. (1999). Effects of inoculation of EPS-producing *Pantoea agglomerans* on wheat rhizosphere aggregation, *Plant Soil*, **211**, 93.
171. Chenu, C. (1989). Influence of a fungal polysaccharide, scleroglucan, on clay microstructures, *Soil Biol. Biochem.*, **21**, 299.
172. Chenu, C. and Guérif, J. (1991). Mechanical strength of clay minerals as influenced by an adsorbed polysaccharide, *Soil Sci. Soc. Am. J.*, **55**, 1076.
173. Chenu, C., Guérif, J. and Jaunet, A. M. (1994).Polymer bridging: a mechanism of clay and soil structure stabilization by polysaccharides. In *Proceedings of XVth World Congress of Soil Science, Acapulco, Mexico*, ISSS, Mexico, p.403.
174. Hallett, P. D. and Young, I. M. (1999). Changes in water repellence of soil aggregates caused by substrate-induced microbial activity, *Eur. J. Soil Sci.*, **50**, 35.
175. Jouany, C. and Chassin, P. (1987). Determination of the surface energy of clay-organic complexes from contact angles measurements, *Colloids Surf.*, **27**, 289.
176. Jouany, C. (1991). Surface free energy components of clay–synthetic humic acid complexes from contact-angle measurements, *Clays Clay Miner.*, **39**, 43.
177. Jańczuk, B., Hajnos, M., Bialopiotrowicz, T., Kliszcs and Biliński, B. (1990). Hydrophobization of the soils by dodecylammonium hydrochloride and changes of the components of its surface energy, *Soil Sci.*, **150**, 792.

178. Capriel, P., Beck, T., Borchert, H. and Härter, P. (1990). Relationships between soil aliphatic fraction extracted with supercritical hexane, soil microbial biomass, and soil aggregate stability, *Soil Sci. Soc. Am. J.*, **54**, 415.
179. Wright, S. F. and Upadhyaya, A. (1996). Extraction of an abundant and unusual protein from soil and comparison with hyphal protein of arbuscular mycorrhizal fungi, *Soil Sci.*, **161**, 575.
180. Wright, S. F. and Upadhyaya, A. (1998). A survey of soil for aggregate stability and glomalin, a glycoprotein produced by hyphae of arbuscular mycorrhizal fungi, *Plant Soil*, **198**, 97.
181. Hu, H., Coleman, D. C., Beare, M. H. and Hendrix, P. F. (1995). Soil carbohydrates in aggrading and degrading agroecosystems—Influences of fungi and aggregates, *Agric. Ecosys. Environ.*, **54**, 77.
182. Bethlenfalvay, G. J., Cantrell, I. C., Mihara, K. L. and Schreiner, R. P. (1999). Relationships between soil aggregation and mycorrhizae as influenced by soil biota and nitrogen nutrition, *Biol. Fertil. Soils*, **28**, 356.
183. Chantigny, M. H., Angers, D. A., Prevost, D., Vezina, L. P. and Chalifour, F. P. (1997). Soil aggregation and fungal and bacterial biomass under annual and perennial cropping systems, *Soil Sci. Soc. Am. J.*, **61**, 262.
184. Molope, M. B., Grieve and Page, E. R. (1987). Contributions by fungi and bacteria to aggregate stability of cultivated soils, *J. Soil Sci.*, **38**, 71.
185. Angers, D. A., Samson, N. and Légère, A. (1993). Early changes in water-stable aggregation induced by rotation and tillage in a soil under barley production, *Can. J. Soil Sci.*, **73**, 51.
186. Haynes, R. J. (1999). Labile organic matter fractions and aggregate stability under short-term, grass-based leys, *Soil Biol. Biochem.*, **31**, 1821.
187. Haynes, R. J. and Beare, M. H. (1997). Influence of six crop species on aggregate stability and some labile organic matter fractions, *Soil Biol. Biochem.*, **29**, 29.
188. Roberson, E. B., Sarig, S., Shennan, C. and Firestone, M. K. (1995). Nutritional management of microbial polysaccharide production and aggregation in an agricultural soil, *Soil Sci. Soc. Am. J.*, **59**, 1587.
189. Sarig, S., Roberson, E. B. and Firestone, M. K. (1993). Microbial activity—soil structure: response to saline water irrigation, *Soil Biol. Biochem.*, **25**, 693.
190. Ma'shum, M. and Farmer, V. C. (1985). Origin and assessment of water repellency of a sandy south Australian soil, *Aust. J. Soil Res.*, **23**, 623.
191. Haynes, R. J. (1993). Effect of sample pretreatment on aggregate stability measured by wet sieving or turbidimetry on soils of different cropping history, *J. Soil Sci.*, **44**, 261.
192. Caron, J., Kay, B. D. and Stone, J. A. (1992). Improvement of aggregate stability of a clay loam with drying, *Soil Sci. Soc. Am. J.*, **56**, 1583.
193. Preston, S., Griffiths, B. S. and Young, I. M. (1999). Links between substrate additions, native microbes, and the structural complexity and stability of soils, *Soil Biol. Biochem.*, **31**, 1541.
194. Roberson, E. B., Sarig, S. and Firestone, M. K. (1991). Cover crop management of polysaccharide-mediated aggregation in an orchard soil, *Soil Sci. Soc. Am. J.*, **55**, 734.
195. Amellal, N., Burtin, G., Bartoli, F. and Heulin, T. (1998). Colonization of wheat roots by an exopolysaccharide- producing *Pantoea agglomerans* strain and its effect on rhizosphere soil aggregation, *Appl. Environ. Microbiol.*, **64**, 3740.
196. Gouzou, L., Burtin, G., Philippy, R., Bartoli, F. and Heulin, T. (1993). Effect of inoculation with *Bacillus polymixa* on soil aggregation in the wheat rhizosphere: preliminary examination, *Geoderma*, **56**, 479.

197. Rao, D. L. N. and Burns, R. G. (1990). Effect of surface growth of blue green algae and bryophytes on some microbiological, biochemical and physical soil properties, *Biol. Fertil. Soils*, **9**, 239.

198. Rogers, S. L. and Burns, R. G. (1994). Changes in aggregate stability, nutrient status, indigenous microbial populations, and seedling emergence following inoculation with *Nostoc muscorum. Biol. Fertil. Soils*, **18**, 219.

199. Robert, M., Hardy, M. and Elsass, F. (1991). Crystallochemistry, properties and organization of soil clays derived from major sedimentary rocks in France, *Clay Miner.*, **26**, 409.

200. Theng, B. K. G., Ristori, G. G., Santi, C. A. and Percival, H. J. (1999). An improved method for determining the specific surface areas of topsoils with varied organic matter content, texture and clay mineral composition, *Eur. J. Soil Sci.*, **50**, 309.

201. Quirk, J. P. and Murray, R. S. (1999). Appraisal of the ethylene glycol monoethyl ether method for measuring hydratable surface area of clays and soil, *Soil Sci. Soc. Am. J.*, **63**, 839.

202. de Jonge, H. and Mittelmeijer-Hazeleger, M. (1996). Adsorption of CO_2 and N_2 on soil organic matter: nature of porosity, surface area and diffusion mechanisms, *Environ. Sci. Technol.*, **30**, 408.

203. Van Olphen, H. and Fripiat, J. J. (1979). *Data Handbook for Clay Minerals and Other Non-metallic Minerals*, Pergamon, Oxford.

204. Theng, B. K. G. (1995). On measuring the specific surface area of clays and soils by adsorption of para-nitrophenol: use and limitations. In *Proceedings of 10th International Clay Conference, Adelaide*, CSIRO, Adelaide, p. 304.

205. Oades, J. M. (1989). An introduction to organic matter in mineral soils. In *Minerals in Soil Environments*, ed. Dixon, S. B. and Weed, S. B., Soil Science Society of America, Madison, WI, p. 89.

206. Feller, C. (1995). *La matière organique dans les sols tropicaux à argile 1:1. Recherche de compartiments fonctionnels. Une approche granulométrique*, Orstom Editions, Paris.

207. Chrétien, J. and Tessier, D. (1988). Influence du squelette sur les propriétés physiques de sols, *Sci. Sol.*, **5**, 255.

208. Bigorre, F. (2000). *Influence de la pédogenèse et de l'usage des sols sur leurs propriétés physiques. Mécanismes d'évolution et éléments de prévision*, Ph.D. Thesis, Université de Nancy I.

209. Roberson, E. B., Chenu, C. and Firestone, M. K. (1993). Microstructural changes in bacterial exopolysaccharides during desiccation, *Soil Biol. Biochem.*, **25**, 1299.

210. Hillel, D. 1980. *Fundamentals of Soil Physics*, Academic Press, New York, 413 pp.

211. Chenu, C., Hassink, J. and Bloem, J. (2001). Short-term changes in the spatial distribution of microorganisms in soil aggregates due to glucose additions. *Biol. Fertil. Soils*, **33**, 349.

2 A Fractal Approach for Interactions between Soil Particles and Microorganisms

NICOLA SENESI
Università di Bari, Italy

LYNNE BODDY
University of Wales, Cardiff, UK

Interactions between Soil Particles and Microorganisms
Edited by P. M. Huang, J.-M. Bollag and N. Senesi. © 2002 John Wiley & Sons, Ltd

1 INTRODUCTION

Fractal geometry provides a powerful approach for quantitative description of complex, highly irregular and random, or disordered, systems [1]. Moreover, it can also be used to describe the processes leading to the formation of such systems and their physical behaviour. Fractal geometry relates to structures which cannot be described in Euclidean whole number dimensions of 1 (straight lines), 2 (flat surfaces) or 3 (volumes), but have fractional dimensions. In two dimensions, structures can have fractal dimensions between 1 and 2 (a completely filled plane), or up to 3 in three dimensions. The degree of irregularity of such structures is independent of scale. Thus, when they are examined under increasing magnification, more and more irregularities come into view and, even though they can be circumscribed within finite planes or volumes, their lengths or areas are infinite when viewed on an infinitesimal scale. 'Ideal' or 'regular' fractal structures exhibit 'self-similarity' (i.e. the structure can be decomposed into smaller copies of itself, so that when any portion of the structure is magnified it will appear identical to a larger part) over all characterisation length scales. Natural structures tend to be self-similar only over a finite range of length scales [2], but they are nonetheless better approximated as fractals than as smooth figures, and are termed 'random' fractals. Many natural objects/systems have been shown experimentally to be fractal, including star constellations, clouds, coastlines, trees, snowflakes, brain circumvolutions, proteins, colloidal aggregates, cellulose, several minerals and clays, limestones and sandstones, sediments, soils, and their organic, mineral and microbial components and physical, chemical and biological properties (e.g. [3–13]).

This chapter introduces the general principles and concepts of fractal geometry and how these can be quantified in soil systems (section 2); the fractal nature of soil organic and mineral particles (sections 3 and 4) and of soil microorganisms (section 5) is described; the ways in which the fractal nature of different components of soil systems affect interactions among and between each other are then considered (sections 6); the chapter ends with some concluding comments and recommendations for future research (section 7).

2 FRACTAL PRINCIPLES AND METHODOLOGY

2.1 NATURAL FRACTALS

Mathematical or nonrandom fractals are scale invariant, i.e. the pattern is the same at all scales (self-similar). Natural, real or random fractals are 'quasi' or 'statistically' self-similar within some finite length scale that is often determined by the characterisation technique used. An object or a process can be classified as fractal when this length scale covers at least one order of magnitude of the property being measured. Fractal structures obey a power law, allowing the fractal dimension, D, to be determined from experimental data:

$$p \propto v^\gamma \tag{1}$$

where p is the property of interest, v is the variable measured and the exponent γ can be related to the fractal dimension, D. For example, in Figure 1 the dark area can be considered as a two-dimensional representation of a natural structure, e.g. a soil particle or microbial colony, whose area is $A(s)$. The area can be

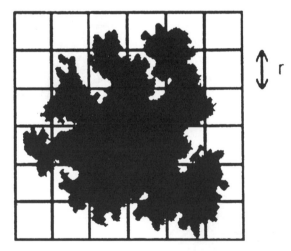

Figure 1. A fractal structure overlaid by a grid of squares. In three dimensions the squares are replaced by boxes. Surface, mass and pore fractal dimensions can be estimated by counting the number of squares occupied by the object. Two dimensional images of three-dimensional objects do not contain the full information about the object, so fractal dimensions of three-dimensional objects need to be estimated in three-dimensions. After [11]. Reproduced by permission of John Wiley & Sons, Ltd

estimated by counting the number of cells of size s^2 that cover the object. For a fractal object

$$A(s) \propto s^{2-D} \qquad (2)$$

By making measurements for different values of s, D can be estimated (section 2.3).

2.2 MASS, SURFACE AND PORE FRACTALS

Environmental particles, microbial colonies, and even patterns of movement of unitary organisms can be characterised in terms of several fractal dimensions [14] (Table 1). The fractal dimension of the surface (boundary/interface) of a solid structure, D_S, is obviously an important characteristic, but many physical properties of solids also depend on the scaling behaviour of the entire solid and of the pore space within. Systems where surface and mass scale similarly are termed 'mass fractal' systems, where surface and pore space scale similarly

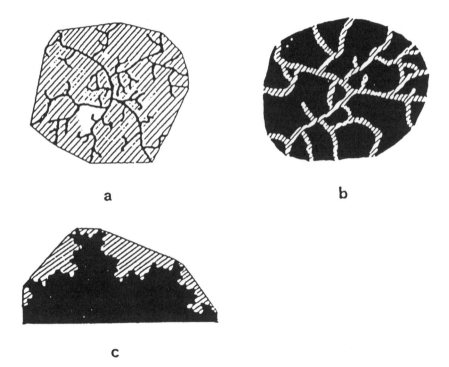

a

b

c

Figure 2. Two-dimensional representations of mass fractal (a), surface fractal (b) and pore fractal (c) structures. After [16]. Reproduced by permission of John Wiley & Sons, Ltd

are termed 'pore fractal', and systems where only the surface is fractal are termed 'surface fractal' (Table 1; Figure 2).

As an example consider the situation where a structure is overlaid by a grid or lattice, and the sites within this can be categorised as either mass sites (occupied sites), surface sites (occupied sites with adjacent empty sites) or pore sites (empty sites in the convex hull of the occupied sites) (Figure 2). The number of x-type (where x is mass, surface or pore) sites within radius, R, from a fixed x-type site, which must be much greater than the distance between nearest neighbour sites, is counted (Figure 3) [15,16]. This quantity, $M_{sites}(R)$, is called the 'mass' in a sphere of radius R, and grows as

Table 1. Classification of fractal objects based on surface (D_S), mass (D_M) and pore (D_P) fractal dimensions. After Pfeifer [14]

Classification	D_S	D_M	D_P	Characteristics
Surface fractal	D^*	$d^\dagger$	d	The interface is fractal, but the actual object and the pore are not fractal
Mass fractal	D	D	d	Interface and the actual object are fractal
Pore fractal	D	d	D	Interface and pore structure are fractal

* D: general fractal dimension.
† d: Euclidean dimension of embedding space (i.e., 1, 2 or 3, respectively, in 1-, 2- or 3-dimensional systems).

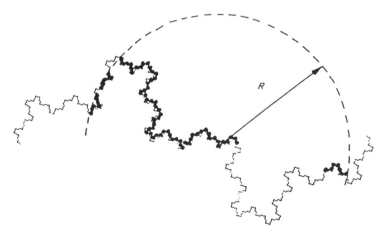

Figure 3. Illustration of $M_{sites}(R)$. The sites are pictured as adsorbed molecules on a surface, but equally well could be monomers in a polymer chain, primary particles of a colloidal aggregate, cells on the surface of a bacterial colony, hyphal tips in a mycelium, etc. After [16]. Reproduced by permission of John Wiley & Sons, Ltd

$$M_{\text{sites}}(R) \propto R^D \tag{3}$$

for increasing R [16].

Each site type gives rise to a mass–radius relation of the form

$$M_{\text{mass}}(R) \propto R^{D_\text{M}} \tag{4}$$

$$M_{\text{surface}}(R) \propto R^{D_\text{S}} \tag{5}$$

$$M_{\text{pore}}(R) \propto R^{D_\text{P}} \tag{6}$$

As D approaches d (Euclidean dimension) the three types of fractal become indistinguishable.

2.3 MODELS AND KINETICS OF FORMATION

Various computer simulations and statistical models have been proposed to explain growth, polymerization, and aggregation processes leading to the formation of fractal structures that closely resemble those found in nature. Attempts have been made to relate the fractal dimension of aggregates to the type and kinetics underlying their aggregation.

Two classical distinct models, particle–cluster aggregation and cluster–cluster aggregation [17], have been proposed to describe aggregation processes. The particle–cluster aggregation model, based on the sequential addition of particles to a growing cluster, is not believed to be feasible in most of the colloidal aggregation processes occurring in nature. Rather, aggregation in natural systems can be better modelled as a cluster–cluster aggregation (CLA) model [17], in which the large number of initially dispersed particles moving randomly in a fluid may come into contact and stick to each other. Small clusters that are thus formed, continue to move, and collide with other particles and clusters to form larger and larger clusters as the aggregation process continues.

Aggregation described by the CLA model can be categorised on the basis of its rate-limiting process as [17]: (a) ballistic CLA, in which particles and clusters follow linear or ballistic trajectories and join at their point of first contact; (b) diffusion-limited CLA (DLCA), where the trajectories that particles and clusters follow are random walks (Brownian motion), and the fractal dimension of generated clusters ranges between 1.75 and 1.80; (c) reaction-limited CLA (RLCA), where a short-range repulsive barrier exists that must be overcome before direct contact and sticking can occur between particles and/or clusters, so that many collisions may be required before an aggregation event can take place with formation of aggregates having a fractal dimension ranging between about 1.9 and 2.1 for $d = 3$, and about 1.5 and 1.6 for $d = 2$ [17, 18].

Aggregation processes are time dependent, thus their kinetic aspects must be considered in addition to geometric aspects. Although the aggregation models described above were originally developed without regard to aggregation kinetics, they can easily be made time dependent. In real systems, two distinct regimes of colloidal aggregation can be considered, i.e., fast and slow aggregation. Each of them has different rate-limiting physics, and the resulting aggregates possess different fractal dimensions and size distributions.

To make these models time dependent, a sticking coefficient, α, is defined which represents the collision efficiency between two particles and/or clusters. The sticking coefficient determines the aggregation rate and depends on the value of the energy barrier that must be overcome to obtain aggregation. The values of α range between 0 and 1, with $\alpha = 0$ or $\alpha = 1$ for 0 or 100% collision efficiency, respectively [19]. When α is large and close to 1 (collision efficiency $\cong$ 100%), the aggregation process is fast and the diffusion rate of particles and/or clusters is the limiting factor. This process is best described by a DLCA model. Under these conditions, loose aggregates, characterised by low fractal dimensions, are formed (Figure 4) [19]. In contrast, when α is small (low collision efficiency), the aggregation rate is slow and the process is controlled mainly by the energy barrier to be overcome to obtain the sticking together of particles and/or clusters. In this case, aggregation is best modelled by an RLCA process, and dense aggregates are formed with high fractal dimensions (Fig. 4) [19]. The sticking probability and aggregation rate can be increased by a decrease of the repulsive component of the energy barrier, e.g., by decreasing the pH of the medium that causes a decrease of the negative charge on the interacting particles and clusters, by increasing the ionic strength of the medium that results in

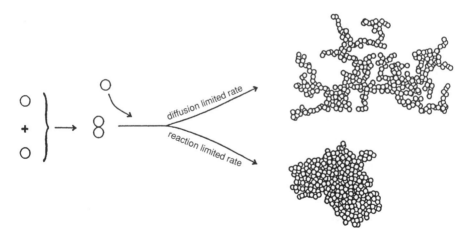

Figure 4. Schematic representation of a diffusion-limited and a reaction-limited cluster–cluster aggregation process leading to aggregates of either low fractal dimension (loose aggregate) or higher fractal dimension (denser aggregate), respectively. Reprinted with permission from [19]. Copyright (1995) American Chemical Society

the screening of particle/clusters charge, and by increasing temperature and/or stirring that provide external energy to the aggregation system.

Most aggregation processes occurring in natural aqueous systems can be described adequately by the simple, limit-case models outlined above. However, several natural aggregation processes are much more complex, and a variety of factors can affect the fractal dimension. For example, the fractal dimension can be substantially reduced by the effects of long-range attractive or repulsive interactions, whereas a number of processes, including restructuring by bending, folding and twisting, and even bond-breaking, which may occur in the often fragile, loosely held aggregates that form in aqueous media, can result in a fractal dimension somewhat larger than that expected from simple models [17]. Deflocculation/dispersion processes to which fragile aggregates may also be subject as a consequence of changing environmental conditions, such as pH, concentration, salinity, etc., can further complicate the interpretation of the fractal dimension measured for these systems. Thus, much care should be used in the interpretation of fractality of aggregation processes in natural systems.

2.4 METHODS FOR ESTIMATING THE FRACTAL DIMENSION

Fractal geometry allows us to determine how objects occupy space, including soil structure, individual soil components and soil organisms. There are a variety of different methods for estimating D, different approaches being suitable for different biotic and abiotic components of soil systems. An embedding three-dimensional space is obviously appropriate for soil, and has been used for inorganic and organic components. However, it has not been used for describing soil microorganisms *in situ* due to difficulties in capturing their images in three dimensions in an opaque substratum. Rather, microrganisms have been constrained to grow on the soil surface.

The most common experimental methods proposed by various authors to estimate the fractal dimension of environmental particles include: (a) visible or laser light scattering, small-angle X-ray scattering (SAXS) and small-angle neutron scattering (SANS), which are based on the measurement of the fraction of the incident beam intensity that is scattered by the sample as a function of the scattering angle, from which the fractal dimension of the object can be obtained by the use of the density autocorrelation function [14,20,21]; (b) turbidimetry, which is based on the measurement of the wavelength dependence of the turbidity, i.e., of the nonscattered or transmitted light that corresponds to the reduction in intensity of the incident beam as the result of light scattering of the sample suspension [22]; (c) static or dynamic vapour adsorption methods, which are based on the measurement of fractal adsorption isotherms and use of various classical adsorption equations rederived for fractal surfaces, which relate the number of adsorbed molecules per unit mass of sorbent, i.e., the monolayer value, to the size of adsorbed probe molecules [23,24]; (d) reaction rate methods, which are based on the measurement of the rate of a surface reaction, such as catalysis, ion exchange, dissolution, and others, on a particle as a function of the number of active sites available, provided that the reaction is not diffusion limited [25]; and (e) transmission and scanning electron microscopy (TEM and SEM) and optical microscopy, in conjunction with a detailed image analysis of the photomicrographs using a procedure based on the density–density autocorrelation function [26] or other image analysis techniques [27].

The methods outlined above have been recently reviewed in some detail in relation to their application to the fractal analysis of soil components and properties [6,11]. Thus, in this chapter only the 'box counting' method, which is the most commonly used method for estimating fractal dimensions of bacterial colonies and mycelial systems, and is well suited to image analysis [13,28–30], will be discussed in detail. This method has recently replaced the 'concentric circles' (two point density–density correlation) method [31] that has been used in several early studies of microbial colonies in agar and soil (e.g. [32,33]), but was found unsuitable because it excludes the extending colony margin from analysis.

The box counting method involves overlaying the microbial image with a series of grids of square boxes of different sizes and counting the number of boxes intersected for each box size. Estimate confidence for filamentous colonies is greatly increased if the counts of intersected boxes are made by grids of each box size at *each* possible starting position [34]. For example, a grid composed of boxes of three pixels side length has nine different starting positions. For a series of boxes of side length s pixels the number of boxes intersected by the set (N) is related to the fractal dimension of the set, D_S, by the power law

$$N(s) = cs^{-D_S} \tag{7}$$

Both interior boxes, which are contained wholly within the fractal set (i.e. contain white pixels only), and border boxes, which contain at least one white pixel and which contain or adjoin at least one black pixel, contribute to the total number of boxes (N) intersected by the set. Thus,

$$N(s) = N_{border}(s) + N_{interior}(s) \tag{8}$$

In practice the image may not be self-similar at all length scales, with departure from the power law occurring at very small or very large box sizes [16,28,29,35]. The characterisation length scale is determined by the limits of the image analysis procedure: the lower limit of resolution is determined by the pixel size of the image, and in mycelial systems this will also be limited to individual hyphae; the upper limit is determined by the largest gap within the image but is usually determined by 25% of the maximum width of the image set [28,35].

The border fractal dimension is estimated by plotting $\log N_{border}(s)$ against $\log s$. Regression analysis of the *linear* portion of this plot gives a gradient of $-D_S$. Similarly, the mass fractal dimension is estimated by regression analysis on the linear portion of a plot of $\log \{N(s) - \frac{1}{2}N_{border}(s)\}$ against $\log s$, giving a gradient of $-D_M$. Employing the subtraction term avoids overestimation of the area of the structure at large box sizes, since border boxes are not entirely filled by the set [27].

3   FRACTAL SOIL ORGANIC MATTER

3.1   INTRODUCTION

Humic substances (HS) and their principal fractions, humic acid (HA), fulvic acid (FA), and humin, are the most abundant, stable, and reactive components of soil organic matter (SOM). These substances occur in soil as a physically and chemically heterogeneous mixture of macromolecules, aggregates, and relatively smaller molecules.

Typical characteristics of HS are: (a) the absence of a discrete structural organization and an intrinsic lack of order at the molecular and supramolecular level; (b) the random assemblage of building blocks, bridges, functional groups, and side chains in the macromolecule; (c) the wide variety of sizes and shapes, implying variable conformation, porosity and compactness assumed by molecular structures in the solid and colloidal states; (d) the occurrence of complex aggregation and dispersion phenomena in aqueous media; and (e) various degrees of irregularity and roughness of exposed surfaces. All these properties play an important role in the binary and ternary interactions that HS experience with mineral and biological components in soil.

Because of these properties, HS represent a typical class of natural compounds that could be described by use of fractal concepts and methodology. In particular, the fractal dimension obtained by fractal analyses of these systems is expected to provide a valuable parameter for the quantitative description of some aspects of the physical and chemical behavior of HS under various environmental conditions.

In recent years fractal geometry has been applied to the study of macromolecular morphology, surface characteristics and aggregation processes of HS by a few research teams in the U.S.A. and Europe as briefly considered below.

3.2   EXPERIMENTS IN SOLUTION STATE

Two sets of experiments were conducted by SANS on solutions of two soil HAs at concentrations of $1-4\,mg\,mL^{-1}$, at pH 5.0 and an ionic strength of $0.1\,mol\,L^{-1}$ NaCl [36–38]. In one experiment HA solutions were analyzed by

SANS at a temperature of 11 °C, after storage at 20 °C for 48 h. In the other experiment, SANS measurements were performed on HA solutions at 4 °C, after storage for 48 h, and at 22 °C after 1, 11, 34 and 60 h. In all cases the measured scattering intensity, $I(q)$, obeyed a power law over one decade of the scattering vector, q, according to:

$$I(q) \propto q^{-\alpha} \tag{9}$$

Neither the HA concentration nor polydispersity appeared to affect the power law scattering observed. The fractal dimension, D, thus coincided with the scattering exponents, α, in equation (9), and could be calculated directly from the slope of the plot $\log I(q)$ versus q. The fractal dimension measured for both HA solutions at 11 °C was $D = 2.3 \pm 0.1$, whereas at 4 °C it was $D = 1.85$. At 22 °C the value of D increased with time from 1.85 (after 1 h) to $D = 2.15$ (after 11 h) to $D = 2.30$ (after 34 h) and to $D = 2.35$ (after 60 h) (Table 2). A D value of 1.85 suggested that the initial aggregation process was of the fast DLCA type yielding loose structures containing large amounts of bound water. The higher D values measured at higher temperature (22 °C) and with increasing time suggested that the loose initial clusters underwent restructuring with formation of more compact structures by a slow RLCA process, with the release of a certain amount of initially bound water molecules. Restructuring and sticking together of HA particles would be favoured by the increased kinetic energy caused by temperature increase, which allowed the energy barrier between negatively charged carboxylate groups of HA particles to be overcome. The authors concluded that both HAs formed aggregates or clusters of average radius of 40–50 nm consisting of building units of radial size ≤ 2.5 nm. In another experiment [39] conducted using SANS, water solutions of two fractions of a soil HA at a concentration of $15 \, \mathrm{mg\,C\,mL^{-1}}$ and at pH 7 provided D values of 2.29 and 2.46 (Table 2), which suggested the existence of relatively compact HA structures.

The SAXS technique has been used to study the fractal nature of aqueous solutions of the largest fraction (obtained by sequential ultrafiltration, nominal size between 50,000 Da and 0.2 μm) of a peat HA at concentrations of 2.0, 6.4, and 9.6 mg mL^{-1} at pH ≈ 9 [40,41], and of the HAs and FAs isolated from a peat and a soil [42]. In all cases the X-ray scattering data obeyed a power law of the type shown in equation (9). Using dynamic light scattering (DLS) and dynamic scaling theory, Ren et al. [40] and Tombácz et al. [41] showed that the HA fraction examined was only weakly polydispersed under the experimental conditions used, and thus its fractal dimension could be calculated directly from the SAXS data without the necessity of correcting for polydispersity effects. Since the exponent α was ≤ 3, the scattering particles could be considered mass fractals with $D_M = \alpha$ [21]. With increasing HA concentration, the D_M value obtained for the dissolved large-size fraction of peat HA slightly increased, from $D_M = 2.09$ to $D_M = 2.23$ [41] (Table 2). At low concentration, a

Table 2. Fractal dimensions of solutions of humic acids (HA) and fulvic acids (FA) from soils and peats measured by different techniques under various conditions

Source and material	Measurement conditions	Measurement technique	Fractal dimension	Reference
Soil HA	HA conc. 1–4 mg mL^{-1}, pH 5.0, ionic strength 0.1 mol L^{-1} NaCl,			
	$t = 11°C$	SANS	$D = 2.3 \pm 0.1$	[36]
	$t = 4°C$, $t = 22°C$ after 1 h	SANS	$D = 1.85 \pm 0.05$	[37]
	$t = 22°C$, after 11 h to 60°C	SANS	$D = 2.15 - 2.35$	[38]
Soil HA, fraction 2	HA conc. 15 mg CmL^{-1}, pH 7.0	SANS	D = 2.29	[39]
Soil HA, fraction 4	HA conc. 15 mg C mL^{-1}, pH 7.0	SANS	$D = 2.46$	[39]
Peat HA, large fraction	HA conc. 2.0, 6.4, 9.6 mg mL^{-1}, pH $\approx$ 9	SAXS	$D_M = 2.09 - 2.23$	[40,41]
Soil FA		SAXS	$D_M = 2.3$	[42]
Soil HA		SAXS	$D_M = 2.6$	[42]
Peat FA		SAXS	$D_M = 2.2$	[42]
Peat HA		SAXS	$D_M = 2.5$	[42]
Soil HA	HA conc. 1 mg mL^{-1}, $t = 4°C$, pH 3, after 4 h pH 6, after 4–28 h	TEM	$D = 1.77 \pm 0.05$ $D = 1.52 - 1.95$	[43]
Soil HA	HA conc. 1 mg mL^{-1}, $t = 4°C$, pH 3, after 10 h pH 6, after 28 h	TEM	$D = 1.72 \pm 0.01$ $D = 1.80 \pm 0.07$	[43]
Soil HA	HA conc. 1 mgmL^{-1}, $t = 4°C$, pH 3, after 10 h pH 6, after 10 h	TEM	$D = 1.79 \pm 0.03$ $D = 1.64 \pm 0.03$	[43]
Peat HA	HA conc. 1 mg mL^{-1}, t = 4°C, pH 3, after 10 h pH 6, after 10 h	TEM	$D = 1.79 \pm 0.05$ $D = 1.55 \pm 0.07$	[43]

loose, sponge-like fractal structure was suggested for the huge (50–100 nm) dynamic units of dissolved HA aggregates that became less porous at higher concentrations. The D_M values of dissolved peat FA and HA and of soil FA and HA were, respectively, 2.2 and 2.5, and 2.3 and 2.6 [42] (Table 2). Similar to the findings of Österberg and collaborators described above [36–38], these values for the mass fractal dimension could be ascribed to fractal aggregates produced by a slow RLCA process in which clusters must overcome repulsive forces, probably due to negative charges on humic particles, before they can form larger aggregates [40,42].

Aqueous solutions of three different soil HAs and a peat HA at a concentration of 1 mg mL^{-1}, at pHs of 3 and 6, and after 4, 10, and 28 h of equilibration at 4°C were investigated for their fractal properties by quantitative image analysis of high resolution TEM micrographs [43]. In all cases examined, straight lines were obtained for the log–log plots of the aggregate mass vs. aggregate size, with a slope corresponding to the fractal dimension of the HA studied. At pH 3 and after 4 to 10 h equilibration, HA aggregates exhibited fractal dimensions ranging from $D = 1.7$ to $D = 1.8$ (Table 2), which were consistent with a fast aggregation process of the DLCA type, i.e., controlled only by diffusion and occurring with high sticking probability between HA particles. At pH 6 and after long equilibration times (up to 28 h), D values of about 1.95 were obtained (Table 2), which suggested that aggregation of HA proceeded by a slow RLCA-type process with low sticking probability, to yield more compact aggregates. At pH 3, protonation of most acidic groups of HA would decrease repulsive negative charges on HA particles, thus inducing fast aggregation by weak attractive forces. At pH 6, repulsive electrostatic forces between largely deprotonated acid groups of HA would predominate thus determining slow aggregation.

3.3 EXPERIMENTS IN SUSPENSION

The fractal behaviour of some soil and peat HAs in aqueous suspension were investigated under different conditions using the turbidimetric technique [44–48]. The dependence of the turbidity, τ, on the wavelength, λ, obeyed a power law of the form:

$$\tau \propto \lambda^\beta \tag{10}$$

where the exponent β could be calculated from the slope of the linear plot log τ versus log λ obtained experimentally [22,26]. The exponent β might assume a value of $\beta < 3$, which implied the existence of a mass fractal of dimension $D_M = \beta$, or a value of $3 < \beta \leq 4$, which reflected the existence of a surface fractal of dimension $D_s = 6 - \beta$, or a value of $\beta = 3$ implying a nonfractal system [21].

Experiments were conducted on aqueous suspensions of four different soil HAs and a peat HA at concentrations of 30–40 mg L^{-1}, at pH values ranging from 3 to 7, at ionic strengths of 1, 5, and 10 mmol L^{-1} NaCl and 1 and 10 mmol L^{-1} CaCl$_2$, and after equilibration times of 2, 4, 8, 16 and 24 h at 25 °C [44–48]. Turbidimetric analysis data showed that the HAs studied might exhibit a mass or a surface fractal behaviour, or a nonfractal nature depending on their origin and experimental conditions.

The most important results obtained in these studies may by summarised as follows. Three of the four soil HAs exhibited a mass fractal behaviour under any condition used, with a D_M value that generally decreased with increasing pH or equilibration time (data for one soil HA are shown in Table 3). The D_M value of these HAs also decreased with increasing ionic strength caused by NaCl, whereas it increased with increasing CaCl$_2$ concentration. The effect of

Table 3. Fractal dimensions of humic acids (HA), fulvic acids (FA), and humins from soils, peats and lignite measured by different techniques in suspension under various conditions and in the solid, powdered state

Source and material	Measurement conditions	Measurement technique	Fractal dimension	Reference
Soil HA	Suspension, HA conc. $30\,mg\,L^{-1}$, $t = 25\,°C$	Turbidimetry		[47]
	pH 3, after 2–24 h		$D_M = 2.77–2.08$	
	pH 4, after 2–24 h		$D_M = 2.68–1.93$	
	pH 5, after 2–24 h		$D_M = 2.68–1.70$	
	pH 6, after 2–24 h		$D_M = 2.04–1.90$	
	pH 7, after 2–24 h		$D_M = 1.78–1.45$	
Soil HA	Suspension, HA conc. $30\,mg\,L^{-1}$, $t = 25\,°C$	Turbidimetry		[47]
	pH 3, after 16–24 h		$D_M = 2.81–2.79$	
	pH 4, after 8–24 h		$D_M = 2.75–2.43$	
	pH 5, after 2–24 h		$D_M = 2.81–2.12$	
	pH 6, after 2–24 h		$D_M = 2.01–1.92$	
	pH 7, after 2–24 h		$D_M = 1.76–1.21$	
Peat HA	Suspension, HA conc. $30\,mg\,L^{-1}$, $t = 25\,°C$	Turbidimetry		[47]
	pH 3, after 2–8 h		$D_S = 2.77–2.61$	
	pH 4, after 2–8 h		$D_S = 2.55–2.83$	
	pH 5, after 2–4 h		$D_S = 2.68–2.87$	
	pH 6, after 2–24 h		$D_S = 2.85–2.66$	
	pH 7, after 2–24 h		$D_S = 1.73–1.45$	
Soil FA	Solid state	SAXS	$D_S = 2.5$	[42]
Soil HA	Solid state	SAXS	$D_S = 2.3$	[42]
Soil humins	Solid state	SAXS	$D_S = 2.9–2.2$	[42,52]
Peat FA	Solid state	SAXS	$D_S = 2.7$	[42]
Peat HA	Solid state	SAXS	$D_S = 2.2$	[42]
Peat humin	Solid state	SAXS	$D_S = 2.9$	[42]
Lignite FA	Solid state	SAXS	$D_S = 2.8$	[42]
Lignite HA	Solid state	SAXS	$D_S = 2.3$	[42]
Soil HA	Solid state	Mercury porosimetry	$D_S = 2.75$	[54]
Soil humin fractions	Solid state, MIBK fractionation	SAXS	$D_S = 3.0–2.2$ $D_S = 3.0–2.3$	[52]

HA concentration on the D_M value was not uniform. One soil HA, however, assumed a nonfractal regime at pH ≤ 4 and after equilibration times $\leq 8\,h$, whereas it assumed a mass fractal regime only at higher pH and longer equilibration times (Table 3), at any concentration and at any ionic strength provided by NaCl. The peat HA exhibited a mass fractal nature after any equilibration time only at pHs 6 and 7, whereas at pH ≤ 5 the peat HA evolved from an initial surface fractal regime to a nonfractal regime, with the crossover between the two regimes occurring after 8 h at pH 3 and 4, and after 4 h at pH 5 (Table 3). Over the entire pH range examined, however, the peat HA behaved as a mass fractal when in the presence of NaCl at any concentration, whereas it behaved as a surface fractal in the presence of $CaCl_2$.

These results were interpreted as follows: (a) a mass fractal system with decreasing D_M values reflected the morphological evolution of HA from collapsed, almost compact structures with slightly rough surfaces to increasingly porous, fragmented and elongated structures with increasingly rough surfaces; (b) a surface fractal regime implied the existence of HA in compact structures with highly corrugated surfaces; and (c) a nonfractal nature suggested compact, space-filled structures with smooth surfaces for HA. These interpretations were supported by corresponding SEM observations on the various HA systems examined [44–48].

The D_M values were also used to identify the possible type of underlying aggregation process occurring in HA suspensions under various conditions. The HA systems that exhibited a D_M value close to 1.78, which generally occurred at pH values close to neutrality, would aggregate according to a DLCA model with high sticking probability and fast kinetics yielding loose, porous aggregates [19,49]. Values of D_M close to 2.05, often exhibited by HA systems at slightly acidic pH values, suggested an RLCA model for aggregation that would occur with low sticking probability and slow kinetics to yield dense, compact aggregates [19,49]. Simulation models more sophisticated than simple DLCA and RLCA models are, however, required to explain experimental values of D_M much lower than 1.78 or much higher than 2.05. Under these conditions, complex restructuring, fragmentation, and/or reconformation processes are suggested to occur during or after aggregation of HA particles. These processes could lead to the formation of either collapsed structures with D_M values much higher than 2.1, or very open, filamentous structures, such as fibrils and/or linear macromolecular chains with D_M values for below 1.7 [50].

3.4 EXPERIMENTS IN THE SOLID PHASE

Solid, powdered samples of HAs, FAs and humins isolated from various soils, a lignite and a peat were studied for their fractal properties by SAXS [42,51,52]. A power law of the form shown in equation (9) was supported in all cases. The power-law exponents, α, ranged between 3 and 4, i.e., the solid samples behaved as surface fractals with $D_S = 6 - \alpha$. The scattering curve of peat humin and some soil humins showed a change in slope attributed to the presence of surface and mass fractal scatterers in these samples [42,52]. The calculated D_S values of FAs increased in the order: soil FAs (2.5) < peat FA (2.7) < lignite FA (2.8) (Table 3). The D_S values of the corresponding HAs were slightly lower, i.e., $D_S = 2.3$ for soil and lignite HAs and $D_S = 2.2$ for peat HA (Table 3) [42,51]. Peat humin exhibited a D_S value of 2.9, whereas soil humins had D_S values ranging from 2.2 to 2.9 (Table 3). The lower D_S values shown by HAs indicated a less corrugated surface for these fractions than for the corresponding FAs and humins. The D_M values were 2.9 for mineral soil humins, thus showing an almost compact, unfractal structure, and 2.2 for organic soil and peat humins, which implied a porous, fragmented structure.

The fractal nature of the bound-HA fraction and the bound-lipid fraction obtained by selective removal of organic matter from four soil humins by the methylisobuthylketone (MIBK) method [53] was also investigated by SAXS using a power law of the form shown in equation (9) [52]. The two humin fractions exhibited a surface and/or mass fractal nature, depending on their origin, with D_S and D_M values ranging from 2.2 to 3.0 and from 2.3 to 3.0, respectively, and not showing any particular trend (Table 3) [52]. Recently, a soil HA sample was analysed in the solid state by mercury intrusion porosimetry and a value of $D_S = 2.75$ was calculated from porosimetry data (Table 3) [54].

4   FRACTAL SOIL MINERALS

4.1   INTRODUCTION

Several physical, chemical and biological processes in soil occur on the surface of clay minerals whose reactivity depends not only on the surface chemical properties, e.g., cation exchange capacity, surface acidity and charge density, but also on the geometry and morphology of the surface [25]. In particular, the surface geometry and morphology determine the extent of the mineral–solution interface that is accessible to a given reactant in reaction processes such as adsorption and desorption, catalysis, weathering, nutrient and pollutant retention and bioavailability, and interaction with the soil liquid phase, plant roots and microorganisms. Minerals with rough and convoluted surfaces will have restricted accessibility, whereas minerals with flat or smooth surfaces will have exposed reactive sites equally accessible.

Thus, it is vital to obtain a quantitative characterisation of the surface geometry and morphology of soil minerals to better understand and evaluate their actual reactivity. A quantitative description of the surface geometry of natural and synthetic minerals in the Euclidean space has proven to be extremely difficult, and information on the mineral surface roughness and irregularity is limited.

In this context, fractal geometry and the use of the surface fractal dimension parameter have been applied successfully to the description of the surface structure of soil minerals at both the particle scale and fine (molecular) scale. The fractal dimension of surfaces ranges between 2 and 3 ($2 < D_S \leq 3$) [24], and also provides a measure of the space filling capacity of the surface [55]. The higher the D_S value, the more irregular and space filling is the surface, the lower the D_S value, the smoother is the surface morphology.

Although a number of limitations and restrictions exist for the application of fractal principles and methodology to the description of natural clay minerals, the advantage of the scaling approach typical of fractal geometry is that, rather than measuring the average properties at a given scale, it allows measurement as the observation scale is changed, thus revealing features that are more 'universal' than those measured at a single scale [55]. In this section, the most relevant results and information obtained by fractal investigation of soil clay minerals of various nature by using different techniques will be briefly reviewed.

4.2 PARTICLE-SCALE STUDIES

The particle-size dependence of the apparent density, ρ_a, was studied for a natural Wyoming montmorillonite clay, and mineralogically purified sepiolite clay and attapulgite clay [56]. Several particle fractions ranging in size (R) from 20 to 1300 μm and roughly spheroidal in shape were measured for their apparent density. The fractions were isolated by mechanical sieving of powders obtained by crushing blocks of clay materials prepared by slow (several weeks) evaporation of clay pastes at room temperature. Experimental mass fractal dimensions, D_M, were calculated from the slope of the log–log plots of ρ_a versus R, according to:

$$\rho_a \propto R^{(D_M-3)} \tag{11}$$

The slope of the log–log plots was extremely small in all cases, i.e., the apparent density was virtually independent of particle size, and D_M values were very close to 3 (Table 4) [56].

The apparent density, $M_V(R)$, the specific surface area, $S_m(R)$, and the internal porosity, $P(R)$, were measured for about 20 granulometric fractions with mesh sizes ranging from 20 to 4500 μm, which were isolated by sieving fragmented and ball milled lumps obtained by slow evaporation of slurries with 10% dry matter of one kaolinite, one sepiolite, one palygorskite and 20 mono-ionic montmorillonites prepared by ion exchange of an Na-exchanged bentonite [57].

Assuming that simple power law relationships exist between the size, R, of macroscopic, roughly spheroidal particles belonging to the same granulometric fraction and their mass, $M(R)$, surface area, $S(R)$, and pore volume, $P(R)$, i.e., $M(R)$, $S(R)$, and $P(R)$ scale as R^{D_M}, R^{D_S}, and R^{D_P}, respectively, the following relations were derived for $M_V(R)$, $S_m(R)$, and $P(R)$ of a bed of $N(R)$ grains [56,58]:

$$M_V(R) = \frac{N(R)M(R)}{V_a} \propto R^{(D_M-3)} \tag{12}$$

Table 4. Fractal dimensions of soil clay minerals measured by different techniques under various conditions

Material	Measurement conditions	Method	Fractal dimension	Reference
Montmorillonite, natural	Several particle size fractions, size range 20–750 μm	Apparent density	$D_M = 2.97 \pm 0.02$	[56]
Sepiolite, purified	Several particle size fractions, size range 30–1300 μm	Apparent density	$D_M = 3.01 \pm 0.05$	[56]
Attapulgite, purified	Several particle size fractions, size range 80–1300 μm	Apparent density	$D_M = 3.02 \pm 0.03$	[56]
Montmorillonite, twenty monoionic samples	Twenty granulometric fractions, size range 20–4500 μm	Apparent density, Specific surface area, Internal porosity	$D_M = 3$ $D_S = 2.4$–3.0 $D_P = 3$	[55,57]
Sepiolite	Twenty granulometric fractions, size range 20–4500 μm	Apparent density, Specific surface area, Internal porosity	$D_M = 3$ $D_S = 3$ $D_P = 3$	[55,57]
Palygorskite	Twenty granulometric fractions, size range 20–4500 μm	Apparent density, Specific surface area Internal porosity	$D_M = 3$ $D_S = 3$ $D_P = 3$	[55,57]
Al_{13}-montmorillonite Sample I sample II	Five probe molecules, size range 0.4–0.10 nm	Monolayer capacity	$D_S = 1.89 \pm 0.09$ $D_S = 1.94 \pm 0.10$	[60]
Georgia kaolinite	Several probe molecules, size range 0.11–0.40 nm^2	Monolayer capacity	$D_S = 2.12 \pm 0.05$	[62]
Russian kaolinite	Several probe molecules, size range 0.11–0.50nm^2	Monolayer capacity	$D_S = 2.40 \pm 0.10$	[62]
Montmorillonite from Wyoming bentonite	Several probe molecules, size range 0.11–0.50 nm^2	Monolayer capacity	$D_S = 2.10$–2.30 ± 0.06	[62]
Georgia kaolinite, KGa-1	Eight probe molecules, size range 0.11–0.46 nm^2	Monolayer capacity Classical BET Fractal BET	$D_S = 2.17 \pm 0.10$ $D_S = 2.13 \pm 0.10$	[63]
Kaolinite, KGa-2	Five probe molecules, size range 0.16–0.56 nm^2	Monolayer capacity SAXS NMR	$D_S = 2.1$ $D_S = 2.0$ $D_S = 1.9$	[66]
Hectorite, SHCa-1	Five probe molecules, size range 0.16–0.56 nm^2	Monolayer capacity SAXS NMR	$D_S = 1.9$ $D_S = 2.8$ $D_S = 2.3$	[66]
Ca-Montmorillonite, STx-1	Five probe molecules, size range 0.16–0.56 nm^2	Monolayer capacity SAXS NMR	$D_S = 2.8$ $D_S = 2.0$ $D_S = 2.2$	[66]

$$S_{\mathrm{m}}(R) = \frac{N(R)S(R)}{N(R)W(R)} \propto R^{(D_{\mathrm{S}}-D_{\mathrm{M}})} \tag{13}$$

$$P(R) = \frac{N(R)P(R)}{N(R)R^3} \propto R^{(D_{\mathrm{P}}-3)} \tag{14}$$

where V_{a} is the apparent volume of the grain bed.

The plots of $\log M_{\mathrm{V}}(R)$ versus $\log R$ had a slope of zero in all cases, so that $D_{\mathrm{M}} = 3$ for all clays studied which thus were not mass fractals [57] (Table 4). The accessible surface area and the open porosity of the grains were both measured by nitrogen adsorption with two different methods, and provided $S_m(R)$ and $P(R)$ values that were independent of the grain size for kaolinite, sepiolite, palygorskite and some monoionic montmorillonites. For most montmorillonite samples, however, neither $S_{\mathrm{m}}(R)$ nor $P(R)$ was independent of the grain size, and linear log–log plots were obtained for $S_{\mathrm{m}}(R)$ versus R and $P(R)$ versus R over almost three decades of R, with slopes leading to D_{S} values for the accessible surface ranging between 2.4 and 3 [57] (Table 4). The D_{P} values obtained for the same samples from the open porosity measuremenrs were in good agreement with the D_{S} values [57], thus most solid montmorillonite clays behave as both surface and open-pore fractals at length scales between a few nm and a few mm and probably beyond that [55].

Fractal dimensions measured for clay particles were tentatively related to the local order of elementary clay sheets and morphomechanic properties of clays [57]. Although the data were scattered, the general trend was that the D_{S} and D_{P} values of various monoionic montmorillonites increased with increasing of the so-called 'coherence length', l_{c}, determined by X-ray diffraction, which is a direct estimate of the thickness of the ordered layered stacks in the solid clay [57].

A relation was also found between the charge of the cation saturating the montmorillonite sample and its capacity to generate stacks of montmorillonite lamellae [55]. In general, highly charged (trivalent) cations that induce a strong local stacking order of the elementary clay sheets led, at the macroscopic scale, to fragile and extensively microcracked clay aggregates with a geometry approaching space filling ($D = 3$). Montmorillonites saturated with divalent cations and sharing intermediate D values (between 2.9 and 2.8) produced intermediate cracking patterns, whereas most monovalent ions that cannot impose strong local order and yielded lower D values led to more plastic and less microcracked aggregates of entagled sheets.

In spite of a relatively large scatter in the data, good relationships were also found between the textural features of solid montmorillonite lumps and the properties of the muds that they form after redispersion in water [55]. In particular, 'stacked' montmorillonites that form thick and well-ordered stacks in the solid state ($D_{\mathrm{S}} = D_{\mathrm{P}} = D_{\mathrm{M}} = 3$) also formed thick quasi-crystals in muds with a small interface area. 'Entangled' montmorillonites characterised by packing of flexible units with D values < 3 in the solid state formed well-dispersed muds with a large solid–liquid interface area. Moreover, muds of 'stacked' montmorillonites, e.g., Ca- and Al-saturated, were shown to have lower plastic viscosity and gel strength (yield stress) than 'entangled' montmorillonites e.g., Na- and Li-saturated.

Van Damme [55] concluded that the fractal character of the accessible interface and of the open porosity of several montmorillonite clays is expected to have important consequences on diffusion–reaction processes occurring at the clay interface, and that this should be taken into account when modelling, for example, controlled release of pesticides or pollutant migration.

4.3 MOLECULAR-SCALE STUDIES

The 'monolayer capacity method' introduced by Pfeifer and Avnir [23] and Avnir et al. [24] has been widely used to characterise, at the molecular scale, the surface fractal behaviour of soil clay minerals. The method is based upon the measurement of the monolayer coverage value, $N_m(\sigma)$, i.e., the number of some probe molecule adsorbed on the sample surface per unit mass of sorbent, as a function of the effective cross-sectional area, σ, of the adsorbed molecules. Because the accessibility of a surface depends on the size of the probe molecules, the fractal dimension of the surface can be estimated from:

$$N_m(\sigma) \propto \sigma^{D_S/2} \tag{15}$$

Generally, the monolayer value, $N_m(\sigma)$, can be calculated from the adsorption isotherms obtained from vapour adsorption experiments conducted with probe molecules of different sizes, σ, by using the classical BET (Brunnauer–Emmett–Teller) equation or a variation of the BET equation [59]. The surface fractal dimension is then obtained from the slope of the log–log plot of $N_m(\sigma)$ versus σ. The use of molecular probes of varying cross-sectional areas allows the evaluation of surface roughness or irregularity of the material at molecular scale. The smallest and largest cross-sectional areas represent the limits of the characterisation scale over which the fractality of the surface can be measured.

Van Damme and Fripiat [60] have applied the method of Pfeifer and Avnir [23], i.e., equation (15), to the fractal analysis of a set of published experimental data [61] on the adsorption of various molecules, including nitrogen, benzene, and three organic molecules, in the size range 0.4–0.10 nm, by two Al_{13}-montmorillonites either freeze-dried or air-dried. The same D_S value held for air-dried and freeze-dried samples of each montmorillonite. The D_S values calculated for the two Al_{13}-montmorillonites were $D_S = 1.89 \pm 0.09$ and 1.94 ± 0.10 (Table 4), indicating that the surface was molecularly smooth.

Sokolowska et al. [62] investigated the fractal surfaces of a Georgia kaolinite, a Russian kaolinite, and a montmorillonite from Wyoming bentonite by the vapour adsorption method using a series of nonpolar and polar probe molecules of size ranging from 0.11 and $0.40\,nm^2$, including nitrogen, water, and various alcohols and hydrocarbons. Adsorption isotherm data were fitted to a fractal BET equation, and equation (15) was used to calculate D_S. For the rigid, anhydrous kaolinites the D_S values were 2.12 ± 0.05 and 2.40 ± 0.1, whereas the expanding-layer montmorillonite featured a D_S value of 2.10–2.30 (Table 4). The D_S values obtained indicated a somewhat corrugated surface for the clay minerals studied, which was attributed to the different level of crystallinity and purity of the mineral.

The surface fractal dimension of a reference kaolinite sample (KGa–1) was measured by adsorption–desorption experiments with eight probe molecules, including nitrogen, water and six hydrocarbons, ranging in size from 0.11 to $0.46\,nm^2$, and by using either the classical BET equation and equation (15), or a fractal BET equation directly [63]. The estimated values of D_S were 2.17 ± 0.10 and 2.13 ± 0.10, respectively, using the classical BET equation and the fractal BET equation (Table 4). The D_S values of the same kaolinite sample similarly calculated from the recommended normalised σ values [64], which are larger than the experimental σ values estimated from the liquid density method for large probes, were $D_S = 1.55 \pm 0.07$ and $D_S = 1.61 \pm 0.06$ [63]. Thus, the calculated fractal dimension values are sensitive to the method used to obtain σ, i.e., the uncertainty associated with the σ value measurements has a major systematic impact on the estimated D_S values. Okuda *et al.* [63] concluded that it is difficult to obtain the effective σ values without adequate information at the molecular level of the probe–surface interactions, such as the orientation/conformation of the probe and its packing density on the surface.

Limitations on the use of the monolayer capacity, or adsorption method for fractal analysis of surfaces include the following: (1) the range of the power law validity is limited to the used set of σ values of the probe molecules; (2) the D_S values obtained actually reflect the effective geometry as 'seen' by the specific set of probe molecules used, and not the 'true' geometry of the surface; and (3) the correct assessment of σ values is difficult, depending on the nature of the molecular interactions between the probe molecule and the surface. Thus, the estimation of the surface fractal dimension from the monolayer sorption method entails several uncertainties, especially when it is applied to complex sorbents such as whole soils.

Two conventional methods, SAXS and monolayer capacity, and a novel method based on nuclear magnetic resonance (NMR) measurements [65] were used to determine and compare the D_S values of three natural reference clays, a kaolinite (KGa-2), a hectorite (SHCa-1), and a Ca-montmorillonite (STx-1) [66]. In the SAXS experiments, data were fitted to a power law of the type shown in equation (9) and $D_S = 6 - \alpha$ was obtained from the slope of the log–log plot of $I(q)$ versus q. In the adsorption experiment, five probe molecules, including nitrogen and four hydrocarbons, ranging in size from 0.16 to 0.56 nm^2, were used to obtain the monolayer capacity by the classical linear BET plot of the adsorption isotherm. The D_S values were obtained from the slope of the log–log plot of equation (15). In the NMR experiment, the magnetization intensity, $m(t)$, is related to the time delay used between pulses, t, by the following power law:

$$m(t) \propto t^{\alpha} \tag{16}$$

where $\alpha = D/6$ [65]. The value of D could thus be calculated from the slope of the linear portion of the log–log plot of $m(t)$ versus t. The D_S values obtained by the three methods were in good agreement only for kaolinite that exhibited D_S values close to 2.0 (Table 4), thus indicating a smooth, planar surface over a length scale of 0.6 to 10 nm [66]. The results for Ca-montmorillonite and hectorite obtained by the three methods agreed only partially (Table 4). However, analysis of these materials by SEM appeared to support the monolayer capacity results. The SEM micrograph of Ca-montmorillonite, which had a D_S value of 2.8 when measured by adsorption, and $D_S = 2.0$ and $D_S = 2.2$ when measured by SAXS and NMR, respectively, showed a pore structure that was more open than that of kaolinite and hectorite, thus exposing more surface for adsorption. The SEM micrograph for hectorite, which exhibited an adsorption D_S value of 1.9, a SAXS $D_S = 2.8$ and an NMR $D_S = 2.3$, showed a compact structure of clay particles with a little surface available for probe molecules. The irregular and more space filling surface of hectorite indicated by the high D_S value (2.8) obtained by SAXS could be attributed to several impurities detected by X-ray diffraction within the sample [66]. The authors concluded that the SAXS and monolayer adsorption methods are better suited for characterising and distinguishing between the surface and mass fractal dimension of clay minerals, whereas NMR measurements are only capable of measuring D_M. Further, the SAXS technique probes the 'overall' interface of the solid material, whereas the adsorption method has the advantage of probing the accessible or open surface, which is of more interest and relevance in describing the surface morphology of clay minerals.

5 FRACTAL SOIL MICROORGANISMS

5.1 INTRODUCTION

A wide variety of microorganisms inhabit soil. In terms of biomass, mycelial fungi considerably outweigh other microorganisms in most soil systems, but that is not to belittle the latter which have complementary roles in carbon and mineral nutrient cycling. In the last two decades soil microbial ecologists have focussed on quantifying microbial biomass. Although results obtained told us 'how much is there' they provided little indication of what an organism is doing (see ref. [67]). Some microorganisms can deploy the same amount of biomass in a variety of different ways (see Section 5.3), and it is important to be able to describe and quantify this, and to relate it to biomass activity and response of the environment.

The growth form of an organism is a major determinant of its ecological role, and organisms can be divided effectively into those having unitary and those having modular life forms [68]. Unitary organisms, represented by higher animals, are generally non-branched, mobile (active) and with a non-iterative, determinate growth pattern. Modular organisms, including bryozoans, corals, ascidians, hydroids, plants, fungi and bacteria, tend to branch, are sessile or passively mobile except during juvenile or dispersal phases, and have iterative, indeterminate growth patterns. Modules may remain attached to the body from which they derive or become detached and function as separate physiological individuals.

Modular organisms are vulnerable to environmental fluctuations by virtue of their lack of mobility, and in the case of microorganisms also because of their organisational simplicity and often external digestion. This is countered by phenotypic plasticity (alteration of genotypic expression by environmental influences), involving physiological and/or morphological changes. Bacteria largely rely on physiological plasticity, though colony characteristics can change (see Section 5.2), while in contrast fungi exhibit considerable morphological (e.g. [69]; see Section 5.3) as well as physiological changes.

Like many naturally irregular structures (e.g. Sections 3 and 4, [1,3,4]), modular organisms are often approximately fractal. Some are only fractal at their boundaries (i.e. surface fractal; see Section 2) having entirely space-filled interiors, while others have gaps in the interiors of the colonies and are mass fractals [28]. Thus, it is appropriate to use both mass fractal (D_M) and surface fractal (D_S) dimensions as descriptors. The use of fractal geometry to describe biomass distribution has revealed small changes which were not detectable by biomass measurement (e.g. [70]). It is far superior to using density—another commonly used measurement which estimates space filling—which is inappropriate because the number of units of length, volume or area identified varies with scale of observation. Linear extent is a commonly used, easy to determine descriptor that is useful to employ in conjunction with D. However, it must be interpreted cautiously; rapid extension is frequently interpreted as indicative of a 'healthy', vigorous organism, but it could be a result of the opposite, being an 'escape' response [71]. Though quantitative descriptors are attractive because of their objectivity, visual observation and qualitative description should be used jointly as microorganisms can sometimes have similar values for D, surface cover and extension rate, but very different branching patterns [13] (Figure 5). A range of qualitative and quantitative descriptors is therefore necessary to obtain the whole picture.

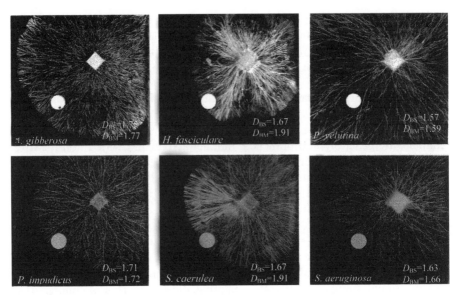

Figure 5. Digital images of mycelial systems of *Agrocybe gibberosa*, *Hypholoma fascicu-lare*, *Phanerochaete velutina*, *Phallus impudicus*, *Stropharia caerulea* and *Stropharia aeruginosa* 28 d after 4 cm^3 beech wood inocula were placed onto non-sterile compacted soil in 24 cm × 24 cm trays. White circles are inert plastic bottle caps. Note interspecific differences in surface and mass fractal dimensions (box-counting) of different species. After [71]. Reproduced by permission of The Mycological Society of America

The fractal nature of modular organisms has been studied at two distinct levels. Firstly, mathematical models (including fractal descriptors) have been proposed to explain how colony development can be related to intra- and extra-cellular substrate concentrations (e.g., for fungi [72–74]; for bacteria [75,76]).

Secondly, fractal geometry has been used to quantify inter-specific and intra-specific, and within-colony (see section 6.3) morphological differences in relation to environment. The morphological forms and modular nature of bacteria and fungi are most easily seen in culture on artificial media, because of the difficulties of making direct observations *in situ* in soil and other opaque environments, nonetheless some studies in natural environments have been made (see section 5.3).

5.2 BACTERIA

In artificial media, colonies of bacteria and yeasts exhibit a variety of morpho-logical patterns in response to environmental conditions, such as the solidity of the medium, nutrient concentration and temperature (e.g. [7,75,77–80]). This is well illustrated by *Bacillus subtilis* that formed circular colonies, colonies with

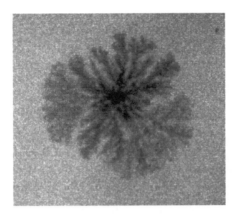

Figure 6. Digital image of a colony of the bacterium *Bacillus subtilis* growing on 0.05% peptone/10.8% agar. Courtesy of J.W.T. Wimpenny and D. Sommerfield, unpublished, 2000

densely branched structure, spreading colonies without any openings, colonies with concentric rings and fractal colonies depending on agar/nutrient concentration (Figure 6) [78]. Fractal colonies tend to form at low nutrient concentration (e.g. [77,78]). On agar, D_M values of about 1.6 have been reported for *Klebsiella ozaenae* [81], and values between 1.7 and 1.8 for *Bacillus subtilis*, *Escherichia coli*, *Salmonella typhimurium*, *Klebsiella pneumoniae* and *Serratia marcescens* [77,82] (Table 5). The ability of many bacteria to swarm is one factor allowing formation of these patterns, and cell movement inside colonies have been demonstrated (e.g. [75]). The complex patterns that develop may partially be accounted for by invoking ideas from pattern formation in non-biological systems, but undoubtedly self-organization is achieved by co-operative behaviour and intricate communication between the individual bacterium and the whole colony [76,84]. The examples given above are all of bacteria which can cause pathological problems in animals; common soil bacteria do not appear to have received much attention, but they are equally likely to form complex colony patterns.

5.3 FUNGI

Mycelial fungi produce fractal colonies on agar (Table 5) [28,32,73,85–93], and in contrast to studies on bacteria most of the fungi examined are common saprotrophs, plant pathogens and mycorrhizal species found in soil. The whole range of fractal dimensions have been reported in two dimensions. Fractal dimensions vary between species, between different genotypes of a species, depending on conditions and with time. Low values of *D* are typically found soon after germination, which then rise. For example, for *Armillaria gallica D*

Table 5. Examples of fractal dimensions of soil microorganisms and some pathogenic bacteria under various conditions

Organism	Growth conditions	Measurement techniques	Fractal dimension*	Reference
Salmonella typhimurium	Agar, 37 °C	Box-counting	$D = 1.73–1.80$	[82]
Salmonella anatum	Agar, 37 °C	Box-counting	$D = 1.77–1.79$	[82]
Escherichia coli	Agar, 37 °C	Box-counting	$D = 1.77–1.81$	[82]
Serratia marcescens	Agar, 37 °C	Box-counting	$D = 1.72–1.85$	[82]
Klebsiella pneumoniae	Agar, 37 °C	Box-counting	$D = 1.80$	[82]
Citrobacter freundii	Agar, 37 °C	Box-counting	$D = 1.79$	[82]
Proteus mirabilis	Agar, 37 °C	Box-counting	$D = 1.76$	[82]
Bacillus subtilis	Low concentration, peptone/agar, 35 °C	Box-counting	$D = 1.73 \pm 0.02$	[78]
Zoogloea ramigera	Rotating test tubes	Length-projected area	$D_S = 1.69 \pm 0.11$	[83]
		Length number scaling	$D_M = 1.79 \pm 0.28$	
	Fermentor	Length-projected area	$D_S = 1.78 \pm 0.11$	[83]
		Length number scaling	$D_M = 2.99 \pm 0.36$	
Saccharomyces cerevisae	Rotating test tubes	Length-projected area	$D_S = 1.92 \pm 0.08$	[83]
		Length number scaling	$D_M = 2.66 \pm 0.34$	
Trichoderma viride	Agar, at different times during colony development	Concentric rings	$D_M = 1.4–2.0$	[32]
Aspergillus oryzae	Agar, different concentrations, at different times during colony development	Box-counting	$D = 1.55–1.8$	[87]
Macrophomina phaseolina	Agar, at different times during colony development	Box-counting	$D = 1.21–1.84$	[90]
Armillaria gallica	Agar, at different times during colony development to 74 h	Box-counting	$D = 1.05–1.20$	[91]
Armillaria ostoyae	Agar, at different times during colony development to 116 h	Box-counting	$D = 1.43–1.57$	[91]
Agrocybe giggerosa	Compressed soil, 20 °C	Box-counting	$D_S = 1.64$ cv 0.77 $D_M = 1.75$ cv 0.72	[29]
Hypholoma fasciculare	Compressed soil, 20 °C	Box-counting	$D_S = 1.80$ cv 0.26 $D_M = 1.89$ cv 0.88	[29]

Table 5. (*continued*)

Organism	Growth conditions	Measurement techniques	Fractal dimension*	Reference
Megacollybia platyphylla	Compressed soil, 20 °C	Box-counting	$D_S = 1.63$ cv 1.04 $D_M = 1.69$ cv 0.92	[29]
Phanerochaete velutina	Compressed soil, 20 °C	Box-counting	$D_S = 1.69$ cv 0.89 $D_M = 1.71$ cv 1.02	[29]
	Compressed soil, $\psi = -0.02$ MPa, up to 29 days, 5 °C		$D_S = 1.29–1.72$ $D_M = 1.54–1.78$	[98]
	10 °C		$D_S = 1.55–1.75$ $D_M = 1.71–1.78$	
	15 °C		$D_S = 1.70–1.83$ $D_M = 1.80–1.88$	
	20 °C		$D_S = 1.64–1.81$ $D_M = 1.78–1.88$	
	25 °C		$D_S = 1.60–1.78$ $D_M = 1.62–1.81$	
Stropharia aeruginosa	Compressed soil, 20 °C	Box-counting	$D_S = 1.74$ cv 0.25 $D_M = 1.82$ cv 0.44	[29]
Stropharia caerulea	Compressed soil, $\psi = -0.02$ MPa, up to 29 days, 5 °C	Box-counting	$D_S = 1.13–1.70$ $D_M = 1.13–1.80$	[98]
	10 °C		$D_S = 1.24–1.72$ $D_M = 1.61–1.83$	
	15 °C		$D_S = 1.24–1.74$ $D_M = 1.72–1.90$	
	20 °C		$D_S = 1.27–1.85$ $D_M = 1.74–1.85$	
	25 °C		$D_S = 1.28–1.45$ $D_M = 1.48–1.61$	
	Compressed soil, 20 °C, up to 31 days $\psi = -0.002$ MPa		$D_S = 1.20–1.44$ $D_M = 1.39–1.69$	[98]
	$\psi = -0.006$ MPa		$D_S = 1.18–1.64$ $D_M = 1.49–1.89$	
	$\psi = -0.02$ MPa		$D_S = 1.23–1.71$ $D_M = 1.51–1.78$	
	$\psi = -0.06$ MPa		$D_S = 1.31–1.59$ $D_M = 1.39–1.62$	
Phallus impudicus	Compressed soil, 20 °C	Box-counting	$D_S = 1.83$ cv 0.84 $D_M = 1.84$ cv 0.91	[29]

* cv, coefficient of variation.

ranged between 1.05 and 1.2 72 h after germination and between 1.43 and 1.57 116 h after germination [91]. It has been suggested that the different colony morphologies formed by mycelia in different situations are produced by operating as non-linear (feedback regulated), hydrodynamic systems with indefinitely expandable (indeterminate) boundaries [94].

Of more direct relevance to soil ecology are the studies that have used fractal dimension to describe/define mycelial systems in soil (Table 5) [13,29, 33,70,85,95–99]. These have mostly used trays of soil that has been compressed to encourage mycelial growth only on the surface, to allow image capture and analysis [29,30], and so consider mycelial development in two dimensions rather than three dimensions. Since many of the fungi examined so far form large mycelial systems at the soil/litter interface this simple model has formed a good starting point. However, a reaction–diffusion mathematical model indicates that patterns of mycelial development will differ in two dimensions from those in three dimensions [74], and experiments have shown that extension rate can be considerably more rapid over surfaces than through soil [100]. Future efforts must, therefore, be directed at obtaining three-dimensional information in non-compacted soil, e.g. by destructive sequential sectioning.

There are inherent differences in the fractal nature of mycelial systems produced by different species. At early stages of outgrowth from resources some species, e.g. *Hypholoma fasciculare*, *Stropharia caerulea* and *S. aeruginosa*, produce surface fractal structures whereas others, e.g. *Agrocybe gibberosa*, *Coprinus picaceus*, *Phallus impudicus*, *Phanerochaete velutina* and *Resinicium bicolor* are mass fractal (Figure 5) [13,29,71]. The former group is characterised by diffuse slowly extending fronts and would be likely to be successful in discovering abundant, relatively homogeneously supplied resources. The latter group, by contrast, is characterised by well-defined, rapidly extending cords (linear organs formed from hyphal aggregates, often with a thickened outer rind and differentiated internally for translocation of water and nutrients) that would be unsuccessful at capitalising on relatively homogeneously supplied nutrients, but successful at discovering large, sparsely distributed organic resources.

Fractal dimensions of fungal colonies change with time (Figures 7 and 8), and in the long term (several/many months) when mycelial cord systems cover large areas (many cm^2) systems that were initially surface fractal become increasingly

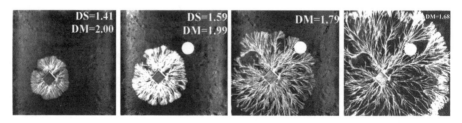

Figure 7. Digital images showing development of a mycelial system of *Hypholoma fasciculare* with time in 24 cm× 24 cm trays of non-sterile soil: from left to right, 11 days, 16 days, 26 days, 44 days. Note differences from left to right in surface (D_S) and mass (D_M) fractal dimensions (box-counting) with time. White circle is an inert plastic bottle cap. Modified from Boddy. From [71]. Reproduced by permission of The Mycological Society of America.

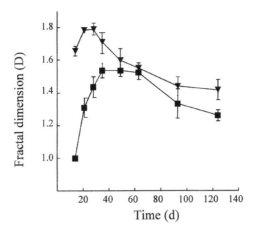

Figure 8. Change with time of mass (▼) and surface (■) fractal dimension (box-counting) of mycelial systems of *Hypholoma fasciculare* and *Phanerochaete velutina*, respectively, grown across non-sterile soil in 24 cm × 24 cm trays. The former is initially border fractal while the latter is mass fractal. Modified from [29] by permission of Bioline

mass fractal (e.g. [13,29]). Non-peripheral regions become more open as minor cords and non-aggregated hyphae regress, leaving a persistent mycelial network behind the actively growing, foraging margin. Changes in mycelial distribution with time may reflect increasing independence of different parts of the growth front (as tips digress with increasing radial extent) from each other (though tangential connections may decrease this effect) and from the original organic food resource [71].

Fractal dimensions of mycelial systems differ depending on the nutrient status of the organic resource from which the system is extending, temperature, water regime and pH, and these have interactive effects [13,71,98]. Different species are affected differently; those fungi whose mycelia tend to be more tightly aggregated into distinct, independent cords (and are mass fractal even in young systems) tend to be least responsive to different abiotic conditions. Soil nutrient status, inorganic composition, encountering organic resources and other organisms also affect mycelial fractal dimension (section 6).

6 FRACTAL INTERACTIONS IN SOIL

6.1 INTRODUCTION

Several physical processes and chemical and biological reactions in soils occur on organic-coated minerals and are affected by the nature and structure of mineral–organic interactions. The development, behaviour and spatial dynamics of microorganisms in soil can be considerably affected by the physical and chemical properties of organic and mineral soil particles and of their organo-mineral associations, and the nature and distribution of organic resources. Further, the habitat space available for organisms in soil is also markedly influenced by the complex interactions occurring between organic and mineral soil phases.

Since most of the interactions described above greatly affect and/or result from the structural and, especially, surface morphology of the components interacting, the application of fractal analysis to these systems is expected to provide some useful insight into the related processes occurring in soil.

6.2 MINERAL–ORGANIC INTERACTIONS

A set of published experimental data [101] on adsorption of nitrogen by some soil organo-clays was submitted to fractal analysis to evaluate the possible surface fractal nature of these materials [60]. The data analysed referred to montmorillonite and hectorite samples pillared with a fixed amount of large organic molecules of variable size, that are homologous series of either monoalkylammonium ions with chain lengths from 0.48 to 1.17 nm, or monoalkyldiammonium ions, with chain lengths from 0.77 to 1.67 nm. The D_S values were calculated from saturation coverage data obtained by measuring with a small yardstick (nitrogen) the remaining free surface of the pillared smectites [60]. The values obtained were $D_S = 2.03 \pm 0.1$, $D_S = 2.03 \pm 0.1$, and $D_S = 1.98 \pm 0.1$ for alkylammonium montmorillonites, alkyldiammonium montmorillonites, and alkyldiammonium hectorites, respectively. Thus, in the range of chain lengths between 0.48 and 1.67 nm, the D_S value of pillared organo-smectites was very close to 2, thus implying not only that their basal surface was molecularly smooth but also that the pillars were homogeneously and regularly, not randomly, distributed on the surface [60].

Fractal analysis of SAXS data was used in conjunction with other measurements to evaluate the surface morphology and properties of the strongly associated organo-mineral soil composite known as humin and of its fractions resulting from sequential removal of the various components of organic matter [52]. Organic matter was removed selectively from humin samples isolated from four different soils using the MIBK method [53]. First, bitumens were removed by Soxhlet extraction of humin with chloroform, thus obtaining a lipid-free humin fraction, then a bound-humic acid fraction and a bound-lipid fraction (discussed in section 4) were removed by MIBK, and, finally, the residual small amount (1–3 %) of organic matter still associated with the mineral matrix was completely removed by bromine oxidation, yielding an unaltered mineral matter fraction. The fractal dimensions of the humin fractions that remain after bitumen removal and complete removal of organic matter by oxidation, i.e., the residual mineral fraction of humin, were compared with those of the corresponding unfractionated humin. The D_S and D_M values of the samples were calculated from the slopes of log–log plots of SAXS data using equation (9) with $D_S = 6 - \alpha$ ($3 < \alpha \leq 4$) and $D_M = \alpha$ ($\alpha \leq 3$) (see sections 3 and 4). The change in slope evident in the plots of the samples originating from three out of four soils examined denoted a transition from mass to surface fractal regime which occurred at different observation scales. The mass and surface fractal nature exhibited by these materials could be attributed to the simultaneous presence of two different classes of grain size and/or different mineral composition in the samples examined [52]. All the materials used displayed a surface fractal behaviour over length scales ranging from ~ 1 to ~ 15 nm, whereas the mass fractal behaviour was observed over a smaller observation range, i.e., from 0.4 to ~ 4.0 nm [52]. Upon removal of organic matter, the D_S values generally decreased from values as high as $D_S = 2.9$ for unfractionated humin to values as low as $D_S \approx 2.0$ for the mineral residue. These results confirmed previous findings of Malekani *et al.* [66] that the mineral components have smooth surfaces, at least over length scales from ~ 1 to ~ 15 nm, and that it is the organic matter coatings of humin that are mainly responsible for its surface roughness [52]. Removal of organic matter, however, did not appear to affect the D_M value of humin fractions originating from soils having low organic matter contents (4.1–4.6 %), whereas the D_M value of humin fractions from an organic rich peat soil increased from $D_M = 2.2$ to $D_M = 2.6$ as organic matter was removed [52]. This result was attributed to the breaking of bonds between particles and fragmentation occurring on oxidation.

Fractal analysis by SAXS of model clay–organic compounds, which were prepared by polycondensation of gallic acid with two clay mineral reference samples, a kaolinite (KGa-2) and a montmorillonite (STx-1), provided results [102] that were consistent with those obtained for the natural organo–mineral samples described above. Gallic acid coatings slightly increased the D_S value of kaolinite (from $D_S = 2.0$ to $D_S = 2.1$), but markedly increased the D_S value of montmorillonite (from $D_S = 2.0$ to $D_S = 2.6$). These results were consistent with the far higher uptake of gallic acid by montmorillonite than that by kaolinite, and confirmed that the degree of surface fractality is dependent on the amount of organic matter present on the mineral [102].

The effect of interaction of varying amounts of a soil HA on the surface fractal dimension of a kaolin sample with grain size $< 2 \mu$m, and consisting of 70 % kaolinite, 18 % mica and 12 % quartz, has been studied by analysis of both water vapour adsorption isotherms using the Frenkel–Hill–Halsey (FHH) type equation, and mercury-intrusion porosimetry data [54]. The unmodified kaolin sample exhibited the highest D_S value (2.76) that dropped to about 2.4 upon its interaction with low amounts of HA (from ~ 0.015 to ~ 0.06 mg mL^{-1}), and remained almost constant, and below the D_S value of original HA ($D_S = 2.75$), when the HA amount added increased up to a value of 16 mg mL^{-1}. The D_S

values obtained from mercury intrusion data were generally slightly lower than those calculated from water vapour adsorption data, but the trends were similar for both sets of data. With changes of amount of HA added to kaolin a trend similar to that of D_S values was observed for the BET surface area and pore size distributions [54]. A remarkable surface coverage (30%) of kaolin with HA occurred for low amounts of HA added, whereas a limited increase of the coverage was obtained with high amounts of HA, which reached a maximum value of about 42% [54]. These results indicated that interaction between mineral surfaces and humic materials dramatically changed the surface morphology and related properties of these soil components with respect to each component considered individually.

Recently, Liu and Huang [103] found that citrate ligands present during the formation of particulate Fe oxides significantly modified the fine-scale morphology, surface geometry, and other surface properties of these materials. In particular, the surface fractal dimension of particulate Fe oxides formed at various concentrations of citrate was investigated by use of atomic force microscopy (AFM). This technique can produce quantitative data in the vertical dimension, i.e., it can yield a quantitative three-dimensional representation (height image) of the surface of insulating soil minerals. Three-dimensional arrays of cubes of varying edge length (ε, in nm) were superimposed on the three-dimensional image so that the cubes of each array completely encompassed the image. The following power law could thus be written:

$$N(\varepsilon) \propto \varepsilon^{-D_S} \qquad (17)$$

where $N(\varepsilon)$ is the number of the cubes intersected by the image and ε is the varying size of the cubes. The D_S value of Fe oxides was then obtained from the negative slope of the log–log plots of the cube intersecting counts, $N(\varepsilon)$, versus the cube size, ε. The AFM images showed that, in the absence of citrate, acicular and cubic crystals of Fe oxides predominantly composed of goethite and maghemite (by XRD analysis) were formed, which exhibited a D_S value of 2.17 ± 0.3. At a citrate concentration of $10\,\mu mol\,L^{-1}$, large lath-shaped crystals of lepidocrocite with a D_S value of 2.15 ± 0.3 were formed. In the presence of $100\,\mu mol\,L^{-1}$ citrate, smaller lath-shaped crystals of lepidocrocite having $D_S = 2.19 \pm 0.3$ were observed. Finally, only irregularly shaped, noncrystalline Fe oxides characterised by a $D_S = 2.24 \pm 0.3$ were formed at the highest citrate concentration ($1000\,\mu mol\,L^{-1}$) [103]. Specific surface area, mean surface roughness and phosphate adsorbed per unit weight of oxides were found to change in the same way as D_S values, i.e., to increase with decreasing Fe oxide crystallinity as citrate concentration increased to 100 and $1000\,\mu mol\,L^{-1}$ [103]. These results suggest that soil oganic matter present during the formation of Fe oxides should have an important effect on surface morphology, geometry and reactivity of these materials in soil.

6.3 INTERACTIONS BETWEEN MICROORGANISMS AND SOIL PARTICLES, ORGANIC MATTER AND NUTRIENT SOURCES

Physical aspects of inorganic components and distribution of large organic resources can considerably affect microbial pattern development in soil. Most attention has focused on pattern development of fungal mycelia, effects of proportion of sand, water potential ψ, soil/litter type, and size and distribution of organic resources all having been demonstrated. For example, the D_S of the saprotrophic, cord-forming, basidiomycete fungus *Resinicium bicolor* growing from pine (*Pinus sylvestris*) wood blocks (2 cm³) varied depending on the proportion of sand in artificial soil/sand mixtures. D_S was similar when growing over soils with 40, 60 or 100% sand, but larger than with 20% sand, which was in turn larger than with no added sand (Figure 9; A.J. Zakaria and L. Boddy, unpublished 2000). Effects of amount of sand were considerably altered by different soil ψ, and degree of soil compaction also affected D_S. Soil ψ, which results not only from the amount of water in a soil but also from the pore size distribution (e.g. [104]) affects D of some fungi, e.g. *Stropharia caerulea* (Table 5), but not others, e.g. *Phanerochaete velutina* [98]. The D_S of *P. velutina* was also unaffected by soil and litter, though mycelial extension rate, mycelial biomass, and time between contact with and emergence from new resources was affected [95]. Clearly, in the last example, space-filling remained constant irrespective of the size of the area searched, and the fractal dimensions exhibited by this fungus are much less variable than for most other species examined [71]. Though the fractal character of the mycelia of these fungi has been well

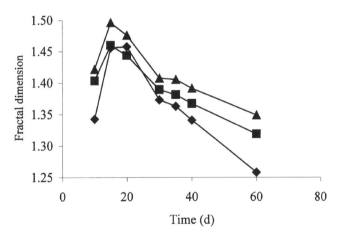

Figure 9. Effect of proportion of sand (◆, 0%; ■, 20%; ▲, 40%) in soil on surface fractal dimension (box-counting) of mycelial systems of *Resinicium bicolor* growing from 2 cm³ wood inocula over the surface of soil, with time. From A. J. Zakaria and L. Boddy, unpublished, 2000

characterised there is, unfortunately, no information available on the fractal properties of any of these soils.

There is little information on the effects of soil inorganic particles on the fractal nature of mycelia, though numerous studies have examined the effects on mycelia (mostly of saprotrophic, cord-forming basidiomycetes) of their encounter with new, solid organic resources (e.g. [13,71,96,99,105]). Some species, e.g. *Hypholoma fasciculare*, respond dramatically to encounter of new resources, mycelium in unsuccessful search directions regressing, leaving a thickened cord between original and new resource; subsequent mycelia egress the new resource [106]. If the new resource contains a relatively much larger nutrient supply than the original resource, complete regression of non-connected mycelium occurs, but if the relative difference between resources is small then some non-connected mycelium remains but D_M is obviously reduced. *P. velutina* mycelia only appear to respond when a newly encountered resource is considerably larger than the inoculum from which the mycelial system is extending [106]. *Stropharia caerulea* exhibited no change in D when systems extending from $0.5 \, \text{cm}^3$ wood inocula encountered resources eight times that size, though when extending from $0.15 \, \text{cm}^3$ *Urtica dioica* rhizomes to $0.5 \, \text{cm}^3$ wood resources D declined for 35 days (though not significantly) [99].

Though *P. velutina* mycelia only appear to respond to encounters with relatively much larger new nutrient resources [106], differences in D were detected when a new resource of the same size as the inoculum ($4 \, \text{cm}^3$) was

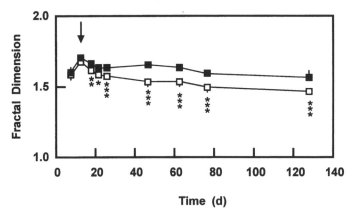

Figure 10. Change with time in mass fractal dimension (box-counting) on two sides of a mycelial system, developing on nutrient poor soil, one of which had an extra resource added after 13 days (■) and the other of which did not (□). Arrow indicates the time of addition of the new resource. Bars are the standard error of the mean. One way ANOVA indicated significant differences (* $P \leq 0.05$, *** $P \leq 0.001$) between means at particular times. Modified from [96]. Reproduced by permission of New Phytologist

placed behind the margin of a mycelial system (13 days old) growing on nutrient poor soil, simulating litter fall [96]. After a further 13 days mycelial 'patches', consisting of fine, highly-branched separate hyphae extending radially, arose from points of origin in cords (i.e. redifferentiation of mature cords occurred), both in systems to which new resources and no resources were added. Patches were temporary structures probably formed in response to nutrient demand for colonising/exploiting the new resource. There were no significant differences in D_S or D_M between controls and systems to which new resources had been added. However, a distinct polarity developed in systems to which new resources were added. D was significantly greater on the side of the system containing the new resource than on the opposite side (Figure 10).

On a smaller scale, *Trichoderma viride* responded dramatically to the presence of a new resource (agar plug containing glucose, $NaNO_3$ and KH_2PO_4) 2 cm away from a germinated spore extending across a cellophane film [85]. There was no growth in the direction away from the new resource, and D was greater in the region between the point of inoculation and the resource compared with a region further away from the new resource.

6.4 EFFECT OF SOIL STRUCTURE ON TRANSPORT PROCESSES AND MICROBIAL DISTRIBUTION

Evidence has already been presented (Section 6.3) indicating that soil structure (in terms of compaction, sand and organic litter content) can influence large scale mycelial behaviour. Structure also undoubtedly has a major influence on the spatial and temporal dynamics of unicellular microbes in soil. This relates to diffusion of respiratory gases, nutrients, movement of microbes and movement of predators in a fractal structure.

With regard to transport, it has been cogently argued [85] that fractal dimension alone may not be sufficient to uniquely define heterogeneity for the purposes of understanding the processes involved. This is because, though D may be related to how a measure of the structure scales with spatial resolution, 'lacunarity', i.e., the magnitude and arrangement of holes in a structure [1], which is concerned with the variability of that measure at a given resolution, may significantly affect processes that occur in and/or on that structure. Crawford et al. [85] demonstrated mathematically that the tortuosity and heterogeneity of a network, such as soil, influence diffusion rate by both D and the fracton dimension [107,108].

The fracton dimension, d, is calculated using a random-walk technique. Beginning at a random point within a pore, a random 'step' is taken, which on a digital image would be to any of the eight surrounding pixel locations. If the new site (pixel) has not been visited, the number of sites visited, S_n, is incremented by 1, and the total number of sites, S_{tot}, is also incremented by 1. If the site has already been visited, S_{tot} but not S_n is incremenmted by 1. The process continues until a boundary is hit or until a designated limit to S_{tot} is reached. The fracton dimension is then calculated from:

$$S_n \propto S_{tot} \overline{\overline{d}}/2 \qquad (18)$$

Classical diffusion equations are, therefore, inappropriate for soil and should be replaced [107] by :

$$\frac{\partial_c}{\partial_t} = \nabla \cdot (D(r) \, \nabla c) \qquad (19)$$

where c is the concentration, r is the distance of the particle from its starting point after time t, and $D(r)$ replaces the constant diffusion coefficient where:

$$D(r) \sim r^{-\theta}; \theta = 2(D_N - \overline{\overline{d}})/\overline{\overline{d}} \qquad (20)$$

where $\overline{\overline{d}}$ is the fracton dimension and D_N is the fractal dimension of the network space. Chemical and biological interactions, viscosity and surface tension effects further complicate the diffusion process, and more research into this general area is clearly needed.

With regard to the soil organisms, the surface area available in a fractal habitat such as soil depends on the scale of the measurement. The habitable space available to organisms of different size can be estimated by using a scale of measurement appropriate to the size of the organism. Obviously smaller organisms have more potential habitable space than larger organisms. This concept has been used to relate body size to population density in soil and other fractal habitats [4,109]. From photographic images of thin sections of soil, soil pores habitable to microarthropods (i.e. pores with an area $> 0.003 \, mm^2$) were analyzed as two-dimensional patches [109]. The fractal dimension of the patch perimeter, D_P, was related to the calculated perimeter, P, and the area, A, by

$$P \propto A^{D_P/2} \qquad (21)$$

The corresponding pore surface dimensions, D_S (calculated as $1 + D_P$), varied only slightly around 2.32. Accordingly, a decrease in body length by an order of magnitude would increase available habitat space approximately fourfold, and also result potentially in a fourfold increase in density of individuals in a given pore area. Extending this to microorganisms, which are a factor of 10^6 or more smaller, obviously implies dramatically greater potential habitat space. Clearly, small pores provide refuge sites for small organisms to which larger predatory organisms cannot gain access. Bacteria tend to reside on the surface of pore walls; thus data on fractal properties of soil structure can give insight into the extent of refuges of particular microbes from particular predators. For example, in a soil with $D = 2.36$, half of the potential habitable space for bacteria has been estimated to be inaccessible to predatory protozoa [85].

As well as affecting habitable space, structure has been shown to differentially affect movement of small and large microbes [85]. Despite lower mobility, larger species can move more rapidly through soil than smaller species because constricting pore necks reduces the effective tortuosity of soils more for larger organisms than for smaller organisms. Fractal geometry has been used to describe and analyse animal movement in heterogeneous environments, though it has been suggested that this should be avoided unless self-similarity can be demonstrated over an appropriate range of scales [110].

In view of the foregoing, soil structure will have a large impact on spatial and temporal dynamics of microbial populations and processes affected by them, e.g. mineral nutrient cycling. Clearly, they can only be understood by taking into account the fractal properties of the environment, because of the effects on habitable space and refuges, movement of nutrients, predators and prey, as well as physiological and biochemical effects.

7 CONCLUSIONS AND RECOMMENDATIONS

Fractal geometry represents a relatively novel and non-conventional approach that can contribute to the description and quantification of heterogeneous systems and processes, such as soil particles and microorganisms, and their interactions, through relatively simple and succint models and the parameterization by a single number, that is, the fractal dimension.

The major weakness of the fractal approach is, however, in the averaging out of any detailed property of the system; thus any interpretation of results can be considered valid only in an average or statistical sense for finite, heterogeneous systems such as soil particles and microorganisms, and their properties and processes. Further, several experimental constraints may influence the fractal analysis of data, including polydispersity of particles, multiple and incoherent scattering processes associated with the use of SAXS and SANS techniques, internal scattering and light absorption in turbidimetry, and others.

Fractal analysis has only recently been applied to soil systems; thus further investigations are needed, for example, to uncover the physical and chemical properties that determine at the molecular level the different fractal natures and fractal dimensions exhibited by organic and mineral soil particles and their interactions. Further studies are also necessary to discover what new insights into the physical, chemical and biological properties and processes of soil particles and microorganisms may be provided by the interpretation of their fractal properties. Finally, research is also needed to provide a better understanding of fractal aspects of interactions of microorganisms with particle surfaces, microbial population dynamics and resource aquisition involving organic and mineral particles in soils.

Some urgent issues that need to be addressed in the application of fractal geometry to soil studies include: (a) the evaluation of the consistency of the fractal dimension values obtained by different techniques applied to the same materials and processes; (b) the precise definition of experimental conditions used in fractal analysis, such as the physical state and concentration of the sample and the pH, temperature and ionic strength of the medium; and (c) the standardization and uniformity of use of fractal principles, equations, models, and methodology, applied to experimental data and their interpretation.

ACKNOWLEDGEMENT

Thanks are expressed to Malcolm F. Wilkins for assistance with preparation of figures.

REFERENCES

1. Mandelbrot, B. B. (1982). *The Fractal Geometry of Nature*, W. H. Freeman, New York.
2. Hastings, H. M. and Sugihara, G. (1993). *Fractals. A User's Guide for the Natural Sciences*, Oxford University Press, Oxford.
3. Richardson, L. F. (1961). The problem of contiguity: an appendix of statistics of deadly quarrels, *Gen. Syst. Yearbook*, **6**, 139.
4. Morse, D. R., Lawton, J. H., Dodson, M. M. and Williamson, M. H. (1985). Fractal dimension of vegetation and the distribution of arthropod body lengths, *Nature (London)*, **314**, 731.
5. Avnir, D. (1989). *The Fractal Approach to Heterogeneous Chemistry. Surfaces, Colloids, Polymers*. John Wiley & Sons, New York.
6. Senesi, N. (1994). The fractal approach to the study of humic substances. In *Humic Substances in the Global Environment and Implications on Human Health*, ed. Senesi, N. and Miano, T. M., Elsevier, Amsterdam, p. 3.
7. Senesi, N. (1996). Fractals in general soil science and in soil biology and biochemistry. In *Soil Biochemistry*, Vol. 9, ed. Stotzky, G. and Bollag, J.-M., Marcel Dekker, New York, p. 415.
8. Senesi, N. (1999). Aggregation patterns and macromolecular morphology of humic substances: a fractal approach. *Soil Sci.*, **164**, 841.
9. Anderson, A. N., McBratney, A. B. and Crawford, J. W. (1998). Applications of fractals to soil studies, *Adv. Agron.*, **63**, 1.
10. Baveye, P., Parlange, J.-Y. and Stewart, B. A. (1998). *Fractals in Soil Science*, CRC Press, Boca Raton, FL.
11. Okuda, I. and Senesi, N. (1998). Fractal principles and methods applied to the chemistry of sorption onto environmental particles. In *Structure and Surface Reac-*

tions of Soil Particles, ed. Huang, P. M., Senesi, N. and Buffle, J., IUPAC Series, John Wiley & Sons, Chichester, p. 77.

12. Pachepsky, Ya. A., Crawford, J. W. and Rawis, W. J. (1999). Fractals in Soil Science, *Geoderma, Special Issue*, **88**, 1.

13. Boddy, L., Wells, J. M., Culshaw, C. and Donnelly, D. P. (1999). Fractal analysis in studies of mycelium in soil, *Geoderma*, **88**, 301.

14. Pfeifer, P. (1987). Characterization of surface irregularity. In *Preparative Chemistry Using Supported Reagents*, ed. Laszlo, P., Academic Press, London, p. 13.

15. Pfeifer, P. (1988). Fractals in surface science: scattering and thermodynamics of adsorbed films. In *Chemistry and Physics of Solid Surfaces*, ed. Vanselow, R. and Howe, R., Springer, Berlin, p. 283.

16. Pfeifer, P. and Obert, M. (1989). Fractal basic concepts and technology. In *The Fractal Approach to Heterogenous Chemistry—Surfaces, Colloids, Polymers*, ed. Avnir, D., John Wiley & Sons, New York, p. 11.

17. Meakin, P. (1991). Fractal aggregates in geophysics, *Rev. Geophys.*, **29**, 317.

18. Brown, W. D. and Ball, R. C. (1985). Computer simulation of chemically limited aggregation, *J. Phys. A*. **18**, L517.

19. Buffle, J. and Leppard, G. G. (1995). Characterization of aquatic colloids and macromolecules. 1. Structure and behavior of colloidal material, *Environ. Sci. Technol.* **29**, 2169.

20. Martin, J. E. and Hurd, J. (1987). Scattering from fractal, *J. Appl. Crystallogr.*, **20**, 61.

21. Schmidt, P. W. (1989). Use of scattering to determine the fractal dimension. In *The Fractal Approach to Heterogeneous Chemistry: Surfaces, Colloids, Polymers*, ed. Avnir, D., John Wiley & Sons, Chichester, p. 67.

22. Horne, D. S. (1987). Determination of the fractal dimension using turbidimetric techniques. Application to aggregation protein systems, *Faraday Discuss. Chem. Soc.*, **83**, 259.

23. Pfeifer, P. and Avnir, D. (1983). Chemistry in noninteger dimensions between two and three: I. Fractal theory of heterogeneous surfaces, *J. Chem. Phys.*, **79**, 3558.

24. Avnir, D., Farin, D. and Pfeifer, P. (1984). Molecular fractal surfaces, *Nature (London)*, **308**, 261.

25. Farin, D. and Avnir, D. (1989). The fractal nature of molecule-surface interactions and reactions. In *The Fractal Approach to Heterogeneous Chemistry. Surfaces, Colloids, Polymers*, ed. Avnir, D., John Wiley & Sons, New York, p. 271.

26. Teixeira, J. (1986). Experimental methods for studying fractal aggregates. In *On Growth and Form*, ed. Stanley, H. E. and Ostrowsky, N., Nijhoff, Dordrecht, The Netherlands, p. 145.

27. Kaye, B. H. (1989). Image analysis techniques for characterising fractal structure. In *The Fractal Approach to Heterogenous Chemistry. Surfaces, Colloids, Polymers*, ed. Avnir, D., John Wiley & Sons New York, p. 55.

28. Obert, M., Pfeifer, P. and Sernetz, M. (1990). Microbial growth patterns described by fractal geometry, *J. Bacteriol.*, **172**, 1180.

29. Donnelly, D. P., Wilkins, M. F. and Boddy, L. (1995). An integrated image analysis approach for determining biomass, radial extent and box-count fractal dimension of macroscopic mycelial systems, *Binary*, **7**, 19.

30. Donnelly, D. P., Boddy, L. and Wilkins, M. F. (1999). Image analysis—a valuable tool for recording and analysing development of mycelial systems, *Mycologist*, **13**, 120.

31. Witten, T. A. and Sander, L. M. (1981). Diffusion- limited aggregation, a kinetic critical phenomenon, *Phys. Rev. Lett.*, **47**, 1400.

32. Ritz, K. and Crawford J. (1990). Quantification of the fractal nature of colonies of *Trichoderma viride*, *Mycol. Res.*, **94**, 1138.
33. Bolton, R. G. and Boddy, L. (1993). Characterisation of the spatial aspects of foraging mycelial cord systems using fractal geometry, *Mycol. Res.*, **97**, 762.
34. Soddell, J. A. and Seviour, R. J. (1994). A comparison of methods for determining the fractal dimension of colonies of filamentous bacteria, *Binary*, **6**, 21.
35. Markx, G. H. and Davey, C. L. (1990). Applications of fractal geometry, *Binary*, **2**, 169.
36. Österberg, R. and Mortensen, K. (1992). Fractal dimension of humic acids, a small angle neutron scattering study, *Eur. Biophys. J.*, **21**, 163.
37. Österberg, R. and Mortensen, K. (1994). Fractal geometry of humic acids. Temperature-dependent restructuring studied by small-angle neutron scattering. In *Humic Substances in the Global Environment and Implications on Human Health*, ed. Senesi, N. and Miano, T. M., Elsevier, Amsterdam, p. 127.
38. Österberg, R., Szajdak, L. and Mortensen, K. (1994). Temperature-dependent restructuring of fractal humic acids: a proton-dependent process, *Environ. Int.*, **20**, 77.
39. Homer, V. J. (1998). Fractal probes of humic aggregation: scattering techniques for fractal dimension determinations. In *Fractals in Soil Science*, ed. Baveye, P., Parlange, J.-Y. and Stewart, B. A., CRC Press, Boca Raton, FL, p. 75.
40. Ren, S. Z., Tombácz, E. and Rice, J. A. (1996). Dynamic light scattering from power-law polydisperse fractals: Application of dynamic scaling to humic acid, *Phys. Rev. E*, **53**, 2980.
41. Tombácz, E., Rice, J. A. and Ren, S. Z. (1997). Fractal structure of polydisperse humic acid particles in solution studied by scattering methods, *Models Chem.*, **134**, 877.
42. Rice, J. A., Tombàcz, E. and Malekani, K. (1999). Applications of light and X-ray scattering to characterise the fractal properties of soil organic matter, *Geoderma, Special Issue*, **88**, 251.
43. Rizzi, F. R., Stoll, S., Senesi, N. and Buffle, J. (2000). A transmission electron microscopy study of fractal properties and aggregation processes of humic acids. (In preparation).
44. Rizzi, F. R., Senesi, N., Acquafredda, P. and Lorusso, G. F. (1994). Applicazione della geometria frattale allo studio morfologico delle sostanze umiche. *Atti XI Conv. SICA*, Cremona 1993, Patron Editore, Bologna. p. 169.
45. Rizzi, F. R., Senesi, N., Dellino, P. and Acquafredda, P. (1994). Effetto della forza ionica e della concentrazione sulla dimensione frattale di un acido umico in sospensione acquosa a diversi pH. *Atti XII Conv. SICA*, Piacenza 1994, Patron Editore, Bologna, p. 143.
46. Senesi, N., Lorusso, G. F., Miano, T. M., Maggipinto, G., Rizzi, F. R. and Capozzi, V. (1994). The fractal dimension of humic substances as a function of pH by turbidity measurements. In *Humic Substances in the Global Environment and Implications on Human Health*, ed. Senesi, N. and Miano, T. M., Elsevier, Amsterdam, p. 121.
47. Senesi, N., Rizzi, F. R., Dellino, P. and Acquafredda, P. (1996). The fractal dimension of humic acids in aqueous suspension as a function of pH and time, *Soil Sci. Soc. Am. J.*, **60**, 1773.
48. Senesi, N., Rizzi, F. R., Dellino, P. and Acquafredda, P. (1997). Fractal humic acids in aqueous suspensions at various concentration, ionic strength, and pH, *Colloids Surf. A*, **127**, 57.
49. Stoll, S. and Buffle, J. (1995). Computer simulations of colloids and macromolecules aggregate formation, *Chimia*, **49**, 300.

50. Wilkinson, K. J., Stoll, S. and Buffle, J. (1995). Characterization of non-colloid aggregates in surface waters: coupling transmission electron microscopy staining techniques and mathematical modelling, *Fresenius J. Anal. Chem.*, **351**, 54.
51. Rice, J. and Lin, J. S. (1993). Fractal nature of humic materials, *Environ. Sci Technol.*, **27**, 413.
52. Malekani, K., Rice, J. A. and Lin, J.-S. (1997). The effect of sequential removal of organic matter on the surface morphology of humin, *Soil Sci.*, **162**, 333.
53. Rice, J. A. and MacCarthy, P. (1989). Isolation of humin by liquid-liquid partitioning, *Sci. Total Environ.*, **81/82**, 61.
54. Sokolowska, Z. and Sokolowski, S. (1999). Influence of humic acid on surface fractal dimension of kaolin: analysis of mercury porosimetry and water vapour adsorption data, *Geoderma, Special Issue*, **88**, 233.
55. Van Damme, H. (1998). Structural hierarchy and molecular accessibility in clayey aggregates. In *Fractals in Soil Science*, ed. Baveye, P., Parlange, J.-Y. and Stewart, B. A., CRC, Boca Raton, FL, p. 55.
56. Ben Ohoud, M., Obrecht, F., Gatineau, L., Levitz, P. and Van Damme, H. (1988). Surface area, mass fractal dimension, and apparent density of powders, *J. Colloid Interface Sci.*, **124**, 156.
57. Ben Ohoud, M. and Van Damme, H. (1990). The fractal texture of swelling clays, *C.R. Acad. Sci., Paris*, **311**, series **II**, 665.
58. Van Damme, H., Levitz, P., Gatineau, L., Alcover, J. F. and Fripiat, J. J. (1988). On the determination of the surface fractal dimension of powders by granulometric analysis, *J. Colloid Interface Sci.*, **122**, 1.
59. Fripiat, J. J., Gatineau, L. and Van Damme, H. (1986). Multilayer physical adsorption on fractal surfaces, *Langmuir*, **2**, 562.
60. Van Damme, H. and Fripiat, J. J. (1985). A fractal analysis of adsorption processes by pillared swelling clays, *J. Chem. Phys.*, **82**, 2785.
61. Pinnavaia, T. J., Tzou, M.-S., Landau, S. D. and Raythatha, R. H. (1984). On the pillaring and delamination of smectite clay catalysts by polyoxo cations of aluminium, *J. Mol. Catal.*, **27**, 195.
62. Sokolowska, Z., Stawinski, J., Patrykiejew, A. and Sokolowski, S. (1989). A note on fractal analysis of adsorption process by soils and soil minerals, *Int. Agrophys.*, **5**, 1.
63. Okuda, I., Johnston, C. T. and Rao, P. S. C. (1995). Accessibility of geometrically-rough (fractal) surfaces of natural sorbents to probe molecules, *Chemosphere*, **30**, 389.
64. McClellan, A. L. and Harnsberger, H. F. (1967). Cross-sectional area of molecules adsorbed on solid surfaces, *J. Colloid Interface Sci.*, **23**, 577.
65. Devreux, F., Boilot, J. P., Chaput, F. and Sapoval, B. (1990). NMR determination of the fractal dimension in silica aerogels, *Phys. Rev. Lett.*, **65**, 614.
66. Malekani, K., Rice, J. A. and Lin, J.-S. (1996). Comparison of techniques for determining the fractal dimensions of clay minerals, *Clays Clay Minerals*, **44**, 677.
67. Ritz, K., Dighton, J. and Giller, K. E. (1994). *Beyond the Biomass*. John Wiley & Sons, Chichester, UK.
68. Andrews, J. H. (1991). *Comparative Ecology of Microorganisms and Macroorganisms*, Springer, New York.
69. Rayner, A. D. M. and Coates, D. (1987) Regulation of mycelial organization and responses. In *Evolutionary Biology of the Fungi*, ed. Rayner, A. D. M., Brasier, C. M. and Moore, D., Cambridge University Press, Cambridge, p. 115.
70. Donnelly, D. P. and Boddy, L. (1998). Developmental and morphological responses of mycelial systems of *Stropharia caerulea* and *Phanerochaete velutina* to soil nutrient enrichment, *New Phytol.*, **138**, 519.

71. Boddy, L. (1999). Saprotrophic cord-forming fungi: meeting the challenge of hetero-geneous environments, *Mycologia*, **91**, 13.
72. Patankar, D. B., Liu, T. C. and Colman, T. A. (1993). Fractal model for the characterization of mycelial morphology, *Biotechnol. Bioeng.*, **42**, 571.
73. Jones, C. L., Lonergan, G. T. and Mainwaring, D. E. (1995). Acid-phosphatase positional correlations in solid–surface fungal cultivation—a fractal interpretation of biochemical differentiation, *Biochem. Biophys. Res. Commun.*, **208**, 1159.
74. Regalado, C. M., Crawford, J. W., Ritz, K. and Sleeman, B. D. (1996). The origins of spatial heterogeneity in vegetative mycelia: a reaction diffusion model, *Mycol. Res.*, **100**, 1473.
75. Kawasaki, K., Mochizuki, A., Matsushita, M. Umeda, T. and Shigesada, N. (1997). Modelling spatio-temporal patterns generated by *Bacillus subtilis*, *J. Theoret. Biol.*, **188**, 177.
76. Golding, I., Kozlovsky, Y., Cohen, I. and BenJacob, E. (1998). Studies of bacterial branching growth using reaction- diffusion models for colonial development, *Physica A*, **260**, 510.
77. Matsuyama, T. and Matsushita, M. (1993). Fractal morphogenesis by a bacterial-cell population, *Crit. Rev. Microbiol.*, **19**, 117.
78. Fujikawa, H. (1994). Diversity of the growth-patterns of *Bacillus subtilis* colonies on agar, *FEMS Microbiol. Ecol.*, **13**, 159.
79. Nakahara, A., Shimada, Y., Wakita, J., Matsushita, M. and Matsuyama, T. (1996). Morphological diversity of the colony produced by bacteria *Proteus mirabilis*, *J. Phys. Soc. Jpn.*, **65**, 2700.
80. Boschke, E. and Bley, T. (1998). Growth patterns of yeast colonies depending on nutrient supply, *Acta Biotechnol.*, **18**, 17.
81. Das, I., Kumar, A. and Singh, U. K. (1997). Nonequilibrium growth of *Klebsiella ozaenae* on agar plates, *Indian J. Chem., Section A*, **36**, 197.
82. Matsuyama, T. and Matsushita, M. (1992). Self-similar colony morphogenesis by Gram-negative rods as the experimental model of fractal growth by a cell popula-tion, *Appl. Environ. Microbiol.*, **58**, 1227.
83. Logan, B. E. and Wilkinson, D. B. (1991). Fractal dimensions and porosities of *Zoogloea ramigera* and *Saccharomyces cerevisae* aggregates, *Biotechnol. Bioeng.*, **38**, 389.
84. BenJacob, E., Shochet, O., Cohen, I., Tenenbaum, A., Czirok, A. and Vicsek, T. (1995). Cooperative strategies in formation of complex bacterial patterns, *Fractals—Interdiscipl. J. Complex Geom. Nature*, **3**, 849.
85. Crawford, J. W., Ritz, K. and Young, I. M. (1993). Quantification of fungal morphology, gaseous transport and microbial dynamics in soil: an integrated frame-work utilising fractal geometry, *Geoderma*, **56**, 157.
86. Matsuura, S. and Miyazima, S. (1993). Colony of the fungus *Aspergillus oryzae* and self-affine fractal geometry of growth fronts, *Fractals*, **1**, 11.
87. Matsuura, S. and Miyazima, S. (1993). Colony morphology of the fungus *Aspergil-lus oryzae*. In *Fractals in Biology and Medicine*, ed. Nonnenmacher, T. F., Losa, G. A. and Weibel, E. R., Birkhauser, Basel, p. 276.
88. Matsuura, S., and Miyazima, S. (1993). Formation of ramified colony of fungus *Aspergillus oryzae* on agar media, *Fractals*, **1**, 336.
89. Matsuura, S. and Tejima, S. (1994). Emerging order in fungal colony patterning, *Forma*, **9**, 289.
90. Mihail, J. D., Obert, M., Taylor, S. J. and Bruhn, J. N. (1994). The fractal dimension of young colonies of *Macrophomina phaseolina* produced from microsclerotia, *Mycologia*, **86**, 350.

91. Mihail, J. D., Obert M., Bruhn, J. N. and Taylor, S. J. (1995). Fractal geometry of diffuse mycelia and rhizomorphs of *Armillaria* species, *Mycol. Res.*, **99**, 81.
92. Baar, J., Comini, B. Elferink, M. O. and Kuyper, T. W. (1997). Performance of four ectomycorrhizal fungi on organic and inorganic nitrogen sources, *Mycol. Res.*, **101**, 523.
93. Ryoo, D. (1999). Fungal fractal morphology of pellet formation in *Aspergillus niger*, *Biotechnol. Techn.*, **13**, 33.
94. Rayner, A. D. M., Ramsdale, M. and Watkins, Z. R. (1995). Origins and significance of genetic and epigenetic instability in mycelial systems, *Can. J. Botany*, **73**, S1241.
95. Abdalla, S. H. M. and Boddy, L. (1996). Effect of soil and litter type on outgrowth patterns of mycelial systems of *Phanerochaete velutina*, *FEMS Microbiol. Ecol.*, **20**, 195.
96. Wells, J. M., Donnelly, D. P. and Boddy, L. (1997). Patch formation and developmental polarity in mycelial cord systems of *Phanerochaete velutina* on a nutrient-depleted soil, *New Phytol.*, **136**, 653.
97. Wells, J. M. Harris, M. J. and Boddy, L. (1998). Encounter with new resources causes polarised growth of the cord- forming basidiomycete *Phanerochaete velutina* on soil, *Microbial Ecol.*, **36**, 372.
98. Donnelly, D. P. and Boddy, L. (1997). Development of mycelial systems of *Stropharia caerulea* and *Phanerochaete velutina* on soil: effect of temperature and water potential, *Mycol. Res.*, **101**, 705.
99. Donnelly, D. P. and Boddy, L. (1997). Resource acquisition by the mycelial-cord-former *Stropharia caerulea*: effect of resource quantity and quality, *FEMS Microbiol. Ecol.*, **23**, 195.
100. Otten, W. and Gilligan, C. A. (1998). Effect of physical conditions on the spatial and temporal dynamics of the soil-borne fungal pathogen *Rhizoctonia solani*. *New Phytol.*, **138**, 629.
101. Barrer, R. M. and Millington, A. D. (1967). Sorption and intracrystalline porosity in organo-clays, *J. Colloid Interface Sci.*, **25**, 359.
102. Malekani, K., Rice, J. A. and Lin, J.-S. (1997). Fractal character of humin and its components, *Fractals*, **5**, 83.
103. Liu, C. and Huang, P. M. (1999). Atomic force microscopy and surface characteristics of iron oxides formed in citrate solutions, *Soil Sci. Soc. Am. J.*, **63**, 65.
104. Griffin, D. M. (1972) *Ecology of Soil Fungi*. Chapman & Hall, London.
105. Boddy, L. (1993). Saprotrophic cord-forming fungi: warfare strategies and other ecological aspects, *Mycol. Res.*, **97**, 641.
106. Dowson, C. G., Rayner, A. D. M. and Boddy, L. (1986). Outgrowth patterns of mycelial cord-forming basidiomycetes from and between woody resource units in soil, *J. Gen. Microbiol.*, **132**, 203.
107. Orbach, R. (1986). Dynamics of fractal networks, *Science*, **231**, 814.
108. Havlin, S. and Ben-Avraham, D. (1987). Diffusion in disordered media, *Adv. Phys.*, **36**, 695.
109. Kampichler, C. and Hauser, M. (1993). Roughness of soil pore surface and its effects on available habitat space of microarthropods, *Geoderma*, **56**, 223.
110. Turchin, P. (1996). Fractal analyses of animal movement: a critique, *Ecology*, **77**, 2086.

3 Interactions of Organic Materials and Microorganisms with Minerals in the Stabilization of Soil Structure

J. A. BALDOCK

CSIRO Land and Water, Australia

1 INTRODUCTION

Mineral soils, with the possible exception of sands found in very dry environments, contain an organic component and support a population of microorganisms. Although generally present as minor components of the total soil mass, living (e.g., plant roots or microorganisms) and nonliving (e.g., particulate debris or molecules) organic materials exert strong controls over many soil physical, chemical and biological properties [1]. The influence that soil organic materials have on these properties is dictated by the composition and activity of soil microorganisms, the chemical and physical nature of the nonliving organic

Interactions between Soil Particles and Microorganisms
Edited by P. M. Huang, J.-M. Bollag and N. Senesi. © 2002 John Wiley & Sons, Ltd

materials present, and/or interaction between the mineral and organic components.

The presence of an active population of microorganisms in mineral soils produces a system in which organic materials are continually being decomposed (mineralized or altered) and new microbial cellular structures and metabolites are synthesized. Soil microorganisms are typically found associated with or near mineral surfaces, creating an environment in which interactions between soil minerals and soil organic materials are common rather than the exception. Such organic–mineral interactions are fundamental to the stabilization of assemblages of soil particles in mineral soils at a range of size scales, especially in soils where inorganic mechanisms of binding soil particles together (e.g., formation of iron oxides or hydroxides) do not exist. Assemblages of soil mineral particles and organic materials are referred to as aggregates, and are important in defining the physical environment of the soil. The extent of soil aggregation directly influences many soil properties including: plant-available water holding capacity, the mechanical strength that must be overcome for roots to penetrate, movement of water, nutrients and salts within the soil profile, and the ease with which a soil can be eroded by wind or water.

In this chapter, the influence of soil microbial activity and organic–mineral interactions on soil structural properties will be addressed. The composition of the soil organic fraction as well as concepts and descriptors of soil structure will first be examined. This will provide a basis for subsequent discussion of the role that interactions of organic materials, microorganisms, and microbial activity with mineral particles have on the development and maintenance of adequate and stable soil structural conditions.

2 ORGANIC COMPONENTS OF MINERAL SOILS

The term soil organic matter (SOM) has been used in various ways to describe all or particular portions of the organic components in soils. Stevenson [2], Schnitzer [3,4], and Baldock and Nelson [1] defined SOM as the total of all organic materials contained within and on soils. Oades[5] excluded charcoal and charred residues, and MacCarthy et al. [6] additionally excluded non-decayed plant and animal tissues, their partial decomposition products and the living soil biomass. Analytical methods used to quantify the content of SOM typically measure soil organic carbon (SOC) contents and use a conversion factor of 1.72 to derive SOM contents. A quantitative discrimination between different forms of SOC is not possible using current analytical methods of SOC determination. Therefore, it is considered most appropriate to define SOC or SOM in its broadest sense, such that it encompasses all of the natural organic materials found in soil and derived from biological sources irrespective of state of decomposition. A series of alternative terms should then be used to

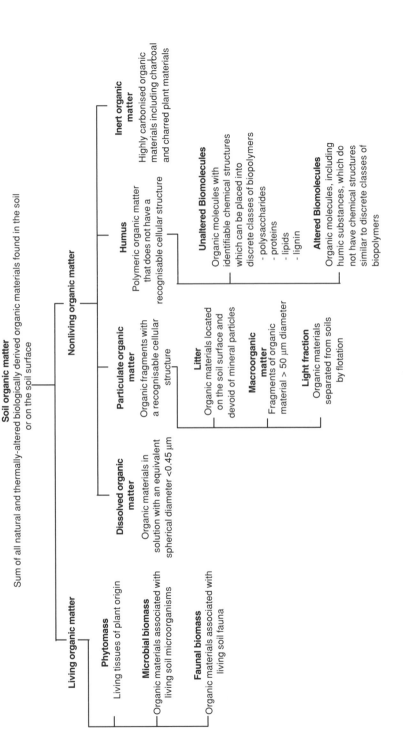

Figure 1. Definitions of soil organic matter and its components (modified from [7])

describe specific components of SOM. In this chapter, the definitions of SOM and its various components presented by Baldock and Nelson [1] will be used and are summarized in Figure 1. SOM therefore consists of a complex hetero-geneous mixture of organic particles and molecules exhibiting numerous extents of decomposition ranging from fresh unaltered through to highly decomposed states. In addition to the complexity imposed by composition and chemical structure, the physical size of the components of SOM vary significantly (Figure 2). In the subsequent paragraphs a brief review of the nature of the various components of SOM that contribute to soil structure will be presented.

The first level of differentiation applied to SOM is whether the organic material is associated with a living organism or is derived from non-living materials. The living components possess the capacity to synthesize enzymes and include plant roots, soil fauna, and soil microorganisms. Although soil fauna have been implicated in the formation and stabilization of soil structure, they are beyond the scope of this chapter. Readers are referred to Lee and Foster [9] for a discussion of the importance of soil animals in defining soil structural properties. Plant roots, filamentous fungi (free living or symbiotic), actinomycetes, and bacteria dominate the living portion of SOM implicated to have a direct influence on soil structural properties.

Nonliving components of SOM include dissolved organic matter (DOM), particulate organic matter (POM), humus, and an 'inert' fraction. DOM is defined as the organic materials in solution that pass through a 0.45 μm filter. DOM represents a small fraction of SOM, but due to its mobility and biological accessibility, it has a dynamic nature and can play an important role in the development and functioning of mineral soils. A detailed review of the proper-ties and importance of DOM in mineral soils was completed by Herbert and Bertsch [10].

POM is defined as fragments of organic material that exhibit a recognizable cellular structure. POM is typically dominated by plant-derived materials but can also contain fungal hyphae, spores, pollen, and faunal skeletons [8,11]. POM serves as a source of energy and nutrients for soil organisms and as a nutrient source for plants. POM has been isolated from soil by particle size fractionation, density fractionation, or a combination of both fractionation methods [12]. Due to methodological constraints associated with isolating small pieces of POM from soil, size fractionation methods typically use a lower size limit of 53 μm for POM as defined by Cambardella and Elliott

Figure 2. Size scales associated with soil mineral particles, organic components, pores and aggregations of mineral and organic components (modified from [8]). The defin-itions of pore size have used those developed by IUPAC (micropores < 2 nm, mesopores 2–50 nm and macropores > 50 nm). Alternatively the pore sizes corresponding to the lower ($\psi_m = -1500$ kPa) and upper ($\psi_m = -100$ kPa) limits of water availability to plants may be used to define the boundaries between the different classes of pore size

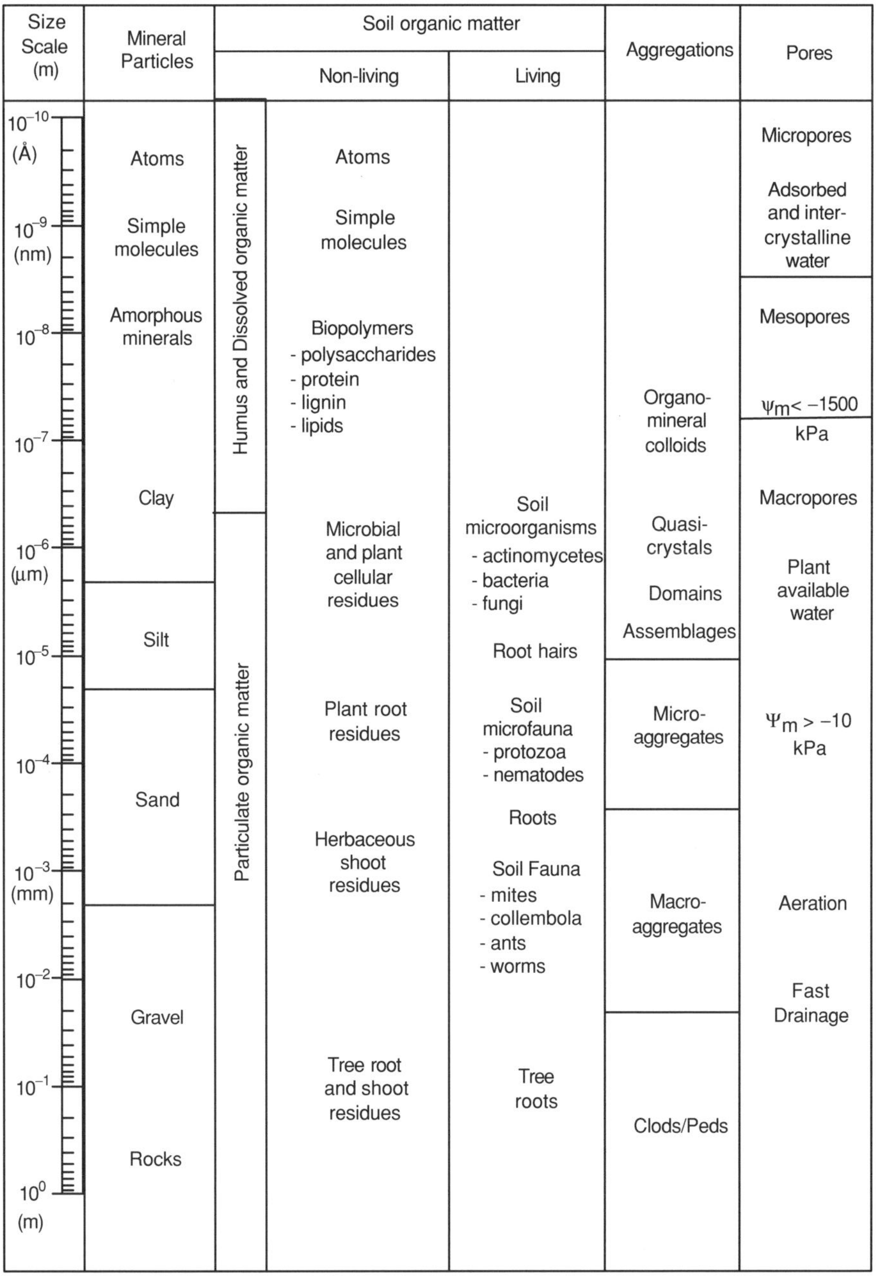

Size Scale (m)	Mineral Particles	Soil organic matter		Aggregations	Pores
		Non-living	Living		
10^{-10} (Å)	Atoms	Atoms			Micropores
10^{-9} (nm)	Simple molecules	Simple molecules			Adsorbed and inter-crystalline water
10^{-8}	Amorphous minerals	Biopolymers - polysaccharides - protein - lignin - lipids		Organo-mineral colloids	Mesopores
10^{-7}					$\Psi_m < -1500$ kPa
10^{-6} (μm)	Clay	Microbial and plant cellular residues	Soil microorganisms - actinomycetes - bacteria - fungi	Quasi-crystals	Macropores
10^{-5}	Silt		Root hairs	Domains Assemblages	Plant available water
10^{-4}	Sand	Plant root residues	Soil microfauna - protozoa - nematodes	Micro-aggregates	$\Psi_m > -10$ kPa
10^{-3} (mm)		Herbaceous shoot residues	Roots Soil Fauna - mites - collembola - ants - worms	Macro-aggregates	Aeration
10^{-2}	Gravel				Fast Drainage
10^{-1}		Tree root and shoot residues	Tree roots	Clods/Peds	
	Rocks				
10^{0} (m)					

Note: Left side spans labeled "Humus and Dissolved organic matter" (upper) and "Particulate organic matter" (lower).

Ψ_m = soil water matric potential

[13,14]. As a result, bacterial and fungal cells, spores, and fragments of organic material having a recognizable cellular structure but a size < 53 μm are operationally not included in the POM fraction. Where the fractionation is based on density (e.g., [15,16]), the nature of the organic materials isolated will be a function of the density of the solution used and the extent of interaction between the organic particles and soil mineral components. In density fractions, it is possible for organic particles < 53 μm to accumulate and for large particles strongly associated with soil minerals to be excluded. The chemical composition of POM fractions > 250 μm and < 1.6 Mg m^{-3} resembles that of fresh undecomposed plant materials [17]. However, as particle size decreases, POM takes on a chemical character more indicative of biologically processed materials (e.g., decreased carbohydrate and increased lignin and aliphatic C contents)[18].

Humus refers to the organic materials remaining in soils after removal of the DOM and POM fractions, but excluding the 'inert' carbonized fraction. Humus consists of a mixture of unaltered and altered biomolecules. Unaltered biomolecules have chemical structures that allow them to be placed into discrete categories of biopolymers (e.g., polysaccharides and sugars, proteins and amino acids, waxes and other lipids, lignin, etc.). Altered biomolecules have chemical structures that have been changed by biological oxidation (enzymatic) or chemical reactions (non-enzymatic) to the extent that the resultant molecules cannot be placed into discrete categories of biopolymers. The altered biomolecules are often referred to as humic substances. Various mechanisms have been proposed to account for the range in type and content of organic functional groups, the presence of many different monomeric species, and the inclusion of biopolymer-like materials in altered biomolecules [2,19–23]. The mechanisms can be placed into one of two categories [1]:

(1) partial biotic biopolymer degradation, where the integrity of the biopolymer is not destroyed, but rather modified by enzymatic degradation such that it forms the backbone of the altered biomolecules, and
(2) abiotic condensation polymerization, in which simple products of biopolymer degradation (phenols, quinones, sugars, and amino acids) repolymerize to form the altered biomolecules.

The two categories of mechanisms are not mutually exclusive since both are dependent upon processes of biological degradation to alter the chemical structure of the unaltered precursor biomolecules. All forms of humus vary in molecular size from small simple monomeric species through to complex large macromolecular structures with variable densities of functional groups.

In traditional studies of SOM, humus was considered to be dominated by humic substances differentiated on the basis of their solubility initially in alkaline and subsequently in acidic solutions. Three types of humic substances were defined: (1) humic acid (soluble in alkali, insoluble in acid), (2) fulvic acid (soluble in alkali and acid), and (3) humin (insoluble in alkali). This fractionation scheme was initially developed because of a need to quantitatively separate soil organic and inorganic materials to enable a selective characterization of the organic components. The fractionation is based entirely on the chemical properties of the organic materials and provides little information related to the biological significance of SOM. As a result of non-quantitative extraction, the potential for producing artifacts, and the development of new analytical techniques that do not require soil organic components to be separated from minerals prior to analysis (e.g., solid-state ^{13}C NMR, various organic mass spectroscopy techniques, and infrared spectroscopy), it is suggested that the extraction and isolation of humic substances be avoided in studies designed to characterize SOM chemistry. Indeed, Nelson and Baldock [24] showed that the chemical composition of organic matter from a range of soil and litter samples could be completely accounted for by common biomolecules found in plants and microorganisms without the need for including a humic molecular structure.

The final component of non-living SOM is the 'inert' fraction that consists of highly carbonized materials typical of charred plant residues or charcoal. The term 'inert' is used to describe this fraction because of its high resistance to biological decomposition. The mineralization of organic C from charcoal, prepared artificially by heating wood, was shown to be much less than that associated with glucose, cellulose or the unheated wood incubated under the same environmental conditions [25]. It should be noted, however, that it is unlikely that the material is completely inert (non-reactive) when considered over long time scales (> 1000 years). Skjemstad and coworkers [26,27] have quantified the content of char/charcoal in soils using a photo-oxidation procedure and solid state ^{13}C NMR. The results of this work and other unpublished data have indicated that char/charcoal can account for 0–60% of the organic C found in soils and that no relationship exists between total organic C and char/charcoal C. The presence of significant quantities of char/charcoal in soils could significantly influence the dynamics of SOC as assessed by measuring total SOC contents.

With the exception of the 'inert' fraction, all forms of non-living organic matter have the capacity to influence soil structure either directly by binding adjacent particles together or indirectly by providing a substrate for soil microorganisms.

3 SOIL STRUCTURE

3.1 DEFINITION AND CONCEPTS

Soil structure is often simply defined as the three-dimensional arrangement of soil particles into solid and void space. Such a definition only reflects one aspect of soil structure: structural form. As discussed by Kay [28] and Kay and Angers

[29], in addition to structural form, soil structure can also be described based on its stability, resiliency, and vulnerability. Soil primary particles may remain as individual discrete structural units, or be bound together into groups of particles referred to as aggregates. Aggregates are separated from one another by failure zones that may exist as either pores of variable diameter or weaknesses within the soil matrix along which preferential fracturing occurs when stresses are applied. The positioning of failure zones within the soil matrix arises in response to the spatial distribution of aggregating agents that bind soil particles together and the distance over which these agents are capable of operating. Structural stability describes the ability of a soil to retain its structural form when exposed to disruptive forces, and is defined by the combined strength of the mechanisms binding primary particles and/or aggregates together. Structural resilience describes the ability of a soil to return to its original structural form through natural processes after exposure to a disruptive force. Resilience may be due to physical processes (e.g., wetting and drying) or biological processes (e.g., root growth) that change soil pore size distribution or the strength of failure zones. Structural vulnerability refers to the combined effects of stability and resilience. Structurally vulnerable soils have low stability and low resiliency and are thus degraded easily on application of a stress and will not recover when the stress is removed.

In the development of soil structure, formation usually precedes stabilization, but the two processes may occur simultaneously (e.g., the formation of a pore by a growing root and stabilization of the pore walls by deposition of organic materials). The mechanisms contributing to the formation and stabilization of soil aggregates are typically different and vary with the size of soil particles involved. In soils where SOM contributes significantly to soil structural properties, its major influence is on structural stabilization. The association of soil primary particles into aggregates can occur at size scales that vary over nine orders of magnitude (Figure 2). Similar variations in the size of SOM from individual molecules through to the large particulate debris associated with plant residues indicate that a variety of potential mechanisms of stabilization must be operative to maintain the entire volume of a mineral soil in a structurally stable state. It is also important to note that stabilization is not static and changes with time and soil conditions (e.g., water content). A continual production of the SOM involved in the stabilization of soil structure is therefore required to maintain soil structural conditions. These last two points will be addressed in more detail later in the chapter.

3.2 MECHANISMS ACCOUNTING FOR STRUCTURAL FORM

Soil mineral fractions are composed of particles that exhibit a range of sizes and mineralogies. The physical behavior of the mineral matrix is dictated by this composition, but may be modified by the presence of organic materials capable of linking mineral particles together into physically stable aggregations. It is important to examine the mechanisms that account for the generation of structural form in soils before examining how this behavior is altered or stabilized by interactions with organic materials or soil microorganisms.

Structural form describes the three-dimensional arrangement of particles within the soil volume. It is created or modified by repositioning soil particles and introducing heterogeneity into the soil matrix through the generation of pores or failure zones. Changes in structural form result when the relative position of neighboring soil primary particles or aggregates within the soil volume is altered. The particles or aggregates may be brought closer together (e.g., compaction or shrinkage), pushed apart (e.g., tillage or swelling), or a combination of both (e.g., formation of a pore by a root forcing its way through pores smaller than its total diameter). Parameters used to monitor change in structural form include bulk density and the corresponding total porosity, and pore size distribution. Although bulk density defines the proportion of the soil volume occupied by particles and pores, significant changes in the distribution of pore sizes within the soil volume can occur with little or no change in bulk density. Pore size distribution, therefore, provides a better description of structural form.

The mechanisms capable of generating or altering structural form can be either abiotic (dependent on soil mineralogical properties) or biotic (dependent on biological processes) and the importance of the mechanisms will vary with soil properties (Table 1). Oades [30] suggested that soils from different textural groups (e.g. sands, loams, and clays) need to be examined separately when structural properties of soil are considered. The distinction between textural groups of soils should not be based on traditional particle size distribution, but rather on their shrink/swell capacity. As soils with a significant shrink/swell capacity dry, shrinkage results in the formation of cracks along failure zones within the soil matrix. With further drying, the process is accentuated and secondary and tertiary cracks form leading to a three-dimensional pattern of pores and failure zones that define aggregate boundaries [31].

Table 1. The importance of abiotic and biotic generation of structural form and biological mechanisms of structural stabilization in soils as defined by clay content (modified from [30])

	Sand (<15% clay)	Loam (15–35% clay)	Clay (>35% clay)
Abiotic generation of structural form	Minimal	Important	Maximal
Biotic generation of structural form	Minimal	Important	Maximal
Biological mechanisms of structural stabilization	Important	Maximal	Minimal

Sands have a pore size distribution and therefore can be viewed as having structural form. However, the structural form of sands can be altered only by changing the packing arrangement of individual grains, because of their low shrink/swell capacity. The presence of small amounts of clay in sand will do little to alter the gross structural form (bulk density or pore size distribution) from that of a pure sand. However, small aggregations of clay particles in the interstices between sand grains will form. In loams, where the content of clay is such that sand or silt grains can be embedded in or surrounded by a matrix of clay, the cohesive properties of the clay and the capacity of clay to shrink/swell will generate a structural form consisting of primary particles and variously sized aggregates separated by failure zones. The generation of structural form in this manner increases with the amount of clay present, the shrink/swell capacity of the clay, and with the extent of wetting and drying.

Maximum development of structural form occurs in clay soils where the mineralogy of the clay fraction is dominated by expanding rather than nonexpanding minerals. The structural form of such soils is defined almost entirely by the shrink/swell capacity of the clay, and may lead to a 'self-mulching' behavior where clay accounts for $> 35\%$ of the total soil mass [32,33]. The surface few centimeters of 'self-mulching' soils are maintained in a friable and granular state. The importance of this abiotic generation of structural form in soils of varying clay content is given in Table 1. The discrete clay content boundaries used in Table 1 should only be taken as indicative since they will also depend on the type of clay present. For example, a different structural form would be expected for a soil containing 40% smetitic clay than for a soil containing 40% kaolinitic clay due to large differences in shrink/swell capacities.

Several components of SOM can contribute to the generation of soil structural form. It is important to note that, unless the SOM component is capable of altering the position of soil pores, particles or aggregates within the soil volume, it does not alter structural form. The growth of roots can directly and indirectly influence soil structural form. As roots grow they create new pores or expand existing pores. During this process adjacent soil particles are either separated or pushed together changing the pore size distribution of the rhizosphere [34,35]. The more extensive the root network and the larger the root diameter, the greater the potential is for altering structural form. Measured influences of root growth on structural form will depend on all environmental factors controlling plant productivity and the allocation of photosynthate below ground, soil strength, and the availability of water and nutrients. An example of the influence of root growth on soil structural form is the creation of macropores that can extend into subsoil by tap roots of Lucerne (*Medicago sativa* L.) [36–40].

Living roots can also indirectly affect structural form by accentuating the rate and extent of soil drying and by creating zones of weakness (wet areas) and strength (dry areas) within the soil volume. Differences in macroporosity of a swelling clay soil resulting from the growth of different grass species were attributed to differences in the amount of grass biomass produced and rates of soil desiccation [41]. Materchera *et al.* [42] proposed that soil cracking induced by the tensile stresses associated with heterogeneous water uptake by plants accounted for the generation of small aggregates around growing roots. Thus, plants that produce large fibrous root systems capable of exploring the soil volume more completely and accentuating the extent of soil wetting and drying cycles, may promote the generation of well-aggregated soil structural forms, provided mechanisms exist to stabilize the structural form produced.

Microorganisms (fungi and bacteria) have little influence on the generation of structural form at a large size scale (> 250 μm). Fungal hyphae have been implicated in the formation of macroaggregates; however, there is little evidence to show that fungi can induce a reorientation of soil particles and an alteration in pore geometry at this size scale. The proliferation of fungal hyphae through soil results in the stabilization or strengthening of preexisting arrangements of pores, individual particles, or small aggregates. At smaller size scales (< 20 μm), microorganisms can alter structural form by reorganizing fine clay particles so that they are aligned parallel to external cellular surfaces [43–45]. Dorioz *et al.* [46] obtained micrographs showing that fungal hyphae compacted and reoriented particles of clay up to 20 μm away from hyphal surfaces. Whether the reorientation of clays resulted from the growth and expansion of fungal cellular structures or the movement of clay towards surfaces of fungal hyphae with water adsorbed by the organisms was not differentiated. The presence of polysaccharide mucilage around bacterial cells may similarly induce a reorientation of clays. Mycorrhizal fungi can indirectly alter structural form through their association with plant roots, by enhancing the extent of the drying phase of wetting and drying cycles.

Although this chapter will not examine the role of soil fauna in the generation of structural form, it should be noted that the activity of soil fauna might contribute significantly. Soil fauna can alter structural form at both large (> 250 μm) and small (< 250 μm) size scales. At large size scales, soil fauna can create pores through their burrowing, compress soil in localized zones around the pores they create, and deposit ingested soil as fecal casts or pellets that have a different structural form to that of the original soil. The compression of soil around the biopores fauna create and the mixing of soil and organic materials in the gut of soil faunal alters the structural form at small scales.

Attempts to quantify the influence of SOM on soil structural form have included measuring the changes in bulk density or pore size distribution across a range of soil types varying in SOM content. Results obtained from such studies have indicated the following:

(1) bulk density tends to decrease in non-swelling soil with increasing SOC content (e.g., [47–49]),

(2) the magnitude of changes in bulk density with increasing SOC content vary across soil types [50],

(3) SOC content is often a significant component of pedotransfer functions that predict pore size distribution (e.g., [51–54]), indicating the importance of SOC in defining the structural form of a soil.

It is important to note, however, that the mere presence of SOM is not responsible for inducing changes in structural form observed by measuring bulk density or pore size distribution. The presence of SOM will typically not alter the position of soil pores, primary particles or aggregates within the soil volume. Rather, the derived relationships result from the ability of SOM to stabilize the structural form generated either abiotically by shrink/swell properties of the soil or biotically by the activity of roots and microorganisms as discussed above.

3.3 STRUCTURAL RESILIENCE

The resilience of soil structure describes the ability of natural processes to regenerate the original structural form after an applied stress is removed (e.g. the ability of a soil to rebound and reform its initial porosity after compression induced by wheel traffic). Resilience is therefore dependent on the presence of the mechanisms capable of generating structural form (described in section 3.2) and the rate and maximum extent that each mechanism expresses itself. The abiotic mechanism for generating structural form is defined by the shrink/swell capacity of the soil. The rate and extent of structural generation depends on the nature of the soil mineral fraction (clay content and mineralogy) and environmental parameters (potential evapotranspiration and rainfall or irrigation). Soils containing > 35 % by mass of expanding clay minerals and subject to rapid wetting and drying cycles will have the most abiotically resilient structure while soils with low contents of nonexpanding clay minerals will exhibit the least abiotic resilience. The capacity to modify the rate or extent of abiotic structural regeneration through management practices is minimal because of the dependence of this mechanism on soil characteristics that are difficult to change over the short term (aside from modifying cation and electrolyte concentrations).

Biotic mechanisms of generating structural form will contribute to the resilience of all soils, but will be most important in soils with medium contents of nonexpanding clay minerals (10–35 % clay by mass). Unlike the abiotic mechanisms, biotic mechanisms for generating structural form can be enhanced by management practices at time scales varying from months to years. For example, removing any nutritional barriers to plant growth by fertilizer application will enhance the amount of root growth and water extraction from soil. Changes in structural form will then result from the direct effect of roots on macroporosity and the indirect effect on shrink/swell capacity. Additionally, different species of plants that have root systems capable of creating the desired changes in structural form could be grown. Examples of this include: (1) planting a grass pasture with an extensive network of fine roots that enhance the uniformity and extent of soil drying, or (2) planting lucerne which will increase the presence of large macropores that extend deep into the soil profile. Microorganisms are not directly involved in soil structural resilience; however, they may have indirect effects. The presence of an active community of microorganisms would (1) increase nutrient availability to plants and enhance root growth, and (2) provide a food source for larger soil animals capable of burrowing and recreating pores.

3.4 STRUCTURAL STABILIZATION

The presence of organic materials and/or the activity of soil organisms is thought to contribute to the stabilization of structural form of most soils. However, organic materials may also destabilize aggregates (see section 3.4.6) In soils where abiotic mechanisms of structural stabilization dominate (e.g., formation of oxides and hydroxides of iron and aluminum in Oxisols or Andisols), organic components and microbial activity still contribute to structural stabilization, but their relative importance is diminished.

Structural stability can be assessed in a number of ways and at several different size scales. Two of the most common methods used are wet sieving and turbidimetry [55–58]. Both methods characterize the stability of soil to the disruptive forces associated with wetting, shattering and abrasion encountered on sieving or shaking soil in water. Wet sieving is used to characterize the stability of aggregates > 250 μm, while turbidimetry is used to determine how effectively clay particles are bound together or into aggregates. Measurements of pore size distribution have also been used to quantify the role of SOM or SOC in stabilizing soil structural form [54,59,60]. Although a single measurement of the pore size distribution in a soil offers little information about the stability of the pore space, a role of SOM can be inferred if pore size distribution measurements are taken through time or across a range of soils varying in SOM content.

In many studies that have examined the role of organic matter in defining the structural condition of soils, no attempts have been made to differentiate the relative importance or role of the different forms of SOM. Although it is important that future experimentation be designed to gain such an understanding, much information pertaining to the general influence of SOM on soil structure can be gleaned from these previous studies. In the next sections of the chapter, the influence of SOM as a whole and the influence of each SOM component, excluding soil fauna, on the stabilization of soil structural form will be discussed.

3.4.1 Entire Soil Organic Fraction

A positive relationship between SOM content and structural stabilization as assessed by measurements of water-stable aggregation has long been recognized (e.g. [61–68]). The stabilizing effect of SOM has been attributed to reduced rates of wetting and an increased resistance to the disruptive forces generated during both the wetting and the subsequent mechanical sieving processes [69–72]. In some of these studies, the prediction of aggregate stability was improved by the inclusion of additional soil properties (e.g., clay and/or iron oxide contents). The inclusion of additional properties undoubtedly arises as a result of an attempt to derive a single relationship that is applicable across soils having a range of textural and mineralogical properties. With increasing clay content, an increasing amount of SOC is required to attain a given level of structural stability [68,73]. Instances where a correlation between SOM content and aggregate stability is not observed may result from one or more of the following:

(1) each component of SOM contributing differently to structural stabilization and the content of each component not being correlated with SOM content,
(2) the existence of an upper limit of SOM or of a SOM component above which no additional increase in stabilization is realized, or
(3) SOM or a component of SOM not being the major agent responsible for structural stabilization.

Williams [63] found that total N content was better correlated with aggregate stability than was SOC content and suggested that this occurred because SOC may include 'inactive' materials that are not involved in stabilizing soil structure. Results from Skjemstad et al. [74], showing that 10–50 % of the total SOC found in six Australian soils existed in the form of charcoal or charred plant residues support this suggestion. For the soils studied by Skjemstad et al. [74], a correlation between total SOC content and structural stability would not exist because char/charcoal carbon would have little influence on the stabilization of soil structure. Rather, the structural stability of these soils would be related to changes in the content of the specific components of SOC involved in stabilization.

The inclusion of SOC content in pedotransfer functions derived to predict the pore size distribution of soils, as defined by the water retention characteristic curve, has also demonstrated the general importance of SOC to the stabilization of structural form (e.g. [51–54]). The pedotransfer function derived by da Silva and Kay [53] is given in equation (1):

$$\theta_v = e^{(4.14+0.68 \ln CL+0.42 \ln OC+0.27 \ln BD)} \psi_m^{(-0.54+0.11 CL+0.02 \ln OC+0.10 \ln BD)} \quad (1)$$

where θ_v = volumetric water content ($m^3 m^{-3}$), ψ_m = matric potential (MPa), CL = clay content expressed as a percentage of total soil mass, OC = organic carbon content expressed as a percentage of total soil mass, and BD = dry bulk density ($Mg\,m^{-3}$). Baldock and Nelson [1] used the equation derived by da Silva and Kay [53] to determine the influence of increasing SOC content on the volume of 0.2–30 μm pores (i.e. the plant available water holding capacity) of 80 red–brown earths (Alfisols) located in South Australia. If SOC content was increased by $0.01\,g\,SOC\,g^{-1}$ soil, the volume of 0.2–30 μm pores increased, with the extent of the increase (ΔWHC) being related to soil clay content (CL) according to equation (2).

$$\Delta WHC = -0.0012\,CL + 0.055 \quad (R^2 = 0.82) \tag{2}$$

Therefore, with increasing clay content a greater change in SOC would be required to stabilize a similar increase in the volume of pores capable of storing plant-available water. This result is consistent with the observation that soils with higher clay contents require more SOC to attain equivalent levels of structural stability [68,73].

Cockroft and Olsson [75] defined coalescence as a distinct form of soil structural degradation that could not be attributed to compaction, slaking or dispersion but resulted in the formation of dense soil layers. Interestingly, the structural stability of the degraded and coalesced soil having a bulk density of $1.4\,Mg\,m^{-3}$ was similar to that of the uncoalesced soil having a bulk density of $0.8\,Mg\,m^{-3}$. Cockroft and Olsson [75] observed that soils having organic C contents greater than $40\,g\,C\,kg^{-1}$ soil did not coalesce but that coalescence increased rapidly as organic C contents fell below $30\,g\,C\,kg^{-1}$ soil across forty different soils.

3.4.2 Plant Roots

The root system of plants distributes itself within the surface layers of mineral soils in the form of a dense network, provided high soil strength or some other limitation does not restrict root growth. Visual and microscopy investigations have indicated that aggregates can be both formed and stabilized in the root rhizosphere and that stabilization can persist after root death [66,76–80]. The ability of root systems associated with growing plants to stabilize soil structural form has been demonstrated in both field and glasshouse studies [43,81–85]. In such studies the stabilization may have resulted from direct or indirect effects of the presence of roots as described earlier and delineated in Table 2.

Quantitative differentiation between the direct and indirect effects of roots on the stability of soil structural form is difficult, but several studies have provided some information. Direct effects of roots on the stabilization of soil structural form include physical entanglement and the exudation of high molecular weight polysaccharide materials capable of binding to mineral surfaces and bridging gaps between adjacent mineral particles. Relationships between root mass or root length and aggregate stability have been observed [82,86,87]. Miller and Jastrow [87] showed, using a path analysis procedure, that the effect of root length was indirect and manifested itself through its influence on the length of root available for infection by mycorrhizal fungi. The correlation between the length of fungal hyphae and geometric mean diameter of aggregates found by Miller and Jastrow [87] was better than that observed between root length and

Table 2. Direct and indirect processes involved in the stabilization of soil aggregates by the growth and presence of living roots

Direct effects	Indirect effects
Physical entanglement by roots and root hairs	Physical entanglement by root associated mycorrhizal fungi
Exudation of mucilaginous materials capable of binding soil mineral particles together	Provision of organic substrates for rhizosphere microorganisms capable of releasing metabolic products with the capacity to bind mineral particles together

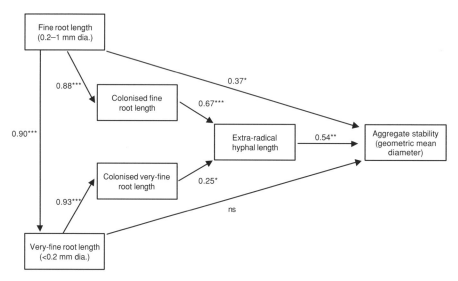

Figure 3. Path model relating the length of roots and associated VAM fungi to aggregate stability as assessed by geometric mean diameter for fine-silty Mollisols and an Alfisol [87]. Arrows indicate causal paths and numeric values are path coefficients indicating the relative strength of each path. Statistical significance of the path coefficients is given by * = $P < 0.05$, ** = $P < 0.01$, and *** = $P < 0.001$. NS denotes a non-significant path coefficient. Reprinted from [87]. Copyright (1990), with permission from Elsevier Science

geometric mean diameter (Figure 3). However, the association of mycorrhizal fungi with roots is not a prerequisite for roots to contribute to structural stabilization. The presence of non-mycorrhizal roots would also contribute to structural stabilization by linking adjacent aggregates together, particularly at size scales > 20 μm.

In addition to the physical enmeshment of aggregates by roots, the organic materials exuded by roots have also been noted to enhance the stability of soil aggregates. Pojasok and Kay [88] observed an increase in the stability of aggregates incubated with exudates collected from bromegrass and maize roots; however, the potential involvement of microbial activity was not excluded. Morel et al. [85] found that aggregate stability was improved by a 'sticking effect' exerted by polysaccharides exuded from maize roots. A direct ability of intact mucilage exuded from maize root tips to stabilize aggregates independent of microbial activity was demonstrated by Morel et al. [89].

Indirect effects of plant roots on structural stabilization include the provision of an organic substrate for soil organisms and the influence that roots have on soil water content. Organic materials capable of acting as a substrate for soil fauna and microorganisms are deposited in soil as roots grow (e.g. exudation of a variety of organic molecules of low and high molecular weight and sloughing off of cellular debris particularly from root caps and root hairs) and when roots die [90]. The distribution of roots within soil ensures that these organic substrates are distributed throughout the soil volume, providing the greatest potential for a general enhancement of structural stability by the activity of the soil microorganisms (see section 3.4.3). The structural stability of soil is influenced by water content and exposure to drying [29]. The ability of roots to accentuate the drying phase of wetting and drying cycles in soil can impact on structural stabilization. It has been proposed that as soils dry, the strength of the association between inorganic particles and organic binding agents increases allowing the mucilaginous materials derived from roots and microbial activity to enhance structural stabilization [81,91].

3.4.3 Soil microorganisms

Enhanced water-stable aggregation has been observed by incubating soils amended with organic substrates (e.g., [92–95]; however, the addition of organic matter in the absence of microbial activity was shown to have little influence on structural stability [96]. The magnitude and longevity of the increased aggregation can be viewed as a function of the nature of the added organic substrate, in particular, the availability of the added organic C to microorganisms (Figure 4). The magnitude of the stabilization increases with increased biological availability, but the duration over which the increased aggregation is maintained decreases. Many studies have observed that the influence of agricultural management on aggregate stability is better correlated with microbial biomass C

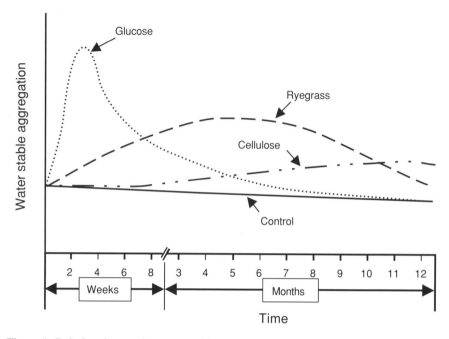

Figure 4. Relative changes in water-stable aggregation after addition of organic materials containing carbon with different biological availability (modified from [66]). The time required for the indicated changes in water stable aggregation to occur will vary with climate in response to the effects that temperature and precipitation have on biological activity

than total soil organic C (e.g., [68,97,98]). In addition, Haynes and Francis [99] found rapid increases in aggregate stability and microbial biomass C in the rhizosphere of grass roots in the absence of changes in SOC. The involvement of soil microorganisms in soil structural stabilization is usually ascribed to the production of fungal hyphae that can enmesh aggregates and extracellular polysaccharide gels that act as glues. The addition of microbial polysaccharides to soils generally increases aggregate stability [96], even in the absence of microbial activity [100]. Microbial activity and the associated synthesis of cellular tissues or metabolic products capable of binding soil particles together are therefore considered an important mechanism of soil structural stabilization.

3.4.3.1 Bacteria Bacteria exist in soils as single cells or in multicell colonies associated with the mineral surfaces of small pores within the soil matrix (Figure 5). Foster [101] found no bacteria on the external surfaces of large aggregates examined with environmental scanning electron microscopy. The apparent absence of bacteria on external aggregate surfaces may result from a combination

(a)   (b)   (c)

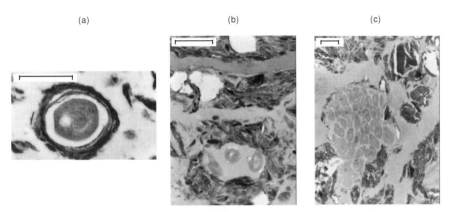

Figure 5. Transmission electron micrographs of microaggregates showing (a) an individual soil bacterium (bar = 0.5 μm), (b) a small colony of soil bacteria (bar = 1.0 μm), and (c) a large colony of soil bacteria (bar = 1.0 μm). From [101]. Reproduced by permission of CSIRO Publishing

of less favorable environmental conditions (lower availability of water), increased exposure to predation by soil fauna, or the burial of bacteria within small aggregates by an adsorption of mineral particles to extracellular polysaccharides surrounding bacterial cells. Larger pore sizes between aggregates than within aggregates and an increased ability of roots to grow in and extract water from larger pores create a drier environment on aggregate surfaces than within aggregates. The activity of soil bacteria is reduced at water potentials of −300 kPa and extinguished at potentials of −1000 kPa [102]. Differences in the chemical composition of organic matter found on aggregate surfaces relative to aggregate interiors [103] indicated a greater biological activity and turnover of soil organic materials at aggregate surfaces relative to aggregate interiors (Table 3).

Changes in pore size distribution can also influence the extent of predation of bacteria by soil fauna because of differences in their physical size. Van der Linden et al. [104] have suggested that bacteria contained in pores < 5 μm and < 30 μm are protected from predation by Protozoa and nematodes, respectively. Killham et al. [105] demonstrated a reduced turnover of glucose-derived microbial C when the glucose was placed in pores < 6 μm rather than pores < 30 μm. These results suggest that bacteria will be more susceptible to predation as pore size distribution tends towards an increased proportion of larger pores. The potential production and effectiveness of bacterial derived organic binding agents (see below) may therefore change with changes in pore size distribution; however, the significance of this effect remains unquantified.

The ability of bacteria to form strong associations with mineral particles in soils is often ascribed to their ability to synthesize and exude mucilaginous

Table 3. Organic carbon contents and chemical properties of organic C isolated from the 0.5 mm exterior layer of soil aggregates and aggregate interiors collected along a north–south transect through the North America prairies [103]

	Organic C content (g C kg^{-1} fraction)	C/N Ratio	Lignin* VSC (g kg^{-1} SOC)	Lignin† (Ac/Ad)$_v$	Neutral Sugars (g C kg^{-1} fraction)	(G + M)/ (X + A)‡
Aggregate interior	22.3	10.1	16.2	0.22	4.38	0.76
Aggregate exterior	20.3	9.5	12.1	0.31	3.90	0.87

* Determined by CuO oxidation
† Ratio of acid to aldehyde forms of the vanyl form of lignin monomers released by CuO oxidation.
‡ Ratio of (galactose + mannose) to (xylose + arabinose) monomer contents.

(a)   (b)

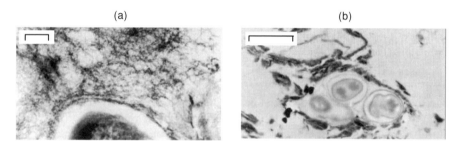

Figure 6. (a) Soil bacterium surrounded by Ru/Os reactive extracellular polysaccharide (bar = 0.1 μm) [101] and (b) a colony of soil bacteria surrounded by a 0.1 μm thick layer of clay and randomly oriented clay embedded in an extracellular polysaccharide layer (bar = 1.0 μm). From [10]. Reproduced by permission of CSIRO Publishing

polysaccharides (e.g. [46,106]). The extracellular polysaccharides produced by bacteria exist as either a discrete layer surrounding individual cells or as a common layer around small colonies [101,107] (Figure 6). As soils dry, the amount of extracellular polysaccharides exuded by bacteria increases and the morphology of the polysaccharide changes from more fibrous to more massive [108]. Dorioz *et al.* [46] found that the major effects of bacterial growth on the organization of concentrated suspensions of pure clays were an adsorption to cell surfaces and a secretion of polysaccharides that penetrated < 1 μm into surrounding clay pores. Addition of polysaccharides to pure clays increased water retention and altered shrinkage/swelling patterns [109]. Adding the bacterial polysaccharide xanthan or dextran to clays enhanced bridging between adjacent clay particles, enabling stronger cohesion and increased water stability [110].

Maintenance of the increased structural stability induced by bacterial extracellular polysaccharide will depend on its subsequent biological availability. Encapsulation of the extracellular polysaccharide and its penetration into clay pores will reduce its accessibility to processes of decomposition and prolong its influence on structural stabilization. Foster [101] observed that extracellular polysaccharides appeared to persist after the lysis of the microorganisms and continued to hold microaggregates together.

3.4.3.2 Filamentous fungi   A role of filamentous soil fungi in the stabilization of soil structural form has long been identified. Harris *et al.* [92] observed that fungal hyphae stabilized soil aggregates in incubation experiments and that the stabilization was enhanced when an organic substrate was added. Vesicular–arbuscular mycorrhizal (VAM) fungi associated with living plants were able to bind smaller aggregates together into larger aggregates stable against wet sieving [43]. Tisdall and Oades [43,66] attributed the stabilization to a physical enmeshment process and the ability of the hyphae to strongly bind to soil particles through the exudation of polysaccharides. The ability of fungi to bind adjacent clay particles together with extracellular polysaccharides was demonstrated by Dorioz *et al.* [46]. Strands of 0.1–1.0 m diameter were observed to interconnect adjacent clay particles, and the extent of bridging increased as the clay/fungus culture aged such that in older cultures (one week) the pore space close to fungal hyphae was filled with extracellular polysaccharide. On exposing the fungal/clay cultures to drying and rewetting, Dorioz *et al.* [46] observed the generation of microfissures and the formation of microaggregates entangled with fungal hyphae. The addition of scleroglucan (an extracellular polysaccharide isolated from the fungus *Sclerotium*) to kaolinite or montmorillonite was found to increase the mechanical strength [111,112] and water stability [112].

Foster [101] observed that all aggregates had fungi and sometimes actinomycetes associated with some part of their surface linking adjacent aggregates together. Contrary to bacteria, Shipton [102] reported that fungi can remain active in soils at very low water potential ($-7200\,kPa$) and are better suited than bacteria to exist in the interpore spaces. Fungi not only have the ability to exude polysaccharide materials capable of binding soil particles together in a manner similar to that of bacteria, but can also stabilize soil structural form through the production of hyphae that can enmesh aggregates.

Several forms of filamentous fungi exist in soils and can contribute to structural stabilization. Saprophytic fungi have been shown to stabilize soil structural form in the presence and absence of added organic substrates, after inoculation and with and without inhibition of other soil microorganisms [92,96,113,114]. Structural stabilization by saprophytic fungi can involve the production and release of extracellular polysaccharides and physical entanglement [46]. Tisdall *et al.* [115] examined the ability of two saprophytic fungi, *Rhizoctonia solani* and *Hyalodendron* sp., to stabilize the clay fraction of a self-mulching soil at two levels of aggregation: $> 2 \,\mu m$ and $< 2 \,\mu m$. After 15 days of incubation, *Rhizoctonia solani* was able to bind a greater proportion of aggregates $< 50 \,\mu m$ into stable aggregates $> 50 \,\mu m$ than was observed in a control treatment where no fungus was present. *Hyalodendron* sp. was found to increase the mean diameter of aggregates $< 2 \,\mu m$ but to have no effect on aggregates $> 2 \,\mu m$. The difference in aggregating behavior between the two saprophytic fungi was possibly related to differences in the length and morphology of the hyphae produced by the two fungi. *Rhizoctonia solani* produced twice the length of hyphae, had thicker hyphal strands, and a calculated surface area 4.5 times that of *Hyalodendron* sp. The potential for stabilization via adsorption of clays to polysaccharides associated with fungal surfaces and by enmeshment by hyphae was therefore greater for the *Rhizoctonia solani*.

The ability of the ectomycorrhizal fungus *Hebeloma* sp and the ericoid mycorrhizal fungus *Hymenoschyphus ericae* to improve the aggregation of clay was also investigated by Tisdall *et al.* [115]. The growth of both fungi increased the proportion of a soil clay found in aggregates $> 50 \,\mu m$, and the mean diameter of particles $< 2 \,\mu m$ was increased by *Hymenoschyphus ericae*. Emerson *et al.* [107] suggested that ectomycorrhizal hyphae were responsible for the increased structural stabilization of soil under forest relative to soil under pastures.

VAM fungi have been implicated in the stabilization of the structural form of soils, sands and clays in field and pot experiments [43,65,87,116,117]. Tisdall and Oades [65], Elliott and Coleman [117], and Miller and Jastrow [87] observed that the aggregation of various soils was related to the length of VAM hyphae in soil. VAM fungi can extend 10–30 mm into the soil from the root surface and the length of VAM hyphae within stable aggregates may reach $50 \,m\,g^{-1}$ aggregate [43,118–120]. VAM fungi also exude polysaccharide materials capable of forming strong associations with soil mineral particles [43]. As noted in section 3.4.2, it is difficult to differentiate the effect of the fungal hyphae from the effects of root systems, but Miller and Jastrow [87] showed that the length of fungal hyphae had the largest direct effect on geometric mean diameter (GMD) of aggregates (Figure 3).

The longevity of the stabilization of soil structural form induced by filamentous fungi depends on the type of fungus involved. Saprophytic fungi die once they have utilized the available substrates present in soil [120], and their stabilization will persist only as long as the residual hyphal structures remain intact. Tisdall and Oades [65] showed that the stabilization induced by the external hyphae of VAM fungi persisted for several months after the death of the infected plant. In addition, while the infected plant is still alive, the symbiotic relationship between VAM fungi and plant roots ensures that the fungi receive a carbon substrate. VAM fungi are therefore not dependent on the continuous presence of an available carbon substrate within the soil matrix to grow and release exudates into the soil matrix.

The amount of VAM hyphae in soil is affected by many soil and plant characteristics including clay content, soil pH, organic matter, nutrient status (especially phosphorus), the presence of fungicides or pesticides, soil fauna, or the presence of other microorganisms [118,120,121]. The morphology of filamentous fungi can also vary significantly both within and between the various types present in soils. Bonfante-Fasolo [122] observed variations in morphology even within one species of VAM fungus. Although it has been established that hyphal length is critical in assessing the ability of fungi to stabilize soil structural form, the importance of other morphological characteristics (e.g., hyphal diameter, cell wall thickness, septate versus non-septate, straight or fan-like) requires more attention. Tisdall [120] suggested that fungi most effective at stabilizing soil aggregates may (a) produce more extracellular mucilage, (b) produce a stickier mucilage, (c) be capable of binding to soil minerals via several mechanisms (e.g., hydrogen bonding or cation bridging), (d) orientate soil particles readily, (e) persist longer in soils, (f) interact more efficiently with plants, (g) produce large lengths of fungal hyphae, or (h) invade soil rapidly and completely.

3.4.4 Particulate Organic Matter

The mode of deposition of POM in soil depends on its source, and is an important factor involved in assessing its ability to affect the stabilization of soil structural form. POM derived from plant shoots is first deposited on the surface. It is then either colonized and decomposed on the soil surface or incorporated into the mineral soil matrix through bioturbation (mixing by the action of soil fauna), tillage or covered by eroded soil. Incorporation into the mineral matrix ensures good contact between shoot-derived POM and soil particles facilitating its colonization by decomposer organisms. However, there will be a lag in the initiation of decomposition as the population of decomposers adapts to the nature of the substrate and increases in numbers. POM derived from roots is deposited within the soil matrix in a zone of enhanced biological activity: the rhizosphere. The sloughing off of root cap cellular debris and the exudation of polysaccharides and other organic materials by growing roots help to ensure that the rhizosphere is biologically primed. Once dead, root POM can be colonized by rhizosphere organisms and decomposed. POM derived from fungal hyphae acts in a manner similar to that of plant root-derived POM.

The structural form of soil can be stabilized by POM through two mechanisms related to its physical properties and its susceptibility to biological decomposition. POM can bridge the failure zones that exist between adjacent stable aggregates. The bridging can result from a combination of binding to aggregate surfaces, penetration through aggregates, being embedded in a portion of an aggregate, or the formation of a network capable of holding groups of aggregates together. The ability of POM to stabilize soil structural form in this manner will be controlled by its physical size and morphology. Long thin particles that branch in different directions such as the remnants of fibrous root systems will be most effective; however, with decreasing particle diameter, POM will become more susceptible to decomposition, and the duration of structural stabilization will decrease.

POM can also enhance the stability of soil structural form by providing a substrate for soil microorganisms that will enhance the production of fungal hyphae and microbial polysaccharides. The magnitude of the stabilization will be a function of the chemical composition of the POM. With increasing carbohydrate (e.g., cellulose) and nitrogen contents the POM will become more biologically available and able to enhance microbial activity. The two mechanisms of structural stabilization by POM have the capacity to operate independently or synergistically to increase the stability of soil structural form.

POM has been divided into various fractions based on its position in the soil matrix and the extent to which it is associated with soil mineral particles [123] as portrayed in Figure 7. The association between POM and the soil mineral matrix is minimal when POM is first added to the soil. POM existing in this state has been termed free POM (Figure 8a) and can be removed from the soil by flotation on a solution of density $1.6\,\mathrm{Mg\,m^{-3}}$ after gentle shaking. The morphology of free POM isolated from an Australian red–brown earth (Alfisol) used for pasture or cereal production was dominated by cylindrical and spherical particles varying in size from 0.1 to 2.5 mm (Figure 8). Free POM has a chemical composition similar to the organic materials from which it was derived suggesting a limited exposure to microbial decomposition (see the ^{13}C NMR spectra in Figure 7). The dominance of carbohydrate structures in free POM [17,123] provides these particles with the potential to become sites of intense biological activity. As the colonization and decomposition of free POM by soil microorganisms proceeds, the concomitant production of metabolic binding agents (fungal hyphae and fungal and bacterial extracellular polysaccharide mucilage) strengthens and extends the interaction between POM and soil mineral particles, resulting in a stabilization of assemblages of soil particles around the decomposing POM. The following two observations support this concept:

Figure 7. The involvement of free and occluded POM fractions in the stability of aggregates in soil and the chemical composition of the POM fractions determined by ^{13}C NMR spectroscopy (d refers to the density of the POM fractions in Mg m^{-3} and C:N is the ratio of organic C to total N) (modified from [123])

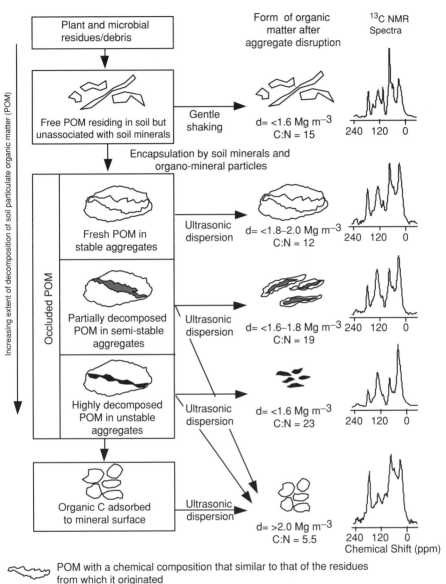

Increasing extent of decomposition of soil particulate organic matter (POM)

Plant and microbial residues/debris

Form of organic matter after aggregate disruption

^{13}C NMR Spectra

Free POM residing in soil but unassociated with soil minerals

Gentle shaking

d= <1.6 Mg m^{-3}
C:N = 15

240 120 0

Encapsulation by soil minerals and organo-mineral particles

Occluded POM

Fresh POM in stable aggregates

Ultrasonic dispersion

d= <1.8–2.0 Mg m^{-3}
C:N = 12

240 120 0

Partially decomposed POM in semi-stable aggregates

Ultrasonic dispersion

d= <1.6–1.8 Mg m^{-3}
C:N = 19

240 120 0

Highly decomposed POM in unstable aggregates

Ultrasonic dispersion

d= <1.6 Mg m^{-3}
C:N = 23

240 120 0

Organic C adsorbed to mineral surface

Ultrasonic dispersion

d= >2.0 Mg m^{-3}
C:N = 5.5

240 120 0
Chemical Shift (ppm)

POM with a chemical composition that similar to that of the residues from which it originated

Partially decomposed POM with a chemical composition partly indicative of the materials from which it was derived

Highly decomposed POM with a chemical composition bearing little resemblance to the residues from which it originated

Organo-mineral complexes with a density > 2.0 Mg m^{-3}

(a)

(b)

 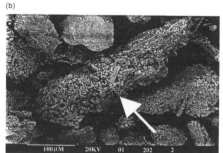

Figure 8. Scanning electron micrographs of (a) 'free' POM extracted from an Australian red–brown earth (Alfisol) with gentle shaking in a solution of density 1.6 Mg m^{-3} and (b) a fragment of POM coated with mineral particles (indicated by the white arrow) Reprinted with permission from [124]. Copyright CRC Press, Boca Raton, Florida

(1) the presence of POM cores around which individual particles and small soil aggregates are bound to produce larger aggregates (Figure 8b) [116,117,123–125], and

(2) an initial incorporation of labeled organic matter derived from plant residues into large soil aggregates [126–128].

POM found within soil aggregates is referred to as occluded POM and can be differentiated on the basis of its extent of interaction with soil mineral particles (Figure 7). When a disruptive force is applied to a soil (e.g., ultrasonic dispersion treatment), a range of soil aggregate structures that can be differentiated on the basis of density are obtained. The highest density fraction (> 2.0 Mg m^{-3}) does not contain occluded POM cores and is composed of either primary mineral particles or small aggregations of primary particles bound together by microbial metabolites (e.g., extracellular polysaccharides) and other organic materials (e.g., humic substances and residual aliphatic materials) adsorbed onto mineral surfaces. Fragments with a density of 1.8–2.0 Mg m^{-3} are associated with occluded POM that has undergone some decomposition but is still capable of supporting an active microbial population. Continued microbial activity results in a production of mucilage and other metabolites that permeate the mineral particles surrounding the POM binding the soil particles together and to the POM. As the decomposition of occluded POM proceeds, the more labile components (carbohydrates and proteins) are utilized by microorganisms leaving an organic core that is more biologically recalcitrant and less able to support an active microbial population. The production of mucilage and metabolites is reduced to the point where it no longer offsets the rate at which these materials are decomposed. Microorganisms then start utilizing alternative organic materials, including the binding agents holding soil mineral particles together and to the occluded POM. As a result, the amount of mineral particles strongly associated with the POM decreases and the density of the aggregates released on application of a stress is reduced. Aggregates having a density of 1.6–1.8 Mg m^{-3} contain occluded POM that has had a significant amount of the microbially available C removed, but still contain enough binding agents to hold some soil particles and the occluded POM together. Aggregates having a density of < 1.6 Mg m^{-3} are derived from the most decomposed occluded POM that is incapable of supporting an active population of microorganisms.

The changes in chemical structure in progressing from the free to occluded POM fractions, as determined by solid-state ^{13}C NMR, are consistent with the increasing extent of biological decomposition proposed to account for variations in density and aggregate structure. A continual reduction in O-alkyl C, normally ascribed to the carbohydrate structures (e.g., cellulose and hemicellulose) is observed. The concomitant increases in aromatic C and alkyl C are indicative of an increase in the more biologically resistant lignin and aliphatic biomolecules typically associated with organic residues in soils [18].

3.4.5 Humus

The humus fraction of SOM has the capacity to adsorb to soil inorganic particles. The extent and strength of adsorption is a function of the properties of both the humus and the inorganic soil matrix. The molecular size and type, content and distribution of functional groups dictate the potential of humus molecules to bind to mineral surfaces. The mineralogy, size, and shape of soil mineral particles dictate the amount of reactive surface area available for adsorption of humus materials and, when considered along with the characteristics of the humus molecules, the strength of adsorption. The observation that peroxide treatment of soils releases large amounts of fine clay and enhances mineral surface areas indicates that the adsorption of organic materials can bind clays or packets of clays together into larger assemblages. Using data from Burford et al. [129] and Turchenek and Oades [130], Oades [116] demonstrated that surface areas of soil clay fractions could be increased by a factor of 2–35 times by removal of SOM.

The polysaccharide fraction is a component of the humus fraction of SOM often implicated in the stabilization of soil structural form. As discussed in sections 3.4.2 and 3.4.3, polysaccharides exuded by both roots and microorganisms are thought to stabilize assemblages of clay particles. Glucuronic, galacturonic or mannuronic acids and other non-sugar acid groups (e.g., pyruvic or succinic) in polysaccharides found in soils [131] provide polysaccharides with the capacity to adsorb strongly to negatively charged clay surfaces through polyvalent cation bridging. Uncharged polysaccharides may form strong linkages via hydrogen bonding or van der Waals forces, because of their typically high molecular weight and numerous sites with the potential to interact with mineral surfaces. In addition to the studies quoted earlier concerning the role of polysaccharide materials in soil structural stabilization, their importance is also demonstrated by the following experimental results:

(1) positive correlations between aggregate stability and polysaccharide content [132]),
(2) extensive presence detected using thin sectioning and selective staining techniques in microscopic studies [133–135], and
(3) a substantial decrease in aggregation induced by treating soils with chemicals that selectively oxidize polysaccharides [136,137].

Roots and microorganisms are typically found adjacent to and in contact with mineral surfaces [90,138]. Therefore, the extracellular polysaccharides exuded by roots and microorganisms have the opportunity to form strong associations with soil minerals. Since extracellular polysaccharides exist as a gel or fibrillae, they are not considered to be mobile within the soil environment. The mobile component involved in the formation of polysaccharide–mineral associations is the mineral phase, typically dispersed clay contained in soil pore water. The dispersed clay moves closer to pore walls as soils dry, and then orients and binds to the surfaces of extracellular mucilage surrounding roots or microbial cells (Figures 5 and 6). The encapsulation of extracellular polysaccharide by mineral particles may account for the observation that not all polysaccharides in soils are available to microbial enzymes and are rapidly decomposed or mineralized. Baldock and Skjemstad [7] discussed the mechanisms involved in the protection of extracellular polysaccharide materials by the mineral matrix of soil.

Mechanisms that increase the biological stability of polysaccharides in soil will prolong any stabilization of soil structural form. The existence of protective mechanisms was demonstrated by the enhanced rate of utilization of polysaccharide materials extracted from soil compared with that of polysaccharides in undisturbed aggregates (e.g., [139]). The chemical properties of polysaccharides can also influence their biological stability. The presence of more than one monomeric component, the presence of 1–3 rather than 1–4 linkages between monomeric components, branching within the polysaccharide structure, or the presence of complex fibrillar structures all tend to enhance biological stability [140].

Little information exists as to the impact of proteins on the stabilization of soil structure. Proteins are one of the most abundant molecules synthesized by soil organisms and can be found as structural proteins, glycoproteins, and enzymes. The large molecular weight, flexibility, and charge characteristics of proteins suggests that they may be able to bind soil particles together in a manner similar to that of polysaccharides. Nelson et al. [141] used solid-state ^{13}C NMR to examine the chemical character of organic matter associated with clay fractions that were relatively easy and difficult to disperse. The SOC found in the clay fraction that was easy to disperse contained a higher proportion of amino acid or protein material. The SOC associated with the clay that was difficult to disperse contained higher proportions of polysaccharide and aliphatic carbon. The chemical structure and amphoteric nature of amino acids and proteins suggest that they can act as dispersants by altering the point of zero charge (PZC = the pH value at which the net total particle charge is zero) of clay minerals, such that it approaches soil solution pH values. As the PZC of the clay–protein complex approaches the soil solution pH, dispersion of the clays is enhanced. Amino acids and proteins can also complex multivalent cations thereby altering the composition of the diffuse double layer surrounding charged clay particles.

The remaining two types of unaltered biomolecules found in soil, lignin and lipids, also have the capacity to affect the stability of soil structural form. The main effect of lignin molecules would be to bridge failure zones between adjacent aggregates at a small scale ($< 50 \, \mu m$). However, partial decomposition products and the recombination of phenolic monomeric species to produce high molecular weight molecules with high contents of acidic functional groups, may allow partially degraded lignin and its decomposition products to bind strongly to mineral surfaces and stabilize structural form.

An interaction of lipids with soil minerals can occur through cation bridging with carboxyl groups (e.g., terminal carboxylic groups found on fatty acids), hydrogen bonding with oxygen-containing functional groups (e.g., alcoholic groups located within alkyl chains), or via van der Waals forces. The association of lipid molecules with mineral surfaces by any of these mechanisms imparts a hydrophobic nature to the resultant mineral–organic complex. The presence of hydrophobic sites within the soil matrix reduces the wettability of specific regions of soil and thereby reduces the magnitude of the disruptive forces associated with wetting and drying cycles (i.e. shrink/swell forces). Enhancing the hydrophobic nature of organic matter located along failure zones in the soil matrix may also enhance the stability of the failure zones and their persistence. Capriel et al. [142] observed a significant correlation between the stability of aggregates to wet sieving and the yield of organic material extracted from the aggregates by a supercritical-hexane extraction procedure. Aliphatic C dominated the extracted organic material, and it was proposed that this aliphatic supercritical-hexane extractable material formed a water-repellent lattice around the soil aggregates. Dinel and coworkers [143,144] demonstrated that the addition of long-chained aliphatic materials to soil increased the stability of aggregates to wet sieving, with the largest increases occurring after addition of acidic or neutral aliphatic fractions. Significant correlations were also noted between the amount of supercritical-hexane extractable material and soil microbial biomass [142], and between the amount of aliphatic carbon added and CO_2 evolution [144], suggesting that soil microorganisms were involved in the stabilization.

The aggregating effect of all forms of humus having functional groups that are negatively charged at pH values typical of those in soils can be enhanced by the presence of polyvalent cations. In the presence of polyvalent cations, carboxylic groups in organic molecules can bind to mineral surfaces through the formation of polyvalent cation bridges with polyvalent cations held either (1) within the crystal lattice at the surface of mineral particles, (2) within amorphous oxide/hydroxide compounds, or (3) in the diffuse double layer of charged mineral particles. The presence of multivalent cations can also change the spatial properties of soil humus molecules, such that they become more condensed and less susceptible to decomposition. Reducing the biological availability of humus enhances the persistence of structural stabilization. The major polyvalent cations present in soils are Ca^{2+} and Mg^{2+} in neutral to alkaline soils, and Fe^{3+} and Al^{3+} in acidic, ferrallitic and andic soils. Adding a source of Ca^{2+} to soils has been shown to reduce the mineralization of organic C derived from ^{14}C labeled substrates in field and laboratory experiments [94,95]. The reduced mineralization of substrate C was accompanied by an increase in the proportion of water stable aggregates $> 250\,\mu m$, and a reduced amount of dispersible clay. The reduction in dispersible clay may have resulted either from an effect of increased Ca^{2+} concentration on flocculation or from a direct effect of Ca^{2+} complexation on the ability of the metabolic products of substrate decomposition to bind soil mineral particles together. The existence of crystals of calcium oxalate and calcium carbonate on surfaces of hyphae [44,145] may provide a localized source of Ca^{2+} directly, and through the release and complexing capability of oxalate, may provide cations of Fe and Al.

Altered biomolecules or humic substances have a capacity to stabilize aggregations of soil primary particles because their high contents of carboxylic and phenolic functional groups facilitate a variety of interactions with soil mineral components (e.g. phyllosilicate clay surfaces, oxides and hydroxides of iron and aluminum and amorphous aluminosilicates) [146]. Humic–mineral complexes with widely different water solubility and chemical and biological stability can be formed in soil environments [147]. The net negative charge associated with humic substances and clay minerals at most soil pH values means that structural stabilization brought about by humic substance adsorption to clay particles occurs via polyvalent cation bridging as discussed above. Humic substances may also bind to mineral surfaces through the weaker but additive hydrogen bonding and van der Waals forces [148] as described previously for polysaccharides. The large molecular size, complexity and high functional group content of humic substances all contribute to defining their ability to bind to and link adjacent soil particles into structurally stable aggregates.

3.4.6 Dissolved Organic Matter

Dissolved organic matter (DOM) in soils consists of a range of organic materials varying in size from individual molecules through to particles with an operationally defined upper size limit of $< 0.45\,\mu m$. At the smaller end of its size scale, DOM consists of individual molecules such as simple sugars, amino acids, monomeric aromatic and phenolic molecules and simple aliphatic acids. At the upper boundary the distinction between dissolved, colloidal and particulate organic materials becomes less well defined and DOM can exist in any of these forms. Despite these variations in composition, DOM in soils represents a dynamic fraction of SOM.

DOM can influence the stability of soil structural form either directly or indirectly. The indirect effect results from the ability of DOM to provide a readily available substrate for microorganisms and the resultant stabilization induced by the activity of soil microorganisms. The direct effect of DOM results from the adsorption of DOM onto soil mineral particles. The adsorption of soluble organic anions decreases the stability of soil structural form through its effect on the dispersion of clay particles[149]. Soluble organic anions can destabilize soils by:

(1) complexing divalent and trivalent cations and removing them from solution,
(2) neutralizing positive sites on the edges of clay minerals,
(3) complexing cations in metal oxides and hydroxides, and
(4) increasing the net negative charge on colloidal surfaces.

Each of these processes results in an increase in the diffuse layer of cations surrounding clay particles and enhances the potential for dispersion. Shanmuganathan and Oades [150] demonstrated that the beneficial effects on the stability of soil structural form induced by the addition of Fe-polycations to soil was negated by the addition of various organic anions. The destabilization induced by the organic anions was related to the amount of organic anion adsorbed by the Fe-polycation/clay surfaces. Citrate, oxalate, and tartrate anions are released into the DOM pool of soil by plants, bacteria and fungi and were among those found by Shanmuganathan and Oades [150] to enhance clay dispersion. Shanmuganathan and Oades [150] also noted a strong dispersive behavior of a soluble fulvic acid. Reid and Goss [81,151] and Reid et al. [152] attributed an increase in the dispersion of clay in soils growing maize plants to the release of soluble organic anions into the soil.

4 AGGREGATE HIERARCHY

Interactions between soil mineral particles, organic matter and microorganisms can occur at many different size scales given the large size range of these materials in soils (Figure 2). It is important when discussing soil structural form and the mechanisms of its stabilization to indicate the size scale being considered, as the potential mechanisms capable of accounting for stabilization vary with aggregate size.

Edwards and Bremner [153] proposed the existence of two size classes of aggregates in soils: macroaggregates > 250 μm diameter and microaggregates < 250 μm diameter. Tisdall and Oades [66] further developed this concept, and Dexter [31], Kay [154], Oades and Waters [11], and Oades [30] extended it into what is now referred to as aggregate hierarchy. The concept of aggregate hierarchy proposes that larger aggregates are composed of more stable smaller aggregates bound together by some mechanism of structural stabilization. Dexter [31] argued that, if aggregate hierarchy exists in a soil, the density, strength and stability of larger aggregates would be less than that of their component smaller aggregates: the 'porosity exclusion principle'. The porosity of larger aggregates is equal to the sum of the internal porosity of the smaller component aggregates plus the porosity between the smaller aggregates. The pores between the small aggregates act as failure zones reducing the stability of the larger aggregates. Thus, when soils exhibiting aggregate hierarchy are exposed to disruptive forces, aggregate structure breaks down progressively from large to small aggregates as the magnitude of the disruptive stress increases. If large aggregates break down directly to primary particles, then the concept of aggregate hierarchy does not apply. The concept of aggregate hierarchy is illustrated in Figure 9 [155].

A level of aggregation often omitted from models is the aggregation of clay particles into stable microstructures referred to as quasicrystals, domains or assemblages as defined by Oades and Waters [11]. In soils where aggregate hierarchy exists, the aggregation of clay forms the basic building blocks from which larger aggregates develop. If clay particles are not aggregated into stable microstructures, they slowly erode from aggregate surfaces resulting in a gradual deterioration of microaggregates and macroaggregates. On this basis, Golchin *et al.* [125] and Jastrow and Miller [155] proposed that three basic levels of structural organization exist in soils containing appreciable contents of clay: (1) the binding together of individual clay particles into stable microstructures 2–20 μm, (2) the binding of clay microstructures into stable microaggregates 20–250 μm, and (3) the binding of microaggregates into stable macroaggregates > 250 μm. Oades and Waters [11] observed the existence of these three levels of aggregation and a stepwise disintegration of macroaggregates to microaggregates to primary particles and clay microstructures with increasing strength of an applied disruptive force for a Mollisol and an Alfisol.

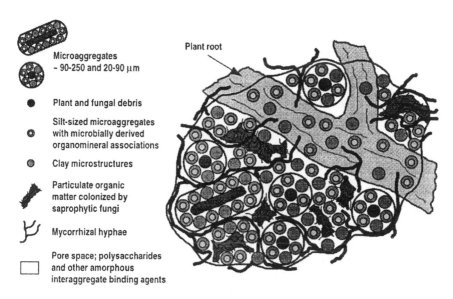

Figure 9. A conceptual model of aggregate hierarchy in soils where organic materials play an important role in the stabilization of aggregates. Reprinted with permission from [155]. Copyright CRC Press, Boca Raton, Florida

However, the disintegration of macroaggregates from an Oxisol, where inorganic cementing agents dominate the stabilization of structural form, resulted in the release of particles < 20 μm with no indication of an intermediate stable microaggregate fraction. Based on these results, Oades [30] speculated that aggregate hierarchy will exist where soils have a long history of being exploited by roots and will not exist in young soils, or soils where inorganic cementing agents are primarily responsible for binding soil particles together. Feller *et al.* [156] found a tendency for aggregate hierarchy to exist in soil dominated by low activity clays (kaolinite) but not in a Vertisol. In addition, Feller *et al.* [156] observed that the size of the stable microaggregate increased with increasing SOC content and decreasing extent of tillage. Bartoli *et al.* [157] also observed that the size of the elementary structural unit in Oxisols, as determined by a fractal analysis, decreased with decreasing SOC content consistent with an ability of SOC to enhance the stabilization of soil structural form.

In the subsequent discussion, the major mechanisms thought to be operative in the stabilization of differently sized aggregates will be addressed. Although discrete size boundaries are suggested for the different levels of aggregation, mechanisms of stabilization will undoubtedly overlap these boundaries.

The extent of aggregation at the smallest size scale, the binding together of clay particles into microstructures, is dictated by the chemical properties of the soil controlling dispersion: the charge density of clay surfaces, the presence or absence of short-range ordered metal oxides, the composition of the cation exchange complex, and the soil solution electrolyte concentration and composition. Significant variations in these characteristics throughout the bulk soil matrix should not occur without anthropogenic inputs (e.g., addition of gypsum or lime), or the passing of enough time to allow pedological process to progress significantly. Consequently, the extent of aggregation of clay in the bulk soil matrix should be relatively homogeneous and should be a function of pedological processes. However, various biological processes may have significant effects that alter clay aggregation and introduce localized heterogeneity into the soil matrix. An example of such a biological process is the exudation of low molecular weight organic acids, such as oxalic acid, that may enhance the dispersion of clays by the mechanisms discussed previously (section 3.4.6).

The adsorption of organic molecules, such as microbially derived polysaccharides and other unaltered and altered biomolecules onto mineral surfaces can enhance the stability of clay microstructures. Adsorption is important to the stabilization of individual clay microstructures and the binding together of clay microstructures and silt particles into small microaggregates with 2–50 μm diameters and densities $> 2.0 \, \text{Mg m}^{-3}$. The high stability of these small microaggregates in some soils is demonstrated by their resistance to the disruptive forces encountered on treatment with ultrasonic energy [11,17]. Electron microscopic studies demonstrated that many microaggregates exist as pieces of fungal hyphae, bacteria or bacterial colonies coated with extracellular polysaccharides and clay minerals [11,158,159]. The distribution of neutral polysaccharides as defined by selective staining [107] is shown in Figure 10. The polysaccharide materials can be seen throughout the matrix but concentrated in pores between clay microstructures.

POM and the metabolic products released during its decomposition become important stabilizing agents at larger size scales: large microaggregates and small macroaggregates [125,155]. Waters and Oades [8] and Oades and Waters [11] demonstrated that POM is often found at the core of microaggregates $< 250 \, \mu\text{m}$. The decomposition of the POM by microorganisms releases polysaccharides and other unaltered and altered biomolecules that have the capacity to bind soil mineral particles and small microaggregates together resulting in the formation of stable microaggregates. While the POM continues to provide a substrate for microorganisms, the production of organic aggregating agents continues and structural stability is maintained. The microbial polysaccharides and other biomolecules are also susceptible to decomposition unless protected by encapsulation with clay particles. The physical size of the microaggregate formed depends on the size of the POM and the amount of binding materials secreted by microorganisms as decomposition proceeds.

Figure 10. Ultrathin section of a surface soil with neutral polysaccharides stained. Many of the pores (0.1 μm diameter) are intensely stained and appear black (indicated by the small arrows). The presence of neutral polysaccharide throughout the soil matrix is indicated by the black dots (bar = 1 μm). From [107]. Reproduced by permission of Soil Science Society of America

Mechanisms of structural stabilization must operate over larger distances in order to bind microaggregates together into macroaggregates where the concept of aggregate hierarchy can be applied to soil. Considering the distances involved, the stabilization of macroaggregates must be related to the presence of POM capable of spanning distances > 100 μm or the existence of a network of fungal hyphae and plant roots that physically enmeshes microaggregates. The death of roots and hyphae growing within and through macroaggregates places organic substrates in a position where its decomposition results in the production of organic agents capable of stabilizing the internal structure of macroaggregates.

5 DYNAMICS OF AGGREGATE TURNOVER

In soils where organic matter provides the main binding agent, aggregation is dynamic because the organic agents responsible for the stabilization of structural form are not inert and are thus subject to decomposition. Therefore, biological activity has the potential not only to stabilize soil structural form through the production of organic materials capable of binding soil particles, but also to destabilize structural form by decomposing organic binding agents. The balance between addition and degradation of the various aggregating agents therefore dictates the level of soil structural stability. A continual addition of organic material to soil is required to ensure that the contents of each type of organic binding agent are adequate to maintain a given level of structural stability in a soil. For this reason, management that provides an ongoing addition of organic materials (e.g., continuous pasture production) has the greatest potential to maintain or enhance soil structural stability. The inclusion of fallow treatments in crop rotations results in a relative destabilization of soil structural form. The influence of practices that return large amounts of organic matter at the end of the growing season on soil structural stability is transient.

Golchin et al. [125] presented a model that can be used to relate the physical and chemical properties of soil organic materials and their distribution and cycling to the stabilization of soil structural form (Figure 11). The model considers the major organic materials contributing to the structural stability of soils to be the free and occluded POM fractions [123] (Figure 7) and the biomolecules synthesized and exuded by roots, mycorrhizal fungi, and other soil microorganisms. The model is most appropriately applied to the aggregation of soils with appreciable clay contents where mechanisms exist to protect organic binding agents from rapid decomposition [7]. For sandy soils, the mechanisms of protection that stabilize organic binding agents against microbial attack are less effective, making aggregation in sands a much more dynamic process. In addition, the larger size of the mineral particles and pores in sandy soils reduces the effectiveness of polysaccharides and other unaltered or altered biomolecules to stabilize structural form. Polysaccharides and biomolecules remain important to the adherence of soil particles to roots and fungal hyphae, and of soil microorganisms to mineral particles. However, they do not have the capacity to span the distances between large primary particles and stabilize the entire soil matrix.

In the proposed model of soil aggregation, freshly deposited POM is typically found in the free POM fraction of SOM. The binding of soil particles to free POM is limited, especially for that derived from plant shoots, and is contained in the free POM fraction having little direct influence on soil structural stability (Stage 1, Figure 11). The physical size and morphology of free POM, and thus the distances over which it can stabilize soil structural units, are defined by its source (e.g., shoot and root architecture) and applied management practices (e.g., tillage). Free POM with a larger size would be expected after production of larger crops such as maize or sugar cane. The distribution of free POM in soil will depend on the mode of deposition of shoot and root residues in the soil. Root residues will be more homogeneously distributed than shoot residues unless shoot residues are well mixed with soil mineral particles by tillage.

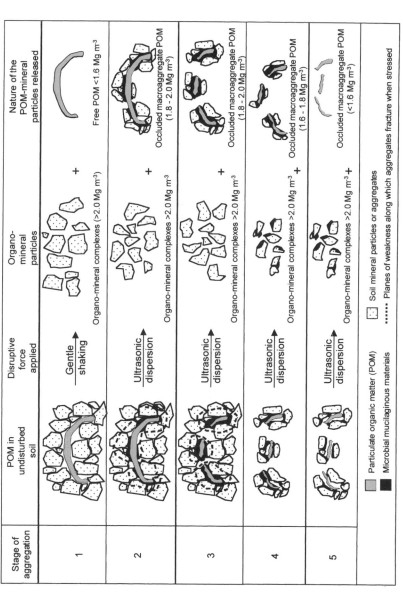

Figure 11. A model depicting the dynamics of soil aggregation and the roles of POM and microbial metabolites in the stabilization of soil aggregates (modified from [125])

With current tendencies towards reducing the amount of tillage used in agricultural systems, the most homogeneous distribution of free POM and potential sites of structural stabilization will occur where plants with extensive fibrous root systems are produced.

Once free POM is colonized by soil microorganisms and decomposition is initiated, individual mineral particles and mineral particles associated with adjacent aggregates adhere to the extracellular mucilage released by decomposer organisms (Stage 2, Figure 11). The free POM becomes occluded within a new macroaggregate. For POM derived from roots or mycorrhizal fungi, the stabilization may have been initiated by an adherence of soil particles to mucilage released as the roots and fungi grew in the soil. As decomposition and the production of microbially derived organic binding agents continues, the stability of the new macroaggregate increases. Concomitantly, the stability of adjacent aggregates will deteriorate as the organic agents responsible for binding their component primary particles together are decomposed. The result of these two processes will be the generation of new failure zones and a new macroaggregate around the decomposing occluded POM (Stage 2, Figure 11). The size of the new macroaggregate will be a function of the size, geometry and mode of deposition of POM. Long pieces of debris derived from roots or fungal hyphae have the potential to span across groups of small microaggregates and pores to form macroaggregates.

Further decomposition of occluded POM results in a decrease in the size of POM through a preferential utilization of exposed POM that has not been encapsulated with mineral materials. The ability of pieces of POM to maintain the stability of the original macroaggregate decreases and the macroaggregate deteriorates into several smaller microaggregates on exposure to disruptive forces (Stage 3, Figure 11). The ability of microaggregate-POM to maintain microaggregate stability then becomes a function of the chemical composition of the POM. Microaggregates containing POM with appreciable contents of carbohydrate remain stable because the POM continues to provide a substrate suitable for microbial decomposition and production of organic biomolecules capable of binding soil particles together. Once the carbohydrates are removed from the microaggregate POM, the more recalcitrant cores are no longer capable of maintaining the integrity of the microaggregate. The microaggregate then breaks up into aggregations of soil particles bound together with remnant microbial mucilage and other unaltered and altered biomolecules (Stages 4 and 5, Figure 11). Continued use of these biomolecules results in a complete deterioration of aggregates to primary particles and clay microstructures with biomolecules adsorbed to primary particle surfaces.

6 SUMMARY

All components of SOM, except for the biologically unavailable char/charcoal fraction, contribute to the stabilization of soil structural form in mineral soils. The diverse range of living and non-living components varying in size, chemical composition, and extent of decomposition (Figure 1) provide SOM with the capacity to stabilize aggregations of soil particles at size scales varying over nine orders of magnitude. Each SOM component may operate across a range of at a given size scales, but primarily in one category among those shown in Figure 2. It is therefore essential to maintain the proper balance of SOM components to ensure a stabilization of the entire soil matrix.

The stabilization imparted by each SOM component may be directly or indirectly related to its presence and content within the soil matrix. Direct effects of SOM components involve a physical enmeshment of primary mineral particles or aggregates, an adsorption of soil mineral particles to the external surfaces of living structures or cells, or an adsorption of organic molecules onto mineral surfaces. In all cases, the organic materials must bridge failure zones existing within the soil matrix to enhance stability. All SOM components can also indirectly affect soil structural stability by providing a substrate for soil microorganisms. During the decomposition of organic substrates, the synthesis of cellular tissues and the exudation of metabolites (e.g., polysaccharide mucilage) will enhance stability.

All components of SOM involved in the stabilization of soil structure are subject to decomposition. Although mechanisms exist that can offer some protection against biological attack (e.g., burial within mineral particles), organic binding agents decompose with time. A continuous input of organic materials, principally through plant production, is therefore required to maintain or enhance structural stability.

REFERENCES

1. Baldock, J. A. and Nelson, P. N. (2000). Soil organic matter. In *Handbook of Soil Science*, ed. Sumner, M., CRC Press, Boca Raton, FL, p. B25.
2. Stevenson, F. J. (1994). *Humus Chemistry. Genesis, Composition and Reactions*, John Wiley & Sons, New York.
3. Schnitzer, M. (1995). Organic–inorganic interactions in soils and their effects on soil quality. In *Environmental Impact of Soil Component Interactions*. Vol. 1. *Natural and Anthropogenic Organics*, ed. Huang, P. M. *et al.*, CRC Press, Boca Raton, FL, p. 3.
4. Schnitzer, M. (2000). A lifetime perspective on the chemistry of soil organic matter, *Adv. Agron.*, **68**, 1.

5. Oades, J. M. (1988). The retention of organic matter in soil, *Biogeochemistry*, **5**, 35.
6. MacCarthy, P., Malcolm, R. L., Clapp, C. E. and Bloom, P. R. (1990). An introduction to soil humic substances. In *Humic Substances in Crop and Soil Sciences: Selected Readings*, ed. MacCarthy, P. *et al.*, Soil Science Society of America, Madison, WI, p. 1.
7. Baldock, J. A. and Skjemstad, J. O. (2000). The role of the soil mineral matrix in protecting natural organic materials against biological attack, *Org. Geochem.*, **31**, 697.
8. Waters, A. G. and Oades, J. M. (1991). Organic matter in water-stable aggregates. In *Advances in Soil Organic Matter Research: The Impact on Agriculture and the Environment*, ed. Wilson, W. S., Royal Society of Chemistry, Cambridge, p. 163.
9. Lee, K. E. and Foster, R. C. (1991). Soil fauna and soil structure, *Aust. J. Soil Res.*, **29**, 745.
10. Herbert, B. E. and Bertsch, P. M. (1995). Characterization of dissolved and colloidal organic matter in solutions: A review. In *Carbon Forms and Functions in Forest Soils*, ed. McFee, W. W. and Kelly, J. M., Soil Science Society of America, Madison, WI, p. 62.
11. Oades, J. M. and Waters, A. G. (1991). Aggregate hierarchy in soils, *Aust. J. Soil Res.*, **29**, 815.
12. Gregorich, E. G. and Ellert, B. H. (1993). Light fraction and maroorganic matter in mineral soils. In *Soil Sampling and Methods of Analysis*, ed. Carter, M. R., Lewis, Boca Raton, FL, p. 397.
13. Cambardella, C. A. and Elliott, E. T. (1992). Particulate soil organic-matter changes across a grassland cultivation sequence, *Soil Sci. Soc. Am. J.*, **56**, 777.
14. Cambardella, C. A. and Elliott, E. T. (1993). Carbon and nitrogen distribution in aggregates from cultivated and native grassland soils, *Soil Sci. Soc. Am. J.*, **57**, 1071.
15. Golchin, A., Oades, J. M. and Skjemstad, J. O. (1994a). Study of free and occluded particulate organic matter in soils by solid state ^{13}C CP/MAS NMR spectroscopy and scanning electron microscopy, *Aust. J. Soil Res.*, **32**, 285.
16. Jastrow, J. D. (1996). Soil aggregate formation and the accural of particulate and mineral-associated organic matter, *Soil Biol. Biochem.*, **28**, 665.
17. Baldock, J. A., Oades, J. M., Waters, A. G., Peng, X., Vassallo, A. M. and Wilson, M. A. (1992). Aspects of the chemical structure of soil organic materials as revealed by solid-state ^{13}C NMR spectroscopy, *Biogeochemistry*, **16**, 1.
18. Baldock, J. A., Oades, J. M., Nelson, P. N., Skene, T. M., Golchin, A. and Clarke, P. (1997). Assessing the extent of decomposition of natural organic materials using solid-state ^{13}C NMR spectroscopy, *Aust. J. Soil Res.*, **35**, 1061.
19. Hatcher, P. G. and Spiker, E. C. (1988). Selective degradation of plant biomolecules. In *Humic Substances and Their Role in the Environment*, ed. Frimmel, F. H. and Christman, R. F., John Wiley & Sons, New York, p. 59.
20. Hedges, J. I. (1988). Polymerization of humic substances in natural environments. In *Humic Substances and Their Role in the Environment*, ed. Frimmel, F. H. and Christman, R. F., John Wiley & Sons, New York, p. 45.
21. Shevchenko, S. M. and Bailey, G. W. (1996). Life after death: lignin–humic relationships reexamined, *Crit. Rev. Environ. Sci. Technol.*, **26**, 95.
22. Bollag, J. M., Dec, J. and Huang, P. M. (1998). Formation mechanisms of complex organic structures in soil habitats, *Adv. Agron.*, **63**, 237.
23. Huang, P. M. (2000). Abiotic catalysis. In *Handbook of Soil Science*, ed. Sumner, M., CRC Press, Boca Raton, FL, pp. B303–B332.
24. Nelson, P. N. and Baldock, J. A. (1998). Using common molecular structures to explain solid-state ^{13}C NMR spectra acquired for natural organic materials. In

Humic Substance Downunder: Understanding and Managing Organic Matter in Soils, Sediments and Waters. Abstracts and Program. CSIRO Publishing, Collingwood, Vic., Australia, p. 173.

25. Baldock, J. A. (1999). Bioavailability and chemical composition of thermally altered *Pinus resinosa* (red pine) wood. In *Ninth Annual V. M. Goldschmidt Conference, LPI Contribution No. 971*, Lunar and Planetary Institute, Houston, USA, p. 16.

26. Skjemstad, J. O., Clarke, P., Taylor, J. A., Oades, J. M. and McClure, S. G. (1996). The chemistry and nature of protected carbon in soil, *Aust. J. Soil Res.*, **34**, 251.

27. Skjemstad, J. O., Taylor, J. A., Janik, L. J. and Marvanek, S. P. (1999). Soil organic carbon dynamics under long-term sugarcane monoculture, *Aust. J. Soil Res.*, **37**, 151.

28. Kay, B. D. (1997). Soil structure and organic carbon: A review. In *Soil Processes and the Carbon Cycle*, ed. Lal, R. *et al.*, CRC Press, Boca Raton, FL, p. 169.

29. Kay, B. D. and Angers, D. A. (2000). Soil structure. In *Handbook of Soil Science*, ed. Sumner, M. *et al.*, CRC Press, Boca Raton, FL, p. A229.

30. Oades, J. M. (1993). The role of biology in the formation, stabilization and degradation of soil structure, *Geoderma*, **56**, 377.

31. Dexter, A. R. (1988). Advances in characterization of soil structure, *Soil Till. Res.*, **11**, 199.

32. Grant, C. D. and Blackmore, A. V. (1991). Self-mulching behaviour in clay soils: its definition and measurement, *Aust. J. Soil Res.*, **29**, 155.

33. Wenke, J. F. and Grant, C. D. (1994). The indexing of self-mulching behaviour in soils, *Aust. J. Soil Res.*, **32**, 201.

34. Guidi, G., Poggio, G. and Petruzelli, G. (1985). The porosity of soil aggregates from bulk soil and soil adhering to roots, *Plant Soil*, **87**, 311.

35. Bruand, A., Cousin, I., Nicoullaud, B., Duval, O. and Bégon, J. C. (1996). Backscatter electron scanning images of soil porosity for analyzing soil compaction around roots, *Soil Sci. Soc. Am. J.*, **60**, 895.

36. Radcliffe, D. E., Clark, R. L. and Sumner M.E. (1986). Effect of gypsum and deep-rooting perennials on subsoil mechanical impedance, *Soil Sci. Soc. Am. J.*, **50**, 1566.

37. Meek, B. D., Rechel, E. A., Carter, L. M. and DeTar, W. R. (1989). Changes in infiltration under alfalfa as influenced by time and wheel traffic, *Soil Sci. Soc. Am. J.*, **53**, 238.

38. Meek, B. D., DeTar, W. R., Rolph, D., Rechel, E. A. and Carter, L. M. (1990). Infiltration rate as affected by an alfalfa and no-till cotton cropping system, *Soil Sci. Soc. Am. J.*, **54**, 505.

39. Mitchell, A., Ellsworth, T. R. and Meek, B. D. (1995). Effect of root systems on preferential flow in welling soil, *Commun. Soil Sci. Plant Anal.*, **26**, 2655.

40. Caron, J., Banton, O., Angers, D. A. and Villeneuve, J. P. (1996a). Preferential bromide transport through a clay loam under alfalfa and corn, *Geoderma*, **69**, 175.

41. Grevers, M. C. J. and de Jong, E. (1990). The characterization of soil macroporosity of a clay soil under ten grasses using image analysis, *Can. J. Soil Sci.*, **70**, 93.

42. Materechera, S. A., Dexter, A. R. and Alston, A. M. (1992). Formation of aggregates by plant roots in homogenized soils, *Plant Soil*, **142**, 69.

43. Tisdall, J. M. and Oades, J. M. (1979). Stabilization of soil aggregates by the root systems of ryegrass, *Aust. J. Soil Res.*, **17**, 429.

44. Foster, R. C. (1988). Microenvironments of soil microorganisms, *Biol. Fert. Soils*, **6**, 189.

45. Chenu, C. (1989). Influence of a fungal polysaccharide, scleroglucan, on clay microstructures, *Soil Biol. Biochem.*, **21**, 299.

46. Dorioz, J. M., Robert, M. and Chenu, C. (1993). The role of roots, fungi and bacteria on clay particle organization. An experimental approach, *Geoderma*, **56**, 179.

47. Anderson S. H., Gantzer, C. J. and Brown, J. R. (1990). Soil physical properties after 100 years of continuous cultivation, *J. Soil Water Conserv.*, **45**, 117.
48. Lal, R., Mahboubi, A. A. and Fausey, N. R. (1994). Long-term tillage and rotation effects on properties of a central Ohio soil, *Soil Sci. Soc. Am. J.*, **57**, 517.
49. Schjonning, P., Christensen, B. T. and Carstensen, B. (1994). Physical and chemical properties of a sandy loam receiving animal manure, mineral fertilizer or no fertilizer for 90 years, *Eur. J. Soil Sci.*, **45**, 257.
50. Manrique, L. A. and Jones, C. A. (1991). Bulk density of soils in relation to soil physical and chemical properties, *Soil Sci. Soc. Am. J.*, **55**, 476.
51. Vereecken, H., Maes, J., Feyen, J. and Darus, P. (1989). Estimating the soil moisture retention characteristic from texture, bulk density and carbon content, *Soil Sci.*, **148**, 389.
52. Williams, J., Ross, P. and Bristow, K. (1992). Prediction of the Campbell water retention function for texture, structure and organic matter. In *Indirect Methods for Estimating the Hydraulic Properties of Unsaturated Soils*, ed. van Genuchten, M.Th. *et al.*, University of California, Riverside, CA, p. 427.
53. da Silva, A. P. and Kay, B. D. (1997). Estimating the least limiting water range of soils from properties and management, *Soil Sci. Soc. Am. J.*, **61**, 877.
54. Kay, B. D., da Silva, A. P. and Baldock, J. A. (1997). Sensitivity of the structure of different soils to changes in organic carbon content: predictions using pedotransfer functions, *Can. J. Soil Sci.*, **77**, 655.
55. Kemper, W. D. and Rosenau, R. C. (1986). Aggregate stability and size distribution. In *Methods of Soil Analysis, Part I. Physical and Mineralogical Methods*, ed. Klute, A., American Society of Agronomy/Soil Science Society of America, Madison, WI, p. 425.
56. Baldock, J. A. and Kay, B. D. (1987). Influence of cropping history and chemical treatments on the water-stable aggregation of a silt loam soil, *Can. J. Soil Sci.*, **67**, 501.
57. Pojasok, T. and Kay, B. D. (1990a). Assessment of a combination of wet sieving and turbidimetry to characterize the structural stability of moist aggregates, *Can. J. Soil Sci.*, **70**, 33.
58. Angers, D. A. and Mehuys, G. R. (1993). Aggregate stability to water. In *Soil Sampling and Methods of Analysis*, ed. Carter, M. R., Lewis Publishers, Boca Raton, FL, p. 651.
59. Hudson, B. D. (1994). Soil organic matter and available water capacity, *J. Soil Wat. Conserv.*, **49**, 189.
60. Emerson, W. W. (1995). Water retention, organic C and soil texture, *Aust. J. Soil Res.*, **33**.
61. Strickling, E. (1950). The effect of soy beans on volume weight and water stability of soil aggregates, soil organic matter content, and crop yield, *Soil Sci. Soc. Am. Proc.*, **15**, 30.
62. Kemper, W. D. and Koch, E. J. (1966). Aggregate stability of soils from Western United States and Canada. *United States Department of Agriculture Technical Bulletin No. 1355*.
63. Williams, R. J. B. (1970). Relationships between the comparison of soils and physical measurements made on them. *Rothamsted Experimental Station Report for 1970, Part 2*, p. 5.
64. Grierson, I. T. (1975). *The effect of gypsum on the chemical and physical properties of a range of red–brown earths*, University of Adelaide, Adelaide, SA, Australia.
65. Tisdall, J. M. and Oades, J. M. (1980). The effect of crop rotation on aggregation in a red-brown earth, *Aust. J. Soil Res.*, **18**, 423.

66. Tisdall, J. M. and Oades, J. M. (1982). Organic matter and water-stable aggregates in soils, *J. Soil Sci.*, **33**, 141.
67. Chaney, K. and Swift, R. S. (1984). The influence of organic matter on aggregate stability in some British soils, *J. Soil Sci.*, **35**, 223.
68. Haynes, R. J., Swift, R. S. and Stephen, R. C. (1991). Influence of mixed cropping rotations (pasture-arable) on organic matter content, water stable aggregation and clod porosity in a group of soils, *Soil Till. Res.*, **19**, 77.
69. Monnier, G. (1965). Action des matières organique sur la stabilité structurale des sols, *Ann. Agron.*, **16**, 327.
70. Quirk, J. P. and Murray, R. S. (1991). Towards a model for soil structural behaviour, *Aust. J. Soil Res.*, **29**, 828.
71. Rashid, V. and Kay, B. D. (1995). Characterizing the rate of wetting: impact on structural destabilization, *Soil Sci.*, **160**, 176.
72. Caron, J., Espindola, C. R. and Angers, D. A. (1996b). Soil structural stability during rapid wetting: influence of land use on some aggregate properties, *Soil Sci. Soc. Am. J.*, **60**, 901.
73. Douglas, J. T. and Goss, M. J. (1982). Stability and organic matter content of surface soil aggregates under different methods of cultivation and in grassland, *Soil Till. Res.*, **2**, 155.
74. Skjemstad, J. O., Taylor, J. A. and Smernik, R. J. (1999). Estimation of charcoal (char) in soils, *Commun. Soil Sci. Plant Anal.*, **30**, 2283.
75. Cockroft, B. and Olsson, K. A. (2000). Degradation of soil structure due to coalescence of aggregates in no-till, no-traffic beds of irrigated crops, *Aust. J. Soil Res.*, **38**, 61.
76. Clarke, A. L., Greenland, D. J. and Quirk, J. P. (1967). Changes in some physical properties of the surface of an impoverished red-brown earth under pasture, *Aust. J. Soil Res.*, **5**, 59.
77. Foster, R. C. and Rovira, A. D. (1976). Ultrastructure of wheat rhizosphere, *New Phytol.*, **76**, 343.
78. Forster, S. M. (1979). Microbial aggregation of sand in an embryo dune system, *Soil Biol. Biochem.*, **11**, 537.
79. Forster, S. M. (1990). The role of microorganisms in aggregate formation and soil stabilization: types of aggregation, *Arid Soil Res. Rehabil.*, **4**, 85.
80. Dormaar, J. F. and Foster, R. C. (1991). Nascent aggregates in the rhizosphere of perennial ryegrass (*Lolium perenne* L.), *Can. J. Soil Sci.*, **71**, 465.
81. Reid, J. B. and Goss, M. J. (1981). Effect of living roots of different plant species on the aggregate stability of two arable soils, *J. Soil Sci.*, **32**, 521.
82. Dufey, J. E., Halen, H. and Frankart, R. (1986). Evolution de la stabiliè structurale du sol sous l'influence des racine de trèfle (*Trifolium pratense* L.) et de ray-grass (*Lollium mutiforum* Lmk.). Observations pendant et après culture, *Agronomie*, **6**, 811.
83. Angers, D. A. and Mehuys, G. R. (1988). Effects of cropping on macroaggregation of a marine clay soil, *Can. J. Soil Sci.*, **68**, 723.
84. Stone, J. A. and Buttery, B. R. (1989). Nine forages and the aggregation of a clay loam soil, *Can. J. Soil Sci.*, **69**, 165.
85. Morel, H. J. L., Guckert, A., Plantureux, S. and Chenu, C. (1990). Influence of root exudates on soil aggregation, *Symbiosis* (*Rehovot*), **9**, 87.
86. Thomas, R. S., Dakessian, S., Ames, R. N., Brown, M. S. and Bethlenfalvay, G. J. (1986). Aggregation of a silty loam soil by mycorrhizal onion roots, *Soil Sci. Soc. Am. J.*, **50**, 1494.
87. Miller, R. M. and Jastrow, J. D. (1990). Hierarchy of root and mycorrhizal fungal interactions with soil aggregation, *Soil Biol. Biochem.*, **22**, 579.

88. Pojasok, T. and Kay, B. D. (1990b). Effect of root exudates from corn and bromegrass on soil structural stability, *Can. J. Soil Sci.*, **70**, 351.
89. Morel, J. L., Habib, L., Plantureux, S. and Guckert, A. (1991). Influence of maize root mucilage on soil aggregate stability, *Plant Soil*, **136**, 111.
90. Russell, R. S. (1977), *Plant Root Systems. Their Function and Interaction with Soil.* McGraw-Hill, London.
91. Caron, J., Kay, B. D. and Stone, J. A. (1992). Improvement of structural stability of a clay loam with drying, *Soil Sci. Soc. Am. J.*, **56**, 1583.
92. Harris, R. F., Chester, G. and Allen, O. N. (1966). Soil aggregate stabilisation by indigenous microflora as affected by temperature, *Soil Sci. Soc. Am. Proc.* **30**, 205.
93. Tisdall, J. M., Cochcroft, B. and Uren, N. C. (1978). The stability of soil aggregates as affected by organic materials, microbial activity and physical disruption, *Aust. J. Soil Res.*, **16**, 9.
94. Muneer, M. and Oades, J. M. (1989b). The role of Ca-organic interactions in soil aggregate stability. II. Field studies with ^{14}C-labelled straw, $CaCO_3$ and $CaSO_4.2H_2O$, *Aust. J. Soil Res.*, **27**, 401.
95. Muneer, M. and Oades, J. M. (1989a). The role of Ca-organic interactions in soil aggregate stability. I. Laboratory studies with ^{14}C-glucose, $CaCO_3$ and $CaSO_42H_2O$, *Aust. J. Soil Res.*, **27**, 389.
96. Lynch, J. M. and Bragg, E. (1985). Microorganisms and soil aggregate stability, *Adv. Soil Sci.*, **2**, 133.
97. Haynes, R. J. and Swift, R. S. (1990). Stability of soil aggregates in relation to organic constituents and soil water content, *J. Soil Sci.*, **41**, 73.
98. Drury, C. F., Stone, J. A. and Finlay, W. I. (1991). Microbial biomass and soil structure associated with corn, grasses and legumes, *Soil Sci. Soc. Am. J.*, **55**, 805.
99. Haynes, R. J. and Francis, G. S. (1993). Changes in microbial biomass C, soil carbohydrate composition and aggregate stability induced by growth of selected crop and forage species under field conditions, *J. Soil Sci.*, **44**, 665.
100. Chaney, K. and Swift, R. S. (1986). Studies in aggregate stability. I. Reformation of soil aggregates, *J. Soil Sci.*, **37**, 329.
101. Foster, R. C. (1994). Microorganisms and soil aggregates. In *Soil Biota: Management in Sustainable Farming Systems*, ed. Pankhurst, C. E. *et al.*, CSIRO, East Melbourne, Vic., Australia, p. 144.
102. Shipton, P. J. (1986). Infection by foot and root rot pathogens and subsequent damage. In *Plant Diseases, Infection, Damage and Loss*, ed. Wood, R. K. S. and Jellis, G. K., Blackwell Scientific, Oxford, p. 139.
103. Amelung, W. and Zech, W. (1996). Organic species in ped surface and core fractions along a climosequence in the prairie, North America, *Geoderma*, **74**, 193.
104. van der Linden, A. M. A., Jeurisson, L. J. J., Van Veen, J. A. and Schippers, B. (1989). Turnover of soil microbial biomass as influenced by soil compaction. In *Nitrogen in Organic Wastes Applied to Soil*, ed. Attansen, J. and Henriksen, K., Academic Press, London, p. 25.
105. Killham, K., Amato, M. and Ladd, J. N. (1993). Effects of substrate location in soil and soil pore-water regime on carbon turnover, *Soil Biol. Biochem.*, **25**, 57.
106. Robert, M. and Chenu, C. (1992). Interactions between soil minerals and microorganisms. In *Soil Biochemistry*, Vol. 7, ed. Stotzky, G. and Bollag, J. M., Marcel Dekker, New York, p. 307.
107. Emerson, W. W., Foster, R. C. and Oades, J. M. (1986). Organo mineral complexes in relation to soil aggregation and structure. In *Interactions of Soil Minerals with Natural Organics and Microbes*, ed. Huang, P. M. and Schnitzer, M., Soil Science Society of America, Madison, WI, p. 521.

108. Roberson, E. B., Chenu, C. and Firestone, M. K. (1993). Microstructural changes in bacterial exopolysaccharides during desiccation, *Soil Biol. Biochem.*, **25**, 1299.

109. Chenu, C. (1993). Clay–organic matter associations: microstructure and swelling and dispersion of clay–polysaccharide complexes. In *Clay Swelling and Expansive Soils, NATO ASI Ser., Phys.*, ed. Baveye, P. and McBride, M. (as referenced by [46]).

110. Chenu, C. (1993). Clay– or sand–polysaccharide associations as models for the interface between micro-organisms and soil: water related properties and microstructure, *Geoderma*, **56**, 143.

111. Chenu, C. and Guérif, J. (1991). Mechanical strength of clay minerals as infuenced by an adsorbed polysaccharide, *Soil Sci. Soc. Am. J.*, **55**, 1076.

112. Chenu, C., Guerif, J., and Jaunet, A. M. (1994). Polymer bridging: a mechanism of clay and soil structure stabilization by polysaccharides. In *15th World Congress of Soil Science, Transactions*, 10–16 July, 1994, Acapulco, Mexico, Sociedad Mexicana de la Ciencia del Suelo, Chapingo, Mexico, p. 403.

113. Molope, M. B., Grive, I. C. and Page, E. R. (1982). Contributions by fungi and bacteria to aggregate stability of cultivated soils, *J. Soil Sci.*, **38**, 71.

114. Kinsbursky, R. S., Levanon, D. and Yaron, B. (1989). Role of fungi in stabilizing aggregates of sewage sludge amended soils, *Soil Sci. Soc. Am. J.*, **53**, 1086.

115. Tisdall, J. M., Smith, S. E. and Rengasamy, P. (1997). Aggregation of soil by fungal hyphae, *Aust. J. Soil Res.*, **35**, 55.

116. Oades, J. M. (1984). Soil organic matter and structural stability: mechanisms and implications for management, *Plant Soil*, **76**, 319.

117. Elliott, E. T. and Coleman, D. C. (1988). Let the soil work for us, *Ecol. Bull.*, **39**, 23.

118. Barea, J. M. (1991). Vesicular–arbuscular mycorrhizae as modifiers of soil fertility, *Adv. Soil Sci.*, **15**, 1.

119. Miller, R. M. (1986). The ecology of vesicular-arbuscular mycorrhizae in grassland shrublands. In *Ecophysiology of VAMycorrhizal Plants*, ed. Safir, G., CRC Press, Boca Raton, FL, p. 135.

120. Tisdall, J. M. (1991). Fungal hyphae and structural stability of soil, *Aust. J. Soil Res.*, **29**, 729.

121. Gianinazzi-Pearson, V. (1989). Vesicular arbuscular endomycorrhizae as determinants for plant growth and survival, *Dev. Soil Sci.*, **18**, 69.

122. Bonfante-Fasolo, P. (1984). Anatomy and morphology of VA mycorrhizae. In *VA Mycorrhiza*, ed. Powell, C. L. and Bagyaraj, D. J., CRC Press, Boca Raton, FL, p. 5.

123. Golchin, A., Oades, J. M., Skjemstad, J. O. and Clarke, P. (1994). Soil structure and carbon cycling, *Aust. J. Soil Res.*, **32**, 1043.

124. Golchin, A., Baldock, J. A. and Oades, J. M. (1997). A model linking organic matter decomposition, chemistry and aggregate dynamics. In *Soil Processes and the Carbon Cycle*, ed. Lal, R. *et al.*, CRC Press, Boca Raton, FL, p. 245.

125. Beare, M. H., Hendrix, P. F. and Coleman, D. C. (1994). Water-stable aggregates and organic matter fractions in conventional and no-tillage soils, *Soil Sci. Soc. Am. J.*, **58**, 777.

126. Puget, P., Chenu, C. and Balesdent, J. (1995). Total and young organic matter distribution in aggregates of silty cultivated soils, *Eur. J. Soil Sci.*, **46**, 449.

127. Angers, D. A., Recous, S. and Aita, C. (1997). Fate of carbon and nitrogen in water-stable aggregates during decomposition of ^{13}C ^{15}N labelled wheat straw, *Eur. J. Soil Sci.*, **48**, 295.

128. Aita, C., Recous, S. and Angers, D. A. (1997). Short-term kinetics of residual wheat straw C and N under field conditions: characterisation by ^{13}C ^{15}N tracing and soil particle size fractionation, *Eur. J. Soil Sci.*, **48**, 283.
129. Burford, J. R., Desphande, T. L., Greenland, D. J. and Quirk, J. P. (1964). Determination of the total specific surface area of soils by adsorption of cetyl pyridinium bromide, *J. Soil Sci.*, **15**, 178.
130. Turchenek, L. W. and Oades, J. M. (1978). Organo-mineral particles in soils. In *Modifications of Soil Structure*, ed. Emerson, W. W. *et al.*, John Wiley & Sons, Chichester, p. 137
131. Chenu, C. (1995). Extracellular polysaccharides: an interface between microorganisms and soil constituents. In *Environmental Impact of Soil Component Interactions. Natural and Anthropogenic Organics*, ed. Huang, P. M. *et al.*, CRC Press, Boca Raton, FL, p. 217.
132. Sanatanatoglia, O. J. and Fernandez, M. (1983). Structural stability and content of microbial gums under different types of management in a soil of the Ramallo series (vertic arguidol), *Ciencia de Suelo*, **1**, 43.
133. Foster, R. C. (1981). Localisation of organic materials *in situ* in ultrathin sections of natural soil fabrics using cytochemical techniques. In *International Workinggroup on Submicrosopy of Undisturbed Soil Materials*, ed. Bisdom, E. B. A., PUDOC Press, Wageningen, The Netherlands, p. 309.
134. Foster, R. C. (1981). Polysaccharides in soil fabrices, *Science*, **214**, 665.
135. Foster, R. C. (1982). The fine structure of epidermal cell mucilages of roots, *New Phytol.*, **91**, 727.
136. Cheshire, M. V., Sparling, G. P. and Mundie, C. M. (1983). Effect of periodate treatment on carbohydrate constituents and soil aggregation, *J. Soil Sci.*, **34**, 105.
137. Baldock, J. A., Kay, B. D. and Schnitzer, M. (1987). Influence of cropping treatments on the monosaccharide content of the hydrolysates of a soil and its aggregate fractions, *Can. J. Soil Sci.*, **67**, 489.
138. Berkeley, R. C. W., Lynch, J. M., Melling, J., Rutter, P. R. and Vincent, B. (1980). *Microbial Adhesion to Surfaces*, Ellis Horwood, Chichester.
139. Cheshire, M. V., Greaves, M. P. and Mundie, C. M. (1974). Decomposition of soil polysaccharide, *J. Soil Sci.*, **25**, 483.
140. Burns, R. G. and Davies, J. A. (1986). The microbiology of soil structure. In *The Role of Microorganisms in a Sustainable Agriculture*, ed. Lopez-Real, J. M. and Hodges, R. D., Academic Press, Hertfordshire, UK, p. 9.
141. Nelson, P. N., Baldock, J. A., Clarke, P., Oades, J. M. and Churchman, G. J. (1999). Dispersed clay and organic matter in soil: their nature and associations, *Aust. J. Soil Res.*, **37**, 289.
142. Capriel, P., Beck, T., Borchert, H. and Härter, P. (1990). Relationship between soil aliphatic fraction extracted with supercritical hexane, soil microbial biomass, and soil aggregate stability, *Soil Sci. Soc. Am. J.*, **54**, 415.
143. Dinel, H., Lévesque, M. and Mehuys, G. R. (1991). Effects of long-chain aliphatic compounds on the aggregate stability of a lacustrine silty clay, *Soil Sci.*, **151**, 228.
144. Dinel, H., Lévesque, P. E. M., Jambu, P. and Righi, D. (1992). Microbial activity and long-chain aliphatics in the formation of stable soil aggregates, *Soil Sci. Soc. Am. J.*, **56**, 1455.
145. Callot, G., Guyon, A. and Mousain, D. (1985). Inter-relations entre argilles de calcite et hyphes myceliens, *Agronomie*, **5**, 209.
146. Haynes, R. J. and Beare, M. H. (1996). Aggregation and organic matter storage in meso-thermal, humid soils. In *Structure and Organic Matter in Agricultural Soils*, ed. Carter, M. R. and Stewart, B. A., Lewis, CRC Press, Boca Raton, FL, p. 213.

147. Schnitzer, M. (1986). Binding of humic substances by mineral colloids. In *Interactions of Soil Minerals with Natural Organics and Microbes*, ed. Huang, P. M. and Schnitzer, M., Soil Science Society of America, Madison, WI, p. 77.
148. Murray, R. S. and Quirk, J. P. (1990). Interparticle forces in relation to the stability of soil aggregates. In *Soil Colloids and their Associations in Aggregates*, ed. DeBoodt, M. F. *et al.*, Plenum, New York, p. 439.
149. Mullins, C. E., MacLeod, D. A., Northcote, K. H., Tisdall, J. M. and Young, I. M. (1990). Hardsetting soils: behavior, occurence and management. In *Advances in Soil Science*, ed. Lal, R. and Stewart, B. A., Springer, New York, p. 37.
150. Shanmuganathan, R. T. and Oades, J. M. (1983). Influence of anions on dispersion and physical properties of the A horizon of a Red-brown earth, *Geoderma*, **29**, 257.
151. Reid, J. B. and Goss, M. J. (1982). Interactions between soil drying due to plant water use and decreases in aggregate stability caused by maize roots, *J. Soil Sci.*, **33**, 47.
152. Reid, J. B., Goss, M. J. and Robertson, P. D. (1982). Relationship between the decreases in soil stability affected by the growth of maize roos and changes in organically bound iron and aluminium, *J. Soil Sci.*, **33**, 397.
153. Edwards, A. P. and Bremner, J. M. (1967). Domains and quasicrystalline regions in clay systems, *Soil Sci. Soc. Am. Proc.*, **35**, 650.
154. Kay, B. D. (1990). Rates of change of soil structure under different cropping systems, *Adv. Soil Sci.*, **12**, 1.
155. Jastrow, J. D. and Miller, R. M. (1997). Soil aggregate stabilization and carbon sequestration: feedbacks through organomineral associations. In *Soil Processes and the Carbon Cycle*, ed. Lal, R. *et al.*, CRC Press, Boca Raton, FL, p. 207.
156. Feller, C., Albrecht, A. and Tessier, D. (1996). Aggregation and organic matter storage in kaolinitic and smetitic tropical soils. In *Structure and Organic Matter in Agricultural Soils*, ed. Carter, M. R. and Stewart, B. A., Lewis, CRC Press, Boca Raton, FL, p. 309.
157. Bartoli, R., Philippy, R. and Burtin, G. (1992). Influences of organic matter on aggregation in oxisols rich in gibbsite and geothite. 1—Structures: the fractal approach, *Geoderma*, **54**, 231.
158. Tiessen, H. and Stewart, J. W. B. (1988). Light and electron microscopy of stained microaggregates: the role of organic matter and microbes in soil aggregation, *Biogeochemistry*, **5**, 312.
159. Jocteur-Monrozier, L., Ladd, J. N., Fitzpatrick, R. W., Foster, R. C. and Raupach, M. (1991). Components and microbial biomass content of size fractions in soils of contrasting aggregation, *Geoderma*, **49**, 37.

4 Impact of Organic Substances on the Formation and Transformation of Metal Oxides in Soil Environments

A. VIOLANTE
University of Naples 'Federico II', Italy

G. S. R. KRISHNAMURTI
CSIRO Land and Water, Australia

P. M. HUANG
University of Saskatchewan, Canada

Interactions between Soil Particles and Microorganisms
Edited by P. M. Huang, J.-M. Bollag and N. Senesi. © 2002 John Wiley & Sons, Ltd

1 INTRODUCTION

Metal oxides are ubiquitous in soils and are of great interest in pedology and soil chemistry, because of their occurrence as weathering products, their specific adsorptive properties for anions and cations, and their impacts on the ecosystem [1–6]. They may exist as crystalline minerals, as short-range ordered or noncrystalline precipitates, which are partly present as coatings on clay minerals and/or humic substances. The short-range ordered aluminum and iron oxides are undoubtedly the most reactive inorganic components of acidic and neutral soils [7,8]. Aluminum and iron are liberated into solution by acidic weathering of minerals and then precipitated either locally or after translocation in soil environments. Humic and fulvic acids as well as organic substances produced by microorganisms and plants (root exudates) are involved in the weathering of primary minerals. Organic compounds play an important role in the hydrolytic reactions of iron and aluminum and in the formation, nature, surface properties, reactivity and transformation of metal oxides. The present chapter deals with the nature and properties of aluminum and iron oxides and the influence of organics on their formation, transformation and surface properties. The importance of these interactions in the fixation and transformation of organic and metal pollutant ions, and on their catalytic role in the abiotic transformation of humic materials is also dealt with.

2 CHEMISTRY OF METAL OXIDES

2.1 ALUMINUM HYDROXIDES AND OXYHYDROXIDES

Aluminum hydroxides crystallize in three polymorphs: gibbsite, bayerite and nordstrandite. A new polymorph, doyleite, has been identified from two localities in Quebec [9], although reservations as to its status have been expressed [2,10]. Gibbsite, nordstrandite, and bayerite are composed of the same fundamental units. The a and b crystallographic axes have the same length. The differences among the $Al(OH)_3$ polymorphs lie in the stacking of octahedral sheets along the c crystallographic axis [3,6,11]. As described by Hsu [3] (see

Figures 7.2 and 7.3 therein), the OH^- groups in one sheet reside directly on the top of those in the next by H-bonding in gibbsite. In bayerite the octahedral sheets are closely packed because the OH^- of one sheet resides in the depression of the next sheet, whereas in nordstrandite, the gibbsite and bayerite arrangements alternate.

Gibbsite is a common product of tropical and subtropical weathering. It is one of the major minerals in many Oxisols and a minor mineral component of many Ultisols [3,12–14]. Oxisols usually have more gibbsite in the silt than in the clay fraction. Gibbsite and the various polymorphs of FeOOH and Fe_2O_3 commonly coexist as the end products of advanced weathering. Goethite and hematite appear earlier and more often than gibbsite in the process of soil genesis [3]. Gibbsite occurs in old and young soils. In the latter case, it occurs when leaching is particularly strong. In soils on volcanic ash, gibbsite is formed in preference to allophane or halloysite when rainfall is very high and drainage effective [15]. In the humid tropics, gibbsite tends to develop in the highlands and kaolinite in the lowlands. Gibbsite is seldom found as a pedogenetic component of soils in temperate and cool climates. In these latter soils, Al mobilized by acidic attack on soil minerals can be precipitated as allophane, imogolite or short-range Al precipitates, and trapped as exchangeable Al or hydroxy Al interlayers in expanding layer silicates. The formation, morphology and reactivity of gibbsite have received great attention [1,3,5].

Nordstrandite and bayerite occur rarely in nature. Their apparent rarity in soils may reflect identification difficulties, due to their low concentration and/or masking by the presence of gibbsite [6]. Bayerite has been recorded in low-grade, thermally metamorphosed sedimentary rocks from Israel, from Hungarian brickclays, and from bauxitic soils from China and Russia [16,17].

Many well-documented occurrences of natural nordstrandite have been reported. The mineral has been identified in microscopic solution cavities in upper Miocene limestone on the island of Guam, in karst limestone Terra Rossa in West Sarawak, Borneo, in Montenegro, in the Dinaric Alps of Croatia, in Hungary, and in Jamaica. Nordstrandite has also been found closely associated with dawsonite nodules in Permian marine strata of the Sydney Basin in New South Wales, Australia, as fillings in thin fissures in dolomitic marlstone of the Green River Formation in northwestern Colorado [16–18 and references therein], in Mont St-Hilaire, Quebec [19], in the bauxites De Haro (La Rioja) in Spain [20], and in Newhaven, Sussex [21]. Electron microscopy has been used to distinguish gibbsite, bayerite and nordstrandite, because particles of these polymorphs can be differentiated with a fair degree of confidence by their morphology [2,3,10,18,22]. The range of shapes and sizes displayed by metal oxides reflects the environment under which they are formed and grown. The habit or morphology is the external shape of the crystal. Presence of additives during the crystal growth usually alters the morphology of

(A)

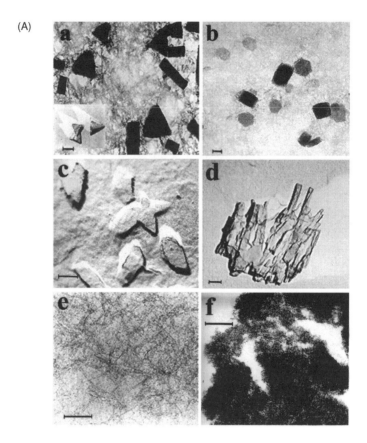

Figure 1. (A) Transmission electron micrographs of Al precipitation products formed in the absence or presence of organic ligands. (a) Bayerite formed at pH 9.0 aged 10 days; (b) gibbsite formed at pH 6.0 aged 2 months; (c) ovoidal particles of nordstrandite and (d) clusters of nordstrandite and gibbsite formed in the presence of montmorillonite (5 mmol g^{-1} clay) at pH 9.0, respectively, at citrate/Al molar ratio of 0.01 and 0.1 (from [17]. Reproduced by permission of The Clay Minerals Society); (e) poorly crystalline boehmite formed at pH 8.2 and tartrate/Al molar ratio of 0.01 aged 10 months at room temperature and (f) noncrystalline Al precipitation product formed at pH 6 and tartrate/Al molar ratio of 0.1 aged 10 years at room temperature. Bar indicates 0.25 μm.

(B)

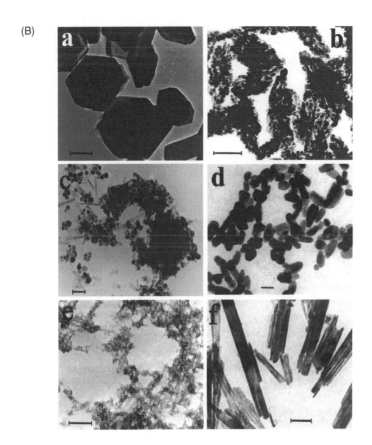

(B) Transmission electron micrographs of Fe precipitation products formed in the absence or presence of organic ligands. (a) Hexagonal plates of hematite formed from ferrihydrite at pH 7.5 and room temperature; (b) ellipsoids of hematite formed from ferrihydrite in the presence of oxalate at pH 6; (c) goethite and maghemite synthesized at pH 8 in the presence of humic acid (Fe/COOH groups molar ratio of 0.1); (d) sub-rounded goethite crystals grown from ferrihydrite in the presence of the reductant cysteine (from [45]. Reproduced by permission of the Royal Society of Chemistry); (e) ferrihydrite freshly precipitated at pH 11.8 (from [52]. Reproduced by permission of the Clay Minerals Society); (f) multidomainic crystals of lepidocrocite prepared by passing air through $FeCl_2$ solutions held at pH 6 (from [49]. Reproduced by permission of the Clay Minerals Society). Bar indicates 0.1 μm

the metal oxides (see later). Bayerite crystals are usually oriented with their c-axis in the plane of the photograph and often have a triangular or barrel-shaped morphology (Figure 1Aa) [10,23]. Gibbsite often shows hexagonal plates, but the crystal growth habit may vary at pH values above that of minimum solubility and/or in the presence of chelating ligands (Figure 1Ab and 1Ad). Elongated hexagonal rods of gibbsite have been observed [10,17,24]. In an Oxisol from Malaysia gibbsite crystallized mostly as nodules, showing typically euhedral and fine and medium silt size crystals [13; their Fig. 6.74]. Synthetic and natural nordstrandite occurs as rectangular plates or in the form of elongated, platy parallelograms, but in the presence of chelating carboxylic acids it crystallizes as roughly rectangular crystals and with the addition of clay to the system, as ovoidal particles (Figure 1Ac) or clusters of acicular crystals (Figure 1Ad), depending on pH [16–18,22].

The two crystalline AlOOH polymorphs that exist in natural environments are boehmite and diaspore. Gibbsite, boehmite and diaspore are the three hydrates of aluminum which are the main constituents of bauxites and laterites, but often boehmite is a principal constituent of some bauxite deposits in which it is sometimes found together with diaspore. As reported in Chapter 11, aluminum hydroxides and oxyhydroxides may form in lateritic soils and bauxites also as a results of biological weathering [25,26]. Boehmite is rare in soils, but Jones et al. [12] found boehmite with calcium carbonate in the coarse and clay fractions of the horizons A_1, B_{21} and B_{22} in a highly weathered soil (Matanzas soil) of Puerto Rico. Because only traces of gibbsite were found in the soil the authors believe that, because of the presence of carbonates, boehmite was formed in lieu of gibbsite.

The structure of boehmite is similar to that of lepidocrocite (see later). Most specimens of boehmite, whether naturally occurring or synthetic, are submicroscopic in size [2,3,27–30]. These poorly crystallized compounds have been and are frequently referred to as *pseudoboehmites* (Figure 1Ae), being considered materials with the same structure of boehmite, but with a poorly ordered and highly defective ion arrangement. Music et al. [31] evaluated the experimental conditions for the preparation of boehmite without the use of Al–organic precursor or organic peptizing agents. Materials that are similar to poorly crystalline boehmite formed in the laboratory have been found in some bauxite deposits, particularly in the area of the Mediterranean Sea [3]. In soils 'noncrystalline alumina of boehmite character' has been described by de Villiers [32] as being present in certain tropical soils from Australia, Zimbabwe, Brazil and Puerto Rico.

Short-range ordered or noncrystalline Al precipitation products without a definite composition and structure (Figure 1Af) easily form and are ubiquitous in soil environments [3,33]. According to Hsu [3] there cannot be a precise differentiation between crystalline and noncrystalline Al hydroxides, which essentially have the same chemical characteristics, differing only in particle size, crystallinity, and reactivity. The noncrystalline Al precipitates dominate the chemical reactions in soils because of their extremely small particle sizes (Figure 1Af) and highly reactive surfaces [1,3,5,6,11,30]. In pure form they are not stable, but in the presence of chelating anions they may remain unchanged for indefinite time (see later). The presence of some accumulated noncrystalline Al hydroxides has been recognized in the upper B horizon of a Podzol in Sweden [34]. Short-range ordered Al precipitates and soluble OH–Al species, often coat crystalline minerals in soils or may be interlayered into the inter-lamellar spaces of vermiculites or smectites, altering the surface properties of these phyllosilicates. Hydroxy–Al interlayered vermiculites and smectites are found in several soil orders, although they tend to be most abundant in Ultisols and Alfisols [35].

In many soils organic matter plays a critical role in pedogenic processes by complexing Al and Fe in surface horizons, translocating them, and depositing them in subsoils [8]. The incorporation of Al into Al–humus complexes and also into 2:1 layer silicates inhibits the formation of other minerals, such as gibbsite and noncrystalline or poorly crystalline aluminosilicates (allophane and imo-golite) in several soils. An extensive literature has developed on the occurrence, synthesis, chemical properties of short-range ordered aluminum silicates and has been reviewed by Wada [36].

2.2 IRON OXIDES AND OXYHYDROXIDES

The Fe oxide content of a soil may vary between one to several hundred $g\,kg^{-1}$, and depends on the Fe content of the parent rock and the maturity of the soil, because as a soil develops the original Fe bearing minerals decompose and most of the Fe is precipitated as pedogenic Fe oxides and the non-Fe phases (Si etc.) are removed [37]. A generalized overview of the occurrence of the different Fe oxides in various soils is summarized in Table 1.

The structure of most of the Fe oxide minerals can be described in terms of hexagonally close packed planes of O atoms stacked one on top of one another with Fe occupying the interstitial octahedral and, in some cases, tetrahedral sites. The stacking of the hexagonally close packed O planes in the third dimension can either be hexagonal or cubic [38]. Often both hexagonal and cubic polymorphs exist. The Fe(III) and Fe(II) ions sit in various octahedral interstices which can be viewed as different assemblages of $[Fe(O, OH)_6]$ octa-hedra. The Fe–Fe interatomic distances across the edge-shared octahedra range from 2.95 to 3.03 Å, whereas those across the corner-shared octahedra range from 3.39 to 3.71 Å. The structural data of Fe oxide minerals has been thoroughly reviewed earlier [4,39,40].

Hematite consists of hexagonally close-packed planes of O atoms with Fe(III) ions occupying the octahedral sites. Only 2/3rd of the possible octa-hedral sites are occupied, giving a diactohedral arrangement. Each Fe(III) is

Table 1. A summary of the occurrence of the different dominant Fe oxides in soils (modified from [4])

Mineral	Major soils
Goethite	Aerobic and anaerobic soils of all regions
Lepidocrocite	Anaerobic, clayey, noncalcareous soils of cooler and temperate regions
Ferrihydrite	Groundwater and stagnant water soils (gleys and pseudogleys) and podzols of temperate and cool regions. Paddy soils
Hematite	Aerobic soils of subtropical, Mediterranean and humid to subhumid tropical regions, laterite and plinthite soils, red mediterranean soils, oxisols, ultisols. Usually absent in soils of temperate and cool regions.
Maghemite	Aerobic soils of the tropics and subtropics
Magnetite	Aquatic and marine sediments, microbial oxidation
Schwertmannite	A poorly ordered Fe-oxyhydroxide mineral occurs in nature in soils, developed under different environments
Green Rust	Reductomorphic soils, a rare occurrence
Akaganite	Rarely occurs in nature. It is found in Cl-rich environments such as hot brines

surrounded by six O atoms and each O atom is shared by four Fe(III) ions. Each $[FeO_6]$ octahedron shares three edges with three neighbouring octahedra and shares a face with an $[FeO_6]$ octahedron of the adjacent layer. The basic habits of hematite are hexagonal plates (Figure 1Ba). However, a number of other morphologies have also been reported. In the presence of citrate or maltose (ligand/Fe = 0.0001–0.001, pH 10–12) hematite grows preferentially along [001] and forms rods, often with fibrous ends [41,42], whereas oxalate (ligand/Fe = 0.001, pH 6–7) causes hematite to grow as small, granular ellipsoids [43] (Figure 1Bb).

Goethite consists of hexagonally close-packed planes of O atoms with Fe(III) ions occupying the octahedral sites. The structure can be described as double chains of octahedra, which are joined to other double chains by sharing corners by Fe-O-Fe and by H bonds. The O atoms are shared by octahedra in two different double chains and the OHs are shared by octahedra in the same chain. The basic morphologies of goethite are acicular needles (Figure 1Bc). Sucrose, at a sucrose/Fe molar ratio of 0.01, led to wedge-shaped outgrowths of goethite on hematite centres possibly due to adsorption of sucrose at the end of the crystals [44], while cysteine (cysteine/Fe = 1.0) at pH 7–8 produced sub-rounded crystals [45] (Figure 1Bd). Aggregates of acicular crystals have been grown by oxidation of Fe(II) in a Na acetate–hydroxylamine solution [46] and in a slightly acid solution with amino-alkyl silicate [47]. Maltose, glucose and citrate promoted epitaxial twins [44].

Like that of goethite, the lepidocrocite structure is described as having double chains of octahedra, with the octahedral chains sharing edges instead of corners as in goethite. The chains are joined to form a corrugated structure, with the H atoms located between these layers held together by H bonds. The basic morphologies of lepidocrocite are lath-like or tabular (Figure 1Bf). In the presence of crystallization inhibitors, such as organics, grassy-type or 'hedge-hog-like' spherulites form [48,49]. Presence of citrate reduced the particle size and rounded the elongated edges of laths [50].

Akaganeite consists of double chains of edge-shared octahedra, which share corners with adjacent chains to give a three-dimensional structure containing tunnels. The tunnels of the structure are thought to be stabilized by Cl^- ions. Other ions besides Cl^- can also be accommodated, such as F^-, NO_3^-, ClO_4^-, Br^-, I^- and SO_4^{2-}. Akaganeite displays two basic morphologies, namely, spindles and rods. Organic ligands may induce novel morphologies of akaganeite, such as large hexagonal plates (dihydroxy-ethylene glycol) [51] and rosette-like polycrystalline aggregates (1-2 ethylene diphosphonic acid) [42].

Both magnetite and maghemite (Figure 1Bc) belong to the inverse spinel structure, which consists of a cubic unit cell of 32 O, forming nearly a cubic-close-packed framework. There are 16 octahedral sites and eight tetrahedral sites. The structural formula is $Y[YX]O_4$ (or $X_8Y_{16}O_{32}$ per unit cell), where $X = Fe(II)$ and $Y = Fe(III)$ with the brackets denoting the cations in octahedral sites. Thus, 1/3 of the Fe occurs as Fe(III) in tetrahedral coordination, 1/3 as Fe(III) in octahedral coordination, and 1/3 as Fe(II) in octahedral coordination.

Ferrihydrite covers a range of poorly ordered compounds whose degree of ordering depends on the method of preparation (rate of hydrolysis) and the time of aging [4,41,52] (Figure 1Be). Neither the formula nor the structure of ferrihydrite has been fully established. Various formulae have been suggested for this poorly ordered ferric hydroxide, such as $Fe_5HO_8\cdot4H_2O$ [53], $5Fe_2O_3\cdot9H_2O$ [54], and $Fe_2O_3\cdot2FeOOH\cdot2\cdot6H_2O$ [55]. The new structure proposed by Eggleton and Fitzpatrick [56,57] includes both octahedral and tetrahedral Fe. Manceau et al. [58], on the basis of EXAFS analysis, question the presence of tetrahedral Fe, whereas, Zhao et al. [59] present evidence of tetrahedral Fe in the structure based on X-ray absorption fine structure (XAFS) analysis. Presence of both octahedral and tetrahedral Fe in the structure was subsequently confirmed [60–62]. The degree of ordering of ferrihydrite is variable and a range of XRD patterns may be obtained. The two extremes of ordering are referred as two-line (0.26 and 0.15 nm) and six-line ferrihydrite (additional lines at 0.221, 0.196, 0.172 and 0.148 nm), because the XRD patterns range from two to six reflections as structural order increases. The two forms of ferrihydrite precipitate under different conditions and the two-line form does not transform to a more ordered ferrihydrite with time [4]. The occurrence and constitution of synthetic and natural ferrihydrite have recently been reviewed in detail [63].

Green rusts have a pyroaurite structure consisting of hexagonally close-packed layers of OH and O of the $Fe(OH)_2$ type, with $Fe(II)$ and $Fe(III)$ in the interstices [64]. The positive charge is balanced by intercalation of anions $(Cl^-, SO_4^{2-}, CO_3^{2-})$ between the layers. Mossbauer and Raman spectroscopies have been used to identify green rust as a specific mineral in a reductomorphic soil in the forest of Brittany, France [65].

Schwertmannite, a poorly crystallized oxyhydroxysulfate, has tunnel structure similar to that of akaganeite. The SO_4^{2-} ion occurs both as a bridging element between Fe atoms lining adjacent walls of the tunnels, and as specifically adsorbed surface component [66,67]. Its ideal formula is $Fe_3O_8(OH)_6(SO_4)$, but may range to $Fe_3O_8(OH)_{4.5}(SO_4)_{1.75}$ depending upon the degree to which tunnel and surface sites are saturated with SO_4.

3 ALUMINUM– AND IRON–ORGANIC COMPLEXES IN SOIL ENVIRONMENTS

Organic matter and metal oxides occur in almost any soil as common products of soil formation. The presence of organics may inhibit the crystallization of Al and Fe oxides because of the formation of Al– and Fe–organic complexes. The presence of organics may influence the nature of the final Al and Fe oxide formed. Further, metal oxides once formed may act as adsorbents for a wide variety of organic agrochemicals, such as biocides and fertilizers. In addition, adsorption may also lead to either immobilization or transformation of the organics.

Soil organic matter consists of a spectrum of material, ranging from compounds that mineralize very easily during the first stage of decomposition to more resistant materials that accumulate as microbial by-products. Soil organic compounds are of two major types: (1) well-defined biochemical substances produced by microorganisms and plants, such as simple, low molar mass organic acids, siderophores, hydroxamate, sugar acids, phenols and phenolic acids, and complex, polymeric phenols, lipids, polysaccharides, nucleic acids and proteins; (2) a series of acidic, yellow to black substances formed by secondary synthesis reactions, which are referred to as humic and fulvic acids [68–70]. Molar mass of organic compounds range from $> 3000\,Da$ for humic acids, $1000–3000\,Da$ for fulvic acids, and $< 1000\,Da$ for low-molar mass organic acids.

Humic and fulvic acids as well as humin materials are regarded as the humic substances. The materials that have recognizable chemical characteristics (carbohydrates, polypeptides, fats, waxes, altered lignin etc.) are considered to be the nonhumic components of soil organic matter. Microorganisms can synthesize the nonhumic materials [68,69,71]. The amounts of nonhumic and humic substances in soil differ. The amount of lipids can range from 2% in forest soil humus to 20% in acid peat soils. Protein may vary from 15 to 45% and carbohydrates from 5 to 25%. Humic substances may vary from 33 to 75%

of the total organic matter with grassland soils having higher quantities of humic acids and forest soils having higher amounts of fulvic acids [7]. For further information see also Chapter 3 (section 2) in this volume.

Complexing organic acids are widely distributed in the environment and are low and variable in their concentrations. In soils they may be produced by leaching from plants, decomposition of litter by microbes and exudation by roots, fungi and other microorganisms.

In the rhizosphere, root exudates comprise both high and low molar mass substances released by the roots. The most important high molar mass compounds are mucilage, polysaccharides and ectoenzymes, whereas the main constituents of the low molecular mass root exudates are carbohydrates, organic acids, amino acids, peptides and phenolics (Table 2). As a rule sugars and organic acids are the predominant compounds in the rhizosphere soil [72–74]. Oxalic, malic, succinic, citric, tartaric and malonic acids are the most abundant aliphatic acids present in the rhizosphere.

Table 2. Some substances detected in the rhizosphere. Reprinted from Marcel Dekker. Inc.

Biomolecules	Components
Sugars	Monosaccharides: glucose, fructose, galactose, rhamnose, ribose, xylose, arabinose, raffinose Disaccharides: sucrose, maltose, oligosaccharide
Acid sugars	Gluconic, glucuronic, galacturonic, 2-ketogluconic acid
Amino acids	Asparagine, α-alanine, glutamine, aspartic acid, leucine/isoleucine, serine, glycine, cystine/cysteine, methionine, phenylalanine, tyrosine, threonine, lysine, proline, tryptophane, β-alanine, arginine, homoserine
Organic acids	Tartaric, oxalic, citric, malic, acetic, propionic, butyric, succinic, fumaric, glycolic, valeric, malonic, salicylic
Fatty acids and sterols	Palmitic, stearic, oleic, linolenic, linoleic acids, cholesterol, campesterol, stigmasterol, sitosterol
Phenolic acids	Gallic, caffeic, vanillic, hydroxybenzoic, ferulic, p-coumaric, tannic acid, tannins
Growth factors	Biotin, thiamine, niacin, pantothenate, choline, inositol, pyridoxine, p-amino benzoic acid, n-methyl nicotinic acid
Enzymes	Phosphatase, invertase, amylase, protease, polygalacturonase
Nucleotides, flavonones	Flavonone, adenine, guanine, uridine/cytidine
Miscellaneous compounds	Auxins, scopoletin, fluorescent substances, hydrocyanic acid, glycosides, saponin (glucosides), organic phosphorus compounds, nematode cyst or egg-hatching factors, nematode attractants, fungal mycelium growth stimulants, mycelium-growth inhibitors, zoospore attractants, spore and sclerotium germination stimulants and inhibitors, bacterial stimulants and inhibitors, parasitic weed germination stimulators

Rhizosphere carbon (C) flow has been estimated to account for a major fraction, up to 40%, of plant primary production. The rhizosphere is a favorable habitat for acid-producing bacteria. The release of exudates is of direct importance to microorganisms living in the rhizosphere, as these organisms feed on the exuded organic material. Microorganisms also release many biomolecules, so that the amount of biomolecules in the rhizosphere is much higher than in the bulk soil. Gluconic, glucuronic, galacturonic and 2-ketogluconic acid are common metabolites of microorganisms. The nature of plants and microorganisms (e.g., species, age), and environmental factors (light, temperature, pH, CO_2, soil solution ionic concentration) affect the nature and the relative amount of organic acids in soil solution [73,74].

Organic acids have an important role in the dissolution and mobilization of elements from minerals [75–78]. The effectiveness of organic acids in the dissolution of soil minerals depends on many factors, such as their chemical composition, concentration and chelating power towards elements (Al, Fe, Ca etc.). Carboxyl and hydroxyl groups are the most important functional groups involved in the reactions of metal ions with organic acids. These functional groups are also present in humic and nonhumic fractions of soil organic matter. Ligands capable of forming a chelate composed of a five- or six-membered ring are the most stable [23] and seem to be particularly resistant to biodegradation.

By forming stable Al– and Fe–organic complexes, humic, fulvic and low molar mass organic acids not only accelerate the decomposition of Al- and Fe-bearing minerals [75,79], but also facilitate the movement of Al and Fe. In fact, when complexed by organic ligands, Al and Fe can be maintained in solution and transported into lower horizons in the soil profile (B and C horizons). According to Vance et al. [80] Spodosol (Podzol) formation is a prime example of the role that organic substances play in the eluviation of Al and Fe during soil formation. Aluminum, iron and organic matter constitute the major amorphous products in spodic horizons. Probably, Al and Fe are complexed in the eluvial horizons (O, A, and E) and translocated downward as soluble complexes with organic substances, with precipitation of the oxide–organic complexes occurring in the lower mineral horizons (B) due to metal saturation, organic polymerization, change of pH, and/or microbial degradation of organic complexes during migration through the profile [77,80,81]. The microbial immobilization process is attributed mainly to decomposition of easily degradable low molecular mass organic acids. Soils forming in volcanic ash (Andisols) lead to the retention and accumulation of organic matter in the surface layer, because of reactions with Al on oxide surfaces [15,80].

The effectiveness of lichens as agents of rock weathering and soil formation has long been recognized [82–84]. The excretion by the mycobiont of low molar mass organic carboxylic acids, such as oxalic, citric, gluconic, and lactic acids, with combined chelating and acidic properties, and the production of slightly water-soluble polyphenolic compounds called 'lichen acids', able to form complexes with the metal cations present in the rock-forming minerals, are phenomena of high local intensity. Due to the abundance of biomolecules, crystallization processes are extremely slow at the rock–lichen interface. Typically, weathering induced by lichens is considered to result in the presence of noncrystalline or poorly ordered secondary products and organo-mineral complexes. Poorly ordered Fe phases have been revealed. Ferrihydrite is likely to be the main component of the pool of short-range ordered iron oxyhydroxides.

The effects of microorganisms on mineral weathering and the mobilization of metals by chelation have been extensively reviewed, respectively, in Chapter 11 and Chapter 5 (Sections 2 and 6) in this issue.

The interactions between plants, soil colloids, organics and microorganisms in the rhizosphere are still very poorly known. Studies have shown that rhizosphere soil has differences in weathering, physical characteristics, and mineralogy compared with bulk soil. Chemical interactions between roots and rhizosphere minerals include precipitation of noncrystalline or short-range ordered Al and Fe oxides, opaline, amorphous silica, and Ca oxalate [85–87]. All the organic substances present in the rhizosphere may interact with Al and Fe released from primary and secondary minerals forming short-range ordered precipitates and organo-mineral complexes. April and Keller [85] also found biomineralization of noncrystalline Al oxides common in the cells of mature root bodies.

Reduction is involved in the weathering of many minerals containing transition metals. The weathering dissolution reactions of some oxides, hydroxides, and oxyhydroxides of transition metals (Mn(III/IV), Fe(III), Co(III), etc.) are greatly enhanced under reducing conditions. Various natural organic compounds have the ability to act as reducing agents in reductive weathering reactions, leading to the formation of polymers (humic substances) and the reduction and dissolution of the metals.

4 INTERACTIONS OF Al SPECIES WITH NATURAL ORGANIC COMPOUNDS

4.1 AQUEOUS CHEMISTRY OF ALUMINUM

The hydrolytic products of Al are important in mineral phase formation and transition, in the mobility of Al in soils and aquatic systems, and in the toxicity of Al to plants and aquatic organisms [88]. In general, the concentrations of Al are low in most natural waters. A median value of $0.4\,\mu$mol $Al\,L^{-1}$ has been reported [5].

The Al^{3+} ion, which is released in natural waters, is octahedrally coordinated with six water molecules and exists as $Al(OH_2)_6^{3+}$ ion. This cation is moderately acidic and its K_1 for the following hydrolytic reaction is 10^{-5} [89]:

$$Al(OH_2)_6^{3+} + H_2O = Al(OH)(OH_2)_5^{2+} + H_3O^+$$

The degree of hydrolysis of $Al(OH_2)_6^{3+}$ increases with increasing solution pH. If only mononuclear Al ions are assumed to equilibrate with aluminous phases, total dissolved Al potentially consists of the following five ion species, namely, Al^{3+}, $AlOH^{2+}$, $Al(OH)_2^+$, $Al(OH)_3^0$ and $Al(OH)_4^-$ (hydrated H_2O molecules are not shown). It is unlikely that $Al(OH)^-$ and $Al(OH)_3^0$ species are of significance in the soil environment [88].

Many scientists have recognized the existence of polynuclear OH–Al complexes in partially neutralized solutions. The hydrolysis of Al to form Al polymers can be represented as:

$$x\ Al^{3+} + y\ H_2O = Al_x(OH)_y^{(3x-y)+} + y\ H^+$$

where $Al_x(OH)_y^{(3x-y)+}$ represents the polymeric Al species. The values of x and y proposed in the literature ranged from $1 \leq x \leq 54$ and $1 \leq y \leq 144$ [3,89 and references therein]. The exact nature of the polymers present in partially neutralized Al solutions is uncertain.

Of all the polynuclear species proposed, those having convincing experimental support are $Al_2(OH)_2^{4+}$, $Al_2(OH)_5^+$, $Al_3(OH)_8^+$, $Al_3(OH)_4^{5+}$, and $Al_8(OH)_{20}^{4+}$, and those of the 'core + links' or 'gibbsite fragment' model, viz., $Al_6(OH)_{12}^{6+}$ through $Al_{54}(OH)_{144}^{18+}$. A more recently proposed model of Al polynucleation includes the tridecameric Al_{13} polynuclear species $AlO_4Al_{12}(OH)_{24}(OH_2)_{12}^{7+}$, usually referred to as the 'Al_{13} polymer' (Figure 2A) and larger condensed units of this basic structure, usually in combination with a minimum of other species, such as the dimer and/or the trimer. The Al_{13} polymer consists of one Al^{3+} at the center, tetrahedrally coordinated to six OH^-, H_2O, or the O^{2-} shared with the Al^{3+} at the center [3,88,90]. The use of ^{27}Al NMR and ^{17}O spectroscopy [90] has provided evidence for the presence of Al_{13} polymers. The resonance peak assigned to the Al_{13} polynuclear species is $\cong 63$ ppm downfield from the monomeric Al species at around 0 ppm (Figure 2B). Other polynuclear species are not detectable with ^{27}Al NMR spectroscopy. The Al_{13} polynuclear species are present in significant amounts only in freshly prepared solutions of OH/Al molar ratios of 1.5 and above and gradually disappear during aging. The possible formation of the Al_{13} tridecamer in natural environments has been emphasized by some researchers. Identification of Al_{13} as the specific cause of polynuclear Al toxicity in chemically simple culture solutions has also been reported [88]. Hunter and Ross [91] reported the presence of Al_{13} in organic horizons of a forested Spodosol. Several investigators,

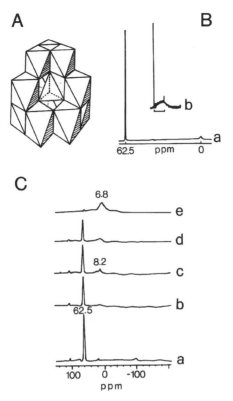

Figure 2. (A) Model of the tridimeric Al_{13} polynuclear species $AlO_4Al_{12}(OH)_{24}$ $(OH_2)_{12}{}^{7+}$. (B) ^{27}Al NMR spectrum of $0.05\,mol\,L^{-1}$ Al solution having a OH/Al molar ratio of 2.25. (A) and (B) from [88]. Reproduced by permission of Soil Science Society of America. (C) ^{27}Al CPMAS NMR spectra of the precipitates after aging for 12 days, from OH–Al solutions aged for 5 days and formed at pH 4.53 and different pyrogallol/Al molar ratios (R). (a) Control at $R = 0$; (b) sample at $R = 0.01$; (c) sample at $R = 0.05$; (d) sample at $R = 0.10$; and (e) sample at $R = 0.50$ [93]

however, have suggested that its presence in soil solutions of agricultural fields is doubtful, because the presence of organic and inorganic ligands in soil solution decreases the concentration of Al_{13} [88,92,93] (Figure 2C). Hiradate *et al.* [94] found that in the KCl extracts from acidic soils, the Al_{13} was not detected at all by ^{27}Al NMR. Whether Al_{13} polynuclear species ever form in natural environments remains an open question.

Polymeric species are transient, metastable intermediates of crystalline $Al(OH)_3$ polymers (mainly gibbsite). Their nature and amount in aqueous systems depend on total Al concentration, OH/Al molar ratio, pH, temperature, nature of anions present, method of preparation and ageing time [3,88]. It is not easy to prove the presence of Al polymers in soil solution because desorption of the highly charged polymers from negatively charged soil colloids into soil solution is difficult.

Very fine noncrystalline or short-range ordered precipitates having some boehmite-like properties may form in partially neutralized Al solutions. Usually the higher the initial OH/Al molar ratio, the greater the amount of noncrystalline precipitates initially formed. These hydroxy-Al solids form quickly, are more soluble than $Al(OH)_3$ polymorphs, are incompletely neutralized and then may contain anions in place of OH^- [3,28,88,89,95–97]. Once these noncrystalline metastable phases have been formed they tend to convert into more stable products but the transformations are governed by many factors (e.g., pH, organic and inorganic ligands, temperature, clay minerals, see later). The initially formed noncrystalline Al precipitate may follow three different pathways (Figure 3): (i) it may very soon organize itself by eliminating anions and H_2O into a crystalline $Al(OH)_3$ polymorph; the formation of nuclei is very important for the subsequent crystallization of $Al(OH)_3$ [1,3,30,88]; (ii) it may dissolve contributing to the formation of monomers and polymers of Al; or (iii) it may convert, particularly in the presence of foreign ligands and at OH/Al ≥ 3 into poorly crystalline boehmite [1,3,30,88,89,95–97] (Figure 3). Poorly crystalline boehmite samples, formed in the presence of various organic and inorganic perturbing ligands, differ significantly in their chemical composition, surface properties and solubility (see later). The nature and solubility of transitional Al precipitation products play a vital role in influencing the subsequent formation of $Al(OH)_3$ polymorphs, which is determined by the rate of its nucleation and

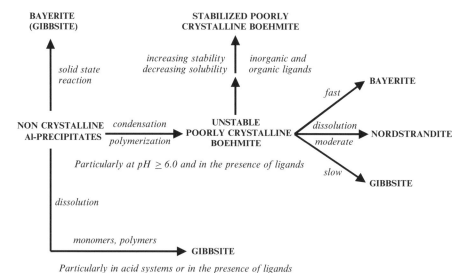

Figure 3. Possible transformation processes for the crystallization of Al hydroxides and oxyhydroxides [modified from 30]

crystal growth. All these transformations may occur in acid as well as in neutral and alkaline environments and probably occur simultaneously, although under given conditions one pathway may be more prominent than the others. The hydrolysis and polynucleation reactions of Al in aqueous solution has been reviewed by Hsu [3] and Bertsch and Parker [88].

The first attempts to determine the various forms of Al in solution were carried out by Okura *et al.* [98] and Turner [95]. Speciation of OH–Al species is important because not all the chemical forms of Al are equally toxic. Organically complexed Al is much less toxic to plants than inorganic Al. Furthermore, organically complexed Al is not toxic to fish and fluoride-complexed Al is more toxic than organic Al, but less toxic than the aquo and hydroxy form [99].

According to Turner [95], in Al salt solutions partially neutralized with a base, the Al extractable after 10 s by 8-hydroxyquinoline is considered to be made up of mainly monomers (Al_m), whereas the Al extracted after 30 min or 3 h is made up of polynuclear (Al_p) ions or noncrystalline solid species (Al_s), respectively. According to Smith [96] mononuclear Al reacts instantaneously with ferron, intermediate polymers react by pseudo first-order kinetics and large polymers or initial solid phase are totally inert or react extremely slowly.

The chemical partitioning of aqueous Al into its inorganic and organic, mononuclear and polynuclear forms has received considerable attention [100]. Kwong and Huang [101] showed that in the presence of citrate the different forms of Al (monomers, polymers and solid species) cannot be determined as in citrate-free solutions. Thus, it is more expedient to divide the Al species on the basis of their reactivity with 8-hydroxyquinoline. Bloom and Erich [99] reported that a contact time of 15 s with 8-hydroxyquinoline before extraction could be used in separating monomeric inorganic from organically bound Al. With a contact time of 15 s a small or relatively large fraction of Al (up to 40%) complexed with simple organic ligands or fulvic acid may be determined. In these studies the anions were introduced prior to neutralization. In other studies anions were allowed to interact with preexisting partially neutralized Al solutions. Jardine and Zelazny [100] reported that distinction of mononuclear and polynuclear Al in solutions, containing various organic and inorganic anions is possible using ferron, if the ligand composition of the solutions is known. Numerous inorganic and organic Al complexes are rapidly degraded by ferron regardless of the ligand concentration and express similar kinetics to that of uncomplexed mononuclear Al. Similarly, mononuclear Al solutions containing $H_2PO_4^-$, F^-, citrate, tartrate, and oxalate with critical ligand/Al ratios < 0.75, 0.75, 0.08, 0.13. 0.08, 0.38 respectively, also react rapidly with ferron and express kinetic reactions similar to that of uncomplexed mononuclear Al. On the contrary, Di Pascale and Violante [102] found that the interaction between phosphate ions and the hydrolytic products of Al leads to the formation of new solid and/or soluble species and retards the rate of removal of Al ions from solution by 8–hydroxyquinoline. Evidently, the sequence of anion addition, the chelating power of ligands for Al species and the ligand/Al molar ratio are important variables influencing organic Al–polynuclear interactions.

Hiradate *et al.* [94] showed that the presence of organically complexed Al (Al_{org}) was also directly detected in KCl extracts from upper horizons by ^{27}Al NMR spectra. The chemical shift of peak of Al_{org} was close to that of the 1:1 complex of Al oxalate.

4.2 PRECIPITATION PRODUCTS OF ALUMINUM: FORMATION

The importance of organic acids as agents for the mobilization, transport and transformation of metal ions in soils and natural waters has been recognized [1,4,5,8]. However, the detailed chemistry of the formation of Al precipitation products in soils remains obscure. Synthesis studies related to pedogenic environments are fundamental to understanding the processes of formation of crystalline and short-range ordered Al hydroxides and oxyhydroxides.

Organic ligands may coordinate with Al hampering the hydrolytic reactions, retarding the crystallization of Al hydroxides and thus affecting the nature of the crystalline Al hydroxides. In slightly acid, neutral and alkaline systems, the transformations during aging of $Al(OH)_3$ suggested by many authors [e.g. 1–3, 103] follow this order (Figure 3):

$$\text{noncrystalline materials} \Rightarrow \text{poorly crystalline boehmites}$$
$$\Rightarrow Al(OH)_3 \text{ polymorphs}$$

Mainly pH, temperature and foreign ligands control the rate of these transformations. Poorly crystalline boehmites, formed at low temperatures and pressures, are very unstable and rapidly convert into crystalline $Al(OH)_3$ polymorphs, when formed in the absence of foreign ligands [3,23,30,103].

Organic ligands delay or inhibit the crystallization of $Al(OH)_3$ to varying degrees. The chemical composition, the molecular structure and the nature of functional groups of each anion control the kinetics of crystallization. Many studies [1,23,28–30] have revealed that the relative effectiveness of organic ligands in retarding or inhibiting $Al(OH)_3$ crystallization is, in general, as follows: acetate < glutarate < succinate = phthalate < glycine < tricarballylate < malonate < acetylacetone < glutamate < aspartate < oxalate < salicylate = malate < tannate < citrate < tartrate. Humic and fulvic acids showed a behavior similar to that of tannate [104,105].

Table 3 lists the precipitation products of Al (crystalline $Al(OH)_3$ polymorphs, poorly crystalline boehmite and/or noncrystalline materials) formed at pH 8.0, 9.0 or 10 and at ligand/Al molar ratios (R) between 0.014 and 0.167. It is well known that at pH ≥ 8.0 and in the absence of foreign ligands, the

Table 3. Aluminum hydroxides and oxyhydroxides formed in the presence of complexing organic acids after 60 days at 20 °C. From [23]. Reproduced by permission of the Clay Minerals Society

Organic acid/Al Molar ratio	Acid								
	Succinic	Glycine	Glutamic	Aspartic	Oxalic	Malic	Salicylic	Citric	Tartaric
Samples aged at pH 8.0									
0.014	B, (G)	B, (G)	–	G	G, (bh)	G, bh	–	bh	bh
0.029	B, (G)	B, (G)	G, N, (B)	bh, G, (N)	bh	bh	bh	A	A
0.050	B, (G)	B, G	G, N	bh, (G)	bh	bh	bh	A	A
0.167	G, N, B	G	bh	A	A	A	A	A	A
Samples aged at pH 9.0									
0.029	B	B, (N)	B, N	N	N, B	N, bh	N, bh	bh	bh
0.050	B	B, N	N	G, N, bh	N, (bh, B)	bh, (N, G)	bh, N, (G)	bh	A
0.167	B, N	N, (bh)	G, N, bh	G, bh	G, bh or bh	bh	bh, (N, G)	A	A
Samples aged at pH 10.0									
0.050	B	B	B	B	B, (bh)	B, N	N, (B)	N	bh
0.167	B	B	B, (bh)	B, (bh)	N, B, (bh)	N, (bh)	N, bh, (G)	bh	A

A, amorphous Al hydroxides; bh, poorly crystalline boehmite; B, bayerite; G, gibbsite; N, nordstrandite; (), small amounts; –, no Al hydroxides or oxyhydroxides detected.

kinetics of crystallization is very fast and bayerite is usually the only polymorph which forms [3,23]. The nature of the precipitates varies greatly with the ligands, the ligand/Al molar ratio and the pH of the systems. The effectiveness of each ligand in inhibiting Al hydroxide formation increased with the increase in its concentration and decreased with the increase of the pH. The data indicate that ligands with a strong affinity for Al inhibit the crystallization of Al hydroxides or oxyhydroxides more than ligands with a moderate or poor affinity for Al. However, the chelating power or affinity of a ligand for Al is not the unique factor which retards the kinetics of $Al(OH)_3$ crystallization. Polydentate and large ligands, such as tannate, fulvate and humate often have a stronger influence in stabilizing noncrystalline materials than those with fewer functional groups or smaller size even if the latter have a stronger affinity for Al than the former. Polydentate and large ligands may not only inhibit crystallization but also promote aggregation of noncrystalline particles.

The presence of organic ligands not only influences the rate of crystallization, but also controls the formation of Al hydroxides and oxyhydroxides. Table 3 and Figure 4 show that at a given pH and ligand/Al molar ratio, the final aluminous products formed range from bayerite to nordstrandite, gibbsite, poorly crystalline boehmite, and noncrystalline materials according to the increasing retarding power of the ligands. Similar results were obtained at a specific pH by increasing the concentration of ligands (Table 3).

A relationship exists between the rate of the crystallization process and the nature of the precipitation products. Usually the lower the rate of crystallization of $Al(OH)_3$, the easier is the formation of gibbsite and nordstrandite over bayerite. In fact in the absence of perturbing ligands or in the presence of low concentrations of ligands with a moderate or poor affinity for Al, the rapid crystallization process leads to the formation of bayerite (Table 3; Figures 1Aa, 3 and 4a). In contrast, critical concentrations of organic ligands with moderate or low affinity for Al significantly retard the crystallization process resulting in the formation of gibbsite even at high pH (Table 3; Figures 1Ad and 4d). In acidic environments gibbsite formation is promoted even in the absence of chelating ligands, because the hydrolytic reactions of Al are usually slow [3,88,97] (Figure 3). When the rate of crystallization is not retarded significantly to favour gibbsite, nordstrandite forms, particularly at pH > 8.0 (Table 3; Figures 1Ac–Ad, and 4b–c). In nature nordstrandite has always been found in alkaline environments.

Nordstrandite crystallizes as euhedral crystals displaying a characteristic tabular habit. In the presence of strongly chelating ligands, the morphology of the crystals of $Al(OH)_3$ polymorphs appears very distorted. In the presence of high concentrations of perturbing ligands, nordstrandite crystals show preferential growth along the crystallographic c – direction [17].

Clay surfaces act as a template in the nucleation of gibbsite and/or nordstrandite, analogous to seeding [16,22,106]. The presence of clays and chelating

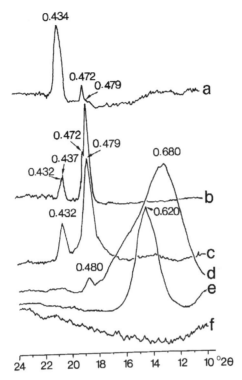

Figure 4. X-ray powder diffractograms of Al precipitation products aged 180 days at pH 8.0 at a carboxylic acid/Al molar ratio of 0.1: (a) bayerite formed in the presence of succinate; (b) mainly norstrandite with bayerite formed in the presence of glycinate; (c) norstrandite formed in the presence of glutamate; (d) poorly crystalline boehmite with low amounts of gibbsite and norstandite formed in the presence of salicylate; (e) pure poorly crystalline boehmite formed in the presence of malate; (f) noncrystalline material formed in the presence of tartrate. The *d* values are in nm. Redrawn from [23]. Reproduced by permission of the Clay Minerals Society

organic ligands, at pH > 8.0, enhances the formation of gibbsite and nordstrandite and inhibits the formation of bayerite and poorly crystalline boehmite [1,16,22]. Clay surfaces not only catalyze the formation of gibbsite and nordstrandite but also influence their morphology in the presence of organic substances. Nordstrandite synthesized in the presence of montmorillonite and citrate at pH > 8.0[16, 18, 22] showed an ill-defined ovoidal outline (usually 0.3–0.5 μm; Figure 1Ac) similar to the 'round pellets' found in soils from West Sarawak [107] on the edge of a sinkhole in limestone where clay may be present. In some synthetic preparations, Violante and Jackson [16,22] found that nordstrandite or nordstrandite and gibbsite crystals nucleated on the clay surfaces and condensed in clusters of weak face to face associations of thin plates (Figure 1Ad). Hathaway and Schlanger [108] and Milton *et al.* [109] reported natural nordstrandite in the form of radiating clusters of platy or bladed crystals. Synthetic preparations confirmed the suggestion of Hathaway and Schlanger [108] that clay surfaces may act as nucleation centers for the growth of radiating clusters of nordstrandite. These authors also noted the presence of several nordstrandite crystals, which were not in contact with clay coatings. These crystals had tabular morphology and would therefore appear to be similar to the nordstrandite observed in synthetic preparations in clay-free systems.

4.3 SHORT-RANGE ORDERED Al PRECIPITATION PRODUCTS: STABILITY

Low molar mass organic ligands, as well as fulvic and humic acids, promote and stabilize the formation of poorly crystalline boehmites over $Al(OH)_3$ polymorphs within certain ranges of optimal concentrations (Figure 3, Table 3) [1,23,28,29,104,105]. Poorly crystalline boehmites in the absence of $Al(OH)_3$ polymorphs have been reported to form under a wide range of pH (from 6.0 to 11.0) in the presence of tartrate, citrate, tannate, malate and salicylate even after 7–15 years of aging [110].

Poorly crystalline boehmite is always formed from an initially noncrystalline material but its crystallization occurs in a few days or a few weeks after the sample preparation. Poorly crystalline boehmites usually show a fibrous morphology (Figure 1Ae), but the degree of fiber development depends on the nature of the ligand, the ligand/Al molar ratio, the pH, the time of aging and the temperature [27–29,111].

Table 4 shows the electron diffraction analysis of Al oxyhydroxides synthesized at pH values ranging from 5 to 8 in the presence of different concentrations of organic ligands and aged 8 months at room temperature. All the samples showed *d* spacings characteristic of boehmite. The relative intensity of the electron diffraction rings was different from one sample to another. Some samples lack some diffraction rings indicating a variation of crystallinity depending on the synthesis conditions such as the nature and concentration of ligands and initial pH. The sample obtained in the presence of citrate at $R = 0.05$ and pH 8.0 has only two very diffused diffraction rings with very weak or weak intensity. This indicates that the sample is very poorly ordered boehmite. These samples also differ significantly in their morphology.

Poorly crystalline boehmite, which very easily forms in synthetic preparations, has been only rarely recognized in soils [32]. Much evidence suggests that poorly crystalline boehmite could often be an unrecognized constituent in many soils. In fact, the identification of poorly crystalline boehmite in the presence of clay minerals is not easy using conventional mineralogical analyses [111,112]. Using XRD analysis, the detection limit of poorly crystalline boehmite in some

Table 4. Electron diffraction analysis of poorly crystalline boehmites formed in the presence of selected organic ligands at different ligand/Al molar ratio (R) and initial pH (Authors' unpublished data, 2000)

Ligand	R	pH	Spacings (nm)										
Aspartic acid	0.1	8.0	0.093 (vw)	0.113 (w)	0.131 (w)	0.144 (s)	0.174 (s)		0.237 (md)		0.282 (m)	0.315 (sd)	0.428 (s)
Tannic acid	0.01	8.0	0.094 (vw)	0.113 (w)	0.131 (vw)	0.144 (md)		0.185 (s)				0.316 (sd)	
Citric acid	0.01	5.0		0.114 (w)		0.144 (s)		0.186 (s)				0.316 (w)	
Citric acid	0.01	6.0		0.113 (w)		0.143 (sd)		0.186 (md)					
Citric acid	0.01	7.0		0.114 (vw)		0.143 (sd)		0.186 (sd)		0.259 (vsd)			
Citric acid	0.01	8.0	(wd)	0.112 (vwd)	0.130 (sd)	0.143	(sd)	0.186 (md)	0.234			0.311 (sd)	
Citric acid	0.02	8.0	(wd)	0.113 (vwd)	0.130 (sd)	0.142	(sd)	0.184 (md)	0.235			0.317 (sd)	
Citric acid	0.05	8.0				0.142 (vwvd)			0.235 (wvd)				
Tannic acid	0.01	8.0	0.093 (vw)	0.112 (m)	0.130 (w)	0.142 (s)		0.185 (s)	0.230 (w)			0.317 (m)	

v, very; s, strong; w, weak; m, middle; d, diffuse.

samples containing montmorillonite or kaolinite was approximately 30–40%, whereas in the samples containing both the minerals, identification of poorly crystalline boehmite was even more difficult. Clearly, conventional approaches are inadequate for identifying poorly crystalline boehmite in the presence of phyllosilicates.

By increasing the ligand/Al molar ratios noncrystalline materials form and remain unchanged for many months or years (Table 3). Noncrystalline Al hydroxides have been found after 7–10 years of aging at 20 °C in samples formed in the presence of citrate, tartrate, and tannate at pH 5–8 and R from 0.1 to 0.5 [110]. The presence of organic (and inorganic) ligands may tremendously retard the transformation of the initially formed noncrystalline Al precipitation products into more stable Al precipitates (Figure 3).

Recently Kawano and Tomita [113] and Kawano et al. [114] have demonstrated that at the earliest weathering stages of K-feldspar or volcanic glass, noncrystalline Al hydroxides, exibiting distinct habits (fibrous, spherical particles or very thin flaky particles), initially formed. Energy dispersive X-ray analysis (EDX) indicated that these materials consist mainly of Al and very small amounts of Si and Fe. According to these authors, these noncrystalline Al hydroxides are similar to those that appear at the initial stages of synthetic experiments [3,29,30], are metastable and are transformed into a stable phase (gibbsite and/or halloysite) as the reaction proceeded. During weathering of volcanic glass the noncrystalline Al hydroxide is transformed into some metastable aluminosilicates [114].

5 INTERACTION OF Fe OXIDES WITH ORGANICS AND MICROORGANISMS

Iron oxides can be precipitated (1) by slow hydrolysis of an Fe(III) salt solution, (2) by addition of base to Fe(III) salt solution and crystallization from ferrihydrite, and (3) by oxidation of Fe(II). The nature of Fe oxide formed depends on the conditions of the synthesis.

5.1 HYDROLYSIS OF Fe(III) SOLUTIONS

The slow hydrolysis of Fe(III) basically follows the same pathway in all solutions with goethite or hematite as the end product; akaganeite is formed in solutions containing Cl^-. When less than stoichiometric amounts of base are added to ferric iron solutions, spherical polycations of 1.5–3.0 nm having similar structure to that of ferrihydrite are formed [54,115–117], regardless of the nature of the anion in the solution and over a wide range of initial OH/Fe ratios [118]. Laths of lepidocrocite may form during early stages of hydrolysis possibly from monomers and dimers. However, once the polycations (ferrihydrite) dominate in the system, goethite is the end product. A general formula $Fe_3O_r(OH)_s^{9-(2r+s)+}$ has been suggested for the polycations and includes Fe-OH-Fe (olation) as well as Fe-O-Fe (oxolation). Although the crystalline end products of this reaction, such as goethite and hematite, are well characterized, the structure of intermediate phases is still to be fully elucidated. An excellent review including the results on the transformation of Fe(III) ions to ferrihydrite to the crystalline end products was recently published [119].

Transformation of ferrihydrite to crystalline components, goethite or hematite is by competing mechanisms: hematite is favored by conditions that promote coagulation of the ferrihydrite particles, whereas goethite forms readily at pHs that promote dissolution of ferrihydrite (review by Cornell and Schwertmann [4]). In laboratory studies, goethite crystallizes first in solutions that are supersaturated with respect to both goethite and hematite [120]. Low temperatures, high soil moisture, and high content of organic matter favor goethite formation and may account for the relative absence of hematite in soils of cool or temperate zones [121].

The effect of organic acids on the transformation of ferrihydrite can be classified broadly into two categories:

(a) They retard the formation of goethite and thus indirectly favor the formation of hematite. This was attributed to the retardation of the dissolution of ferrihydrite, the primary requisite for the formation of goethite, by blocking specific dissolution sites by absorption of the ligands [44,122]. Citrate and simple sugars such as glucose and maltose had a strong effect, while lactate and sucrose had a weak effect.

(b) They favour the formation of hematite directly and accelerate the transformation of ferrihydrite to hematite. When organic acids are adsorbed at the ferrihydrite surface; OH and COOH can delay ferrihydrite crystallization or act (oxalate) as a template, inducing hematite formation [43,123–125]. This effect decreases with increasing pH (pH > 7) where goethite is favoured. However, the goethite/hematite ratio in that pH range varied with the type of organic acid adsorbed: lactic > mesotartaric > citric > L-tartaric [41]; a similar effect was observed with fulvic acid [126]. Within the pH range of 4 through 10, tendency to form crystalline Fe oxides increased with increasing pH. Hematite/goethite formation was more inhibited when Fe(III)–FA complexes were used instead of ferrihydrite–FA mixtures.

The data reported suggest that, compared with nonhydroxy carboxylic acids (e.g., formic, acetic, oxalic, and malonic acids), organic acids with adjacent hydroxyl and carboxyl groups (e.g., citric, tartaric, and malic acids) have a stronger inhibitory effect on crystallization of Fe oxides [41]. This is attributed to a strong affinity of adjacent functional groups of Fe atoms at the surface of ferrihydrite.This proposed mechanism is in accordance with the observation that the surface area of natural ferrihydrite decreases with increasing content of organic C [127].

Akaganeite (β-FeOOH) is a typical solid product of slow hydrolysis of Fe(III) chloride solutions [128], and the product of oxidation of hydrothermal brines [129]. It has been suggested that akaganeite had an important role in prebiotic formation of organic substances on earth. The size and morphology of akaganeite can be modified by the addition of organic molecules [42,125,130–132]. The presence of sodium polyethanol sulfonate modified the morphology of the akaganeite particles from different shapes of submicron particles, such as needles, stars, X- and T-shaped particles to large spherical particles [133], and this was attributed to stereochemical factors. Influence of urotropin on the Fe(III) hydrolytic products was studied by Saric *et al.* [134]; the nature of the products (goethite, hematite, akaganeite) depended on the concentrations of $FeCl_3$, urotropin, pH and time of aging.

Hydrolysis of Fe(III) chloride solutions containing SO_4 ions results in a poorly crystallized oxyhydroxysulfate of Fe (schwertmannite) with a tunnel structure similar to that of akaganeite [66,67]. Because of its abundance in nature and high surface reactivity, this compound should play an important role in regulating the solubilities of both major and trace elements in surface waters impacted by acid mine drainage. Schwertmannite was observed to be a product of formation at various stages of acid mine drainage treatment [135]. An akaganeite tunnel structure was also reported during hydrolysis of Fe(III) nitrate solutions resulting in a new Fe(III) oxyhydroxynitrate, $FeO(OH)_{1-x}(NO_3)_x$ with $0.2 < x < 0.3$, with NO_3 ions in the tunnels [136].

5.2 OXIDATION OF Fe(II) SOLUTIONS

In an aqueous weathering environment, Fe is commonly mobilized as Fe(II) and the Fe oxides can be produced from Fe(II) solutions by oxidation followed by hydrolysis. The nature of the Fe oxide formed is governed by the pH, the rate of oxidation, the temperature, and the concentration of Fe^{2+} and by complexing ligands. Once formed, Fe oxides can be redissolved through either microbial reduction or the presence of complexing organic ligands.

Microbial reduction occurs whenever O_2 needed for metabolic oxidation of biomass by aerobic soil organisms becomes deficient, such as in poorly drained soils. The microbial reduction of Fe oxides is by far the most important process in the Fe cycle in soils and landscapes, besides the Fe oxides formed from primary lithogenic Fe. Aerobic soils are more widespread than anaerobic soil in the pedosphere, so that the Fe oxides once formed remain stable. Many species of microorganisms (e.g., *Enterobactriaceae*, *Bacillaceae*, *Clostridiae*) are capable of reducing Fe oxides (review by Ottow and Glathe [137]). As heterotrophic organisms, they depend on available biomass so that redoxomorphosis in soils is limited to a depth in which biomass is available for metabolic oxidation. Fe(III) ions at the oxide surface are reduced to Fe(II) by electrons produced during metabolic oxidation of organic compounds, bound to an enzymatic system at the outer surface of a bacterial cell. This process is facilitated if the organism is in contact with an Fe oxide particle [138]. The resultant Fe(II) will be in solution and will be reoxidized when O_2 is again reintroduced or moved to zones of higher Eh (higher O_2 concentration) and precipitated as Fe(III) oxide. The highly reactive Fe oxide surface, such as ferrihydrite may be redissolved within a few days through bacterial reduction (Figure 5) [139,140].

The photoreduction of Fe(III) to Fe(II) and the presence of high amounts of Fe(II) in aquatic systems has also been attributed to a variety of electron donors such as EDTA and organic ligands such as oxalate and formate [141–143].

5.2.1 Kinetics of Fe(II) Oxidation

Oxidation of Fe(II) solutions has been investigated extensively [144–151] It obeys first order kinetics with respect to $[Fe^{2+}]$, and second order with respect to pH [144], and is retarded by organic ligands [152]. The rate constant of Fe(II) oxidation decreases in the order: oxalate > acetate > tartrate > citrate [153]. The effect of ligands can be explained by assuming that the Fe(II)–ligand complex is the reactive species and the relative stability of Fe(II)–ligand complex controls the rate of Fe(II) oxidation. A higher value of the Fe(II)–ligand stability constant thus leads to slower kinetics and a corresponding lower value of the rate constant of Fe(II) oxidation (Figure 6). The inhibition of Fe(II) oxidation by organic compounds can be explained as follows [154]:

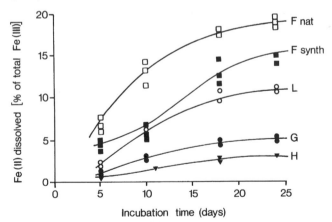

Figure 5. Reduction of various Fe oxides by *Corynebacterium ferrireductans*. Ferrihydrite (natural-Fnat, synthetic-Fsyn), synthetic goethite (G), hematite (H), and lepidocrocite (L). Modified from [139]. Reproduced by permission of Urban & Fischer Verlag

$$Fe(II) + O_2 \rightarrow Fe(III) \qquad (a)$$
$$Fe(II) + organic \rightarrow Fe(II)-organic \qquad (b)$$
$$Fe(II)-organic + O_2 \rightarrow Fe(III) \qquad (c)$$
$$Fe(III) + organic \rightarrow Fe(II) + oxidized\ organic \qquad (d)$$
$$Fe(II) + O_2 \rightarrow Fe(III) \qquad (e)$$

The rates of reactions (c) and (d) are dependent on both pH and the nature of the organics. In the presence of citrate, at high concentrations (citrate/Fe molar ratio > 0.1), the Fe(II) is stabilized in the system as an Fe(II)–citrate complex [155].

5.2.2 Pathways of Formation of Fe Oxides

As Fe(II) is oxidized in neutral or slightly basic solutions it passes through intermediate green solution complexes or solid green rusts, respectively. These green products are composed of both Fe(II) and Fe(III) held together by ol- and oxo-bridges formed during the consumption of OH^-. Green solution complex I, $[Fe(II)_2Fe(III)O_x(OH)_y]^{(7-2x-y)+}$, is formed in Cl^- solutions while green solution complex II, $[Fe(II)Fe(III)O_x(OH)_y]^{(7-2x-y)+}$, is formed in SO_4^{2-} solutions [156–158]. The Fe(II)/Fe(III) ratio in the green rusts is the same as that of green solution complexes in ClO_4^- solutions [159]. Solid green rusts are greenish-blue crystalline compounds of alternating layers of $[(Fe^{2+}, Fe^{3+})_3(OH, O)_8]$ and $(anion^-)_2$ with the Fe(II)/Fe(III) varying between 3.6 to 0.8 including anions as the integral part of a predominantly hexagonal structure [160–162]. The layers are postively charged due to their Fe(III) content and the charge is balanced by the anions Cl, SO_4, CO_3, NO_3, ClO_4, Br, and I. The composition of green rust varies depending on the anion in the interlayer. The anions SO_4, NO_3, ClO_4 accommodate water in the interlayers [163]. Upon oxidation, these transform to goethite, and/or lepidocrocite in slightly acidic and near neutral conditions [160,164–169]

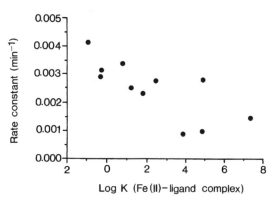

Figure 6. Relationship between rate constant of Fe(II) oxidation at a ligand/Fe molar ratio of 0.1 and the stability constant ($\mu = 0.01, 25\,^{\circ}C$) of Fe(II)–ligand complexes. Redrawn from [153]

Lepidocrocite formation appears to be restricted to the cases of rapid oxidation, either through green solution complexes or green rusts (Figure 7). Because of rapid oxidation and incomplete deprotonation of green rusts, lepidocrocite has a disordered cubic structure of oxygens and hydroxyls. Lepidocrocite can also be formed directly during oxidation without the presence of intermediate compounds [159]. This polymerization process can be visualized as:

$$FeOH^+ \rightarrow Fe(OH)^{2+} \rightarrow \begin{array}{c} OH \\ \diagdown \\ \diagup \\ OH \end{array} Fe\text{-}O\text{-}Fe(OH)^+$$

$$\downarrow \text{Precipitation}$$

$$\gamma\text{-FeOOH}$$

However, the nature of the intermediate phases is yet to be elucidated.

Lepidocrocite is less stable than its polymorph goethite. The transformation of lepidocrocite to goethite is through solution phase rather than topotactic [170,171]. The rate controlling step can be either the dissolution of lepidocrocite or the formation of small spherical polycations or ferrihydrite particles which link together in linear arrays and age into goethite [115–117,162]. The forma-

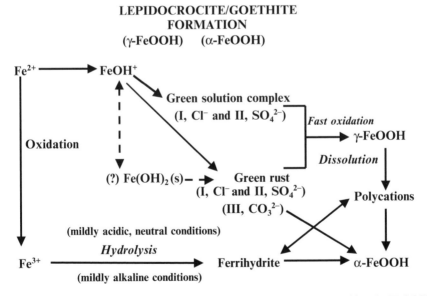

Figure 7. Pathways of formation of lepidocrocite (γ-FeOOH)/goethite (α-FeOOH) through oxidation of Fe(II) solutions

tion of goethite either through the intermediate green compounds or through ferrihydrite is as shown in Figure 8.

Magnetite is formed by the initial dehydration and subsequent slow oxidation of the green rusts in alkaline conditions [64,159]. Misawa *et al.* [159] argue that it would be difficult for $Fe(OH)_{2(solid)}$ which has a hexagonal structure to convert directly to magnetite (Fe_3O_4), which has a cubic close-packed structure. Two of three $Fe(OH)_2$ molecules must be oxidized with dehydration, deprotonation, and reorganization. The magnetite that forms eventually oxidizes into maghemite (Figure 9). Magnetite is easily synthesized by oxidation of Fe(II) solutions at pH 8 under the same conditions where green rust and lepidocrocite form at pH 6–7 [4].

If the rate of oxidation is extremely high (by H_2O_2 or exposure of $Fe(OH)_2$ to air) feroxyhyte (δ'-FeOOH) and δ-FeOOH are formed [159,161,172,173], the latter from $Fe(OH)_2$ and the former from green rust. At pH 12, well-crystallized δ-FeOOH is formed but as the pH of the system drops, the product becomes increasingly less ordered leading to the formation of feroxyhyte [173]. The oxidation is assumed to be topotactic or solid-state transformation between two solid phases of close-packed structures of oxygens [161],

GOETHITE FORMATION
(α-FeOOH)

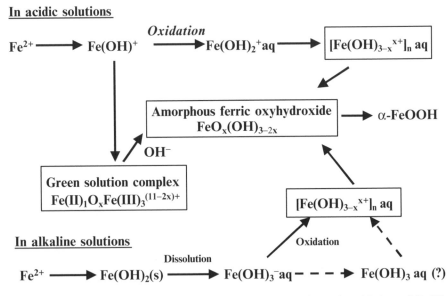

Figure 8. Pathways of formation of goethite (α-FeOOH) through oxidation of Fe(II) solutions

$$2Fe(OH)_2 + \frac{1}{2}O_2 \rightarrow 2(\delta\text{-FeOOH}) + H_2O$$

δ-FeOOH is not observed in nature. Chukrov *et al.* [174], however, reported finding a disordered form, which they called δ'-FeOOH and proposed the mineral name feroxyhyte.

The influence of complexing ligands on the rate of Fe(II) oxidation and the resulting oxidation products has been investigated [153]. The concentrations of common organic ligands in soil solution are of the order of $10^{-3} - 10^{-5}$ mol L^{-1} [1,68]. The precipitation products formed from the oxidation of Fe(II) perchlorate solution at pH 6 is predominantly goethite with small amounts of lepidocrocite and poorly crystalline Fe oxide. The presence of ligands, such as acetate, oxalate, citrate, tartrate, phosphate and silicate decreases the rate of Fe(II) oxidation and the rate of Fe(III) hydrolysis, thus influencing the nature of the precipitation product (Figure 10). At a ligand/Fe molar ratio = 0.1 (Fe(II) = 0.01 mol L^{-1}), the formation of goethite (acetate), lepidocrocite (oxalate) and poorly crystalline Fe oxide (tartrate) is favoured, whereas citrate stabilizes Fe(II) in solution, and thus no precipitation of Fe oxide was observed. The formation of lepidocrocite was attributed to the modification of the oxygen coordination in the edge-sharing Fe(O,OH) octahedra by the Fe(II) ligand complexes.

MAGNETITE/MAGHEMITE FORMATION
(Fe_3O_4) $(\gamma\text{-}Fe_2O_3)$

Slightly acid or slightly alkaline conditions

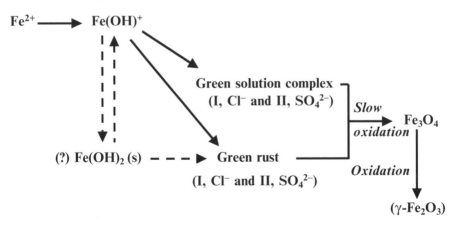

Figure 9. Pathways of formation of magnetite (Fe_3O_4)/maghemite $(\gamma\text{-}Fe_2O_3)$ through oxidation of Fe(II) solutions

Krishnamurti and Huang [155,175] extended their study to investigate the influence of citrate, at different citrate/Fe molar ratios (R), on the kinetics of Fe(II) oxidation and the resulting hydrolytic products of Fe(III) under both mildly acidic (pH 6.0) and alkaline (pH 7.5) conditions. The retardation of the rate of Fe(II) oxidation inhibits the nucleation of goethite and maghemite and promotes the formation of lepidocrocite especially at $R = 0.001$, and $R = 0.01$, respectively (Table 5). The laths of lepidocrocite were observed to be elongated along the c-axis. The citrate anion possibly exerts significant influence on the oxygen coordination and plays a positive role in the way the double rows of Fe(O,OH) octahedra are linked during crystallization of the precipitation products. Further, the Fe–citrate complexation possibly modified the dehydration of the intermediate green rusts, which was a prerequisite for the maghemite formation at slightly alkaline pH conditions. This possibly resulted in the formation of lepidocrocite (Figure 11).

Most organics tend to accumulate in the poorly drained soil environments [68,80]. Even small amounts of citrate can promote and stabilize lepidocrocite, which is present in reductomorphic and poorly drained soils [121]. Citrate and phenolic compounds are dominant in the root exudates, which reduce Fe(III) of the soil minerals to Fe(II) [176]. Fe(II) released to soil solution may be transformed to lepidocrocite during oxidation. Lepidocrocite was reported to be the dominant constituent in the root channels of soil profiles away from the roots, whereas goethite was observed close to the root channels due to the presence of CO_2, a product of root respiration [177].

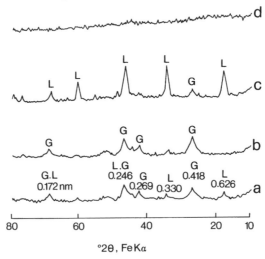

Figure 10. X-ray powder diffractograms of hydrolytic products of Fe(II) oxygenation formed at pH 6.0. (a) $0.01 \, mol \, L^{-1}$ ferrous perchlorate; and in the presence of organic ligands (initial ligand/Fe(II) molar ratio = 0.1): (b) acetate, (c) oxalate, and (d) tartrate. L, lepidocrocite; G, goethite. Redrawn from [153]

Table 5. Rate constants of Fe(II) oxidation and nature of Fe oxides formed in absence and presence of citrate in ferrous perclorate–NaOH system at pH 6.00 and 23.5 °C. From [155]. Reproduced by premission of the Clay Minerals Society

Initial citrate/Fe molar ratio	Rate constant $(min^{-1}) \times 10^4$	Dominant Fe-minerals	XRD* 020 peak of lepidocrocite	
			WHH ±0.02 (°20)	Area (mm^2)
0	41.3 ± 2.0	G, L	1.40	40 ± 5
0.0003	28.6 ± 3.5	L, G	1.00	97 ± 5
0.0005	26.9 ± 1.8	L	1.00	124 ± 12
0.0010	24.4 ± 3.0	L	0.80	263 ± 3
0.0050	20.2 ± 1.8	L	1.10	82 ± 4
0.0060	19.1 ± 1.5	L	1.20	25 ± 3
0.0080	16.8 ± 0.6	L	1.20	10 ± 3
0.0100	15.2 ± 0.7	PC	n.a.†	n.a.
0.1000	7.6 ± 0.7	no ppt	n.a.	n.a.

* X-ray powder diffraction data; WHH width at half height; L, lepidocrocite; G, goethite; PC, poorly crystalline.
† n.a., not applicable.

5.3 DISSOLUTION OF IRON OXIDES

5.3.1 Complexation

The solubility of a compound is determined by the free energy of dissolution. In general, the solubility of Fe oxides is low. This means that, except under extreme pH conditions, Fe oxides maintain a very low level of total Fe in solution. In pH range of 6–10, and in the absence of complexing ligands total Fe in solution is $< 10^{-6}$ mol L^{-1}. The formation of soluble complexes of Fe(II) or Fe(III) with ligands such as citrate and oxalate increases the solubility of Fe oxides. Fe oxide precipitates from a 0.001 mol L^{-1} Fe(III) solution at pH 2–3; 0.01 mol L^{-1} citrate will keep Fe(III) as a soluble complex up to pH 7.6 [4]. Fe oxide crystallizes from 0.01 mol L^{-1} Fe(II) solution at pH 5–7; however, low amounts of citrate (0.001 mol L^{-1}) do not allow Fe(II) to oxidize and keep Fe(II) in solution as a soluble complex [175]. Organic molecules with oxygen donor groups that form mononuclear bidentate surface chelates (oxalate, salicylate, catechol, and citrate) are very dissolution active. These surface ligand orbitals shift electron density toward the central Fe(III) ion and labilize its binding to oxygen lattice atoms (*trans* effect) [178]. The general reaction for ligand-promoted dissolution of Fe oxides may be visualized as follows [179]: ligand is first adsorbed on the surface of Fe oxide weakening the Fe-O bonds to neighboring atoms, and then leading to the detachment of Fe(III) complex.

Mobilization of Fe from Fe oxides by siderophores is of great importance in natural systems [180,181]. The active components of siderophores were identified as phenolates and hydroxamates which can form stable Fe(III) complexes

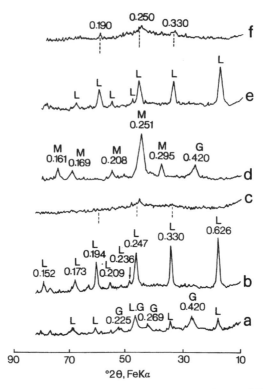

Figure 11. X-ray powder diffractograms of hydrolytic products of Fe(II) oxygenation formed in 0.01 mol L^{-1} ferrous perchlorate system at pH 6.0 after reaction time of 120 min after aging in suspension for 9 months: (a) in the absence of citrate; and in the presence of citrate at an initial citrate/Fe(II) molar ratio of (b) 0.001 and (c) 0.01; and at pH 7.5 after reaction time of 60 min after aging in suspension for 9 months: (d) in the absence of citrate, and in the presence citrate at an initial citrate/Fe(II) molar ratio of (e) 0.01 and (f) 0.1. L, lepidocrocite; G, goethite; M, maghemite. The d values are in nm. Redrawn from [175]. Reproduced by permission of the Clay Minerals Society

(pK up to 32) and thus mobilize Fe from Fe(III) compounds [182–184]. These higher Fe concentrations are important for the supply of Fe to plant roots which excrete organic acids at the soil–root interface.

5.3.2 Reduction

Reduction of structural Fe(III) to Fe(II) is one of the most important dissolution mechanisms, which can be mediated both biotically and abiotically. The important electron donors result from metabolic oxidation of organic compounds. A number of reductants such as ascorbic acid [179], hydroquinone [185], fructose, sucrose and fulvic acid were studied. For details see Cornell and Schwertmann [4].

Photochemical reduction of Fe(III) to Fe(II) was demonstrated with ligands such as oxalate, citrate and thioglycolate with hematite [186], lepidocrocite [187] and goethite [188]. The formation and diurnal fluctuations of Fe(II) in surface waters was attributed to photochemical reduction [141].

No data are available on the relative importance of different dissolution processes [189–191]. The results suggest that reduction, particularly when assisted by complexation, will be the main mechanism for Fe transport in global ecosystems. The Fe(III) reduction, which can take place either aerobically or under microaerophillic or anoxic conditions may have an important influence on the iron geochemistry and magnetic properties of soils, sediments and water (review by Lovely [192]).

5.4 INTERACTION OF IRON OXIDES WITH MICROORGANISMS

Mineral dissolution processes at the interface with living microorganisms are of general occurrence and a variety of dissolution phenomena are observed. Microsites around living organisms always exhibit precipitates of poorly crystallized compounds. Jackson and Keller [193] observed noncrystalline Fe oxides under lichens in Hawaii, in an environment that ordinarily would produce crystalline Fe oxides. The presence of ferrihydrite under lichens occurring on basalt has also been reported [194]. Iron oxide precipitation has frequently been observed near the glaciers at the surface of rocks or more generally in running water, where bacteria such as *Gallionella* and *Leptothrix* are found [195], and in drainage trenches [127,196]. Fungi, lichens and roots can also concentrate Fe [197]. Iron oxidation and accumulation occur in the epidermal cells of rice roots, and goethite or lepidocrocite are identified around the roots [177].

The Fe–microbe interactions can be divided broadly into two fundamental processes, viz., microbial oxidation and microbial reduction. Microbial oxidation may be either nonenzymatic (indirect), which takes place in aerated environments at pH > 5.2 (utilization of Fe–organic chelates by *bacilli* and *pseudomonads*, [198]), or enzymatic (direct). The rod-shaped *Thiobacillus ferrooxidans* and curved rod *Leptospirillum* grow well at pH 2.5–3.5, and can exist with Fe(II) as their only source of energy and CO_2 as their sole source of C [199].

Certain microorganisms are capable of oxidizing Fe^{2+} in aqueous systems [200], and some of them, e.g., *Thiobacillus ferroxidans*, are adapted to high acid conditions (pH < 3). In such conditions biotic oxidation is dominant because abiotic Fe^{2+} oxidation at such low pH conditions is extremely slow. A typical abiotic oxidation product in an acid mine drainage area is schwertmannite, found under divergent pedogenic environments [66,67]. The biotic formation of Fe oxides can also proceed through metabolic oxidation of organic ligands which complex with Fe(II) and Fe(III). After oxidation of the organic ligand, the released Fe undergoes hydrolysis and polymerization to form magnetite [201–207].

Formation of biominerals follows one of two pathways [208]: it can be induced by modification of the chemical environment or through provision of an organic support or surface. Regardless of the pathway of formation, the iron oxide formed by microbial oxidation is usually ferrihydrite [174] which is found on the surface of bacterial or algal cells. Iron is always localized at the periphery of the bacteria [209] and always occurs in a short-range ordered form, usually as ferrihydrite. The poorly crystalline nature of Fe oxide formed by the biotic pathway may be attributed to a high oxidation rate and/or the presence of organics as inhibitors for the formation of crystalline Fe oxides.

With the exception of hematite, all the major Fe oxides are found in living organisms. Like other biominerals, Fe oxides have various homeostatic functions: they participate in iron metabolism, act as magnetic navigational devices, and provide support, hardness and density in structures such as teeth [210]. The field of biomineralization has received widespread interest recently and was reviewed recently by Frankel and Blakemore [206], Skinner and Fitzpatrick [211], and Cornell and Schwertmann [4]. However, the understanding of the *in vivo* mechanisms of biomineralization is still at the descriptive stage and it is in this that direction further research should be concentrated.

Some specific examples of interaction of Fe oxides with anaerobic microorganisms are described in Chapter 11 of this issue.

6 INFLUENCE OF ORGANICS ON MIXED Fe–Al GELS

The influence of Al on the crystallization process of Fe-oxides and on the formation of Al-substituted goethite, hematite and ferrihydrite has been reported in detail [4,121,212]. Many studies have demonstrated that in soil environments, especially acidic conditions and at certain critical Fe/Al ratios, Fe and Al ions may interact with each other forming, particularly in the presence of organic substances, short-range ordered mixed oxides of different chemical composition, size, nature and reactivity. Rengasamy and Oades [213] showed evidence that polymerization of Al and Fe(III) in mixed solutions $(OH/Al + Fe \leq 2.5)$ favors the formation of Al–Fe copoly-cations rather than a mixture of separate Al and Fe species. Goh et al. [214] found that, after prolonged ageing of mixed Al–Fe systems (initial Fe/Al molar ratio $(R_i) \leq 1$), crystalline $Al(OH)_3$ and mixed noncrystalline Fe–Al oxide or highly Al-substituted noncrystalline Fe oxides formed at pH 6–7. More recently, it has been demonstrated that the initially formed mixed Fe–Al products are metastable and slowly convert, depending on R_i, through different soluble and short-range ordered species into more stable Al or Fe oxides, mainly gibbsite and hematite [215–218]. The distribution of Fe and Al in soluble or solid phases of different sizes depends on R_i, pH and ageing period. Usually, the higher R_i, the higher the amounts of Fe + Al present in soluble or colloidal particles. The high stability of mixed Fe–Al species has been demonstrated. In unmixed Al or Fe solutions gibbsite or hematite and/or goethite crystallized in a few days or weeks at pH 5.0 and 50 °C, whereas in the Fe–Al samples, the formation of well crystallized Fe oxides is not evident at $R_i = 0.1-4$ even after 32–120 days at 50 °C. Samples formed at pH values ranging from 4.0 to 10.0 and R_i of 1 or 2.5 showed considerable differences in the mineralogy of the precipitates after 60 days of aging at 50 °C. The samples formed at $R_i = 1$ contained ferrihydrite at pH 4.0, ferrihydrite + gibbsite at pH 5.0–7.0, and hematite + $Al(OH)_3$ polymorphs + ferrihydrite at pH 9.0–10.0. The samples formed at $R_i = 2.5$ have greater quantities of ferrihydrite. Large amounts of $Al + Fe$ (25–85%) are solubilized from these samples by ammonium oxalate [216].

The addition of increasing concentrations of citrate to the gel suspensions formed at pH 5 or 8.5 promotes the formation of short-range ordered materials. Citrate added initially to Fe–Al solutions (citrate/Fe + Al = 0.1 and Fe/Al = 1) completely inhibits the formation of crystals even after 135 days at 50 °C [216].

The effect of time, pH, organic ligands and temperature on the stability of hydroxy Fe(III, II)–Al–montmorillonite complexes formed at Fe/Al molar ratios (R_i) ranging from 0.1 to 10 was also investigated [216,217,219]. Upon ageing for long periods of time (up to 120 days at 25–50 °C), the degree of interlayering in the hydroxy Fe(III or II)–Al–montmorillonite complexes is always higher than for samples containing only Al ($R_i = 0$) or Fe ($R_i = \infty$) and it is greater in the complexes with R_i ranging from 0.5 to 4. The amounts of Fe and Al extractable from the complexes by oxalate solution decrease steadily with time. However, even after prolonged ageing, large quantities of Al and Fe are solubilized by oxalate from the mixed Fe–Al– montmorillonite complexes formed at pH 5.0 with $R_i > 0.5$ (34–78% of Fe + Al present in the samples). The presence of strongly chelating organic ligands greatly increases the stability of hydroxy Fe–Al–montmorillonite complexes (A. Violante, unpublished data).

7 EFFECT OF ORGANIC COMPOUNDS ON SURFACE CHEMISTRY OF OXIDES

7.1 SURFACE CHARGE AND SPECIFIC SURFACE AREA

The nature of the surfaces of pure oxide minerals in contact with aqueous solutions is pH dependent [3,4,11]. The metal ions at the surface of an oxide complete their ligand shell with hydroxyls in aqueous solutions. The hydroxylation of the surface is followed by the adsorption of H_2O molecules through H-bonding. The amount of adsorbed H_2O increases with decrease in crystal size and increase in surface area [220]. At the hydroxylated or hydrated surface, positive or negative charge is developed by adsorption or desorption of protons (H^+) and/or hydroxyl ions (OH^-) or dissociation of surface species. Such reactions occur at all surface sites and they are likely to be anionic, neutral and/ or cationic over a range of pH values. The pH values for the point of zero charge (*pzc*) of the pure Fe and Al oxides tend to be either close to or greater than neutrality. The pH of the *pzc* of Fe oxides ranges between 7 and 9, without any marked difference between the various mineral forms. Hsu [3] reported that the *pzc* of Al oxides ranges from pH 8 to 9.2.

The surface charge characteristics of the minerals in soil environments are altered from those of pure minerals by adsorption and precipitation reactions and can be regenerated by proton adsorption:

$$M\text{-}OH_2 + A^- \rightarrow M\text{-}A + H_2O$$
$$M\text{-}A + H^+ \rightarrow M\text{-}AH^+$$

When the chemical interaction between the mineral surface (M) and anion (A) cannot be described by electrostatic forces, the process is known as specific adsorption and can be viewed as a chemical surface reaction involving ligand exchange and formation of covalent bonds between the metal cations of the oxides and the anion. Organic ligands with a strong affinity for Al and Fe (oxalate, citrate, tartrate and malate) appear to be strongly adsorbed as inner-sphere complexes, and adsorption occurs through a ligand exchange mechanism. The polydentate functionality of many organic compounds is an important factor in forming a strong bond at the surface. Fulvic and humic acids are adsorbed mainly by ligand exchange. Monodentate ligands (acetate, benzoate and formate) are weakly adsorbed on oxide surfaces. Adsorption of organic compounds usually increases by decreasing pH.

Adsorption of anions via ligand exchange results in a shift in the *pzc* of the oxide to a more acid value. The surface charge characteristics of Fe oxides formed from the oxidation of Fe(II) solutions in the presence of citric acid was recently investigated by Liu and Huang [50]. The *pzc* of the Fe oxides decreases from 7.0 to 3.9 with increase in the initial citrate/Fe(II) molar ratio from 0 to 0.1, which is attributed to increasing coprecipitation and adsorption of citrate and partial neutralization of the surface positive charge.

Table 6. Selected physicochemical properties of the Fe oxides. Reprinted from [221], Copyright (1995), with permission from Elsevier Science

Sample	Specific surface $(m^2\,kg^{-1}) \times 10^{-3}$	Organic C $(g\,kg^{-1})$	PZC	Excess charge* $(cmol_c\,kg)$
FAC†	344	0	9.20	27.80
FPC‡	205	5.3	3.65	−14.79

* In the $0.1\,mol\,L^{-1}$ $NaNO_3$ solution at pH 7.
† The Fe oxide prepared in the absence of citric acid.
‡ The Fe oxide prepared in the presence of citric acid.

Crystalline Fe and Al oxides have specific surface of $15–70\,m^2\,g^{-1}$, whereas short-range ordered Fe and Al precipitates may have specific surface areas even greater than $300\,m^2g^{-1}$ (see Table 1 in Chapter 7 in this issue). Multifunctional organic compounds associated with Al and Fe play an important role in both distortion of Al and Fe precipitates and promotion of aggregation of the reaction products. The specific surface of Al oxides formed in the presence of chelating ligands (tannic or tartaric acid) increases with increasing the initial ligand/Al molar ratio from 0 to 0.01 for tannate and to 0.1 for tartrate. Further increase in the ligand/Al molar ratio decreases the specific surface of the precipitates due to the ability of the polydentate anions to promote aggregation of the precipitates above certain critical concentrations [33,87 and references therein]. Probably, structural distortion and aggregation occur simultaneously but aggregation may be more prominent when the amount of certain ligands in the solid is sufficiently high. Xue and Huang [221] also found that the presence of citrate lowers the specific surface of ferrihydrite (Table 6). They reported that the lower surface area of ferrihydrite formed in the presence of citrate might be indicative of citrate modifying the small ferrihydrite particles into a rigid network of aggregates. Susser and Schwertmann [127] reported that the specific surface of natural ferrihydrite decreases with increasing content of natural organic carbon and increases after H_2O_2 treatment.

7.2 ANION AND CATION RETENTION

7.2.1 Anion Sorption

Metal oxides adsorb a wide variety of organic compounds, which include natural organics and xenobiotics [1,4,7,11,80,87,200,222].

In soil environments, depending on the mechanisms of their interfering reactions, organic acids, tannins, humic and fulvic acids have a dual role of both promoting and hindering anion retention by the Al or Fe precipitation products. Organic acids such as malic, citric, aspartic, oxalic and tannic acid promote the formation of active sites for the sorption of phosphate by distorting the structure of precipitation products of Al and enhancing their specific surface. Stability of the short-range ordered structure of the precipitates with a large specific surface, formed in the presence of critical concentrations of some biomolecules, contributes to development of a high anion retention capacity of organo-mineral complexes [1,223,224].

However, competition between strongly adsorbing anions for an adsorbing surface is well documented. Organic compounds compete for common sites on Al and Fe oxides and inhibit the adsorption of other organic and inorganic ligands. Nagarajah et al. [225] found that the competitive ability of carboxylic acids in preventing phosphate adsorption on gibbsite and goethite is in the order citrate > oxalate > malonate > tartrate > acetate. Lopez-Hernandez et al. [226],

Fox *et al.* [227], Violante *et al.* [228] and Violante and Gianfreda [87] have demonstrated that chelating organic ligands strongly reduced phosphate adsorption by variable charge minerals and soils. Citrate and tartrate suppressed the adsorption of phosphate on ferrihydrite [229], which is believed to be one of the mechanisms by which plant roots, which excrete simple and di-carboxylic acids, mobilize adsorbed phosphate and improve their phosphate supply [230]. The inhibiting effect of various low molecular mass organic acids and phenols on phosphate adsorption by goethite has been demonstrated by Klees (cited by Guggenberger and Haider in Chapter 7 of this issue; see their Table 5). Vice versa, phosphate suppresses the adsorption of mugineic acid on ferrihydrite, thereby impeding the release of Fe to plant roots by siderophores [182].

Inskeep [231] and Liu *et al.* [232] have demonstrated that some carboxylic acids as well as humic and fulvic acids, tannic and gallic acids prevent sulfate adsorption on iron oxide. Oxalate, tartrate, gluconate, malate and thiocyanate cause more reduction than others in sulfate adsorption. Maximum reduction in phosphate and sulfate adsorption occurs when organic compounds react with oxides before the addition of phosphate and sulfate.

The fate of pesticides and other xenobiotics is strongly affected by adsorption in soils. Metal oxides act as sorbents for pesticides such as 2,4-D, quinmerac [233,234]. The adsorption of 2,4-D was strongly competitive with that of phosphate and citrate [235].

7.2.2 Cation Sorption

Fe and Al oxides selectively adsorb divalent cations even at solution pH values lower than the *pzc* of metal oxides. Both the metal oxides and humic matter are efficient scavengers of heavy metals. Divalent transition and heavy metal cations, both of which are often sorbed as inner-sphere complexes, are more strongly adsorbed than alkaline earth cations. Spectroscopic studies have given evidence that the surface of alumina acts as a bidentate ligand in binding Cu^{2+} [7,11,236]. The mechanism of metal ion association with hydrous oxide surfaces involves an ion exchange process in which the adsorbed cations replace bound protons. Specifically adsorbed cations raise the value of the *pzc* of oxides. pH affects adsorption of metal cations, either by changing the number of sites available for adsorption or by changing the concentration of cation species $(Me^{2+}, MeOH^+, Me(OH)_2)$ that are preferentially adsorbed. The following affinity series for freshly precipitated Fe and Al have been reported [237,238] as Pb > Cu > Zn > Ni > Cd > Co > Sr > Mg for Fe oxide, and Cu > Pb > Zn > Ni > Co > Cd > Mg > Sr for Al oxide. Manceau *et al.* [239] have reported that below a critical pH the percentage of Pb(II) sorbed by metal oxides varied in the order Mn > Fe > Al.

Organic compounds may either hinder or promote the sorption of metal cations, depending, respectively, on whether the metal complexes which they form remain in solution or are themselves sorbed by the mineral, and on whether sorbed complexes are bound more or less strongly than free cations [238,240,241]. Furthermore, adsorption of chelating organic anions can increase the negative charge of surfaces and hence increase adsorption of metal cations. Organic chelating anions may promote metal adsorption on oxides of Fe and Al by forming stable surface–metal–ligand complexes. Some organic compounds have no net effect on metal sorption, at least under certain conditions (e.g. within a given pH range). Many authors also found that in systems containing organic ligands metal adsorption is often enhanced at low pH (pH $<$ pzc) and depressed at high pH (pH $>$ pzc) [11,236,238 and references therein].

The ligand/metal ratio seems to be critical in determining whether the adsorption of a metal is favored or prevented [236,238,242]. Zhou $et\ al.$ [242] found that oxalate promoted the adsorption of Cu on bayerite and goethite at pH 4.5 and oxalate/Cu molar ratios $<$ 5; in the presence of increasing concentrations of oxalate, Cu adsorption firstly increased, reached a maximum at a critical oxalate/Cu molar ratio and then decreased.

The importance of pollutant heavy metal ions bound to the Fe– and/or Al–organic complexes to the availability of the metals was recently brought out using the sequential extraction speciation scheme [243]. The metal–organic complex-bound Cd was found to be important in the assessment of plant available Cd in the temperate soils of Saskatchewan, Canada [244,245], tropical soils of Kenya [246] and in the Mediterranean soils of South Australia [247]. Specifically, the fulvic acid fraction of the metal–organic complexes was observed to play an important role in the assessment of plant available Cd [247,248].

Owing to its high surface area and affinity to many ions, poorly crystalline Fe oxide has been used to adsorb many pollutant ions and elements. [249–252].

Cleaning of sewage using Fe oxides has been developed by CSIRO in Melbourne [253]. Magnetite powder, positively charged at an acid pH, was used to attract negatively charged organics. The Fe–organic associations are then separated and removed using a magnetic field and the organics subsequently desorbed under alkaline conditions

An overview of cation contaminant associations with soil colloids (including Fe and Al oxides) is presented in Chapter 12 of this volume.

7.3 SORPTION OF BIOPOLYMERS

The adsorption of biopolymers (proteins, toxins, DNA, RNA, polysaccharides), bacteria and viruses on clay minerals has been studied in detail and critically reviewed [254–256]. In contrast, the fixation of proteins on metal oxides and organo-mineral complexes has received attention only in the last decade [87,257].

The immobilization of proteins on variable charge minerals apparently proceeds through ligand exchange whereby the carboxylate groups of proteic molecules displace the hydroxyl or water groups from the surfaces of metal oxides or OH–Al (–Fe)–clay complexes. Protein molecules strongly compete with strongly chelating organic anions (citrate) or phosphate for sorption sites of metal oxides on variable charge soils [87 and references therein]. Recently, it has been demonstrated that increasing concentration of citrate, oxalate and phosphate reduce the adsorption of phosphatase and albumin on ferryhidrite, goethite and selected acidic Chinese soils, i.e. Oxisols and Ultisols (H. Huang and A. Violante, unpublished data).

Studies on the residual activity of selected enzymes immobilized on noncrystalline Al oxides and $Al(OH)_x$–montmorillonite (chlorite-like) complexes have been carried out. Adsorbed invertase, urease and phosphatase display a consistent reduction of their specific activities compared with the free enzymes. Adsorption of enzymes on metal oxides also modifies their stability towards thermal and proteolytic denaturation. Very active complexes have been obtained by the interaction of enzymes with OH–Al or OH–Fe species, organic compounds and clay minerals [87].

8 CATALYTIC ROLE OF METAL OXIDES IN THE FORMATION OF HUMIC SUBSTANCES

Even though the formation of humic substances has long been regarded as a biological process, recent research has shown that humic substances may also be formed by the heterogeneous catalytic effect of metal oxides in soils [33]. Oxides and hydroxides of Fe catalytically oxidize phenols and phenolic acids. The catalytic oxidative powers of iron oxides and oxyhydroxides are in the order: hydrohematite > maghemite > lepidocrocite > hematite. In soils, poorly ordered iron oxides with large surface area provide active sites for the catalytic reaction to occur. The importance of hydrous oxides of Fe in promoting the formation of phenolic polymers seems to be related to the nature of phenolic compounds [258].

The catalytic role of aluminum oxides in the abiotic formation of humic polymers was suggested from the results of Seebald and Schunack [259] and Wang et al. [260]. Recently, Krishnamurti et al. [93] have shown the role of Al_{13} tridecamer, the hydrolyzed species of Al, in the abiotic formation of humic-like polymers from phenolic acids. The ESR spectrum of the product formed at Al/ phenolic acid molar ratio = 0.50 showed a single symmetrical line devoid of any fine splitting, indicating the presence of free radicals. The free radicals have a g value of 2.0031 with a line width of 7.2 G, indicative of semiquinones, the major free radicals normally observed in the humic acid fractions of soil organic matter [261]. The g value and the line width are in close agreement with the values reported for the humic acids produced from phenols [262] and for those polymerized through natural clays [260]. Further confirmation of the formation of humic substances was obtained from the ^{13}C CPMAS NMR spectrum of the sample [93].

The role of metal oxides in abiotic transformation of organic compounds has been reviewed by Huang [33,263].

The research data so far available indicate that Fe and Al components of soil have a vital role in the abiotic formation of humic substances through the polymerization of phenolic compounds. However, more research is needed using advanced sophisticated instruments, such as ESR, and pyrolytic CPMAS NMR, to understand the mechanism and the role of metal oxides in the abiotic formation of humic substances.

9 CONCLUSIONS AND FUTURE PROSPECTS

Soil organic matter is important in the formation and transformation of both iron and aluminum oxides/oxyhydroxides, as revealed by intensive studies done during the last three decades. Organic substances strongly interfere both in the formation of Fe(III) oxides, either by hydrolysis of Fe(III) or by oxidation of Fe(II) solutions, and in the transformation of ferrihydrite. They also have an important effect on the nature, morphology and surface properties of Al and Fe hydroxides and oxyhydroxides.

Organic complexing anions delay or inhibit, at certain critical concentrations, the crystallization of Al or Fe oxides, favoring the formation of short-range ordered precipitates. Organic anions adsorbed or coprecipitated with Al and Fe have a great influence on the adsorption of anions, cations, nutrients, pollutants and biopolymers.

The fine scale morphology and surface geometry of these oxides formed under different ionic environments and the impact on their surface chemistry with reference to the dynamics, transformation and fate of nutrient and pollut-ant ions in terrestrial environments warrants in-depth future research.

The interaction of organic ligands with soluble monomeric and polymeric species of Al in natural environments is also extremely important. Speciation of OH–Al species free or complexed with organic ligands is important because not all the chemical forms of Al are equally toxic.

The abiotic formation of humic substances through catalytic polymerization of phenolic compounds by metal oxides deserves close attention. The functional groups of HAs synthesized by polymerization of phenolic compounds as cata-lyzed by Al_{13} tridecamer species resemble those of natural HA. The ESR spectra are devoid of hyperfine splitting indicating the formation of semiqui-nones. Availability of advanced instrumentation facilities such as CPMAS NMR, ESR, STM/AFM should advance the knowledge on metal oxide–humic acid interactions.

REFERENCES

1. Huang, P. M. and Violante, A. (1986). Influence of organic acids on crystallization and surface properties of precipitation products of aluminum. In *Interactions of Soil Minerals with Natural Organics and Microbes*, ed. Huang, P. M. and Schnitzer, M., SSSA Spec. Publ. No. 17, Soil Science Society of America, Madison WI, p. 159
2. Wefer, K. and Misra, C. (1987). *Oxides and hydroxides of aluminum*, Alcoa Technical Paper 19, 92 pp.
3. Hsu, P. H. (1989). Aluminum hydroxides and oxyhydroxides. In *Minerals in Soil Environments*, ed. Dixon, J. B. and Weed, S. B., 2nd edn, Soil Science Society of America, Book Ser. no.1, Madison, WI, p. 331.
4. Cornell, R. M. and Schwertmann, U. (1996). *The Iron Oxides: Structure, Properties, Reactions, Occurrences and Uses*, VCH, Weinheim, 573 pp.
5. Sposito, G. (1996). *The Environmental Chemistry of Aluminum*, 2nd edn, CRC Press, Lewis Publishers, London, 464 pp.
6. Kampf, N., Scheinost A. C. and Schultze D. G. (2000). Oxides minerals. In *Handbook of Soil Science*, ed. Sumner, M. E., CRC Press, Boca Raton, FL, p. F-125.
7. Sparks, D. L. (1995). *Environmental Soil Chemistry*, Academic Press, San Diego, CA, pp 267.
8. Sumner, M. E. (ed.). (2000). *Handbook of Soil Science*, CRC Press, Boca Raton, FL, pp. 2048.
9. Chao, G. Y., Baker, J., Sabina, A. P. and Roberts, A. C. (1985) Doyleite, a new polymorph of $Al(OH)_3$, and its relationship to bayerite, gibbsite and nordstrandite, *Can. Miner.*, 23, 21.
10. Rodgers, K. A. (1993). Routine identification of aluminium hydroxide polymorphs with the laser Raman microprobe, *Clay Miner.*, 28, 85.
11. Goldberg, S., Davis, J. A. and Hem, J. D. (1996). The surface chemistry of aluminum oxides and hydroxides. In *The Environmental Chemistry of Aluminum*, 2nd edn, ed. Sposito, G., Lewis Publishers, Boca Raton, FL, p. 271.
12. Jones, R. C., Hudnall, W. H. and Sakai, W. S. (1982). Some highly weathered soils of Puerto Rico. Part 2: Mineralogy, *Geoderma*, 27, 75.
13. Beinroth, F. H., Eswaran, H., Uehara, G. and Reich, P. F. (2000). Oxisols. In *Handbook of Soil Science*, ed. Sumner, M. E, CRC Press, Boca Raton, FL, p. E-373.
14. West, L. T. and Beinroth, F. H. (2000). Ultisols. In *Handbook of Soil Science*, ed. Sumner, M. E., CRC Press, Boca Raton, FL, p. E-358.
15. Kimble, J. M., Ping, C. I., Sumner, M. E. and Wilding L. P. (2000). Andisols. In *Handbook of Soil Science*, ed. Sumner, M. E., CRC Press, Boca Raton, FL, p. E-209.
16. Violante, A. and Jackson, M. L. (1979). Crystallization of nordstrandite in citrate systems and in the presence of montmorillonite. In *Proceedings of the VI International Clay Conference*, Oxford 1978, *Develop. Sedimentol.*, 27, 517.
17. Tait J. M., Violante, A. and Violante, P. (1983). Co-crystallization of gibbsite and bayerite with nordstrandite, *Clay Miner.*, 18, 95.
18. Violante, P., Violante, A. and Tait, J. M. (1982). Morphology of nordstrandite, *Clays Clay Miner.*, 30, 431.
19. Chao, G. Y. and Baker, J. (1982). Nordstrandite from Mont St-Hilaire, Quebec, *Can. Miner.*, 20, 77.
20. Pardo, S. E. M., Gardillo, J. R., Gallego, M. R. and Ramos, J. D. M. (1985) Nordstrandita, $Al(OH)_3$, en la bauxitas De Haro (La Riòja), *Bol. Soc. Esp. Min.*, 83.
21. Wilmot, R. D. and Young, B. (1985). Aluminite and other aluminium minerals from Newhaven, Sussex: the first occurrence of nordstrandite in Great Britain, *Proc. Geol., Ass.* 96, 47.

22. Violante, A. and Jackson, M. L. (1981). Clay influence on the crystallization of $Al(OH)_3$ polymorphs in the presence of citrate, sulfate or chloride, *Geoderma*, **25**, 199.

23. Violante, A. and Violante, P. (1980). Influence of pH, concentration and chelating power of organic anions on the synthesis of aluminum hydroxides and oxyhydroxides, *Clays Clay Miner.*, **28**, 425.

24. Schoen, R. and Roberson, C. E. (1970). Structures of aluminum hydroxide and geochemical implications. *Am. Miner.*, **55**, 43.

25. Ehrich, H. L., Wickert, L. M., Noteboom, D. and Doucet, J. (1995). Weathering of pisolitic bauxite by heterotrophic bacteria. In *Biohydrometallurgical Processing*, Vol. I, ed. Vargas, T., Jerez, C. A., Wiertz, J. V. and Toledo, H., University of Chile, p. 395.

26. Ehrich, H. L. and Wickert, L. M. (1997). Bacterial action on bauxites in columns fed with full-strength and dilute sucrose-mineral salts medium. In *Biotechnology and Mining Environment. Proceedings 13th Annu. Meet*, ed. Lortie, L., Bédard, P. and Gould, W. D., *BIOMINET*, Natural Resources Canada, SP 97–1, Ottawa, Canada, p. 74.

27. Tettenhorst, R. and Hofmann, A. (1980). Crystal chemistry of boehmite, *Clays Clay Miner.*, **28**, 373.

28. Violante, A. and Huang, P. M. (1984). Nature and properties of pseudoboehmites formed in the presence of organic and inorganic ligands, *Soil Sci. Soc. Am J.*, **48**, 1193.

29. Violante, A. and Huang, P. M. (1985). Influence of inorganic and organic ligands on precipitation products of aluminum, *Clays Clay Miner.*, **33**, 181.

30. Violante, A. and Huang, P. M. (1993). Formation mechanism of aluminum hydroxide polymorphs, *Clays Clay Miner.*, **41**, 590.

31. Music, S., Dragcevic, D. and Popovic, S. (1995a). Formation of boehmite via precipitation from aqueous solutions, *Mater. Lett.*, **24**, 59.

32. de Villiers, J. M. (1969). Pedosesquioxides—composition and colloidal interactions in soil genesis during the Quaternary. *Soil Sci.*, **107**, 454.

33. Huang, P. M. (1995). The role of short range ordered mineral colloids in abiotic transformation of organic components in the environment. In *Environmental Impact of Soil Component Interactions II Metals. Other Inorganics, and Microbial Activities*, ed. Huang, P. M., Berthelin, J., Bollag, J.-M., McGill, W. B., Page, A. L., CRC Press, Boca Raton, FL, p. 135.

34. Karltun, E., Bain, D. C., Gustafsson, J. P., Mannerkoski, H., Murad, E., Wagner, U., Fraser, A. R., McHardy W. J. and Starr, M. (2000). Surface reactivity of poorly-ordered minerals in podzol B horizons, *Geoderma*, **94**, 265.

35. Barnhisel, R. I. and Bertsch, P. M. (1989) Chlorite and hydroxy-interlayered vermiculite and smectite. In *Minerals in Soil Environments*, ed. Dixon, J. B. and Weed, S. B., Soil Science Society of America, Madison, WI, p. 729.

36. Wada, K. (1989). Allophane and imogolite. In *Minerals in Soil Environments*, ed. Dixon, J. B. and Weed, S. B., Soil Science Society of America, Madison, WI, p. 1051.

37. McFadden, L. D. and Hendricks, S. B. (1988). Changes in the content and composition of pedogenic iron oxyhydroxides in a chronosequence of soils in Southern California, *Q. Res.*, **23**, 189.

38. Fasiska, E. J. (1967). Structural aspects of the oxides and oxydehydrates of iron, *Corros. Sci.*, **7**, 833.

39. Lindsley, D. H. (1976). The crystal chemistry and structure of oxide minerals as exemplified by the Fe-Ti oxides. In *Oxide Minerals. Reviews in Mineralogy 3*, ed. Rumble III, D., Min. Society of America, Book Crafters, Chelsea, MI, p. L1.

40. Eggleton, R. A., Schulze, D. G. and Stucki, J. W. (1988). Introduction to crystal structures of iron containing minerals. In *Iron in Soils and Clay Minerals*, ed. Stucki, J. W., Goodman, B.A and U. Schwertmann, Nato ASI Ser. 217, D. Reidel, Dordrecht, The Netherlands, p.141.

41. Cornell, R. M. and Schwertmann, U. (1979). Influence of organic anions on the crystallization of ferrihydrite, *Clays Clay Miner.*, **27**, 402.

42. Reeves, N. J. and Mann, S. (1991). Influence of inorganic and organic additives on the tailored synthesis of iron oxides, *J. Chem. Soc. Faraday Trans. I*, **87**, 3875.

43. Fischer, W. R. and Schwertmann, U. (1975). The formation of haematite from amorphous iron (III)-hydroxide, *Clays Clay Miner.*, **23**, 33.

44. Cornell, R. M. (1985). Effect of simple sugars on the alkaline transformation of ferrihydrite into goethite and haematite, *Clays Clay Miner.*, **33**, 219.

45. Cornell, R. M., Schneider, W. and Giovanoli, R. (1991). Preparation and characterization of colloidal α-FeOOH with a narrow size distribution, *J. Chem. Soc. Faraday Trans. I*, **87**, 869.

46. Ardizzone, S. and Formaro, L. (1985). Hydrothernal preparation of goethite crystals, *Surf. Technol.*, **26**, 269.

47. Bye, G. C. and Howard, C. R. (1971). An examination of the nitrogen adsorption of the thermal decomposition of pure and silica doped goethite, *J. Appl. Chem. Biotechnol.*, **21**, 324.

48. Brauer, G. (1982). *Handbuch der Preparativen Anorganischen Chemie*, Band 3, F. Enke, Stuttgart.

49. Cornell, R. M. and Giovanoli, R (1988) Acid dissolution of akaganite and lepidocrocite: the effect on crystal morphology, *Clays Clay Miner.*, **36**, 385.

50. Liu, C. and Huang, P. M. (1999). Atomic force microscopy and surface characteristics of iron oxides formed in citrate solutions, *Soil Sci. Soc. Am. J.*, **63**: 65.

51. Nightingale, E. R. and Benck, R. F. (1960). Precipitation of crystalline iron(III) oxide from homogeneous solution, *Anal. Chem.*, **32**, 566.

52. Cornell, R. M. and Giovanoli, R (1985) Effect of solution conditions on the proportion and morphology of goethite formed from ferrihydrite, *Clays Clay Miner.*, **33**, 424.

53. Towe, K. M. and Bradley, W. F. (1967). Mineralogical constitution of colloidal 'hydrous ferric oxides', *J. Colloid Interface Sci.*, **24**, 384.

54. Chukrov, F. V., Zoyagin, B. B., Gorshkov, A. I., Yermilova, L. P. and Balashova, V. V. (1973). Ferrihydrite, *Int. Geol. Rev.*, **16**, 1131.

55. Russell, J. D. (1979). Infrared spectroscopy of ferrihydrite: evidence for the presence of structural hydroxyl groups, *Clay Miner.*, **14**, 109.

56. Eggleton, R. A. and Fitzpatrick, R. W. (1989). New data and a revised structural model for ferrihydrite, *Clays Clay Miner.*, **36**, 111.

57. Eggleton, R. A. and Fitzpatrick, R. W. (1990). New data and a revised structural model for ferrihydrite. A reply, *Clays Clay Miner.*, **38**, 335.

58. Manceau, A., Combes, J. M. and Calas, G. (1990). New data and a revised structural model for ferrihydrite: Comment, *Clays Clay Miner*, **38**, 331.

59. Zhao, J., Huggins, F. E., Feng, Z. and Huffman, G. P. (1994). Ferrihydrite: Surface structure and its effects on phase transformation, *Clays Clay Miner.*, **42**, 737.

60. Drits, V. A., Sakharov, B. A., Salyn, A. L. and Manceau, A. (1993). Structural model for ferrihydrite, *Clay Miner.*, **28**, 185.

61. Manceau, A. and Drits, V. A. (1993). Local structure of ferrihydrite and ferroxyhite by EXAFS spectroscopy, *Clay Miner.*, **28**, 165.

62. Manceau, A. and Gates, W. P. (1997). Surface structural model for ferrihydrite, *Clays Clay Miner.*, **45**, 448.

63. Jambor, J. J. (1998). Occurrence and constitution of natural and synthetic ferrihydrite, a widespread iron oxyhydroxide, *Chem. Rev.*, **98**, 2549.
64. Bernal, J. D., Dasgupta, D. R. and Mackay, A. L. (1959). The oxides and hydroxides of iron and their structural interrelationships, *Clay Miner. Bull.*, **4**, 15.
65. Trolard, F., Genin, J-M. R., Abdelmoula, M., Bourrie, G., Humbert, B. and Herbillon, A. (1997). Identification of a green rust mineral in a reductomorphic soil by Mossbauer and Raman spectroscopies, *Geochim. Cosmochim. Acta*, **61**, 1107.
66. Bigham, J. M., Schwertmann, U., Carlson, L. and Murad, E. (1990). A poorly crystallized oxyhydroxysulfate of iron formed by bacterial oxidation of Fe(II) in acid mine waters, *Geochim. Cosmochim. Acta*, **54**, 2743.
67. Bigham, J. M., Schwertmann, U., Traina, S. J., Winland, R. L., and Wolf, M. (1996) Schwertmannite and the chemical modeling of iron in acid sulfate waters, *Geochim. Cosmochim. Acta*, **60**: 2111.
68. Stevenson, F. J., (1994). *Humus Chemistry.Genesis, Composition, Reactions*, 2nd edn, John Wiley and Sons, New York, NY 498.
69. Piccolo, A., (1996). Humus and soil conservation. In *Humic Substances in Terrestrial Ecosystems*, ed Piccolo, A., Elservier, Amsterdam, p. 225
70. Schulten, H. R. and Leinweber, P. (2000). New insights into organic-mineral particles: composition, properties and models of molecular structure, *Biol. Fertil. Soils*, **30**, 399.
71. Schnitzer, M. and Schulten, H. R. (1995). Analysis of organic matter in soil extracts and whole soils by pyrolysis-mass spectrometry, *Adv. Agr.*, **55**, 168.
72. Rovira, A. D. (1969). Plant root exudates, *Bot. Rev.*, **35**, 35.
73. Lynch, J. M., ed. (1990). *The Rhizosphere*, John Wiley & Sons, Chichester, pp 460.
74. Marschner, H. (1995). *Mineral Nutrition of Higher Plants*, 2nd edn, Academic Press, London, pp 889.
75. Tan, K. H. (1986). Degradation of soil minerals by organic acids. In *Interactions of Soil Minerals with Natural Organics and Microbes*, ed. Huang, P. M. and Schnitzer, M., Spec. Publ. No. 17. Soil Science Society of America, Madison WI, p. 1.
76. Robert, M. and Chenu, C. (1992). Interactions between soil minerals and microorganisms. In *Soil Biochemistry*, ed. Stotsky, G. and Bollag, J.-M., Marcell Dekker, New York, p. 307
77. Lundstrom, U. S., van Breemen, N. and Bain, D. (2000). The podsolization process. A review, *Geoderma*, **94**, 91.
78. Churchman, G. J. (2000). The alteration and formation of soil minerals by weathering. In *Handbook of Soil Science*, ed. Sumner, M. E., CRC Press, Boca Raton, FL, p. F-3–F76.
79. Robert, M. and Berthelin, J. (1986). Role of biological and biochemical factors in soil mineral weathering. In *Interaction of Soil Minerals with Natural Organics and Microbes*, ed. Huang, P.M and Schnitzer, M., SSSA Spec. Publ. 17, Soil Science Society of America, Madison, WI, p. 453
80. Vance, G. F., Stevenson, F. J. and Sikora, F. J. (1996). Environmental chemistry of aluminum–organic complexes. In *The Environmental Chemistry of Aluminum*, ed. Sposito, G., CRC Press, Lewis, Boca Raton, FL, p.169.
81. Mokma, D. L. and Evans, C. V. (2000). Spodosols. In *Handbook of Soil Science*, ed. Sumner, M. E., CRC Press, Boca Raton, FL, p. E-307.
82. Jones, D. (1988). Lichens and pedogenesis. In *Handbook of Lichenology*, ed. Galun, M., CRC Press, Boca Raton, FL, p. 109.
83. Wilson, M. J. (1995). Interactions between lichens and rocks: a review, *Cryptogamic Bot.*, **5**, 299.
84. Adamo, P. and Violante, P. (2000) Weathering of rocks and neogenesis of minerals associated with lichen activity, *Appl. Clay Sci.*, **16**, 229.

85. April, R. and Keller, D. (1990a). Interactions between minerals and roots in forest soils. *Proc 9th Int. Clay Conf.*, ed. Farmer, V. C. and Tardy, Y., *Sci. Geol. Mem.*, **85**, 89.

86. April, R. and Keller, D. (1990b). Mineralogy of the rhizosphere in forest soils of the eastern United States, *Biogeochemistry*, **9**, 1.

87. Violante, A. and Gianfreda, L. (2000). Role of biomolecules in the formation of variable-charge minerals and organo-mineral complexes and their reactivity with plant nutrients and organics in soil. In *Soil Biochemistry*, Vol. 10, ed. Bollag, J.-M. and Stotzky, G., Marcell Dekker, New York, p. 207.

88. Bertsch, P. M. and Parker, D. R. (1996). Aqueous polynuclear aluminum species. In *The Environmental Chemistry of Aluminum*, ed. Sposito, G., 2nd edn, Lewis Boca Raton, FL, p. 117.

89. May, H. M. and Nordstrom, D. K. (1991). Assessing the solubilities and reaction kinetics of aluminous minerals in soils. In *Soil Acidity*, ed. Ulrich, B. and Sumner, M. E., Springer-Verlag, Berlin p. 125.

90. Akitt, J. W. (1989). Multinuclear studies of aluminum compounds, *Prog. NMR Spectrosc.*, **21**, 1.

91. Hunter, D. and Ross, D. S. (1991). Evidence for a phytotoxic hydroxy-aluminum polymer in organic soil horizons, *Science*, **251**, 1056.

92. Krishnamurti, G. S. R., Wang, M. K. and Huang, P. M. (1999a). Role of tartaric acid in the inhibition of the formation of Al_{13} tridecamer using sufate precipitation, *Clays Clay Miner.*, **47**, 658.

93. Krishnamurti, G. S. R., Wang, M. K. and Huang, P. M. (1999a). Pyrogallol inhibition of Al_{13} tridecamer formation and the synthesis of humic substances. *Proceedings 5th ICOBTE, Vienna, July 11–15, 1999*, International Society for Trace Element Research, Vienna, Austria, p. 742.

94. Hiradate, S., Taniguchi, S. and Sakurai, K. (1998) Aluminum speciation in aluminum-silica solutions and potassium chloride extracts of acidic soils, *Soil Sci. Soc. Am. J.*, **62**, 630.

95. Turner, R. C. (1969). Three forms of aluminum in aqueous system determined by 8–quinolinolate extraction method, *Can J. Chem.*, **47**, 2521.

96. Smith, R. W. (1971). Relations among equilibrium and nonequilibrium aqueous species of aluminum hydroxy complexes, *Am. Chem Soc. Adv. Chem.*, **106**, 250.

97. Hsu, P. H. (1988). Mechanisms of gibbsite crystallization from partially neutralized aluminum chloride solution, *Clays Clay Miner.*, **36**, 25.

98. Okura, T., Goto, K. and Yotuyanagi, T. (1962). Forms of aluminum determined by an 8-quinolinolate extraction method, *Am. Chem.*, **34**, 581.

99. Bloom, P. R. and Erich, M. S. (1996). The quantitation of aqueous aluminum. In *The Environmental Chemistry of Aluminum*, 2nd edn, ed. Sposito G., Lewis Publisher, New York, Inc, Boca Raton, FL, p. 1.

100. Jardine, P. M. and Zelazny, L. W. (1989). A speciation method for partitioning mononuclear and polynuclear aluminum using ferron. In *Environmental Chemistry and Toxicology of Aluminum* Chapter 2., ed. Lewis, T. E., Lewis Pub, Michigan,., p 19.

101. Kwong, N. K. K. F. and Huang, P. M. (1977). Influence of citric acid on the hydrolytic reactions of aluminum. *Soil Sci. Soc. Am. J.*, **41**, 692.

102. Di Pascale, G. and Violante, A. (1986). Influence of phosphate ions on the extraction of aluminum by 8-hydroxyquinoline from OH-Al suspensions, *Can. J. Soil Sci.*, **66**, 573.

103. Bye, G. C. and Robinson, J. G. (1974). The nature of pseudoboehmite and its role in the crystallization of amorphous aluminum hydroxide, *J. Appl. Chem. Biotechnol.*, **24**, 633.

A. VIOLANTE, G. S. R. KRISHNAMURTI AND P. M. HUANG   181

104. Kodama, H. and Schnitzer, M. (1980). Effect of fulvic acid on the crystallization of aluminum hydroxides, *Geoderma*, **24**, 195.
105. Singer, A. and Huang, P. M. (1990). The effect of humic acid on the crystallization of precipitation products of aluminum, *Clays Clay Miner.*, **38**, 47.
106. Nagy, K. L., Cygan, R. T., Hanchar, J. M. and Sturchio, N. C. (1999) Gibbsite growth kinetics on gibbsite, kaolinite, and muscovite substrates: Atomic force microscopy evidence for epitaxy and an assessment of reactive surface area, *Geochim. Cosmochim. Acta*, **63**, 2337.
107. Wall, J. R. D., Wolfenden, E. B., Beard, E. H. and Deans, T. (1962). Nordstrandite in soil from West Sarawak, Borneo, *Nature (London)*, **196**, 264.
108. Hathaway, J. C. and Schlanger, S. O. (1962). Nordstrandite from Guam, *Nature (London)*, **196**, 265.
109. Milton, C., Dwornik, E. J. and Finkelman, R. B. (1975). Nordstrandite Al(OH)$_3$, from the Green River Formation in Rio Blanco County, Colorado, *Am. Miner.*, **60**, 285.
110. Violante, A., Gianfreda, L. and Violante, P. (1993). Effect of prolonged aging on the transformation of short-range ordered aluminum precipitation products formed in the presence of organic and inorganic ligands, *Clays Clay Mine.*, **41**, 353.
111. Violante, A. and Huang, P. M. (1994). Identification of pseudoboehmite in mixtures with phyllosilicates, *Clay Miner.*, **29**, 351.
112. Kawano, M. and Tomita, K. (1997). Experimental study on the formation of zeolites from obsidian by interaction with NaOH and KOH at 150 and 200 °C, *Clays Clay Miner.*, **45**, 365.
113. Kawano, M. and Tomita, K. (1996). Amorphous aluminum hydroxide formed at the earliest weathering stages of K-feldspar, *Clays Clay Miner.*, **44**, 672.
114. Kawano, M. Tomita, K. and Shinohara, Y. (1997). Analytical electron microscopic study of the noncrystalline products formed at early weathering stages of volcanic glass, *Clays Clay Miner.*, **45**, 440.
115. Murphy, P. J., Posner, A. M. and Quirk, J. P. (1976a). Characterization of partially neutralized ferric nitrate solutions, *J. Colloid Interface Sci.*, **56**, 270.
116. Murphy, P. J., Posner, A. M. and Quirk, J. P. (1976b). Characterization of partially neutralized ferric chloride solutions, *J. Colloid Interface Sci.*, **56**, 284.
117. Murphy, P. J., Posner, A. M. and Quirk, J. P. (1976c). Characterization of partially neutralized ferric perchlorate solutions, *J. Colloid Interface Sci.*, **56**, 298.
118. Schneider, W. and Schwyn, B. (1987). The hydrolysis of iron in synthetic, biological, and aquatic media. In *Aquatic Surface Chemistry*, ed. Stumm, W., Wiley Interscience, New York, p. 167.
119. Schwertmann, U., Friedl, J. and Stanjek, H. (1999). From Fe(III) ions to ferrihydrite and then to hematite, *J. Colloid Interface Sci.*, **209**, 215.
120. Hsu, P. H. and Wang, M. K. (1980). Crystallization of goethite and haematite at 70 °C, *Soil Sci. Soc. Am. J*, **44**, 143.
121. Schwertmann, U. and Taylor, R. M. (1989). Iron oxides. In *Minerals in Soil Environments*, ed. Dixon, J. B. and Weed, S. B., 2nd edn, SSSA Book Series No. 1., Soil Science Society of America, Madison, WI, p. 379.
122. Cornell, R. M., Giovanoli, R. and Schindler, P. W. (1987). Effect of silicate species on the transformation of ferrihydrite into goethite and haematite in alkaline media, *Clays Clay Miner.*, **35**, 21.
123. Cornell, R. M. and Schwertmann, U. (1979). Influence of organic anions on the crystallization of ferrihydrite, *Clays Clay Miner.*, **27**, 402.
124. Cornell, R. M. and Schindler, P. W. (1980). Infrared study of the adsorption of hydroxycarboxylic acids on α-FeOOH and amorphous Fe(III) hydroxide, *Colloid Polym. Sci.*, **258**, 1171.

125. Cornell, R. M. (1987). Comparison and classification of the effects of simple ions and molecules upon the transformation of ferrihydrite into more crystalline products, *Z. Pflanzenernahr. Bodenkd.*, **150**, 304.
126. Kodama, H. and Schnitzer, M. (1977). Effect of fulvic acid on the crystallization of Fe(III) oxides, *Geoderma*, **19**, 279.
127. Susser, P. and Schwertmann, U. (1983). Iron oxide mineralogy of ochreous deposits in drain pipes and ditches, *Z. Kulturtech. Flurbereinig*, **24**, 389.
128. Maeda, H. and Maeda, Y. (1996). Atomic force microscopic studies for investigating the smectitic structures of colloidal crystals of β-FeOOH, *Langmuir*, **12**, 1446.
129. Holm, N. G., Wadsten, T. and Dowler, M. J. (1982). β-FeOOH (akaganeite) in Red Sea brine, *Estud. Geol. (Madrid)*, **38**, 367.
130. Music, S., Orehovec, Z., Popovic, S. and Czako-Nagy, I. (1994). Structural properties of precipitates formed from hydrolysis of Fe^{3+} ions in $Fe_2(SO_4)_3$ solutions, *J. Mater. Sci.*, **29**, 1991.
131. Music, S., Santana, G. P., Smit, G. and Garg, V. K. (1999). [57]Fe Mossbauer, FTIR and TEM observations of oxide phases precipitated from concentrated $Fe(NO_3)_3$ solutions, *Croatica Chim. Acta*, **72**, 87.
132. Gotic, M., Popovic, S., Ljubesic, N. and Music, S. (1994). Structural properties of precipitates formed by hydrolysis of Fe^{3+} ions in aqueous solutions containing NO_3^- and Cl^- ions, *J. Mater. Sci.*, **29**, 2474.
133. Music, S., Gotic, M. and Ljubesic, N. (1995). Influence of sodium polyenathol sulphonate on the morphology of β-FeOOH particles obtained from the hydrolysis of a $FeCl_3$ solution. *Mater. Lett.*, **25**, 69.
134. Saric, A., Music, S., Nomura, K. and Popovic, S. (1998). Influence of urotropin on the precipitation of iron oxides from $FeCl_3$ solutions, *Croatica Chim. Acta*, **71**, 1019.
135. Karathanasis, A. D. and Thompson, Y. L. (1995). Mineralogy of iron precipitates in a constructed acid mine drainage wetland, *Soil Sci. Soc. Am J.*, **59**, 1773.
136. Schwertmann, U, Friedl, J. and Pfab, G. (1996). A new iron(III) oxyhydroxynitrate, *J. Soil State Chem.*, **126**, 336.
137. Ottow, J. C. G. and Glathe, H. (1971). Isolation and identification of iron-reducing bacteria from gley soils, *Soil Biol. Biochem.*, **3**, 43.
138. Munch, J. C. and Ottow, J. C. G. (1983). Reductive transformation mechanism of ferric oxides in hydromorphic soils, *Ecol. Bull.*, **35**, 383.
139. Fischer, W. R. and Pfanneberg, T. (1984). An improved method for testing the rate of iron(III) oxide reduction by bacteria, *Zbl. Mikrobiol.*, **139**, 169.
140. Fischer, W. R. (1987). Standard potentials (Eo) of iron(III) oxides under reducing conditions, *Z. Pflanzenernahr. Bodenkd.*, **150**, 286.
141. McKnight, D. M., Kimball, B. A. and Bencala, K. E. (1988). Iron photoreduction and oxidation in an acidic mountain stream, *Science*, **240**, 637.
142. Siffert, C. and Schulzberger, B. (1991). Light-induced dissolution of haematite in the presence of oxalate: A case study, *Langmuir*, **7**, 1627.
143. Pehkonen, S. O., Siefert, R., Erel, Y., Webb, S. and Hoffman, M. R. (1993). Photoreduction of oxyhydroxides in the presence of important atmospheric organic compounds, *Environ. Sci. Technol.*, **27**, 2056.
144. Stumm, W. and Lee G. F. (1961). Oxygenation of ferrous iron, *Ind. Eng. Chem.*, **53**, 143.
145. Ghosh, M. M. (1976). Oxygenation of ferrous iron in highly buffered waters. In *Aqueous Environmental Chemistry of Metals*, ed. Rubin, A. J., Ann Arbor Science, Ann Arbor, MI, p. 193.
146. Tamura, H., Gato, K. and Nagayama, M. (1976). Effect of anions on the oxygenation of ferrous iron in neutral solutions, *J.Inorg. Nucl. Chem.*, **38**, 113.

147. Sung, W. and Morgan, J. J. (1980). Kinetics and products of ferrous iron oxygenation in aqueous systems, *Environ. Sci. Technol.*, **14**, 561.
148. Davidson, W. and Seed, G. (1983). The kinetics of the oxidation of ferrous iron in synthetic and natural waters, *Geochim. Cosmochim. Acta*, **47**, 67.
149. Roekens, E. J. and Van Grieken, R. E. (1983). Kinetics of iron (II) oxidation in sea water of various pH, *Mar. Chem.*, **13**, 195.
150. Millero, F. J., Sotolongo, S. and Izaguirre, M. (1987). The oxidation kinetics of Fe(II) in sea water, *Geochim. Cosmochim. Acta*, **51**, 793.
151. Von Gunten, U. and Schneider, W. (1991). Primary products of oxygenation of iron(II) at an oxic/anoxic boundary: nucleation, agglomeration and ageing, *J. Colloid Interface. Sci.*, **145**, 127.
152. Liang, L., McNabb, J. A., Paulk, J. M., Gu, B. and McCarthy, J. F. (1993). Kinetics of Fe(II) oxygenation at low partial pressures of oxygen in the presence of natural organic matter, *Environ. Sci. Technol.*, **27**, 1864.
153. Krishnamurti, G. S. R. and Huang, P. M. (1990). Kinetics of Fe(II) oxygenation and the nature of hydrolytic products as influenced by ligands, *Sci. Geol. Mem.*, **85**, 195.
154. Theis, T. L. and Singer, P. C. (1974). Complexation of Fe(II) by organic matter and its effect on iron (II) oxygenation, *Environ. Sci. Technol.*, **8**, 569.
155. Krishnamurti, G. S. R. and Huang, P. M. (1991). Influence of citrate on the kinetics of Fe(II) oxygenation and the formation of iron oxyhydroxides, *Clays Clay Miner.*, **39**, 28.
156. Kiyama, M. (1969). Commentary experiments on the formation of Fe_3O_4 precipitates from aqueous solutions, *Bull. Inst. Chem. Res. Kyoto Univ.*, **47**, 607.
157. Misawa, T., Hashimoto, K. and Shimodaira, S. (1973). Formation of Fe(II)-Fe(III) green complex on oxidation of ferrous iron in neutral and slightly alkaline sulphate solution, *J.Inorg. Nucl. Chem.*, **35**, 4167.
158. Misawa, T., Hashimoto, K., Suetaka, W. and Shimodaira, S. (1973). Formation of Fe(II)-Fe(III) green complex on oxidation of ferrous iron in perchloric acid solution, *J.Inorg. Nucl. Chem.*, **35**, 4159.
159. Misawa, T., Hashimoto, K. and Shimodaira, S. (1974). The mechanism of formation of iron oxyhydroxides in aqueous solutions at room temperature, *Corros. Sci.*, **14**, 131.
160. Feitknecht, W. and Keller, G. (1950). Uber die dunkelgrunen Hydroxyverbindungen des Eisens, *Z. Annorg. Allg. Chem.*, **262**, 61.
161. Feitknecht, W. (1959). Uber die oxydation von festen Hydroxyverbindungen des Eisens in wassinrigen Losungen, *Z. Elektrochem.*, **63**, 34.
162. Schwertmann, U. and Thalmann, H. (1976). The influence of Fe(II), Si and pH on the formation of lepidocrocite and ferrihydrite during oxidation of aqueous $FeCl_2$ solutions, *Clay Miner.*, **11**, 189.
163. Lewis, D. G. (1997). Factors influencing the stability and properties of green rusts, *Adv. EcoGeol.*, **30**, 345.
164. Derie, R. and Ghodsi, M. (1972). Contribution a l'etude de la formation des sesquioxydes de la Fe(III) monohydrates par aeration de gels d'hydroxyde de Fe(II). *Ind. Chim. Belge*, **37**, 731.
165. Detournay, J., Ghodsi, M. and Derie, R. (1974). Etude cinetique de la formation de goethite par aeration de gels d'hydroxyde ferreux, *Ind. Chim. Belge*, **39**, 695.
166. Detournay, J., Ghodsi, M. and Derie, R. (1975). Influence de la temperature et de la presence des ions etrangers sur la cinetique et le mecanisme de formation de la goethite en milieu aqueux, *Z. Annorg. Allg. Chem.*, **412**, 184.
167. Detournay, J. Derie, R. and Ghodsi, M. (1976). Etude de l'oxydation par aeration de $Fe(OH)_2$ en mileue chlorure, *Z. Annorg. Allg. Chem.*, **427**, 267.

168. Taylor, R. M. (1984). Influence of chloride on the formation of iron oxides from Fe(II) chloride. Effect of (Cl) on the formation of lepidocrocite and its crystallinity, *Clays Clay Miner.*, **32**, 175.
169. Schwertmann, U. and Fechter, H. (1994). The formation of green rust and its transformation to lepidocrocite, *Clay Miner.*, **29**, 87.
170. Schwertmann, U. and Taylor, R. M. (1972). The transformation of lepidocrocite to goethite, *Clays Clay Miner.*, **20**, 151.
171. Schwertmann, U. and Taylor, R. M. (1972). The influence of silicate on the transformation of lepidocrocite to goethite, *Clays Clay Miner.*, **20**, 159.
172. Feitknecht, W., Hani, H. and Dvorak, V. (1969). The mechanism of the transformation of δ-FeOOH to α-Fe$_2$O$_3$. In *Reactivity of Solids*, ed. Mitchell, J. W., DeVries, R. C. and Roberts, R. W., John Wiley & Sons, New York., p. 237.
173. Carlson, L. and Schwertmann, U. (1980). Natural occurrence of feroxyhyte (δ'-FeOOH). *Clays Clay Miner.*, **28**, 272.
174. Chukrov, F. V., Zoyagin, B. B., Yermilova, L. P. and Gorshkov, A. I. (1976). Mineralogical criteria in the origin of marine iron-manganese nodules, *Miner. Deposits*, **11**, 24.
175. Krishnamurti, G. S. R. and Huang, P. M. (1993). Formation of lepidocrocite from Fe(II) solutions: Stabilization by citrate, *Soil Sci. Soc. Am. J.*, **57**, 861.
176. Vempati, R. K., Kollipara, K. P., Stucki, J. W. and Wilkinson, H. T. (1990). *Agron. Abs.*, ASA, Madison, WI, p.353.
177. Fitzpatrick, R. W., Taylor, R. M., Schwertmann, U. and Childs, C. W. (1985). Occurrence and properties of lepidocrocite in some soils of New Zealand, South Africa and Australia. *Aust. J. Soil Res.*, **23**, 543.
178. Stumm, W. (1995). The inner-sphere surface complex: A key to understanding surface reactivity. In *Aquatic Chemistry. Interfacial and Interspecies Processes*, ed. Huang, C. P., O'Melia C. R. and Morgan, J. J., American Chemical Society, Washington, DC, pp. 1–3
179. Stumm, W., Furrer, G., Wieland, E. and Zinder, B. (1985). The effect of complex-forming ligands on the dissolution of oxides and aluminosilicates. In *The Chemistry of Weathering*, ed. Drever, J. I., D. Reidel, Dordrecht, The Netherlands, p. 55
180. Emery, T. (1978). The storage and transport of iron. In *Metals in Biological Systems*, ed. Sigel, H., Marcel Dekker, Basel, p. 77.
181. Marschner, H., Romheld, V. and Kissel, M. (1986). Different strategies in higher plants in mobilization and uptake of iron, *J. Plant. Nutr.*, **9**, 695.
182. Watteau, F. and Berthelin, J. (1990). Iron solubilization by mycorrhizal fungi producing siderophores, *Symbiosis*, **9**, 59.
183. Watanabe, S. and Matsumoto, S. (1994). Effect of monosilicate, phosphate, and carbonate on iron dissolution by mugeneic acid, *Soil Sci. Plant Nutr.*, **40**, 9.
184. Hersman, L., Lloyd, T. and Sposito, G. (1995). Siderophore-promoted dissolution of hematite, *Geochim. Cosmochim. Acta*, **59**, 3327.
185. LaKind, J. S. and Stone, A. T. (1989). Reductive dissolution of goethite by phenolic reductants, *Geochim. Cosmochim. Acta*, **53**, 961.
186. Waite, W. D. (1986). Photoredox chemistry of colloidal metal oxides. In *Geochemical Processes at Mineral Surfaces*, ed. Davis, J. A.and Hayes, K. F., ACS Symp. Ser. 323, American Chemical Society, Washington, DC, p. 426.
187. Waite, W. D. and Morel, F. M. M. (1984). Photoreductive dissolution of colloidal iron oxide: effect of citrate, *J. Colloid Interface Sci.*, **102**, 121.
188. Cornell, R. M. and Schindler, P. W. (1987). Photochemical dissolution of goethite in acid/oxalate solution, *Clays Clay Miner.*, **35**, 347.

189. Zinder, B., Furrer, G. and Stumm, W. (1986). The coordination chemistry of weathering: II. Dissolution of Fe(III) oxides, *Geochim. Cosmochim. Acta*, **50**, 1861.
190. Stumm, W. and Fuhrer, G. (1987). The dissolution of oxides and aluminum silicates: Examples of surface-coordination-controlled kinetics. In *Aquatic Surface Chemistry*, ed. Stumm, W., John Wiley & Sons, New York, p. 197.
191. Banwart, S., Davies, S. and Stumm, W. (1989). The role of oxalate in accelerating the reductive dissolution of haematite by ascorbate, *Colloids Surf.*, **39**, 303.
192. Lovely, D. R. (1995). Microbial reduction of iron, manganese, and other metals, *Adv. Agron.*, **54**, 175.
193. Jackson, T. A. and Keller, W. D. (1970). A comparative study of the role of lichens and 'inorganic' processes in the chemical weathering of recent Hawaiian Lava flows, *Am. J. Sci.*, **269**, 446.
194. Jones, D., Wilson, M. J. and Tait, J. M. (1980). Weathering of a basalt by *Pertusaria corallina. Lichenologist*, **12**, 277.
195. Ivarson, K. C. and Sojak, M. (1978). Microorganisms and ochre deposits in field drains of Ontario, *Can. J. Soil Sci.*, **58**, 1.
196. Houot, S., Cestre, T. and Berthelin, J. (1984). Origine du fer et conditions de formation du colmatage ferrique. Etudes de differentes situations en France.. In *Proc. 11th Congr. Inter. Irrigations and Drainage, Fort Collins, CO*, p. 151
197. Chen, C., Dixon, J. B. and Turner, F. T. (1980). Iron coatings on rice roots. Morphology and modes of development, *Soil Sci. Soc. Am. J.*, **44**, 1113.
198. Kullman, K. H. and Schweisfurth, R. (1978). Iron-oxidizing rod-shaped bacteria. (II) Quantitative study of metabolism and iron oxidation using iron (II) oxalate, *Allg. Mikrobiol.*, **18**, 321.
199. Harrison, A. P.Jr. (1984). The acidophilic thiobacilli and other acidophilic bacteria that share their habitat, *Annu. Rev. Microbiol.*, **38**, 265.
200. Schwertmann, U., Kodama, H. and Fisher, W. R. (1986). Mutual interactions between organics and iron oxides. In *Interaction of Soil Minerals with natural Organics and Microbes*, ed. Huang, P. M. and Schnitzer, M., SSSA Spec. Pub. 17, Soil Science Society of America, Madison, WI, p. 223.
201. Blakemore, R. P., Short, K. A., Bazylinski, D. A., Rosenblatt, C. and Frankel, R. B. (1985). Microaerobic conditions are required for magnetite formation within *Aquasprillum magnetotacticum. Geomicrobiol. J.*, **4**, 53.
202. Bazylinski, D. A., Frankel, R. B. and Jannasch, H. W. (1988). Anaerobic magnetite production by a marine, magnetotactic bacterium, *Nature (London)*, **334**, 518.
203. Moskowitz, B. M., Frankel, R. B., Flanders, P. J., Blakemore, R. P. and Schwartz, B. B. (1988). Magnetic properties of magnetotactic bacteria, *J. Magn. Magn. Mater.*, **73**, 273.
204. Moskowitz, B. M., Frankel, R. B., Bazylinski, D. A., Jannasch, H. W. and Lovely, D. R. (1989). A comparison of magnetite particles produced anaerobically by magnetotactic and dissimilatory iron-reducing bacteria, *Geophys. Res. Lett.*, **16**, 665.
205. Fassbinder, J. W. E., Stanjek, H. and Vali, H. (1990). Occurrence of magnetic bacteria in soil, *Nature (London)*, **343**, 161.
206. Frankel, R. B. and Bakemore, R. P. (ed.) (1990). *Iron Biominerals*. Plenum Press, New York.
207. Fassbinder, J. W. E. and Stanjek, H. (1993). Occurrence of bacterial magnetite in soils from archeological sites, *Archeol. Pol.*, **31**, 117.
208. Lowenstam, H. A. (1981). Minerals formed by organisms, *Science*, **211**, 1126.
209. Houot, S. (1984). *Mise en evidence des principaux mechanismes de formation du colmatage par le fer des reseaux de drainage agricol en France*, These, INAPG, Paris.

210. Williams, R. J. P. (1991). Biominerals and homeostasis. In *Iron Biominerals*, ed. Frankel, R. B. and Bakemore, R. P. Plenum Press, New York, p. 7.
211. Skinner, H. G. W. and Fitzpatrick, R. W. (ed.) (1992). *Biomineralization Processes of Iron and Manganese*, Catena Verlag, Cremlingen-Destedt, pp. 432.
212. Schwertmann, U. (1985). The effect of pedogenic environments on iron oxide minerals, *Adv. Soil Sci.*, **1**, 172.
213. Rengasamy, P. and Oades, J. M. (1979). Interaction of monomeric and polymeric species of metal ions with clay surfaces. IV Mixed systems of aluminium and iron (III). *Aust. J. Soil Res.*, **17**, 141.
214. Goh, T. B., Huang, P. M., Dudas, M. J. and Pawluk, S. (1987). Effect of iron on the nature of precipitation products of aluminum, *Can. J. Soil Sci.*, **67**, 135.
215. Colombo, C. and Violante, A. (1996). Effect of time and temperature on the chemical composition and crystallization of mixed iron and aluminum species, *Clays Clay Miner.*, **44**, 113.
216. Violante, A., Colombo, C., Cinquegrani, G., Adamo, P. and Violante, P. (1998). Nature of mixed iron and aluminum gels as affected by Fe/Al molar ratio, pH and citrate, *Clay Miner.*, **33**, 511.
217. Krishnamurti, G. S. R., Violante, A. and Huang, P. M. (1995). Influence of Fe on the stabilization of hydroxy-Al interlayers in montmorillonite. In *Clays Controlling the Environment*, ed. Churchman, G. J., Fitzpatrick, R. W. and Eggleton, R. A., *Proc. 10th Int. Clay Conf. Adelaide, Australia*, CSIRO Publishing, Canberra, Australia, p. 183.
218. Schwertmann, U., Friedl, J., Stanjek, H. and Schulze, D. G. (2000). The effect of Al on Fe oxides. XIX. Formation of Al-substituted hematite from ferrihydrite at 25 °C and pH 4 and 7, *Clays Clay Miner.*, **48**, 159.
219. Colombo, C. and Violante A. (1997). Effect of ageing on the nature and interlayering of mixed hydroxy Al–Fe–montmorillonite complexes, *Clay Miner.*, **32**, 55.
220. Schwertmann, U., Cambier, P. and Murad, E. (1985). Properties of goethite of varying crystallinity, *Clays Clay Miner.*, **33**, 369.
221. Xue, J. and Huang, P. M. (1995). Zinc adsorption-desorption on short-range ordered iron oxides as influenced by citric acid during its formation, *Geoderma*, **64**, 343.
222. Zeltner, W. A., Yost, E. C., Machesky, M. L., Tejedor-Tajedor, M. I. and Anderson, M. A. (1986). Characterization of anion binding on goethite using titration colorimetry and cylindrical internal reflection-Fourier transformed inrared spectroscopy. In *Geochemical Processes at Mineral Surfaces*, ed. Davis, J. A.and Hayes, K. F., ACS Symp. Series 323, American Chemical Society Washington, DC, p. 142.
223. Violante, A. and Huang, P. M. (1989) Influence of oxidation treatments on surface properties and reactivity of pseudoboehmites formed in the presence of organic ligands, *Soil Sci. Soc. Am. J.*, **53**, 1402.
224. De Cristofaro, A., He, J. Z., Zhou, D. and Violante, A. (2000). Sorption of phosphate and tartrate on hydroxy-aluminum-oxalate coprecipitates, *Soil Sci. Soc. Am. J.*, **64** 1347.
225. Nagarajah, S., Posner, A. M. and Quirk, J. P. (1970). Competitive adsorption of phosphate with polygalacturonate and other organic anions on kaolinite and oxide surfaces. *Nature (London)*, **228**, 83.
226. Lopez-Hernandez, D., Siegert, G. and Rodriguez, J. V. (1986). Competitive adsorption of phosphate with malate and oxalate by tropical soils, *Soil Sci. Soc. Am. J.*, **50**, 1460.

227. Fox, T. R., Comerford, N. B. and McFee, W. W. (1990). Phosphorus and aluminum release from a Spodic horizon mediated by organic acids, *Soil Sci. Soc. Am. J.*, **54**, 1763.
228. Violante, A., Colombo, C., and Buondonno, A. (1991). Competitive adsorption of phosphate and oxalate by aluminum oxides, *Soil Sci. Soc. Am. J.*, **55**, 65.
229. Earl, K., Syers, J. and McLaughlin, R. (1979). Origin of the effect of citrate, tartrate and acetate on phosphate sorption by soils and synthetic gels, *Soil Sci. Soc. Am. J.*, **43**, 674.
230. Gerke, J., Romer, W. and Jungk, A. (1994). The excretion of citric and malic acid by proteoid roots of *Lupinus albus* L.; effects on soil solution concentrations of phosphate, iron, and aluminum in the proteoid rhizosphere in samples of an oxisol and a luvisol, *Z. Pflanzenernahr. Bodenkd.*, **157**, 289.
231. Inskeep, W. P., (1989). Adsorption of sulfate by kaolinite and amorphous iron oxide in the presence of organic ligands, *J. Environ. Qual.*, **18**, 379.
232. Liu, F., He, J., Colombo, C. and Violante, A. (1999). Competitive adsorption of sulfate and oxalate on goethite in the absence or presence of phosphate, *Soil Sci.*, **164**, 180.
233. Watson, J. R., Posner, A. M. and Quirk, J. P. (1973). Adsorption of the herbicide 2,4-D on goethite, *J. Soil Sci.*, **24**, 503.
234. Schwandt, H., Kogel-Knabner, I., Stanjek, H. and Totsche, K. (1992). Sorption of an acidic herbicide on synthetic iron oxides and soils: sorption isotherms, *Sci. Total Environ.*, **123/124**, 121.
235. Madrid, L. and Diaz-Barrientos, E. (1991). Effect of phosphate on the adsorption of 2,4-D on lepidocrocite, *Aust. J. Soil Res.*, **29**, 15.
236. McBride, M. B. (1989). Reactions controlling heavy metal solubility in soils, *Adv. Soil Sci.*, **10**, 1.
237. Kinniburgh, D. G. and Jackson, M. L. (1976). Adsorption of alkaline earth, transition and heavy metal cations by hydrous oxides gels of iron and aluminum, *Soil Sci. Soc. Am. J.*, **40**: 796.
238. Jackson, T. A. (1998). The Biogeochemical and Ecological Significance of Interactions between Colloidal Minerals and Trace Elements. In *Environmental Interactions of Clays*, ed. Parker, A. and Rae, J. E., Springer, Berlin, p. 93.
239. Manceau, A., Charlet, L., Boisset, M. C., Didier, B. and Spadini, L. (1992). Sorption and speciation of heavy metals on hydroxides of Fe and Mn oxides. From macroscopic to microscopic, *Appl. Clay Sci.*, **7**, 201.
240. Pickering, W. F. (1980). Cadmium retention by clays and other soil or sediment components. In *Cadmium in the Environment. Part I. Ecological Cycling*, ed. Nriagu, J. O., John Wiley & Sons, New York. p. 365.
241. Tam, S-C., Chow, A. and Hardley, D. (1995). Effects of organic component on the immobilization of selenium on iron oxyhydroxide, *Sci. Total Environ.*, **164**, 1.
242. Zhou, D., De Cristofaro, A., He, J. Z. and Violante, A. (1999). Effect of oxalate on adsorption of copper on goethite, bayerite and kaolinite: In *Clays for our future, Proc. 11th Int. Clay Conf., Ottawa, Canada, 1997*, ed. Kodama H., Mermut, A. R. and J. K., Torrance, p. 523.
243. Krishnamurti, G. S. R., Huang, P. M., Van Rees, K. C. J., Kozak, L. M. and Rostad, H. P. W. (1995). Speciation of particulate-bound cadmium of soils and its bioavailability, *Analyst*, **120**, 659.
244. Krishnamurti, G. S. R., Huang, P. M., Kozak, L. M., Rostad, H. P. W. and Van Rees, K. C. J. (1997). Distribution of cadmium in selected soil profiles of Saskatchewan, Canada: Speciation and bioavailability, *Can. J. Soil Sci.*, **77**, 613.

245. Krishnamurti, G. S. R., Huang, P. M., Van Rees, K. C. J., Kozak, L. M. and Rostad, H. P. W. (1997). Differential FTIR study of pyrophosphate extractable material of soils: implication in Cd-bonding sites and availability. In *Contaminated Soils*, ed. Prost, R., *Proc. 3rd ICOBTE, May 15–19, 1995*, Paris, France, INRA, Paris, CD-ROM 012pdf. 1–10.

246. Onyatta, J. O. and Huang, P. M. (1999). Chemical speciation and bioavailability index of cadmium for selected tropical soils in Kenya, *Geoderma*, **91**, 87.

247. Krishnamurti, G. S. R. and Naidu, R. (2000). Speciation and availability of cadmium in selected surface soils of South Australia, *Aust. J. Soil Res.* **38**, 991.

248. Krishnamurti, G. S. R. and Huang, P. M. (2001). The nature of organic matter of soils with contrasting cadmium phytoavailability, *Proc. 10th IHSS Conf., 20–25 Sep. 1998*, Adelaide, Australia. (in press)

249. Pierce, M. L. and Moore, C. B. (1982). Adsorption of arsenite and arsenate on amorphous iron hydroxide, *Wat. Res.*, **16**, 1247.

250. Mark, A., Merrill, D. T., McLearn, M. E., Winston, S., Fames, J., Kobayashi, S. and Martin, W. J. (1988). Trace elements including As, Be, Cd, Cr, Cu, Pb, Mo, Ni and Zn, can be removed from coal-fired power plant wastewaters by Fe oxyhydroxide adsorption. *49th Proc. Int. Water Conf. Eng. Soc. West Pa.*, p. 361.

251. Appleton, A. R., Papelis, C. and Leckie, J. O. (1989). Adsorptive removal of trace elements from coal fly-ash waste waters onto iron oxyhydroxide, *Proc. 43rd Purdue Industrial Waste Conf. 1988*, p. 375.

252. Carpenter, C., Sucui, D. and Wikoff, W. (1990). Sodium sulfide/ferrous sulfate metals treatment for hazardous waste minimization, *Proc. 44th Purdue Industrial Waste Conf., 1989*, p. 617.

253. Dayton, L. (1993). Magnets are attractive for quicker sewage cleaning, *New Scientist* (19 June 1993) p. 20.

254. Theng, B. K. G. (1979). *Formation and Properties of Clay–Polymer Complexes*. Elsevier, New York, p. 362

255. Stotzky G. (1986). Influence of soil mineral colloids on metabolic processes, growth, adhesion, and ecology of microbes and viruses. In *Interactions of Soil Minerals with Natural Organics and Microbes*, ed. Huang, PM and Schnitzer, M. Spec Publ No 17, Soil Science Society of America, Madison, WI, p. 305.

256. Venkateswarlu G and G Stotzky. (1992). Binding of the protoxin and toxin proteins of *Bacillus thuringiensis* subsp kurstaki on clay minerals, *Curr. Microbiol.*, **25**, 225.

257. Fusi, P. G., Ristori, G., Calamai, L. and Stotzky, G. (1989). Adsorption and binding of protein on 'clean' (homoionic) and 'dirty' (coated with Fe oxyhydroxides) montmorillonite, illite, and kaolinite, *Soil Biol Biochem.*, **21**, 911.

258. Shindo, H. and Huang, P. M. (1984). Catalytic effects of manganese (IV), iron (III), aluminum and silicon oxides on the formation of phenolic polymers, *Soil Sci. Soc. Am. J.*, **48**, 927.

259. Seebald, H. J. and Schunack, W. (1972). Reaktionen en Aluminiumoxiden, 4. Mit Umsetzungen von Acetophenon on Aluminiumoxide, *J. Chromatogr.*, **74**, 129.

260. Wang, T. S. C., Wang, M. C., Ferng, Y. L. and Huang, P. M. (1983). Catalytic synthesis of humic substances by natural clays, silts, and soils, *Soil Sci.*, **135**, 350.

261. Schnitzer, M. and Levesque, M. (1979). Electron spin resonance as a guide to the degree of humification of peats, *Soil Sci.*, **127**, 140.

262. Schnitzer, M., Barr, M. and Hortenstein, R. (1984). Kinetics and characteristics of humic acids produced from simple phenols, *Soil Biol. Biochem.*, **16**, 371.

263. Huang, P. M. (1990). Role of soil minerals in transformations of natural organics and xenobiotics in soil. In *Soil Biochemistry*, Vol. 6, Chap. 2, ed. Bollag, J-M. and Stotzky, G., Marcel Dekker, New York, p. 29

5 Microbial Mobilization of Metals from Soil Minerals under Aerobic Conditions

E. KUREK

University of Maria Curie-Sklodowska, Lublin, Poland

Interactions between Soil Particles and Microorganisms
Edited by P. M. Huang, J.-M. Bollag and N. Senesi. © 2002 John Wiley & Sons, Ltd

1 INTRODUCTION

The elemental composition of the mineral fraction of soil is influenced by the composition of parent rock from which it is formed [1]. The breakdown of rocks to form soil(s) is part of a biogeochemical cycle encompassing biological, geological and chemical processes [2]. At present, there is no doubt that microorganisms are important agents involved in metal mobilization from minerals.

Microbes can dissolve minerals by direct (enzymatic) or indirect action under aerobic and anaerobic conditions. When oxidized metal compounds such as Fe(III), Mn(IV) or As(V) act as terminal electron acceptors (dissimilatory reduction), anaerobic respiration becomes an example of direct dissolving action under anaerobic conditions [3,4] (see chapter 11). Oxidation of either ferrous iron or sulfur entities of metal sulfides to obtain energy is an example of direct dissolving action under aerobic conditions [5]. Indirect dissolution of minerals can be the result of microbial activity connected with production of organic and inorganic acids, and oxidizing agents which can influence soil conditions including changes in pH and redox potential. Metals can also be mobilized from geological sources by complexation with end products of microbial metabolism or with metal-sequestering agents (siderophores) produced to acquire microelements, necessary to synthesize enzymes, vitamins and cofactors [6,7]. Volatilization of metal (metalloid) or biomethylated metal compounds from the soil into the atmosphere can be a mechanism of detoxification for toxic elements such as As, Se and Hg [8].

Mineral diagenesis (transformation of one mineral into another) can be an indirect effect of aerobic and anaerobic microbial metabolism. The formation of a new mineral can be the result a chemical reaction between a product of microbial dissolution of a mineral and appropriate cations present in the environment (see chapter 11).

2 BIOLOGICAL LEACHING OF MINERALS BY ACIDOPHILIC IRON AND SULFUR OXIDIZERS

Metal sulfide minerals occur in nature as major or minor constituents of various ores, or are associated with deposits of other fossils such as coals or phosphorites. They are also constituents of wastes resulting from mining and recovery of metals from ores and burning of lignite.

As a result of microbial metabolism, insoluble metal deposits or metals contained in wastes are converted into soluble metal sulfates. Acid sulfate soils as well as acid mine drainage, enriched with heavy metals, can be formed causing serious environmental problems. These soils have to be remediated before being used for plant cultivation because the yield of plants grown in such soils is usually low and the biomass contains large amounts of toxic heavy metals. Conversely, bioleaching can also be beneficial, as in microbial mining of uranium, copper, zinc, lead, cobalt, nickel, bismuth and antimony from low-grade ores, removal of constituents that interfere with the extraction of metal elements (e.g. in the extraction of gold from pyrite and arsenopyrite containing ores), depyritization of coal, and decontamination of metal-polluted soils, wastes and sediments [9].

2.1 METAL SULFIDE MINERALS

The minerals mentioned below include the most common in nature, whose biological dissolution processes are the best studied.

Pyrite (FeS_2) occurs in a wide variety of geological settings and is one of the most common sulfide minerals in nature. Marcasite has the same chemical composition, yet it is crystallographically different. The structure of marcasite is orthorhombic while that of pyrite is isometric, which results in a lower stability of marcasite than that of pyrite. Pyrite is the most prevalent form present in coals and is associated with many ores including those containing zinc, copper, uranium, gold and silver [10].

Pyrite can vary significantly in grain size and morphology, depending upon the environment in which it is found. This mineral can occur in the form of acicular crystals, coarse-grained masses, euhedral forms, framboids and poly-framboids, octahedral and pyritohedral crystals. Arora *et al.* [11] reported that pyrite isolated from lignite coal consisted of porous and non-porous irregular grains. In morphological studies of pyrite isolated from Pennsylvanian-age shale in Missouri (USA), Ainsworth and Blanchar [12] described three pyrite groups: (1) pyrite with a smooth surface, including octahedral, cubic and pyritohedral, (2) conglomerates with irregular surface, composed of many cemented particles and, (3) framboids in which cemented crystals formed a smooth sphere. Framboid and polyframboid pyrites are more reactive than conglomeritic pyrite, due to their high specific surface area and high porosity.

Pyrrhotite ($Fe_{1-x}S$) is an important component of many ore deposits and may in some instances be the only significant sulfide present in mining waste. It has been suggested that pyrrhotite oxidizes at rates of 100 times faster than those measured for pyrite under abiotic conditions [13]. As a result, the mineral may play an important role in acid production, especially in the early stage of weathering. The structure as well as stoichiometry of pyrrhotite is varied. The monoclinic form is usually the most iron-deficient (FeS) and may be the most reactive chemically. Intermediate and end-member phases with stoichiometries of Fe_7S_8 to Fe_9S_{10} have orthorhombic or hexagonal crystal structures. These phases may occur as mixtures with monoclinic pyrrhotite [14]. Though in varying amounts, iron compounds are always associated with metallic sulfide ores and often with oxide uranium ores, as well.

Chalcopyrite ($CuFeS_2$), which forms tetrahedral crystals, is the most abundant Cu mineral found in nature. *Chalcocite* (Cu_2S) occurs as framboids, and is the second most abundant Cu mineral in nature. *Covellite* (CuS) with hexagonal crystals is another Cu mineral frequently found in ores [15].

Arsenic-containing minerals are relatively common in many sulfide ores. *Arsenopyrite* (FeAsS) and pyrite are associated with recalcitrant gold ore deposits [5]. *Sphalerite* (zinc blende ZnS) is the most common zinc mineral in high-grade zinc ores [16]. *Galena* (PbS) is also relatively common as a minor mineral constituent in sulfide ore materials [17]. In nature, sulfide minerals sometimes form polymetallic ores, such as the complex Cu–Co–Zn ore in the Keretti mine, Outokumpu, Finland, which contains chalcopyrite, pentlandite, pyrite, pyrrhotite and sphalerite [18].

2.2 LEACHING MICROORGANISMS

The studies of cultures isolated from natural leaching sites and industrial leaching operations have indicated that diverse bacterial species can participate in this process. However, aerobic mesophilic chemolithotrophic bacteria, belonging to genera *Thiobacillus* (*T. ferrooxidans, T. thiooxidans*) and *Leptospirillum ferrooxidans* are considered as the most important when bioleaching takes place below 40 °C. Obligate autotrophic species from these genera are highly acidophilic (optimal pH 1.5–2.0) and grow optimally at a temperature of 25–35 °C [19].

T. ferrooxidans obtains energy from oxidation of either ferrous iron or reduced sulfur compounds (S(0), sulfides), and can rapidly decompose mineral sulfides in pure cultures. Meanwhile *T. thiooxidans* can gain energy only from oxidation of reduced sulfur compounds, and *L. ferrooxidans* can only use ferrous iron oxidation as an energy source. Atmospheric CO_2 is the carbon source for all of them. In leaching operations and acid mine waters, these three most common chemolithotrophic bacteria occur as a consortium. Their combined action results in rapid oxidative leaching of sulfide minerals [20]. In heaps their distribution is limited to the top meter of ore, with the highest number in the upper half-meter zone, and similar depth profiles for their distribution were also found in mine tailings [21]. Anaerobic oxidation of elemental sulfur coupled with ferric iron reduction by *T. ferrooxidans* has been demonstrated; however, no biological oxidation of ferrous iron under anaerobic conditions by thiobacilli has been shown so far [22,23]. Nevertheless the recent studies by Benz *et al.* [24] and Straub *et al.* [25] indicate that the ability to oxidize iron under anaerobic conditions (at approximately pH 7) with nitrate as an electron acceptor appears to be a widespread feature among mesophilic denitrifying bacteria (see chapter 11). A study of microbial diversity in uranium mine waste heaps indicates that irrespective of the sampling depth, moderately acidophilic thiobacilli such as *T. neapolitanus, T. novellus* or *T. intermedius*, were at least as abundant as *T. ferrooxidans*. In samples from 3–4 m below the surface, these thiobacilli species were a hundred times more frequent than the former. However, the source of their substrate remains unclear because the uranium ore contained only pyrite as a nutrient source, which is known to be non-degradable by these bacteria [26]. Acidophilic bacteria which oxidize reduced sulfur compounds but not ferrous iron such as obligate chemolitho-trophic *T. thiooxidans* and the mixotroph *T. acidophilus*, may also enhance leaching by producing sulfuric acid [27].

Chemolithotrophic iron- and sulfur-oxidizing acidophilic bacteria share mineral leaching environments with a range of heterotrophic microorganisms. Some data have indicated that heterotrophic acidophilic bacteria belonging to the *Acidiphilium* genus may enhance the rate of sulfide mineral oxidation by autotrophs. This can occur because heterotrophic bacteria metabolize organic materials which accumulate in leaching liquors and inhibit iron oxidizers [5]. A novel group of mesophilic heterotrophic acidophiles has recently been described. These bacteria are able to oxidize ferrous iron to ferric, but in contrast to chemolithotrophic acidophiles, they require organic carbon to grow. One strain isolated by Johnson *et al.* [28], which had filamentous morphology similar to bacteria from genera *Sphaerotilus* and *Leptothrix*, grew and oxidized iron in ferrous sulfate–yeast extract medium, but could not oxidize pyrite in yeast extract amended media. Bacelar-Nicolau and Johnson [29] were successful in isolating heterotrophic bacteria from acid mine drainage, which, when grown in pyrite-containing media supplemented with 0.02% (w/v) of yeast extract, oxidized pyrite. One strain (T-24) promoted rates of mineral dissolution similar to those observed with iron-oxidizing autotroph *T. ferrooxidans*. A mixed culture of the three isolates and sulfur-oxidizing autotroph *T. thiooxidans* promoted pyrite dissolution in yeast extract-free medium. *T. thiooxidans*, like other autotrophic acidophiles, released organic materials into culture media and these materials could be used by heterotrophic acidophiles as C source. Phylogenetically, strains isolated by these authors were distinct from other acidophilic bacteria. Basing upon 16S rRNA sequence analysis, they appear to belong to more than a single species. The name '*Ferromicrobium acidophilus*' has been proposed for strain T-23. Other isolates were more closely related to moderately thermophilic iron oxidizer *Acidimicrobium ferrooxidans*. They required yeast extract as a carbon source in order to catalyze the oxidative dissolution of pyrite when grown in pure cultures. The requirement of these organisms for organic carbon is minimal, and in a natural environment, it can be well satisfied by the organic carbon originating from indigenous to sulfide minerals autotrophic acidophiles with which they usually grow in consortium, as well as from extraneous sources (i.e. soil leachates).

Mesophilic iron- and sulfur-oxidizing acidophiles are active at temperature as low as 2 °C, but truly psychrophilic isolates have not been found [30–32]. The optimum temperature for their growth is usually reported to be 28–33 °C. The maximum growth temperature (T_{max}) determined by Niemela *et al.* [33] for pure (*T. ferrooxidans*, *T. thiooxidans*) and mixed cultures of acidophilic thiobacilli was within the range 36.1–43.6 °C.

Oxidation of sulfides is an exothermic reaction resulting in a temperature rise in leach dumps to 50–80 °C. Such elevated temperatures are likely to hamper the growth and activity of common mesophilic bacteria, indicating that the thermophilic ones can play an important role in metal solubilization [34].

A moderately thermophilic bacterium, called *Thiobacillus* TH, capable of growth on pyrite, chalcopyrite and similar ores at temperatures of 50 °C was isolated [5]. Tuovinen *et al.* [35] isolated a moderately thermophilic consortium composed of iron- and sulfur-oxidizing acidophiles in which *Thiobacillus caldus* (strain KU) was one of the predominant organisms. This mixed culture oxidized arsenopyrite faster at 45 °C than did *T. ferrooxidans* at 22 °C. *T. caldus*, capable of oxidizing sulfur optimally at 45 °C, appears to be a moderately thermophilic equivalent of *T. thiooxidans*. Comparison of 16S rRNA sequences from *T. caldus* and *T. thiooxidans* indicates that the two bacteria are phylogenetically closely related [36]. The most common moderately thermophilic iron-oxidizing microorganism is a Gram-positive spore-forming bacterium belonging to the genus *Sulfobacillus*. *S. thermosulfidooxidans* has been reported to be the most efficient iron-oxidizing moderate thermophile, and its growth and ability to oxidize iron were stimulated in the presence of yeast extract and CO_2-enriched air [37]. *Acidimicrobium ferrooxidans* is a moderate thermophile, growing well even in the absence of an elevated concentration of CO_2, but its ability to oxidize iron is lower than that of *S. thermosulfidooxidans* even when grown in CO_2-enriched air. Mixed cultures of the two bacteria are efficient oxidizers of iron in the absence of added CO_2 [37].

Extreme acidophilic thermophilic archaebacteria, belonging to the *Sulfolobus* (*S. acidocaldarius*) and *Acidianus* (*A. brierleyi*) genera, are capable of oxidizing ferrous iron and metal sulfides in extreme environments, such as hot acid springs at temperatures above 50 °C and pH of 1–2 [38]. Konishi *et al.* [39] reported that the specific growth rate of *A. brierleyi* on pyrite at 65 °C was about four times higher than that of the mesophilic bacterium *T. ferrooxidans* at 30 °C, while the growth yields on pyrite for the two microbes were approximately equal to one another in magnitude. This indicates that the thermophile was more effective in pyrite leaching.

Information about composition of bacterial populations inhabiting natural leaching sites and industrial leaching operations was obtained from culture studies. However, most of the microorganisms occurring in soil are non-culturable.

Progress in studies on ecology of bioleaching microorganisms was made by amplifying of 16S rRNA genes from the total DNA extracted from environmental samples with polymerase chain reaction (PCR) techniques. Using these techniques Pizarro *et al.* [40] found the microbial population in a heap leaching system of copper ore to be a mixture of different bacteria. However, *T. ferrooxidans* was a dominant bacterium at high ferrous-iron concentration. At low ferrous-iron concentration, *T. thiooxidans* and *L. ferrooxidans* were also present. It has been reported, however, that *T. ferrooxidans* occurs in very low numbers in stirred tank reactors in which the bacterial population is dominated by *L. ferrooxidans*, *T. thiooxidans* or *T. caldus* [20,41]. Using similar techniques to measure species-dependent 16S and 23S rDNA intergenic spacing, negligible numbers of *T. ferrooxidans* and a dominance of *Leptospirillum* and *T. thiooxidans* have been reported in copper heap leaching environments operated under highly acidic conditions [42].

2.3 MECHANISMS OF DISSOLUTION OF SULFIDE MINERALS

The mechanism of microbial metal sulfide oxidation has been most extensively studied using pyrite as the metal sulfide and *T. ferrooxidans* as a leaching organism. Because of the ability of *T. ferrooxidans* to oxidize both iron and sulfur entities, the results of pyrite leaching experiments suggested that two mechanism—direct metabolic reactions and indirect metabolic reactions— could be involved in this sulfide dissolution.

2.3.1 Direct Attack Mechanism

Due to the insolubility of metal sulfides, a direct contact between the bacterium and the sulfide surface is required for the direct attack mechanism to operate. Moreover, attachment of bacteria to mineral is probably mediated by a thin organic film existing between the cell outer membrane of attached bacteria and the surface of sulfide mineral. There are suggestions that attached *T. ferrooxidans* multiplies on the surface of pyrite using pyritic sulfur as nutrient, that pyrite is directly attacked by oxidizing bacteria, and that bacterially catalyzed oxidation of sulfur and iron by O_2 occurs [10]. Observations by scanning electron microscopy of pyrite particles inoculated with *T. ferrooxidans* revealed corrosion, cracks and small pits on its surface. Specific surface areas of pyrite— before and after bioleaching—were 1.1 and 1.6 m^2g^{-1}, respectively [43]. Studies by Bennett and Tributsch [44] indicated that chemical processes taking place on a pyrite surface due to the oxidation by bacteria occurred mainly in the region of contact between bacteria and pyrite, and that the bacterial distribution on the pyrite surface was probably strongly dependent on the crystal structure and diversity in crystal orders of the pyrite.

The events taking place during the direct microbial attack on pyrite are presented below [10,45]. The prerequisite for bacterial oxidation of the mineral is chemical dissociation of insoluble metal sulfide from the mineral surface:

$$FeS_2 \xrightarrow{\text{abiotic}} Fe^{2+} + S_2{}^{2-} \tag{1}$$

The equilibrium conditions of sulfide mineral dissociation are affected by ferric ion and protons in medium. The concentration of H^+ is very important since protons can chemically break surface bonds, shifting electron levels of many sulfides into a range where they can more directly interact with the bacterial metabolic system.

Disulfide anion (equation 1) is immediately bound by bacterial enzymes and oxidized to sulfate:

$$S_2^{2-} + 4O_2 \xrightarrow{\text{biotic}} 2SO_4^{2-} \tag{2}$$

The effectiveness of leaching is correlated with the activity of some key enzymes involved in the metabolism of inorganic sulfur compounds including sulfite reductase, thiosulfate oxidase and rhodanase [45].

If *T. ferooxidans* is involved iron cation is oxidized:

$$4Fe^{2+} + O_2 + 4H^+ \xrightarrow{\text{biotic}} 4Fe^{3+} + 2H_2O \tag{3}$$

Oxidation external to the cell membrane ferrous ion is favored by high acidity.

The sum of bacteria-catalyzed reactions of sulfur and iron oxidation by O_2 operates as follows:

$$4FeS_2 + 15O_2 + 2H_2O \xrightarrow{\text{biotic}} 2Fe_2(SO_4)_3 + 2H_2SO_4 \tag{4}$$

2.3.2 Indirect Attack Mechanism

This attack mechanism does not require a direct contact between the bacterium and sulfide surface. It is based upon chemical pyrite (metal sulfides) oxidation by Fe(III), which is a more efficient pyrite oxidant than by O_2 over a pH range of 2–9. The faster rate of pyrite oxidation by Fe(III) than by O_2 according to Luther [46] is due to the fact that Fe(III) can bind chemically to the pyrite surface. Fe(III) has a vacant orbital, and binds to the pyrite surface via sulfur to form a persulfido brigde (Fe-S-S-Fe$(H_2O)_5(OH)^{2+}$, a somewhat intermediate transition state. Through this bridge an electron can be transferred from the highest occupied molecular orbital of S(II) to the lowest unoccupied molecular orbital of Fe(III). The process requires that all electron transfers would involve the reduced sulfur of pyrite and the oxidant adsorbed on the pyrite surface.

Pyrite is oxidized according to reaction:

$$FeS_2 + Fe_2(SO_4)_3 \xrightarrow{\text{abiotic}} 3FeSO_4 + 2S^0 \tag{5}$$

The ferrous iron and S(0) produced by this reaction (equation 5) are oxidized by *T. ferrooxidans* to Fe(III) (regeneration of oxidant) and acid. Oxidation reactions may take place on the pyrite surface or in interstitial solutions [10]. Oxidation of Fe(II) by dissolved O_2 is a limiting step in the oxidation of pyrite. The oxidation rate of Fe(II) to Fe(III) by O_2 in the presence of *T. ferrooxidans* is directly proportional to the concentration of bacteria, O_2 and H^+, yet below pH 2.2 it is independent of pH value. The Michealis–Menten equation can be used to explain the rate of biological Fe(II) oxidation.

In nature, all sulfide minerals contain iron, though in varying amounts, and are oxidized by bacterial consortia consisting of organisms capable of oxidizing either the iron or the sulfur moieties or both. In this situation, it is not easy to establish which of the two mechanisms is operating during particular mineral dissolution at a natural leaching site, or which is more important.

Taking into consideration their own research results, sulfur chemistry and ecological, and biochemical observations of other authors, Sand *et al.* [47] have presented a model for bacterial leaching of metal sulfides and concluded this process to proceed only *via* indirect attack mechanism (Figure 1). Premises for this conclusion are given in the following: leaching bacteria grow and attach to the surface of mineral sulfides. The attachment of *T. ferrooxidans* as well as *L. ferrooxidans* is not accidental. A chemotactic mechanism is involved with Ni(II), Fe(II), Cu(II) ions as positive attractants for *L. ferrooxidans*, while thiosulphate is an additional positive attractant for *T. ferrooxidans*. The two leaching bacteria are able to recognize sites on the mineral sulfide surface where electrochemically controlled processes may cause dissolution and liberate ferrous ions and thiosulphate, so that the cells attach to those sites where electrochemical dissolution processes make the substrates available. Chemical analysis has demonstrated that isolated extracellular polymeric substances of *T. ferrooxidans* and *L. ferrooxidans* contained between 0.5 and 5% of tightly bound ferric iron, and that these iron compounds were not removed by any washing procedure. Thus, the cells of the two leaching bacteria contain a considerable supply of Fe(III), even if the mineral is devoid of them. In any event, the attachment is mediated by exopolymers containing ferric iron. The nature of these polymeric compounds and their interaction with the mineral sulfide surface is unknown. After attachment, the ferric ions start attacking the mineral sulfide. As a result, ferrous iron hexahydrate and thiosulphate are generated. In case of *T. ferrooxidans*, both compounds are used by the cells. With *Leptospirillum*, the degradation of thiosulphate progresses in purely chemical reactions (oxidation by ferric iron and effect of high proton concentrations in this acidic environment). Consequently, polythionates should be produced by chemical oxidation, resulting in a reduction of ferric iron. Tetra- and pentathione were detected in cultures of *L. ferrooxidans* oxidizing pyrite (besides the end product—sulfate). With *T. ferrooxidans*, ferrous iron is reoxidized to ferric iron, and thiosulphate may either be oxidized to sulfate, or to polythionate granules in the periplasm. According to this model, the presence of thiosulphate accounts for the abundance of moderately acidophilic thiobacilli in uranium mine waste heaps, since this compound is the preferred substrate for these bacteria.

However, the model presented above can be fitted to many experimental data though the deduction by the authors is probably too rigorous. There is experimental evidence for a direct attack mechanism operating during sulfide

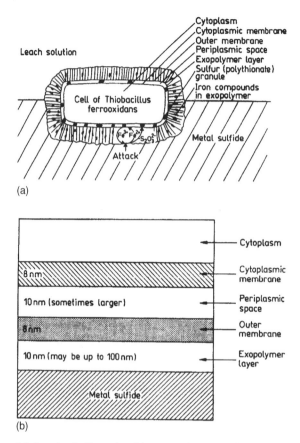

Figure 1. A model for the indirect leaching attack mechanism as catalyzed by a cell of *Thiobacillus ferrooxidans* attached to a metal sulfide [47]. **(a)** A general description of the bacterial leaching attack exaggerating the exopolymer layer in order to show its importance. **(b)** The correct dimensions of the cellular structures where leaching biochemistry takes place. Note: the attack takes place in the exopolymer layer between the cell and the metal/sulfide and is mediated by the iron compounds in the exopolymer layer (at least primarily). Thiosulfate is an intermediate compound; for clarity the ensuing reactions were omitted. From [47]. Reproduced by permission of Springer-Verlag

oxidation. Garcia *et al.* [17] demonstrated that iron-free ZnS could be oxidized by *T. thiooxidans*; however, sphalerite oxidation by *T. ferrooxidans* was three fold more efficient. Pistario *et al.* [48] reported *T. thiooxidans* to enhance the oxidation of synthetic ZnS by *T. ferrooxidans*, providing evidence for the involvement of the direct mechanism in sulfide oxidation, as *T. thiooxidans* oxidizes only inorganic S compounds. The bioleaching of natural pyrite by

Figure 2. Scanning electron micrographs of pyrite particles after various degrees of leaching using *Acidianus brierleyi*. An initial pyrite-liquid loading ratio $= 10 \, kg \, m^{-3}$, an initial total cell concentration $= 1.07 \times 10^{13}$ cells m^{-3}, and an initial ferric iron concentration $= 0.0 \, kg \, m^{-3}$; (**a**) initial pyrite particles; (**b**) after 30% pyrite leaching; (**c**) after 80% pyrite leaching. From [39]. Reproduced by permission of John Wiley & Sons, Ltd

Acidianus brierleyi was found to take place with a direct attack by adsorbed cells on the surface of pyrite (Figure 2), while chemical leaching of pyrite by ferric iron was insignificant [39].

In nature, both mechanisms probably contribute to leaching of metals from sulfide minerals, and there is no uniform mechanism of mineral dissolution in biological leaching, as suggested by the results of Ahonen and Tuovinen [18] on bioleaching of complex Cu–Co–Zn ore with mixed bacterial culture obtained by enrichment of mine water samples, using sulfide ore material from mine site as a substrate. The leaching rates of Co, Cu, Ni and Zn from this ore (containing chalcopyrite, pentlandite, pyrite–pyrrhotite and sphalerite) responded differently in each case to the redox potential and the concentration of ferric iron in the solution. Furthermore, in leaching complex sulfide ores, electrochemical interactions between sulfide minerals may also play an important role.

2.4 BIOLEACHING OF METALS FROM SOIL MINERALS, SEDIMENTS AND SEWAGE SLUDGE

T. ferrooxidans has been found to play a major role in genesis of acid sulfate soils by oxidizing pyritic minerals [49]. Acid sulphate soils with elevated As concentration that developed from a pyritic parent material were identified in Canada [50]. Seleniferous soils may also develop from selenide minerals incorporated into pyrite [51].

Soils permanently or temporarily flooded with water (wetlands, marshes, fresh-water or seawater sediments) exist in nature. Flooding of the soil and respiration of aerobic microbes result in formation of anaerobic conditions. In soils located in the neighborhood of mining industries or power plants using pyrite-containing coal or in soils periodically exposed to brackish or seawater containing organic matter and sulphate, dissimilatory reduction of sulfate occurs leading to the formation of significant amounts of sulfide. Sulfide is immediately precipitated with cations of heavy metals present in the environment. Insoluble sulfides retained within wetlands or sediments do not cause significant environmental problems. However, when aerated as the result of lowering the water level (e.g. drainage) or by dredging, they are oxidized by microorganisms to sulfates. Oxidation of sulfidic minerals results in acidification. Acid sulfate soils formation is observed in areas periodically inundated with salt water [9].

River bottoms have to be dredged regularly to maintain shipping channels. When Hamburg harbor sediments were excavated and disposed on land, the pH decreased within a few days from circum-neutral to pH < 4 accompanied by an increase in concentration of Cd and Zn in the aqueous phase [52].

Organic topsoil and mineral soil are removed when deposits of ores and coal are exploited by open-pit mining. This spoil material containing metal sulphides is transported to permanent disposal sites and heaped into spoil banks. Leachate is formed from the metal sulfides mainly pyrite. It often happens that other accompanying metals are mobilized, and a population of indigenous microorganisms develop inside the heap [53]. Analysis of leachate from a 10-year old spoil bank at Litov (Northern Bohemia, Czech Republic) revealed elevated concentrations of Al, As, Cd, Cu, Cr, Fe, Ni, Zn at a pH of 6.5. A reason for the relatively high pH of this leachate was the consumption of produced acidity by the pH buffering system inside the spoil [9]. Spoil bank leachates are usually less concentrated, and having a higher pH than acid mine drainage, they cause fewer environmental problems.

Bioleaching procedure can be applied in remediation of metal sulfide-contaminated sediments or soils, and for leaching heavy metals from the sludge produced by wastewater treatment plants. The reduced sulfur content in municipal sludges or river sediments can be insufficient to reach a desirable low pH for optimal bacterial action and high metal extraction. Hence an addition of ferrous iron or elemental sulfur, substrates for thiobacilli, is usually required. Bioleaching of heavy metals (Cd, Cr, Cu, Ni, Pb, Zn, Mn, Co and As) from deposited oxidized river sediments supplemented with elementary sulfur for indigenous sulfur-oxidizing bacteria, was found to be an effective treatment. About 62% of the total contents of the tested metals were removed by percolation leaching after 120 days [54].

Solubilization of metal sulfides contained in wastewater sludge using a biological process required pre-acidification to a pH < 4, addition of ferrous iron, and inoculation with *T. ferrooxidans.* Although this bacterium was responsible for oxidation of iron and acidification of the sludges, the activity of populations of diverse indigenous bacteria was necessary to make the growth of *T. ferrooxidans* and initiation of the bioleaching process possible. By coupling of 16S rDNA gene amplification by PCR and resolution of the amplification products by denaturing gradient gel electrophoresis, it was reported that this bacterial population consisted of more than 20 different units undetected by cultivation techniques [55].

2.5 ACID MINE DRAINAGE

Acid mine drainage (AMD) is a specific type of wastewater resulting from oxidation of sulfide minerals contained in exposed ore bodies, ore tailings, and spoil materials originating from metals and coal mining [10].

When exposed to oxygen and moisture, sulphide minerals undergo oxidation via abiotic and microbiologically mediated reactions. Iron sulphides (pyrite, marcasite and pyrrhotite) are the most abundant sulphidic minerals, and the possibility of oxidation of both iron and sulphur moieties makes them the main source of acid mine drainage [56].

Pyrite contained in mining waste is first chemically oxidized by atmospheric O_2 producing H^+, SO_4^{2-} and Fe(II) (equation 6). Produced this way, Fe(II) can be further chemically oxidized by O_2 into Fe(III) (equation 7) which in turn hydrolyzes into iron hydroxide and releases additional amounts of acid into the environment (equation 8) [57]. When the pH in the vicinity of a pyrite surface drops below 3.5 (due to acid production), oxidation of the pyrite by Fe(III) becomes the main mechanisms for acid production and pyrite oxidation (equation 9). There is evidence that Fe(III) can oxidize pyrite at much higher rate than O_2 [58]. At low pH, reaction (7) is catalyzed and accelerated by a factor greater than 10^6 by an acidophilic, chemoautotrophic, iron oxidizing bacterium *T. ferrooxidans.* Therefore, this bacterium is considered to be primarily responsible for rapid pyrite oxidation in mine waste at low pH by indirect attack mechanisms (Figure 3).

$$FeS_2 + \frac{7}{2}O_2 + H_2O \xrightarrow{\text{abiotic}} Fe^{2+} + 2SO_4^{2-} + 2H^+ \tag{6}$$

$$Fe^{2+} + \frac{1}{4}O_2 + H^+ \xrightarrow{\text{abiotic,biotic}^1} Fe^{3+} + \frac{1}{2}H_2O \tag{7}$$

^{1}at low pH

$$Fe^{3+} + 3H_2O \xrightarrow{\text{abiotic}} Fe(OH)_3 + 3H^+ \tag{8}$$

(direct oxidation)   **(indirect oxidation)**

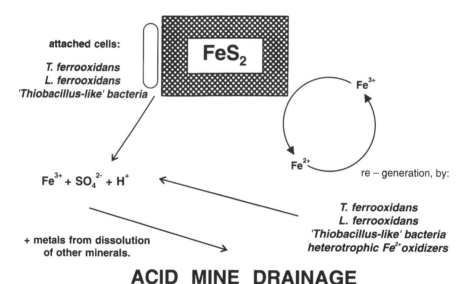

ACID MINE DRAINAGE

Figure 3. Bacterial oxidation of pyrite (FeS$_2$) by the direct and indirect attack mechanisms, and its central position in the formation of acid mine drainage. Reprinted from [56]. Copyright (1995), with permission of Elsevier Science

The role Fe(III) generated at pH below 3.5, mostly by the activity of *T. ferrooxidans* in oxidation of pyrite, is demonstrated as follows:

$$FeS_2 + 14Fe^{3+} + 8H_2O \xrightarrow{abiotic} 15Fe^{2+} + 2SO_4^{2+} + 16H^+ \qquad (9)$$

Acid mine drainage is hostile to most forms of aquatic life. The majority of bacterial species found in unpolluted streams are killed on exposure to AMD, yet microorganisms are dominant forms of life in AMD [59]. Most forms of life inhabiting AMD, in which temperatures generally range between 10 and 35 °C, are either acidophilic or acid tolerant.

Besides acidophilic, iron-oxidizing chemolitothrophs (*T. ferrooxidans* and *Leptospirillum ferrooxidans*), heterotrophic iron oxidizers are also numerous in AMD. Heterotrophic microbes play an important role by metabolizing organic materials potentially toxic to autotrophic iron-oxidizing bacteria [29]. Some of the microorganisms (40% of isolates) growing in AMD are capable of reducing ferric iron (amorphous and crystalline) compounds to ferrous iron [60]. This process results in mobilization of iron as well as other metals which may be associated with ferric iron deposits. At higher temperature, the microflora inhabiting a highly acidic metalliferous environment is quite distinct. These 'iron bacteria' are phylogenetically more closely related to Gram – positive heterotrophic bacteria than to *T. ferrooxidans*. Archaebacteria belonging to *Sulfolobus*, *Acidianus*, *Metallosphera* and *Sulfurococcus* are indigenous to AMD with temperatures higher than 60 °C [56].

Rainwater percolating through sulphide-containing deposits becomes highly acidic (as low as pH 2) and enriched with soluble iron, manganese, aluminum, and sometimes also with heavy metals such as Pb, Hg, Cd, Zn [10]. The major region of acid generation within a deposit such as a waste heap is restricted to the aerobic zone, which is generally the outer layer of the heap (approximately 1 m deep). Subsurface waters passing through disused mines in which appreciable amounts of sulphides are contained, and to which air has access, can become highly acidic and metalliferous. When reappearing at the surface on top soil or flowing into groundwaters, the subsurface waters can bring severe environmental pollution problems [61].

It has been estimated that wet cleaning of coal in the USA each year generated wastes containing at least 10 million tonnes of pyrite [62]. In Poland $1.046 \times 10^8 \, m^3 \, year^{-1}$ of drainage water is brought to the surface with a total content 807 t of zinc, 67 t of lead and 17 t of cadmium from the Zinc Mining and Processing Complex at Bukowno alone [63].

The procedures used to avoid production of AMD at the source involve suppression of the bacterial sulfur- and iron-oxidizing activities or prevention of pyrite oxidation. Other technologies aim at abating harmful effects of AMD on the soil and groundwater and most often involve passing AMD through constructed wetlands or neutralization by addition of alkaline materials.

At-source strategies for decreasing activity of thiobacilli include stimulation of microbial competition by enriching the environment with nutrients that favor other microbes growing faster than them, and inactivation of acid-producing bacteria by biocides [10]. A treatment of pyrite (sulfide minerals) with surfactants is promising as a source prevention method for solution of the AMD problem. Bioleaching can be suppressed by preventing the bacteria attaching to surface sulfidic minerals because attachment was found to be necessary for initiation of the microbial oxidation. Addition of E-30 (sodium paraffinsulfonate) to percolate of a pilot ore damp successfully inhibited the growth and activity of thiobacilli. The specific E-30 concentration was experimentaly determined as $0.016 \, mg \, cm^{-2}$ of the ore surface [64]. The treatment of pyrite with primary fatty acid amine surfactants, ArmacT and Armac HrR (Akzo Chemicals Ltd.) in the presence of *T. ferrooxidans* reduced the rate of both chemical and biological oxidation of pyrite making the pyrite surface highly hydrophobic [65]. Anionic surfactants such as sodium dodecyl sulphate (SDS) have been demonstrated to be effective in controlling the activities of *T. ferrooxidans* and *L. ferrooxidans* in the laboratory, and have been also used in the field with some success [66]. However, this strategy requires the quantification of the specific surface area of sulfide minerals since applied surfactant concentration should ensure full coverage of the mineral surface. There is also an extra problem of water pollution resulting from the presence of the detergents themselves. Another at source strategy is reduction of pyrite oxidation by indirect microbial attack [Fe(III) as oxidant] by addition of phosphate which can precipitate Fe(III) as insoluble $FePO_4$ or $FePO_4 \cdot 2H_2O$ [67].

On the other hand, passing AMD and spoil bank leachates through constructed wetlands—the so-called passive technology or biological mitigation— is another approach to their treatment [10,56].

A wetland is usually composed of two zones: an oxidation zone which is vegetated with aquatic plants (*Sphagnum* sp., *Typha latifolia, Phragmites australis*) and a reduction zone which is a sedimentation zone rich in sulfate-reducing (SBR), denitryfing and Mn-reducing bacteria. Biological activity in wetlands results in the wastewater neutralization and removal of metals by precipitation and adsorption in wetland sediment. Reductions of iron and sulphate have been identified as microbially driven alkali-generating processes. SBR convert SO_4^{2-} contained in AMD into H_2S, that reacts with heavy metals into insoluble precipitates. No extra organic matter is required as a substrate for sulfate and iron reduction, since wetlands are self-supporting due to the activity of photosynthesizing plants growing in them. Therefore over 400 wetlands have been constructed to treat AMD, some of which have appeared highly successful. Thus passing through a constructed wetland AMD from the large Five Tunnel Mine in Idaho Spings, Colorado, USA decreased the concentration of heavy metals (Cu, Pb, Zn, Cd) by 94–99% and increased the pH from 2.9 to 6.0 [56].

The major advantage of wetlands is their low costs to set up and maintain in comparison with conventional chemical treatment. Drawbacks include a large surface required for higher AMD flows and concerns relating to the diffuse spread and long term stability of the metal deposited [68]. Moreover a packed-bed column with manure, compost or similar materials rich in organic matter used as a packing material, or bioreactor can be an alternative to wetlands for AMD treatments [9,68].

At active mine sites, AMD is usually neutralized by limestone treatment. However, metals such as iron and manganese must be oxidized before being precipitated as stable insoluble compounds. Bacteria such as *T. ferrooxidans*, mostly responsible for AMD generation, are used to oxidize ferrous iron as an alternative to the addition of chemical oxidants before neutralization of ADM [56].

3 MOBILIZATION OF METALS BY CHELATION

Microorganisms require some metals for their metabolism and growth. For most microbes, essential ones include Fe, Mg, Mn, Mo, Co, Cu, Zn, Ni and V. They may serve as constituents of cytochromes, specific enzymes, growth factors, or enzyme cofactors [69]. In most soils, the only source of these metals happens to be corresponding oxides, hydroxides or sulfides. Because of the insolubility or very low solubility of these compounds, the metals are unavailable for uptake. Therefore, microorganisms have evolved various powerful systems to overcome solubility problems. In this respect, the high affinity-system for Fe uptake is a particularly well studied one [70].

3.1 IRON AVAILABILITY AND REQUIREMENTS OF MICROBES

Although iron is the fourth most abundant element in the Earth's crust, it is largely unavailable for uptake aerobically. In soils, iron is mostly found as a constituent of oxyhydroxide polymers of general composition $FeOOH$ (e.g. goethite, hematite), which are stable and have very low solubility at neutral pH (see chapter 4). The concentration of dissolved iron for optimal microbial growth is in a range from 10^{-8} to 10^{-6} mol L^{-1}, while the free Fe(III) concentration in the aqueous phase of soil under aerobic conditions is approximately 10^{-17} mol L^{-1}, i.e. far below the level required for microbial growth [71]. Thus, acquisition of external iron by microbes has to involve solubilization of Fe(III) oxide or other ferric minerals by one of the following mechanisms: chelation, reduction and protonation [7].

3.2 HIGH AFFINITY SYSTEM FOR IRON UPTAKE

For scavenging iron from the environment most free-living, as well as symbiotic, microorganisms produce specific chelators collectively termed *siderophores* [7,72,73]. They have been defined as extracellular low molar mass compounds (500–1000 Da) having a high affinity for iron(III), whose biosynthesis is regulated by the iron level in the environment and whose function is to supply iron to the cell. Some also show a high affinity for other multivalent metals and form complexes with them, though with a lower formation constants [74].

Microbial siderophores may be classified by the nature of their Fe(III) chelating ligands such as hydroxamates, catechols, carboxylates or mixed ligands (Table 1). Despite their considerable structural variety, all iron is chelated in hexadentate fashion [76] (Figure 4). The universal chrome azurol S (CAS) method developed by Schwyn and Neilands [79] has made it easy to detect siderophores, regardless of their structure.

Hydroxamate siderophores are very water soluble and less affected by changes in Ph than catechol siderophores (produced only by bacteria), which are also less soluble [7].

Table 1. Ligands and stability constants of some siderophore–Fe(III) complexes (for structure see Figure 1)

Siderophore*	Ligand*	Stability† constant (log K)	Producer*
Ferrichrome	Hydroxamate	29.1	*Aspergillus, Neurospora, Penicillium, Ustilago, Actinomyces, Streptomyces*
Aerobactin	Hydroxamate and hydroxycarboxylate	22.5	*E. coli, Klebsiella pneumoniae*
Ferrioxamine B	Hydroxamate	30.6	*Nocardia, Micromonospora, Streptomyces pilous, Actinomyces*
Rhodotorulic acid	Hydroxamate	31.2	*Rhodotorula, Sporobolomyces, Leucosporidum*
Enterobactin (Enterochelin)	Catechol	52.0	*E. coli, Salmonella typhimurium, Aerobacter aerogenes*
Parabactin	Catechol and 2–hydroxyphenyloxazoline	52.0	*Paracoccus denitryficans*
Agrobactin	Catechol and 2-hydroxyphenyloxazoline	52.0	*Agrobacterium tumefaciens*
Pseudobactin	Catechol, hydroxamate and α-hydroxy acid	52.0	*Pseudomonas* B10

* [74]
† [75]

Figure 4. Structures of representative siderophores. (**a**) ferrichrome, (**b**) ferrioxamine B and (**c**) rhodotorulic acid, (**d**) Schizokinen, (**d**) aerobactin, (**e**) parabactin, (**e**) agrobactin and (**f**) enterobactin, (**g**) Pseudobactin. (a), (b) and (c) from [77], reproduced by permission of Elsevier Science; (d), (e) and (f) from [75], reproduced by permission of American Institute of Biological Sciences and (g) reprinted with permission from [78], copyright (1986) American Chemical Society

a

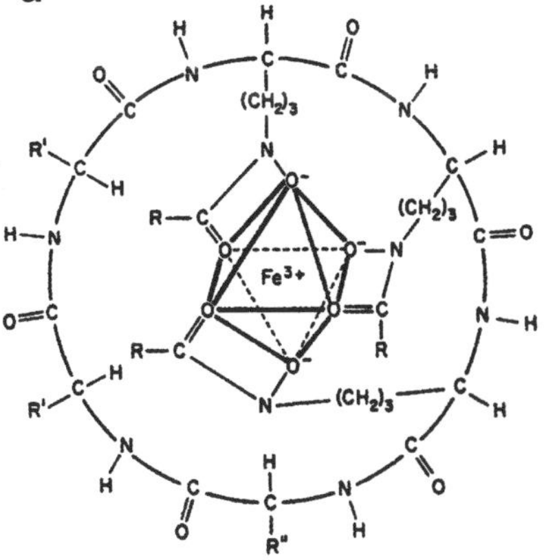

Ferrichrome: R = CH_3—
R' = R'' = H—
Ferrichrome A: R = HOOC—CH_2 / H_3C > C=C < H
R' = H—
R'' = $HOCH_2$—

b

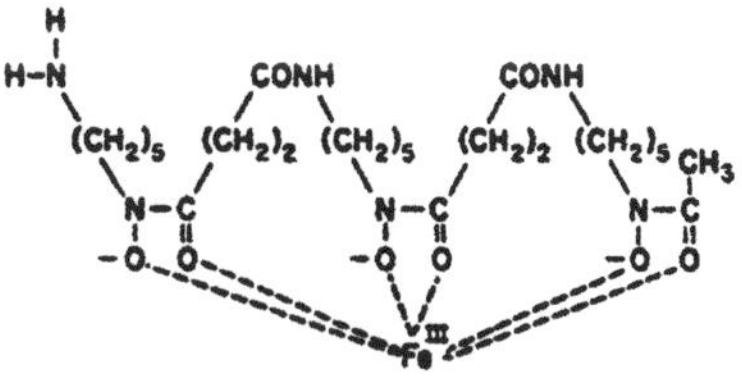

Ferrioxamine B

c

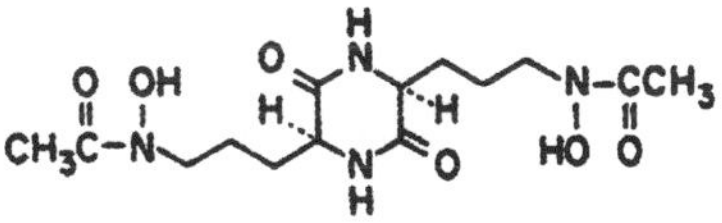

Rhodotorulic Acid

d

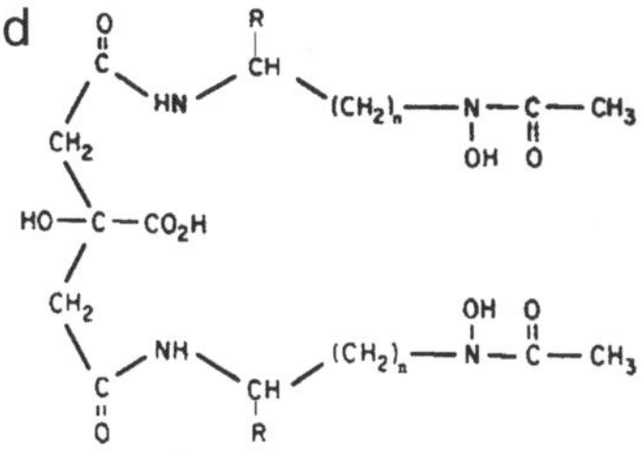

	R	n
Schizokinen:	H	2
Aerobactin:	COOH	4

e

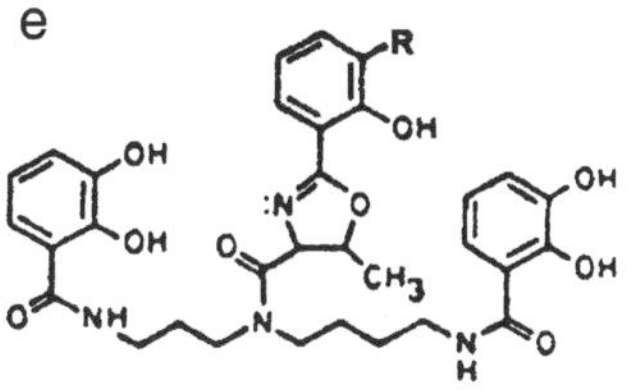

Parabactin (R=H)
Agrobactin (R=OH)

f

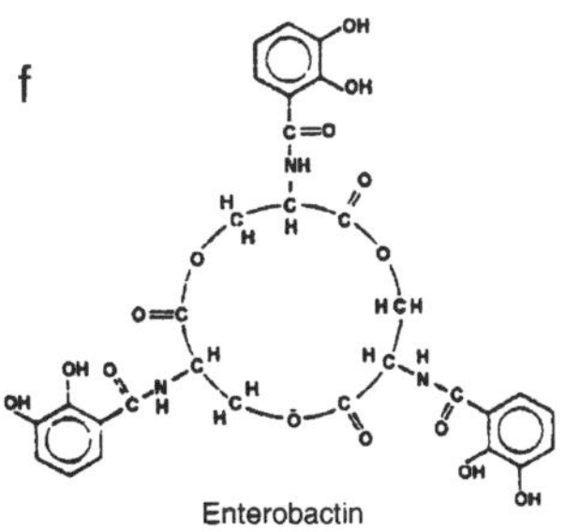

Enterobactin

g

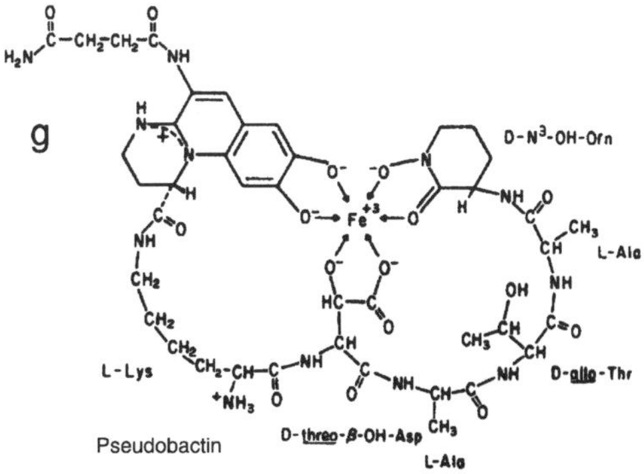

Pseudobactin

Besides chelators (siderophores), the high-affinity system of iron uptake induced by iron starvation includes outer-membrane specific receptors for ferrisiderophores and a system of active transport of iron into the cell. Outer-membrane receptors confer specificity of the iron transport system (for each ferrisiderophore, a specific receptor is found in the outer membrane). A set of periplasmic and inter-membrane associated proteins, known as permease, is

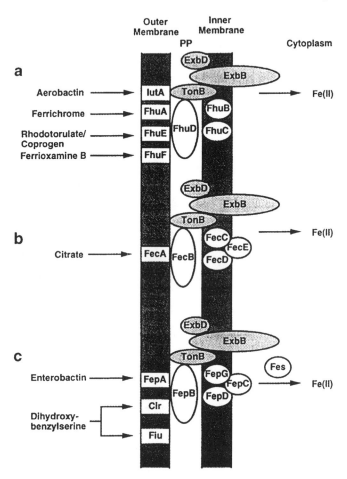

Figure 5. Schematic diagram of iron-uptake systems in *Escherichia coli*. Inner and outer membranes are shaded. PP=periplasm. Rectangles on outer membrane indicate receptors for complexes of corresponding siderophore–Fe(III). Elipses marked in periplasm and on inner membrane represent protein components of permease systems. Light shading indicates the three proteins known to be common to all three [hydroksamate (**a**), citrate (**b**) and catechol (**c**)] transport systems. Proteins are not drawn to scale. From [7]. Reproduced with permission from Annual Reviews (www.Annual Reviews. org)

required for translocation of ferrisiderophore into the cell [70] (Figure 5). Several microbes produce more than one siderophore, and most Gram-negative bacteria produce three to nine extra outer-membrane proteins that act as a receptors for ferrisiderophores.

In *E. coli*, all known siderophore systems are negatively regulated by the Fur (*ferric uptake regulation*) repressor—cytoplasmic protein (17 kDa)—coded by *fur* gen. At low Fe(II) concentration, the Fur protein has a weak affinity for operator DNA, but at high Fe(II) concentration, the Fur repressor binds tightly to operator DNA. In addition to iron, several other divalent cations including Mn, Co and Cu bind to Fur [80].

3.3 SIDEROPHORE-MEDIATED SOLUBILIZATION OF MINERALS

High values of formation constants for siderophore–Fe(III) complexes result in a very effective release of iron from insoluble ferriminerals commonly found in the soil and effective transport of the solubilized iron into the cell. The formation constant for enterobactin (catechol type siderophore produced by *E. coli*) is the highest to have ever been measured for an iron complex, $K_f = 10^{52}$ [74]. Pure cultures of the ectomycorrhizal fungus *Suillus granulatus* mobilized iron from goethite, which was attributed to chelation by hydroxamate siderophores produced by this organism [73].

Azotobacter vinelandii strain producing two catechole siderophores—azoto-chelin and azotobactin—was able to mobilize and use iron from the following minerals: marcasite (FeS_2), vivianite ($Fe_3(PO_4)_2 \cdot 8H_2O$), olivine ($(Mg, Fe)_2 SiO_4$), hematite (Fe_2O_3), siderite ($FeCO_3$), goethite ($FeOOH$), pyrite (FeS_2) and illite. When an *Agrobacterium tumefaciens* strain that was unable to use these minerals as an iron source was cocultured with *A. vinelandii*, the number of living *Agrobacterium* cells increased considerably (by more than eight fold when marcasite was used as an iron source) [81].

Other metals readily form complexes with siderophores when their size and charge are similar to those of Fe(III). For example, aluminum forms trivalent cations with similar charge and radius as iron(III), and therefore has a high affinity for siderophores. Al(III) binds less tightly to hydroxamate ligands than does iron(III), but has a similar affinity for catechole ligands [82]. In iron-deficient cultures of *Bacillus megaterium* ATCC 19213, siderophore formation mobilized aluminum from its hydroxides or phosphates [83]. Some actinides, e.g. plutonium(IV), by virtue of charge densities similar to that of Fe(III) also have a high affinity for hydroxamate and catechol type siderophores [6,84]. Two other metals—copper and molybdenum—have an appreciable affinity for siderophores. Copper(II) has high affinity for the both types of siderophores, such as to form stable complexes with them. Molybdenum forms complexes with catechol and hydroxamate siderophores, but the affinity constants are not high [74]. When grown in a culture medium deficient in Mo and Fe,

A. vinelandii produces several phenolate compounds that complex Mo [85]. This type of chelating compound was also found in *B. thuringiensis* cultures. Formation constants for complexes of hydroxamate siderophore desferrioxamine B with Ca(III), Al(III) and In(III) are in the range of 10^{20}–10^{28} [86], and those for catechole type pseudobactin with Zn(II), Cu(II) and Mn(II) range between 10^{17} and 10^{22} [87].

It was found that interaction of crystalline orthorhombic $CdHPO_4$ with alcaligin E, the phenolate type of siderophore produced by heavy metal-resistant *Alcaligenes eutrophus* strain CH24, resulted in the disappearance of the orthorhombic crystals and monoclinic 'desert-rose' like structures were formed [88] (Figure 6).

Mean concentration of siderophores in the soil was estimated to be 12 nmol L^{-1} desferrioxamine methanesulfonate equivalents taking into consideration values in the range of 2.7–34 nmol L^{-1}, determined in samples of various soils [77].

Although soil microbes produce a variety of siderophores under laboratory conditions, only schizokinen (hydroxamate siderophore) produced by *Bacillus* and *Anabaena* sp. has been purified directly from soil samples [89].

3.4 MOBILIZATION OF METALS BY OTHER MICROBIAL METABOLITES

Besides highly specific complexing compounds (siderophores), other microbial metabolites can also chelate metals mobilizing them from soil minerals. Unlike iron(III), copper(II) has high affinity for aminoacids and small peptides, making it more soluble [74].

Some organic acids secreted by microorganisms have a dual effect on solubilization of minerals—by lowering pH and complexation. The degree of complexation depends on the particular organic acid involved (number and proximity of carboxyl groups), the concentration and type of metal and the pH of the soil solution [90]. Malate, citrate and oxalate, all show a high affinity for trivalent metals such as Al(III) and Fe(III), which are most readily mobilized by organic acids in most soils (Table 2). These acids can also release Mn from insoluble synthetic MnO_2 [91].

The rate of Fe dissolution mediated by organic acids is dependent on crystaline structure of iron minerals and pH. Amorphous $Fe(OH)_3$ can be much more readily dissolved than well-defined goethite, Fe_2O_3 and Fe_3O_4. Citrate-mediated Fe dissolution is extremely pH dependent with mobilization potential decreasing with an increasing pH. Organic acids can rapidly form stable complexes with Fe at soil pH <6.5 [92].

Organic acid metabolites such as oxalic, isocitric, citric, succinic, hydrobenzoic and coumaric acids produced by a variety of heterotrophic microorganisms have been found involved in the solubilization of uranium oxide from granitic rock [6].

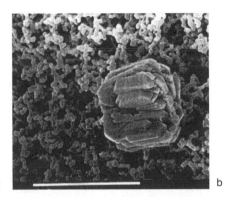 

a   b

Figure 6. Scanning electron micrographs: **(a)** orthorhombic $CdHPO_4$ crystals formed when 8 mM $CdCl_2$ was added to iron-limited Schatz lactate medium; **(b)** showing monoclinic 'desert-rose'-like structures formed when *Alcaligenes eutrophus* CH34, Sid⁻ derivative was grown for 4 days in this culture medium in the presence of alcaligin. The bar represents 10 μm. From [88]. Reproduced by permission of Nature Publishing Group

Table 2. Stability constants (log K) of organic acid–metal complexes at stoichiometry of the complex metal:ligand ratio 1:1 [91]

Metal	Malate	Citrate	Oxalate
Fe(III)	7.1	11.5	7.7
Al(III)	6.0	7.9	6.1
Cu(II)	3.3	5.9	6.2
Zn(II)	3.3	4.7	4.9
Mn(II)	2.2	3.7	4.0
Ca(II)	2.7	4.9	3.2
Mg(II)	1.6	4.8	3.4
Cd(II)*	–	3.8	–
Pb(II)*	–	4.1	–

* [6]

In general, organic acids appear to be able to induce a two- to four fold increase in mineral dissolution rate in comparison with rainwater alone. However, this is highly dependent on mineral type, pH, Al content of the mineral and organic acid type [90,93].

Certain α-keto acids generated by amino acid deaminases (phenylpyruvic, indolylpyruvic, α-ketoisovaleric, α-ketoisocapronic) show an affinity for Fe (III). However, α-keto acids do not form very stable complexes with ferric iron, as compared with hydroxamate siderophores, but their stability is high enough to be determined using chrome azurol S (CAS) assay [94].

4 METAL VOLATILIZATION

Volatilization of metals (and metalloids) is frequently considered as a mechanism of detoxification of polluted soil, since organic compounds of some trace metals are volatile and less toxic than inorganic ones and can easily pass from the soil into the atmosphere [8].

4.1 ARSENIC

In the soil, arsenic, very toxic for all biota, can originate from parent rocks from which soil is formed as well as from anthropogenic sources. The concentration of As in sedimentary rocks is much higher (1.7–400 mg kg^{-1}) than in igneous ones (1.5–30 mg kg^{-1}). The most common and widespread mineral form of arsenic is arsenopyrite (FeAsS); orpiment (As$_2$S$_3$ and realgar (AsS) are also quite common. In native soils, the As concentration does not generally exceed 15 mg kg^{-1}. However, the As concentration in contaminated soil can be greater than 20,000 mg kg^{-1}. Arsenic is a natural component of Pb, Zn, Cu and Au ores. Mining and smelting processes of non-ferrous metal ores usually result in contamination of soil, water and air. Use of pesticides, herbicides and some fertilizers in agriculture becomes another source As contamination in soil [50,95,96].

Several microorganisms (*Thiobacillus ferrooxidans, Sulfolobus acidocaldarius, Pseudomonas arsenitoxidans*) can use arsenic in arsenopyrite, orpiment and reaglar as sole source of energy oxidizing it to arsenite and arsenate. Heterotrophic arsenite-oxidizing bacteria that have been identified include *Alcaligenes fecalis, A. eutrophus* strain 280 and *Pseudomonas putida* strain 18 [96] (see chapter 6 for microbial As redox transformations).

Under aerobic conditions, some fungi such as *Scopulariopsis brevicaulus, Penicillium* sp., *Gliocladium roseum*, yeast *Candida humicola* and bacteria (*Proteus* sp., *E. coli, Flavobacterium -Cytophaga* group, *Pseudomonas* sp., *Corynebacterium* sp. and some species of the genera *Nocardia, Achromobacter, Aeromonas, Enterobacter* and *Alcaligenes*) are capable of arsenic biomethylation. Both arsenic oxyanions can be substrates for biomethylation, except that arsenite is methylated faster than arsenate [97]. Carbonium ion (CH$_3$$^+$) from *S*-adenosylomethionine is the source of methyl groups in microbial production of methylarsenicals (trimethyl- or dimethylarsine). There is little or no detailed information on genetics and biochemical regulation of metal(loid) methylation in microorganisms [8]. As less toxic and less bound to cell compounds than inorganic forms of this metalloid, methylarsenicals diffuse from the organisms through the cell wall and are lost from the soil into the atmosphere [95]. Microbial methylation of As(III) compounds can be used as a method for bioremediation of contaminated soils [8].

4.2 SELENIUM

Selenium is an essential microelement for most organisms, though very toxic at higher concentration. This metalloid occurs in nature in distinct minerals, e.g. ferroselite ($FeSe_2$) and challomenite ($CuSeO_3 \cdot 2H_2O$), is also often a natural component of uranium deposits and sulfide minerals (chalcopyrite, pyrite, galene and pyrrhotite) whose weathering can be a source of this element in the soil [96]. Commercial uses of its inorganic compounds and burning of fossil fuels can be other sources of increased selenium concentration in the soil [8]. The solubility of minerals, microbiologically mediated oxidation–reduction reactions, methylation and volatilization all are potential processes that control Se concentration, mobility and toxicity in the soil. Selenium can exist in four different oxidation states: selenide [Se(II)], elemental selenium [Se(0)], selenite [Se(IV)] and selenate [Se(VI)]. Its two oxyanion forms are soluble but elemental Se and metal selenides are insoluble and nontoxic. Amorphous and black crystaline Se(0) are formed as end products of dissimilatory reduction of Se oxyanions when they act as terminal electron acceptors [98] (see chapters 6 and 11), yet information concerning microbially mediated mobilization of Se from minerals is very scarce. Microbial oxidation of CuSe in laboratory experiments involving acidophilic bacterium *T. ferrooxidans* has been reported [96]. The reaction may be written:

$$CuSe + 2H^+ + 0.5O_2 \rightarrow Cu^{2+} + Se^0 + H_2O$$

Elemental Se has been reported to be oxidized aerobicly to oxyanions by heterotrophs *Micrococcus selenicus* isolated from mud [96], a strain of *Bacillus megaterium* isolated from the soil [99] and by a group of autotrophs [96]. Nothing is known about the physiological role of this process in the metabolism of these bacteria. Selenium is subjected to methylation and volatilization. Gaseous Se species detected in the atmosphere include $(CH_3)_2Se$ (DMSe) and dimethyldiselenide $(CH_3)_2Se_2$. DMSe can be produced in natural conditions both enzymaticaly and nonenzymaticaly. Due to its limited water solubility and high vapor pressure, DMSe volatilizes rapidly [98].

Se-methylating microorganisms are part of the natural soil population under aerobic conditions. They include fungi (e.g. *Alternaria alternata*) and bacteria. Volatile and gaseous compounds have been found to be produced in soil, and in sedimentary samples supplemented with inorganic selenium compounds [100]. Environmental factors that affect and stimulate microbial production of DMSe include addition of carbon sources, alkaline pH values and oxidative down to moderately reduced conditions. The mechanism of selenium biomethylation appears similar to that found for arsenic with carbonium ion from *S*-adenosylomethionine [8,98].

Besides dissimilatory reduction of selenate to Se(0) (see chapters 6 and 11), microbial volatilization of selenium is another method for bioremediation of selenium-contaminated soil, sediments and waters. This treatment consists of stimulating the activity of Se-methylating microorganisms naturally occurring in the contaminated environment by addition of nutrients (carbon source), aeration (ploughing and irrigation), optimizing moisture and elevating temperature [101].

4.3 MERCURY

Mercury is the only environmental toxicant whose input from natural sources into the environment is greater than that from anthropogenic sources. It is the only metal to occur in nature in a liquid state at ambient temperature [102]. The most common form of mercury to be found in nature is elemental mercury whose natural sources include weathering of cinnabar deposits and volcanic and geothermal emissions. Monovalent and divalent forms normally do not exist as ionic species, but are usually associated with inorganic or organic compounds [103].

Mercuric sulfide (HgS) occurs as red hexagonal (α-form cinnabar) or black cubic crystals (β-form, metacinnabar). Metacinnabar crystals are mainly produced by microbial production of H_2S which reacts with Hg compounds. HgS is very stable as to its turnover in the Earth's pool in terms of millions of years. Natural events and mining of cinnabar release free Hg from the sulfide ore. Trace amounts of mercury occur in most rocks with high sulfide-containing minerals. This mercury is released to the soil during weathering [104].

Some amounts of mercury can be released from sulfide minerals by microbial activity. Hg-resistant *T. ferrooxidans* strain has been reported to produce Hg(0) from red cinnabar mixed with pyrite in 1:1 ratio (w/w); however, it could not grow chemolithotrophically on mercury sulfide [105]. The relatively high vapor pressure of elemental mercury makes it degass from the lithosphere and hydrosphere [106]. Its main transformations involve oxidation/reduction and methylation/demethylation, which can result in mercury species of higher mobility and solubility. Volatilization of Hg(0) is considered to be the most common mercury detoxification mechanism in bacteria. Microbes transform inorganic Hg compounds to volatile Hg(0) by means of the enzyme mercuric reductase coded by plasmid genes (*mer A*) [106,107].

Under anaerobic conditions, mercuric ion undergoes microbial methylation into methylmercury compounds [108]. In the reaction of Hg(II) with methylcobalamine two products can be formed: salts of the methylmercuric ion (CH_3Hg^+)—which are hydrophobic and more (200-fold) toxic than Hg(II)—and the final product—volatile dimethylmercury—which is harmless and volatilizes. In air it is photolyzed to methane and elemental mercury, or oxidized by hydroxyl radicals [8,103,109,110].

5 MOBILIZATION OF METALS DUE TO MICROBIOLOGICALLY MEDIATED CHANGES IN ENVIRONMENTAL CONDITIONS

Weathering of minerals and mobilization of constituent metals can be enhanced by indirect effects of microbial activity. The growth of microorganisms can affect environmental conditions through changes of redox potential and pH as a result of oxygen depletion and secretion of metabolites (acids and reductants) by aerobes in zones with limited aeration. Microbial decomposition and transformation of native and added organic matter may also affect the redox status of the environment and enrich it with complexing agents.

Both Mn and Fe oxides sometimes interact rapidly with a variety of different organic reductants such as organic acids, phenolic compounds, quinones and others, all of which are biological products and many of which may be found free in anoxic environments such as wetlands and sediments. These reductants can rapidly reduce both metals abiotically, especially at low pH [111,112], achieving what might otherwise be attributed to microbiologically catalyzed reduction. However, as field studies have indicated, only insignificant amounts of manganese can be reduced nonenzymatically. Iron reduction even in marine wetlands and sediments rich in a strong reductant such as H_2S, is primarily biologically catalyzed [4,113].

Addition of oxalate or hydroxylamine to the soil in which microbial activity was inhibited with antibiotics resulted in notable chemical reduction of Fe(III), whereas ethanol, glucose, formic acid and formaldehyde reduced Fe(III) to a lesser degree. However, even after the addition of as much as 0.4 g of oxalate to 4 g of soil, the amount of Fe(III) reduced over an incubation period of 10 days was less than 15 % of the Fe(III) reduced in untreated soils in which microbial iron reduction was not inhibited. Acetic and butyric acids—two common fermentation products in anaerobic environments—did not reduce Fe(III) even at pH 4.4 [114].

Organic acids of low molar mass produced by microorganisms can significantly change pH in the surrounding environment and enhance solubilization of minerals. The concentration of citric acid produced by mycorrhizal plants was demonstrated to be significantly higher than that produced by nonmycorrhizal plants, and likely to reduce soil pH by 0.5–1.0 units [115]. Jongmans *et al.* [116] have provided direct evidence that the ectomycorrhizal mycelia are able to penetrate and colonize mineral microsites. These authors reported fungal hyphae to penetrate aluminosilicate minerals physically by means of strongly complexing acid exudates and create pores to further enhance weathering rates of the mineral.

Fungi such as *Aspergillus niger* and *Penicillium simplicissimum* are known to produce large amounts of organic acids such as citric and oxalic, which are involved in solubilization of rock phosphate ores and release of metals associated with them. They form a wide range of complexes with metals including magnesium, manganese, copper, iron and calcium (Table 2). This ability of fungi has a potential for leaching metals from industrial waste material as well as from low-grade ores. The ability of *P. simplicissimum* to produce large amounts of citric acid was utilized to extract zinc from industrial filter dust [117,118]. Likewise, *Pseudomonas cepacia* and *Erwinia herbicola*, Gram-negative soil bacteria, are highly efficient solubilizers of poorly soluble mineral phosphates such as fluoroapatite (rock phosphate ore), and mobilize metals associated with the minerals. Solubilization is a result of acidification of the periplasmic space by direct oxidation of glucose or other aldose sugars producing gluconic acid and 2-ketogluconic acid. These acids bring acidification of a region adjusted to the cell or colony. Gluconic acid is weak but capable of dissolving oxides, hydroxides and carbonates of polyvalent cations, forming water-soluble complexes with these cations [119]. As reported by Bipp *et al.* [120], at concentrations of 0.5–5%, this compound could be used as an effective agent for removing heavy metals bound to or included in the matrix of waste incineration residues. Leaching of metals with sugar acids was found to be particularly effective in alkaline media in which gluconic acid forms a very stable complex with heavy metals [120].

Mobilization of metals such as Zn and Cu contained in the soil have been shown to be usually less affected by changes in redox potential in the surrounding environment than by humic acids which strongly complex them at pH levels around 4.5–5.0 [121]. Meanwhile mobilization of tungsten from its natural mineral form (wolframite—iron and/or manganese tungstate) as an indirect result of microbial activity was demonstrated in laboratory cultures by Chashchina and Lyalikova [122]. These authors indicated that under aerobic conditions bacteria were able to accelerate tungsten mobilization from the surface of polished pieces of wolframite in liquid media. Some tungsten mobilization was noted even in the absence of bacteria in the mineral acid media—9K at pH 2.3 and Leathens of pH 3.5 without iron and manganese added—and in neutral (pH 7.0) van Veen medium containing organic compounds. In the uninoculated acid mineral media, mobilization of tungsten (as tungstic acid) was an effect of low pH, yet in the organic medium it was the effect of complexation with organic ligands contained in the medium (mono-, di- and tri-carboxylic acids). In the presence of bacteria *T. ferrooxidans* or *Arthrobacter siderocapsulatus* an enhanced leaching of tungsten, iron and manganese from wolframite was observed. During leaching, new minerals precipitated on the surface of wolframite. The pH of the media tended to drop slightly in the presence of bacteria, whereas it remained fairly constant in the uninoculated controls. The tungsten solubilized by the bacteria, especially by *A. siderocapsulatus*, was believed to be held in the solution as a result of complexation by some of the organic acids formed by the bacteria.

6 CONCLUSIONS

Metal ions play a variety of very important roles in microbial metabolism. They can be used as electron acceptors, a source of energy; they are also constituents of cytochromes, and essential compounds of specific enzymes, growth factors and enzyme cofactors. Because minerals are often the only source of physiological essential metals in the soil, in order to survive, microbes have developed various strategies to mobilize them from insoluble sources. The consequence of microbial activity involved in mineral dissolution is their significant role in metal cycling in the biosphere. Chemical mineral weathering, as a part of the soil formation process, can be enhanced by microbial activity by a factor as high as 10^6. So, microbially mediated mineral dissolution processes can promote soil mineral diagenesis, influence the nutrient status of the soil, quantity and quality of metal ions in surface and subsurface waters, as well as the formation of atmospheric gases. As to ecological and economic effects, microbial dissolution of minerals, depending on environmental conditions, can be harmful or beneficial.

The activity of iron and sulfur oxidizers can result in the formation of acidic sulfate soils of low agricultural value because the yield of plants grown in such soils is usually low, and the biomass contains large amounts of toxic heavy metals. Another environmental problem relates to formation of acid main drainage—enriched with sulfate and toxic heavy metals—hostile to most forms of aquatic life. However, the activity of this physiological group of microbes can be beneficial for the ecosystem or economy.

Microbial mining (or bioleaching) is a term used for extraction of metals or removal of constituents interfering with the extraction of metal elements from ores through mediation of microorganisms. This technology of copper recovery from sulfide ores has been practiced on an industrial scale since the late 1950s [123]. At present, it is used in the extraction of copper from low grade ores (containing less than 0.5% copper) and uranium, and it applies to the recovery of gold from gold-bearing arsenopyrite ores. However, this technology can also be applied to ores containing sulfides of zinc, lead, cobalt, nickel, bismuth and antimony [5]. It has been estimated that at least 15% of the world production of copper is obtained by bioleaching [123]. Deposits of low-grade copper ores (chalcopyrite, chalcocite, covellite) are bioleached *in situ*. Higher-grade ores are most often treated *ex situ*, in a similar process called heap leaching. Bioleaching of uranium *in situ* has been practiced since the 1960s [20]. Uranium occurs in ores as insoluble tetravalent oxides and such ores usually contain some pyrite ingredients. Dissolution of uranium from ores is a result of oxidation of U(IV) to soluble hexavalent uranyl ion by *T. ferrooxidans* activity.

Since the 1980s bioleaching has been a part of the technological process for the extraction of gold from recalcitrant ores [20]. In those gold is encased in a matrix of pyrite and arsenopyrite, and for this reason cannot be solubilized by the cyanidation process usually applied [5]. Bioleaching, in which chemolitho-trophic bacteria (thiobacilli) oxidize (dissolve) arsenopyrite, is used to expose gold for cyanide extraction. After this pretreatment, more than 95% of the gold is recovered, depending on the mineral composition of the ore and the extent of treatment [20].

Biomining has become a very significant part of industrial microbiology given relatively low operating costs. All tested *T. ferrooxidans* and *L. ferrooxidans* strains contain *nif* genes [124], and are able to grow in very nutrient-poor solutions. Aeration of a suitable ore in water is usually sufficient to satisfy their essential growth requirements. Air provides carbon (CO_2), nitrogen (N_2) and electron acceptor (O_2) sources while the ore is a source of energy, and the growth medium the source of water. Microorganisms are probably incapable of fixing N_2 when growing in an aerobic environment. Therefore, growing in highly aerated tank reactors, small amounts of ammonium sulphate and potassium phosphate are added to ensure sufficient nutrient supply. Because biomining bacteria tolerate many metal ions—from moderate up to high levels—and only few other organisms are able to grow in inorganic acidic environment, sterility is not necessary [20].

At present, the leaching potential of indigenous sulfur-oxidizing bacteria is also verified as a treatment for removal of heavy metals (Cd, Cr, Cu, Ni, Pb, Zn, Mn, Co and As) from contaminated river sediments and wastewater sludge. Bioleaching with elemental sulfur as a substrate to thiobacilli was found to be better than the method with sulfuric acid added for solubilization of all the metals tested [54]. During coal combustion sulfur dioxide is produced, which is a cause of acid rains. Inorganic sulfur contained in coal consists mainly of pyrite, generally associated with coal fields all over the world. Coal desulfurization/depyritization can be performed by abiotic as well as microbiological oxidation of sulfur entity of pyrite [125]. Microorganisms effective in leaching metal sulfide ores containing pyrite appeared to be suitable in coal desulfurization.

Organic acids of low molar mass produced by free-living heterotrophs as well as ectomycorrhizal fungi are a significant factor involved in the processes of bioremediation of soils contaminated with heavy metals or radionuclides, and removal of these elements from industrial waste materials [126].

Application of new molecular techniques, such as analysis of DNA extracted from environmental samples, has shown natural sites enriched with soluble metals inhabited by physiological diverse microorganisms often in consortia, including autotrophs as well as heterotrophs. Knowledge of the interactions between microbes living in such sites and the effect of environmental factors on their activity may be helpful in limiting the adverse effects of microbes on the soil.

REFERENCES

1. Kimbrough, D. E., Cohen. Y., Winer, A. M., Creelman, L. and Mabuni, C. (1999). A critical assessment of chromium in the environment, *Crit. Rev. Environ. Sci. Technol.*, **29**, 1.
2. Douglas, S. and Beveridge, T. J. (1998). Mineral formation by bacteria in natural microbial communities, *FEMS Microbiol. Ecol.*, **26**, 79.
3. Ahmann, D., Roberts, A. L., Krumholz, L. R. and Morel, F. M. M. (1994). Microbe grows by reducing arsenic, *Nature* (London), **371**, 750.
4. Nealson, K. H. and Saffarini, D. (1994). Iron and manganese in an anaerobic respiration. Environmental significance, physiology and regulation, *Annu. Rev. Microbiol.*, **48**, 311.
5. Rawlings, D. E. and Silver, S. (1995). Mining with microbes, *Biotechnology*, **13**, 773.
6. Francis, A. J. (1998). Biotransformation of uranium and other actinides in radioactive waste, *J. Alloys Compounds*, **271/273**, 78.
7. Guerinot, M. L. (1994). Microbial iron transport, *Annu. Rev. Microbiol.*, **48**, 743.
8. Gadd, G. M. (1993). Microbial formation and transformation of organometallic and organometalloid compounds, *FEMS Microbiol. Rev.*, **11**, 297.
9. Tichý, R., Lens, P., Grotenhuis, J. T. C. and Bos, P. (1998). Solid state reduced sulfur compounds: environmental aspects and bioremediation, *Crit. Rev. Environ. Sci. Technol.*, **28**, 1.
10. Evangelou, V. P. and Zhang, Y. L. (1995). A review: Pyrite oxidation mechanisms and acid mine drainage prevention, *Crit. Rev. Environ. Sci. Technol.*, **25**, 141.
11. Arora, H. S., Dixon, J. B. and Hessner, L. R. (1978). Pyrite morphology in lignitic coal and associated strata of east Texas, *Soil Sci.*, **125**, 151.
12. Ainsworth, C. C. and Blanchar, R. W. (1984). Sulfate production rates in pyritic Pennsylvania – aged shales, *J. Environ. Qual.*, **13**, 93.
13. Nicholson, R. V. and Scharer, J. M. (1994). Laboratory studies of pyrrhotite oxidation kinetics. In *Environmental Geochemistry of Sulfide Mine-wastes*, Short Course Handbook, Vol. 22, ed. Blowes, D. W. and Jambor, J. L., Mineralogical Association of Canada, Ottawa, Ont., p. 163.
14. Bhatti, T. M., Bigham, J. M., Vuorinen, A. and Tuovinen, O. H. (1995). Biological leaching of sulfides with emphasis on pyrrhotite and pyrite. In *Biotechnology for Sustainable Development*, ed. Malik, K. A., Nasim, A. and Khalid, A. M., National Institute for Biotechnology and Genetic Engineering, Faisalabad, Pakistan, p. 299.
15. Dixon, J. B. and Weed, S. B. (eds) (1989). *Minerals in Soil Environments*, 2nd edn, Soil Science Society of America, Madison, WI.
16. Garcia, O. Jr., Bigham, J. M. and Tuoavinen, O. H. (1995). Sphalerite oxidation by *Thiobacillus ferrooxidans* and *Thiobacillus thiooxidans*, *Can. J. Microbiol.*, **41**, 574.
17. Garcia, O. Jr., Bigham, J. M. and Tuovinen, O. H. (1995). Oxidation of galena by *Thiobacillus ferrooxidans* and *Thiobacillus thiooxidans*, *Can. J. Microbiol.*, **41**, 500.
18. Ahonen, L. and Tuovinen, O. H. (1995). Bacterial leaching of complex sulfide ore samples in bench-scale column reactor, *Hydrometallurgy*, **37**, 1.
19. Rawlings, D. E. (1997). Mesophilic, autotrophic, bioleaching bacteria: description, physiology and role. In *Biomining: Theory, Microbes and Industrial processes*, ed. Rawlings, D. E. and Landes R. G., Springer, Berlin, p. 229.
20. Rawlings, D. E. (1998). Industrial practice and the biology of leaching of metals from ores. The 1997 Pan Lab Lecture *J. Ind. Microbiol. Biotechnol*, **20**, 268.
21. De Silo'niz, M. I., Lorenzo, P. and Perera, J. (1991). Distribution of oxidizing bacteria activities and characterization of bioleaching, related microorganisms in a uranium mineral heap, *Microbiological SEM, 7, 82.*

22. Sugio, T., Tsujita, Y., Hirayama, K., Inagaki, K. and Tano, T. (1988). Mechanism of tetravalent manganese reduction with elemental sulfur by *Thiobacillus ferrooxidans*, *Agric. Biol. Chem.*, **52**, 185.

23. Sugio, T., Tsujita, Y., Inagaki, K. and Tano, T. (1990). Reduction of cupric ions with elemental sulfur by *Thiobacillus ferrooxidans*, *Appl. Environ. Microbiol.*, **56**, 693.

24. Benz, M., Brune, A. and Schink, B. (1998). Anaerobic and aerobic oxidation of ferrous iron at neutral pH by chemoheterotrophic nitrate-reducing bacteria, *Arch. Microbiol.*, **169**, 159.

25. Straub, K. L. and Buchholz-Cleven, B. E. E. (1998). Enumeration and detection of anaerobic ferrous iron-oxidizing, nitrate-reducing bacteria from diverse European sediments, *Appl. Environ. Microbiol.*, **64**, 4846.

26. Schippers, A., Hallmann, R., Wentzien, S. and Sand, W. (1995). Microbial diversity in uranium mine waste heaps, *Appl. Environ. Microbiol.*, **61**, 2930.

27. Norris, P. R. and Kelly, D. P. (1982). The use of mixed microbial cultures in metal recovery. In *Microbial Interaction and Communities*, Vol. 1, ed. Bull, A. T. and Slater, J. H., Academic Press, London, p. 443.

28. Johnson, D. B., Ghauri, M. A. and Said, M. F. (1992). Isolation and characterization of an acidophilic heterotrophic bacterium capable of oxidizing ferrous iron, *Appl. Environ. Microbiol.*, **58**, 1423.

29. Bacelar-Nicolau, P. and Johnson, D. B. (1999). Leaching of pyrite by acidophilic hetrotrophic iron-oxidizing bacteria in pure and mixed culture, *Appl. Environ. Microbiol.*, **65**, 585.

30. Berthelot, D., Leduc, L. G. and Ferroni, G. D. (1993). Temperature studies of iron-oxidizing autotrophs and acidophilic heterotrophs isolated from uranium mines, *Can. J. Microbiol.*, **39**, 384.

31. Berthelot, D., Leduc, L. G. and Ferroni, G. D. (1994). The absence of psychrophilic *Thiobacillus ferrooxidans* and acidophilic heterotrophic bacteria in cold, tailings effluents from a uranium mine, *Can. J. Microbiol.*, **40**, 60.

32. Leduc, L. G., Trevors, J. T. and Ferroni, G. D. (1993). Thermal characterization of different isolates of *Thiobacillus ferrooxidans*, *FEMS Microbiol. Lett.*, **108**, 189.

33. Niemela, S. I., Sivela, C., Luoma, T. and Touvinen, O. H. (1994). Maximum temperature limits for acidophilic, mesophilic bacteria in biological leaching systems, *Appl. Environ. Microbiol.*, **60**, 3444.

34. Murr, L. E. and Brierley, J. A. (1978). The use of large-scale test facilities in studies of the role of microorganisms in commercial leaching operations. In *Metallurgical Applications of Bacterial and Related Microbiological Phenomena*, ed. Murr, L. E., Torma, A. E. and Brierley, J. A., Academic Press, New York, p. 491.

35. Tuovinen, O. H., Bhatti, T. M., Bigham, J. M., Hallberg, K. B., Garcia, O. Jr. and Lindström, E. B. (1994). Oxidative dissolution of arsenopyrite by mesophilic and moderately thermophilic acidophiles, *Appl. Environ. Microbiol.*, **60**, 3268.

36. Hallberg, K. B. and Lindström, E. B. (1994). Characterization of *Thiobacillus caldus* sp. *nov*, a moderately theromphilic acidophile, *Microbiology*, **140**, 3451.

37. Norris, P. R. (1997). Thermophiles and bioleaching. In *Biomining: Theory, Microbes and Industrial Processes*, ed. Rawlings, D. E. and Landes, R. G., Springer, Berlin, p. 247.

38. Norris, P. R. and Parrot, L. (1986). High temperature, mineral concentrate dissolution with *Sulfolobus*. In *Fundamental and Applied Biohydrometallurgy*, ed. Lawrence, R. W., Branion, R. M. R. and Ebner, H. G., Elsevier, Amsterdam, p. 355.

39. Konishi, Y., Yoshida, S. and Asai, S. (1995). Bioleaching of pyrite by acidophilic thermophile *Acidianus brierleyi*, *Biotechnol. Bioeng.*, **48**, 592.

40. Pizarro. J., Jedlicki, E., Orellana, O., Romero, J. and Espejo, R. T. (1996). Bacterial populations in samples of bioleached copper ore as revealed by analysis of DNA obtained before and after cultivation, *Appl. Environ. Microbiol.*, **62**, 1323.

41. Rawlings, D. E. (1995). Restriction enzyme analysis of 16S rRNA genes for the rapid identification of *Thiobacillus ferrooxidans*, *Thiobacillus thiooxidans* and *Leptospirillum ferrooxidans* strains in leaching environments. In *Biohydrometallurgical Processing*, Vol. II, ed. Jerez, C. A., Vargas, T., Toledo, H. and Wiertz, J. V., University of Chile Press, Santiago, p. 9.

42. Vasquez, M. and Espejo, R. T. (1997). Chemolitotrophic bacteria in copper ores leached at high sulfuric acid concentration, *Appl. Environ. Microbiol.*, **63**, 332.

43. Mustin, C., Berthelin, J., Marion, P. and Donato, P. (1992). Corrosion and electrochemical oxidation of pyrite by *Thiobacillus ferrooxidans*, *Appl. Environ. Microbiol.*, **58**, 1175.

44. Bennett, J. C. and Tributsch, H. (1978). Bacteria leaching patterns on pyrite crystal surface, *J. Bacteriol.*, **134**, 310.

45. Tuovinen, O. H., Kelley, B. C. and Groudev, S. N. (1991). Mixed cultures in biological leaching processes and mineral biotechnology. In *Mixed Cultures in Biotechnology*, ed. Zeikus, J. G. and Johnson, E. A., McGraw-Hill, New York, p. 373.

46. Luther, G. W. III. (1987). Pyrite oxidation and reduction: molecular orbital theory consideration, *Geochem. Cosmochem. Acta*, **51**, 3193.

47. Sand, W., Gerke, T., Hallmann, R. and Schippers, A. (1995). Sulfur chemistry, biofilm and the (in) direct attack mechanism—a critical evaluation of bacterial leaching, *Appl. Microbiol. Biotechnol.*, **43**, 961.

48. Pistario, M., Curutchet, G., Donato, E. and Tedesco, P. (1994). Direct zinc sulphide bioleaching by *Thiobacillus ferrooxidans* and *Thiobacillus thiooxidans*, *Biotechnol. Lett.*, **16**, 419.

49. Ivarson, K. C. and Sojak, M. (1978). Microorganisms and ochre deposits in field drains of Ontario, *Can. J. Soil Sci.*, **58**, 1.

50. Smith, E., Naidur, R. and Alston, A. M. (1998). Arsenic in the soil environment: A review, *Adv. Agron.*, **64**, 149.

51. Howard, J. H. (1977). III Geochemistry of selenium: formation of ferroselite and selenium behavior in the vicinity of oxidizing sulfide and uranium deposits, *Geochim. Cosmochim. Acta*, **41**, 1665.

52. Maass, B. and Miehlich, G. (1988). Die Wirrkung des Redoxpotentials auf die Zusammnesetzung der orenlösung in Hafenschlicks-feldern, *Mitt Dtsch. Bodekunde. Ges.*, **56**, 289.

53. Bradshaw, A. D. (1993). Natural rehabilitation strategies. In *Integrated Soil and Sediment Research: A Basis for Proper Protection*, ed. Eijsackers, H. J. P. and Hamers, T., Kluwer, Dordrecht, The Netherlands, p. 577.

54. Seidel, H., Ondruschka, J., Morgenstern, P. and Stottmeister, U. (1998). Bioleaching of heavy metals from contaminated aquatic sediments using indigenous sulfur-oxidizing bacteria: a feasibility study, *Wat. Sci. Technol.*, **37**, 387.

55. Fournier, D., Lemieux, R. and Couillard, D. (1998). Genetic evidence for highly diversified bacterial populations in wasterwater sludge during biological leaching of metals, *Biotech. Lett.*, **20**, 27.

56. Johnson, D. B. (1995). Acidophilic microbial communities: candidates for bioremediation of acidic mine effluents, *Int. Biodeter. Biodegrad.*, 41.

57. Nordstrom, D. K. (1982). Aqueous pyrite oxidation and the consequent formation of secondary iron minerals. In *Acid Sulfate Weathering: Pedogeochemistry and Relationship to Manipulation of Soil Minerals*, ed. Hossner, L. R., Kittrick, J. A. and Fanning, D. F., Soil Science Society of America, Madison, WI.

58. Singer, P. C. and Stumm, W. (1970). Acid mine drainage: rate-determining step, *Science*, **167**, 1121.
59. Wortman, A. T., Voelz, H., Lantz, R. C. and Bissonnette, G. K. (1986). Effect of acid mine water on *Escherichia coli*: structural damage, *Curr. Microbiol.*, **14**, 1.
60. Johnson, D. B. and McGinness, S. (1991). Ferric iron reduction by acidophilic heterotrophic bacteria, *Appl. Environ. Microbiol.*, **57**, 207.
61. Johnson, D. B. (1995). Mineral cycling by bacteria: iron bacteria. In *Microorganisms and the Maintenance of Biodiversity: Microbial Diversity and Ecosystem Function*, CAB International, Egham, UK, p. 137.
62. Blessing, N. V., Lackey, J. A. and Spry, A. H. (1975). *Minerals and Environment*, ed. Jones, E. J., Institution of Mining, London, p. 341.
63. Suschka, J. (1993). Effects of heavy metals from mining and industry on some rivers in Poland: an already exploded chemical time bomb? *Land Degrad. Rehab.*, **4**, 387.
64. Ondrushka, G. and Glombitza, F. (1993). Inhibition of natural microbiological leaching processes. In *Proc. Int. Conf. Contaminated Soil*, ed. Arendt, F., Annokei, G. J., Bosman, R. and van der Brink, W. J., Kluwer, Dordrecht, The Netherlands, p. 1195.
65. Nyavor, K., Egeibor, N. O. and Fedorak, P. M. (1996). Suppression of microbial pyrite oxidation by fatty acid amine treatment, *Sci. Tot. Environ.*, **182**, 75.
66. Dugan, P. R. (1987). Prevention of formation of acid drainage from high-sulfur coal refuse by inhibition of iron-oxidizing and sulfur-oxidizing bacteria. 2. Inhibition in run of mine refuse under simulated field condition, *Biotechnol. Bioeng.*, **29**, 49.
67. Spotts, E. and Dollhopf, D. J. (1992). Evaluation of phosphate materials for control of acid production in pyritic mine overburden, *J. Environ. Qual.*, **21**, 627.
68. Rose, P. D., Boshoff, G. A., van Hille, R. P., Wallace, L. C. M., Dunn, K. M. and Duncan, J. R. (1998). An integrated algal sulphate reducing high rate ponding process for the treatment of acid mine drainage wastewaters, *Biodegradation*, **9**, 247.
69. Streyer, L. (1995). *Biochemistry*, 4th edn, W. H. Freeman, New York
70. Briat, J. F. (1992). Iron assimilation and storage in prokaryotes, *J. Gen. Microbiol.*, **138**, 2475.
71. Neilands, J. B. (1991). A brief history of iron metabolism, *Biol. Metals*, **4**, 1.
72. Dilworth, M. J., Carson, K. C., Giles, R. G. F., Byrne, L. T. and Glenn, A. R. (1998). *Rhizobium leguminosarum* bv *viciae* produces a novel cyclic trihydroxamate siderophore, vicibactin, *Microbiology*, **144**, 781.
73. Haselwandter, K. (1995). Mycorrhizal fungi: siderophore production, *Crit. Rev. Biotechnol.*, **15**, 287.
74. Hider, R. C. (1984). Siderophore mediated absorption of iron. In *Structure and Bonding*, Vol. 58, Springer, Berlin, p. 25.
75. Moody, M. D. (1986). Microorganisms and iron limitation, *BioScience*, **36**, 618-622.
76. Höfte, M. (1993). Classes of microbial siderophores. In *Iron Chelation in Plants and Soil Microorganisms*, ed. Barton, L. L. and Hemming, B. C., VCH Press, San Diego, CA, p. 3.
77. Castignetti, D. and Smarrelli, J. Jr. (1986). Siderophores, the iron nutrition of plants, and nitrate reductase, *FEBS Lett.*, **209**, 147.
78. Buyer, J. S., Wright, J. M. and Leong, J. (1986). Structure of pseudobactin A214, a siderophore from a bean-deleterious *Pseudomonas*, *Biochemistry*, **25**, 5492.
79. Schwyn, B. and Neilands, J. B. (1987). Universal chemical assay for the detection and determination of siderophores, *Anal. Biochem.*, **160**, 47.
80. Braun, V. and Hantke, K. (1991). Genetics of bacteria iron transport. In *Handbook of Microbial Iron Chelates*, ed. Winkelmann, G., CRC Press, Boca Raton, FL, p. 107.

81. Page, W. J. and Dale, P. L. (1986). Stimulation of *Agrobacterium tumefaciens* growth by *Azotobacter vinelandii* ferrisiderophores, *Appl. Environ. Microbiol.*, **51**, 451.
82. Garrison, J. M. and Crumbliss, A. L. (1987). Kinetics and mechanism of aluminum (III) siderophore ligand exchange: mono(deferriferrioxamine B) aluminum(III) formation and dissociation in aqueous acid solution, *Inorg. Chim. Acta*, **138**, 61.
83. Hu, X. and Boyer, G. L. (1996). Siderophore-mediated aluminum uptake by *Bacillus megaterium* ATCC 19213, *Appl. Environ. Microbiol.*, **62**, 4044.
84. Brainard, J. R., Strietelmeier, B. A., Smith, P. H., Langston-Unfeker, P. J., Barr, M. E. and Ryan, R. R. (1992). Actinide binding and solubilization by microbial siderophores, *Radiochim. Acta*, **58/59**, 357.
85. Duhme, A. K., Hider, R. C. and Khodr, H. (1996). Spectrophotometric competition study between molybdate and Fe(III) hydroxide on N,N'-bis(2,3-dihydroxybenzoyl)-L-lysine, a naturally occuring siderophore synthesized by *Azotobacter vinelandii*. *BioMetals*, **9**, 245.
86. Evers, A., Hancock, R. D., Martell, A. E. and Motekaitis, R. J. (1989). Metal ion recognition in ligands with negatively charged oxygen donor groups. Complexation of Fe(III), Ga(III), In(III), Al(III) and other highly charged metal ions, *Inorg. Chem.*, **28**, 2189.
87. Chen, Y., Jurkevitch, E., Bar-Ness, E. and Hadar, Y. (1994). Stability constants of pseudobactin complexes with transition metals, *Soil. Sci. Soc. Am. J.*, **58**, 390.
88. Gilis, A., Corbisier, P., Baeyens, W., Taghavi, S., Margeay, M. and van der Lelie, D. (1998). Effect of the siderophore alcaligin E on the bioavailability of Cd to *Alcaligenes eutrophus* CH34, *J. Ind. Microbiol. Biotech.*, **20**, 61.
89. Akers, H. A. (1983). Isolation of the siderophore schizokinen from soil of rice fields, *Appl. Environ. Microbiol.*, **45**, 1704.
90. Jones, D. L. and Kochian, L. V. (1996). Aluminum-organic acid interactions in acid soils 1. Effect of root-derived organic-acids on the kinetics of Al dissolution, *Plant Soil*, **182**, 221.
91. Jones, D. L. (1998). Organic acids in the rhizosphere-a critical review, *Plant Soil*, **205**, 25.
92. Jones, D. L., Darrah, P. R. and Kochian, L. V. (1996). Critical-evaluation of organic-acid mediated iron dissolution in the rhizosphere and its potential role in root iron uptake, *Plant Soil*, **180**, 57.
93. Lundström, U. S. and Öhman, L. O. (1990). Dissolution of feldspars in the presence of natural organic solutes, *J. Soil. Sci.*, **41**, 359.
94. Drechsel, H., Thieken, A., Reissbrodt, R., Jung, G. and Winkelmann, G. (1993). α-Ketoacids are novel siderophores in the genera *Proteus*, *Providencia* and *Morganella* and are produced by amino acid deaminasas, *J. Bacteriol.*, **175**, 2727.
95. Tamaki, S. and Frankenberger. W. T. (1992). Environmental biochemistry of arsenic, *Rev. Environ. Contam. Toxicol.*, **124**, 79.
96. Ehrlich, H. L. (1996). *Geomicrobiology*, 3rd edn, Marcel Dekker, New York.
97. Honschopp, S., Brunken, N., Nehrkorn, A. and Bruenig, H. J. (1996). Isolation and characterization of a new arsenic methylating bacterium from soil, *Microbiol. Res.*, **151**, 37.
98. Masschelyen, P. H. and Patrick, W. H. Jr. (1993). Biogeochemical processes affecting selenium cycling in wetlands, *Environ. Toxic. Chem.*, **12**, 2235.
99. Sarathchandra, S. U. and Watkinson, J. H. (1981). Oxidation of elemental selenium to selenite by *Bacillus megaterium, Science*, **211**, 600.
100. Thompson-Eagle, E. T., Frankenberger, W. T. and Karlson, V. (1989). Volatilization of selenium by *Alternaria alternata*, *Appl. Environ. Microbiol.*, **55**, 1406.

101. Thompson-Eagle, E. T. and Frankenberger, W. T. (1992). Bioremediation of soils contaminated with selenium, *Adv. Soil Sci.*, **17**, 261.

102. Stein, E. D., Cohen, Y. and Winer, A. M. (1996). Environmental distribution and transformation of mercury compounds, *Crit. Rev. Environ. Sci. Technol.*, **26**, 1.

103. Baldi, F. (1997). Microbial transformation of mercury species and their importance in the biogeochemical cycle of mercury. In *Metal Ions in Biological Systems*, Vol. 34, ed. Sigel, A. and Sigel, H., Marcel Dekker, New York, p. 213.

104. Bailey, E. H., Clark, A. L. and Smith, R. M. (1972). Mercury. In *United States Mineral Resources. US Geological Survey Professional Paper* 820, Goverment Printing Office, Washington, DC, p. 401.

105. Baldi, F. and Olson, G. J. (1987). Effects of cinnabar on pyrite oxidation by *Thiobacillus ferrooxidans* and cinnabar mobilization by a mercury-resistant strain, *Appl. Environ. Microbiol.*, **53**, 772.

106. Silver, S. and Walderhaug, M. (1992). Ion transport. In *Encyclopedia of Microbiology*, ed. Lederberg, J., Academic Press, San Diego, CA.

107. Chang, J-S., and Law, W-S. (1998). Development of microbial mercury detoxification processes using mercury-hyperresistant strain of *Pseudomonas aeruginosa* PU21, *Biotechnol. Bioengin.*, **57**, 467.

108. Seigneur, C., Wrobel, J. and Constantinou, E. (1994). A chemical kinetic mechanism for atmospheric inorganic mercury, *Environ. Sci. Tech.*, **28**, 1589.

109. Choi, S-C., Chase, T. Jr. and Bartha, R. (1994a). Metabolic pathways leading to mercury methylation in *Desulfovibrio desulfuricans* LS, *Appl. Environ. Microbiol.*, **60**, 4072.

110. Choi. S-C., Chase, T. Jr. and Bartha, R. (1994b). Enzymatic catalysis of mercury methylation by *Desulfovibrio desulfuricans* LS, *Appl. Environ. Microbiol.*, **60**, 1342.

111. Stone, A. T., Godtfredsen, K. L. and Deng, B. (1994). Sources and reactivity of reductants in aquatic environments. In *Chemistry of Aquatic Systems: Local and Global Perspectives*, ed. Bidoglio, G. and Stumm, W., Kluwer, Dordrecht.

112. Jones, J. G., Gardener, S. and Simon, B. M. (1984). Reduction of ferric iron by heterotrophic bacteria in lake sediments, *J. Gen. Microbiol.*, **130**, 45.

113. Lovley, D. R., Phillips, E. J. and Lonergan, D. J. (1991). Enzymatic versus non-enzymatic mechanisms for Fe(III) reduction in aquatic sediments, *Environ. Sci. Technol.*, **25**, 1062.

114. Lovley, D. R. (1987). Organic matter mineralization with reduction of ferric iron: a review, *Geomicrobiol. J.*, **5**, 375.

115. Wickman, T. (1996). *Weathering assessment and nutrient availability in coniferous forests*, Ph.D. thesis, Department of Civic and Environmental Engineering Royal Institute of Technology, Stockholm, Sweden.

116. Jongmans, A. G., van Breemen, N., Lundström, U., Finlay, R. D, van Hees, P. A. W., Giesler, R., Melkerud, P-A., Olsson, M., Srinivasan, M. and Unestam, T. (1997). Rock-eating fungi: a true case of mineral plant nutrition? *Nature*, **389**, 682.

117. Sayer, J. A., Raggett, S. L. and Gadd, G. M. (1995). Solubilization of insoluble metal compounds by soil fungi: development of a screening method for solubilizing ability and metal tolerance, *Mycol. Res.*, **101**, 987.

118. Gharieb, M. M., Sayer, J. A. and Gadd, G. M. (1998). Solubilization of natural gypsum (CaSO$_4$ · 2H$_2$O) and the formation of calcium oxalate by *Aspergillus niger* and *Serpula himantioides. Mycol. Res.*, **102**, 825.

119. Goldstein, A. H. (1995). Recent progress in understanding the molecular genetic and biochemistry of calcium phosphate solubilization by Gram negative bacteria, *Biol. Agr. Hortic.*, **12**, 185.

120. Bipp, H-P., Wunsch, P., Fischer, K., Bieniek, D. and Kettrup, A. (1998). Heavy metal leaching of fly ash from incineration with gluconic acid a molasses hydrolysate, *Chemosphere*, **36**, 2523.
121. Waller, P. A. and Pickering, W. P. (1992). Effect of time and pH on the lability of copper and zinc sorbed on humic acid particules, *Chem. Speciation. Bioavailability*, **4**, 29.
122. Chashchina, N. M., Lyalikova, N. N. (1989). Role of bacteria in transformation of tungsten minerals, *Mikrobiologiya*, **58**, 122.
123. Agate, A. D. (1996). Recent advances in microbial mining, *World J. Microbiol. Biotechnol.*, **12**, 487.
124. Norris, P. R., Murrell, J. C. and Hinson, D. (1995). The potential for diazotrophy in iron- and sulphur-oxidizing acidophilic bacteria, *Arch. Microbiol.*, **164**, 294.
125. Olsson, G., Pott, B-M., Larsson, L., Holst, O. and Karlsson, H. T. (1995). Microbial desulfurization of coal and oxidation of pure pyrite by *Thiobacillus ferrooxidans* and *Acidianus brierleyi*, *J. Ind. Microb.*, **14**, 420.
126. Entry, J. A., Vance, N. C., Hamilton, M. A., Zabowski, D., Watrud, L. S. and Adriano, D. C. (1996). Phytoremediation of soil contaminated with low concentration of radionuclids, *Wat. Air Soil Pollut.*, **88**, 167.

6 Interactions of Bacteria and Environmental Metals, Fine-grained Mineral Development, and Bioremediation Strategies

J. S. McLEAN, J.-U. LEE AND T. J. BEVERIDGE
University of Guelph, Canada

Interactions between Soil Particles and Microorganisms
Edited by P. M. Huang, J.-M. Bollag and N. Senesi. © 2002 John Wiley & Sons, Ltd

1 INTRODUCTION

Bacteria have adapted to almost every conceivable environment... from heavy metal-laden acidic mine wastes to thermal hot springs and marine vents, and the deep subsurface. Researchers in many disciplines have realized that bacteria undoubtedly play a dominant role in the speciation, fate and transport of metals, metalloids and radionuclides in the environment and, eventually, the global geochemical cycling of these elements. The implications of bacterial mineralization and metal accumulation/transformation are far reaching. Unfortunately, geochemical modeling of metal speciation and transport is only beginning to include bacteria as geochemically active surfaces. The physical and chemical characteristics of bacteria, such as their large surface area-to-volume ratio, serve to increase the metal binding capacity of their charged surfaces leading to precipitation and formation of mineral phases on their cell walls or other surface polymers. In addition, the diversity of ways in which they transform and accumulate metals makes them ideal candidates to immobilize toxic metals from contaminated environments. Mann [1] has listed the subdisciplines that encompass the topic of bacterial interactions with metals; these include biochemistry, geomicrobiology, toxicology, biomineralization, ore deposition, biofouling in industry and transportation, biocorrosion, the leaching of ores, mineral prospecting, metal recovery, and bioremediation of metal contamination.

Bacteria have been shown to sorb a variety of dissolved metals, reduce or oxidize transition metals, and participate in the formation and growth of authigenic minerals on their surfaces. Certain species of microorganisms have developed highly complex mechanisms to derive energy from metal redox reactions or to deal with toxic metal species at high concentrations through enzymatic or non-enzymatic processes. Since bacteria play a predominant role in the global geochemical cycling of metals and because these effects are generally relevant to soil science, mining industry and bioremediation of metal contamination, fundamental research into metal–microbe interactions has been the focus of many studies. Consequently, a variety of reviews devoted to describing these interactions have been published. This chapter is designed to give an overview of metal–microbe interactions, bioformation of mineral phases, and current and potential metal bioremediation applications.

2 CHARACTERISTICS OF BACTERIA

2.1 GENERAL UBIQUITY AND SIZE

Prokaryotes come in a variety of shapes that maximize their surface areas, and are typically 1.0–$5.0\,\mu m$ in length and 0.5–$1.0\,\mu m$ in width [2]. The high surface area-to-volume ratio of individual cells combined with a high surface charge density allows bacteria to concentrate metals very effectively. The surfaces of bacteria vary in that they may have several different layers or a combination of layers, each of which may possess metal binding sites. Some bacteria may have only a cell wall in contact with the aqueous environment, whereas others can generate additional closely associated layers such as capsules and S-layers or more loosely associated materials such as sheaths and 'slimes'. In the soil environment, microbial communities (such as biofilms) composed of many different bacterial cells in a matrix of variably charged exopolysaccharides, can serve to increase the metal binding capacity above that of planktonic cells [3].

It is estimated that bacteria comprise $\sim 10^{18}$ g of global living biomass with 10^6–10^9 bacteria per gram of soil [4]. Typical soil genera (e.g., *Arthrobacter*, *Bacillus* and *Actinomycetes*) may make up only 5% of the culturable types in soils. The number of bacteria present in any given environment may, however, be of less importance than their relative level of activity and the role certain genera play in the cycling of nutrients and metals. As mediators of chemical phase transitions, certain bacterial species direct major processes in soils (e.g., denitrifiers, methanogens, etc.). The available space between soil particles and colloids provides a habitat for bacteria and allows them to interact with the surrounding aqueous phase. Interfaces between biofilms and soil particles are highly complex regions where the speciation and fate of metals can be influenced by the active and passive functions of indigenous microbial communities. At the soil particle–bacteria interface, mineral dissolution may occur releasing metals into the aqueous phase where they will be more bioavailable or reform into secondary mineral phases [5]. As with many soil particles, the net charge on the surfaces of bacteria at circumneutral pH is negative due to the ionization of carboxyl and phosphoryl groups which act as a sink for metals at the soil solution–bacteria interface (Figure 1). If the cells are planktonic or are bound to colloids, they contribute to the overall mobility of metal in groundwater. Alternatively, if the cells are growing as a biofilm on larger soil particles they may sequester metals in more immobile and less bioavailable forms.

2.2 SURFACE LAYERS AND BINDING SITES OF BACTERIA

Eubacteria can be divided into gram-positive and gram-negative forms which are profoundly different in their cell wall structure. Each type of cell wall results in a variable number of metal binding sites which control complexation constants and total metal accumulating capacity. The following sections describe some general characteristics of gram-positive and gram-negative cell walls and other surface layers in relation with their metal binding capacity.

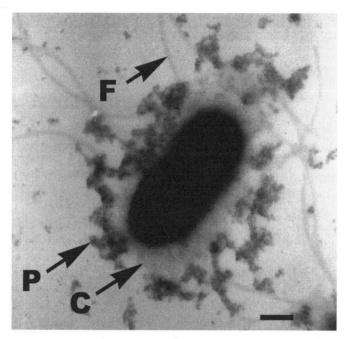

Figure 1. Unstained whole mount of a peritrichous flagellated (F) chromium-reducing bacterium with fine grained Cr(III) precipitates (P) on its capsule (C). Reduction of Cr(VI) and precipitation of Cr(III) mineral phases is mediated by this organism isolated from a Cr-contaminated groundwater. Chromium precipitates were identified by energy dispersive X-ray spectrometry (EDS). Scale bar = 500 nm. From [139b]. Reproduced by permission of American Society for Microbiology

2.2.1 Gram-positive Cell Walls

These cell walls are usually 20–50 nm thick (Figure 2) and are 40–90% comprised of peptidoglycan, a molecule containing linear chains of the disaccharide N-acetylglucosamine-β-1,4-N-acetylmuramic acid (2). The different chains are cross linked by short peptide stems emanating from the N-acetylmuramyl moieties creating a porous network which is highly permeable to molecules of molecular weights between 1,200 and 70,000 Da. Secondary polymers, such as teichoic or teichuronic acids, are cross linked into the peptidoglycan network through the N-acetylmuramyl residues and are intercalated throughout the inter-peptidoglycan spaces (2). The cell wall of *Bacillus subtilis*, as one type of gram-positive bacterium, has been well characterized and its metal sequestering properties have been identified in laboratory studies [6,7]. Carboxylates (peptidoglycan, teichuronic acid and protein) and phosphates (teichoic acid) are distributed on the cell surface as the dominant ionizable chemical groups at pH ~ 7 [7–9]. Therefore, the electronegative character of the cell wall

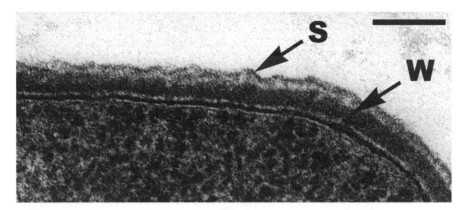

Figure 2. Thin section of a Gram-positive cell wall (W) which has an S-layer (S) on top of it. This wall belongs to a common soil bacterium named *Bacillus thuringiensis*. Scale bar = 50 nm

at circumneutral pH arises from the contribution of free carboxyl and phosphoryl groups within the wall as well as the negatively charged secondary polymers (teichoic and teichuronic acids) attached to the peptidoglycan matrix.

The enormous extent of metal accumulation by gram-positive bacteria is also enhanced by their cell architecture. The outermost surface of the wall is pocketed with numerous depressions [10] and this structure increases the surface area-to-volume ratio of the cell [11], making it possible to bind huge amounts of metal ions. The overall metal binding capacity of gram-positive cell walls is usually greater than that of gram-negative cell walls [12] and they are consequently used in a variety of applications as biosorbents [13].

2.2.2 Gram-negative Cell Walls

The structure of gram-negative cell walls is complex and consists of inner and outer bilayer membranes separated by a thin layer of peptidoglycan (Figure 3). Although a thick layer of peptidoglycan is responsible for most metal binding in gram-positive cells, gram-negative peptidoglycan is not believed to exert the same influence. It is thin (ca. 2 nm) and makes up only 10 % or less of the wall mass. In addition, it is surrounded by a dense periplasm [14,15] and is shielded from the external environment by the outer membrane [16].

However, there is a variety of negatively charged lipopolysaccharides (LPS), phospholipids and proteins in the outer membrane. Since the lipids are asymmetrically arranged so that virtually all LPS is on the outer face of the membrane, this molecule is a primary metal-binding site. The net electronegative

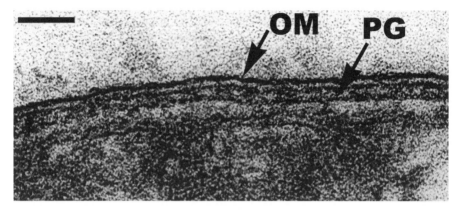

Figure 3. Thin section of a Gram-negative cell wall showing the outer membrane (OM) and peptidoglycan layer (PG)(see ref. [2] for more details). This wall also belongs to a common soil bacterium named *Pseudomonas aeruginosa*. Scale bar = 50 nm

surface charge arises mainly from the ionization of several phosphate and carboxylate groups in the LPS as well as the phosphate groups of the major phospholipid (phosphidylethanolamine) at the inner face of the membrane [12,17,18].

2.2.3 Capsules/Slimes and S-layers

Capsules and slimes are found on some species of bacteria although they are hard to examine by conventional electron microscopical methods. The differentiation between capsules and slimes is that one (capsule) is physically attached to the bacterial surface whereas the other (slime) is not [19]. The presence of these loose hydrated structures may serve as a concentrating filter for dissolved metal ions. The fabric of most capsules consists of a sequence of sugar residues (or their derivatives) linked together forming a layer that can extend from 0.1 to 10 μm from the surface into the environment [20]. In fact the presence of toxic metals may induce the production of a capsule on a bacterium which then binds metals on acidic functional groups for detoxification. Depending on the chemical properties of a particular capsule, the estimates of thermodynamic metal binding constants vary with the type of organic polymer present as well as with the geochemical conditions which control the availability of reactive sites [20].

Certain bacteria surround themselves in S-layers (Figure 2) which are paracrystalline surface arrays composed of a single species of protein (or glycoprotein) that self-assembles to form the intact structure [21]. Most S-layers impart a neutral charge to the cell making them more hydrophobic. Usually S-layers do not precipitate metal ions, but an unique cyanobacteria, *Synechococcus* strain

GL24, with a hexagonally arranged S-layer on which available carboxyl groups are located, can mediate the formation of calcite or gypsum from natural waters [22]. Patches of the S-layer are shed once they become mineralized to prevent the cell's total encasement in mineral and premature death [23].

2.3 BIOFILMS AND EXOPOLYSACCHARIDES

In natural settings, including soil environments, a community of bacteria may form a biofilm on solid surfaces as a way to obtain nutrients in oligotrophic environments and to protect themselves from a variety of unfavorable conditions such as dehydration or chemical toxicity [24,25]. Biofilms are commonly found in natural environments as coatings around particles that occur in soils, lakes, rivers, and in the deep subsurface. (Indeed, one can imagine that the pore spaces within the subsurface are biofilm-coated channels where the dilute dissolved load of nutrients and metals in groundwater is available for bacteria as they grow at the water–solid interface.) The physiology of bacteria in biofilms is profoundly different from that of their planktonic counterparts [26].

The bacteria in a biofilm are bound together in a matrix of hydrous exopolysaccharides (EPS) which have a variety of binding sites similar to those of

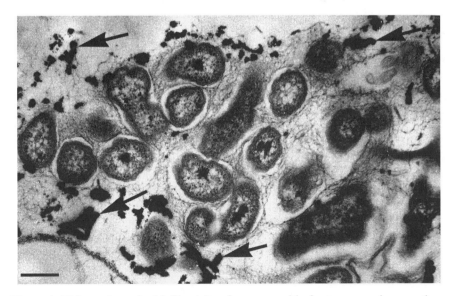

Figure 4. Thin section of a biofilm taken from a granitic fracture zone in a northern Ontario mine. The biofilm is precipitating dissolved iron from the water seeping through the fracture into amorphous iron oxide mineral phases (arrows). The fibrillar material connecting bacteria is a jelly-like material called the exopolymeric matrix which is often made up of extracellular polysaccharide (EPS). Iron oxide mineral phases were identified by EDS. Scale bar = 500 nm

capsules and slimes (Figure 4). A biofilm of *P. aeruginosa* can increase the total metal uptake above planktonically grown counterparts in the bulk fluid phase [3].

Internal pH and redox potential gradients are formed by the differential diffusion of nutrients, oxygen and metabolic products within the biofilm (see also Chapter 11). These gradients can be monitored at micrometer-scale levels by using microelectrodes [e.g., 26,27] or fluorescent chemical probes [26], and have shown that chemical microenvironments exist within a biofilm which are intrinsically different from those in bulk fluid phases. Biofilms are no longer regarded as homogenous smooth matrices containing bacteria. They are highly structured containing heterologous communities surrounded by microenvironments of their own making. All of these complex, variable and dynamic features have a bearing on metal binding and mineralization.

3 BIOMINERALIZATION

3.1 OBSERVING METAL–MICROBE INTERACTIONS

Because bacteria and their fine-grained minerals are so small, microscopy is often used to discern metal–microbe interactions. A variety of techniques can be used; each has its own advantages and disadvantages, and often a combination of methods is employed to obtain information on structure and the physico-chemical properties of samples.

Transmission electron microscopy (TEM) has amongst the highest resolving power and is an essential tool for visualizing the association between minerals and bacteria [28]. TEM associated with energy dispersive X-ray spectrometry (EDS) can help identify mineral phases by giving compositional analyses of adsorbed metals. Crystalline phases can be determined by selected area electron diffraction (SAED). Scanning confocal laser microscopy (SCLM) and environmental scanning electron microscopy (ESEM), which allow the observation of natural hydrated samples, are also powerful imaging systems that are increasing the understanding of biofilm structure and the spatial deposition of metals within them [29,30]. The reader is referred to Beveridge *et al.* [28] for an extensive compilation of some of the newest methods of analysis for total metal and metal species in biological samples, and for the molecular biological methods in environmental microbiology.

3.2 MODELING METAL BINDING AND BIOMINERALIZATION

From a geochemical standpoint, the reactive layers of a bacteria provide active interfacial sites for adsorption and complexation of dissolved aqueous metal species. Research with soil constituents such as the mineral phases (e.g., clays and oxyhydroxides) and soil organics often disregard the potential of soil bacterial populations to interact with metal ions (for detailed discussion on the interactions between microorganisms and minerals in soil environments, see Chapter 11). Bacteria can also alter soil organics by using them as nutrients thereby altering their stable concentration. Unfortunately, thermodynamic equilibrium modeling of aqueous metal species cannot account for the diversity of microorganisms, their active metabolism, and the microenvironments they create. As bacteria change the chemistry of their environment and actively or passively sequester metals, they may control the speciation and bioavailability of these elements. Processes that are involved in metal–microbe interactions include adsorption, complexation, precipitation, oxidation, reduction, methylation and demethylation.

Attempts to model the interaction of metals with bacterial cells require that surface functional groups must be adequately defined. Elucidation of the active metal binding sites of the *B. subtilis* cell wall revealed the contributions of carboxyl, phosphoryl, hydroxyl and amine groups [7,8]. The amine group is positively charged and may be involved with the binding of negatively charged anions such as silicates (SiO_4^{2-}) which can then sorb metal ions [31]. The other groups are negatively charged at neutral pH with deprotonation constants (pK_a) of around 4.8 for carboxyl, 6.9 for phosphoryl and 9.4 for hydroxyl groups [32]. Conditional stability constants have been obtained for a variety of metals with the carboxyl and phosphoryl sites of the *B. subtilis* cell wall [32]; however, these values may vary with ionic strength, bacterial species and the metal involved [33].

Metals prone to hydrolysis in solution bind strongly to cells. Under favorable geochemical conditions the binding of metals may lead to further metal complexation and the formation of precipitates. The specific mineral phases formed, however, may not be continuously controlled by the cells. The resulting mineral is a function of the cell surface, the microenvironment around the cell, and the type and abundance of the negatively-charged counterions (e.g., sulfate, sulfide, phosphate, carbonate or silicate ions). The proposed mechanism of precipitation by bacterial surfaces involves a two-step process, the first being a stoichiometric binding of metal cations with the charged surface functional groups, and the second being a nucleation reaction for precipitation [6]. Homogeneous and heterogeneous nucleation reactions proceed as the microenvironment around the cell exceeds the activation energy barrier (i.e., the energy to inhibit the spontaneous formation of insoluble precipitates [34]). The process of precipitate growth will continue as long as the microenvironment is saturated with respect to the mineral phase. Depending on how the surface is modeled, though, the continuum from surface complexation to precipitation may occur on the surface under saturated or undersaturated conditions [34]. The resulting solid phase may undergo phase transitions with the degree of crystallinity controlled by such factors as time, dehydration and the total metal concentration in the bulk solution. For example, amorphous ferric oxides may age to more crystal-line phases and, during this aging, incorporate other trace metals into their lattice structure.

Bacterial surfaces promote mineral precipitation by reducing the interfacial portion of the overall activation energy barrier and by acting as heterogeneous nucleation templates of similar atomic spacing [35]. This reaction appears to conform, qualitatively, with the basic principles of equilibrium thermodynamics [33,36]. The quantification of metal binding and precipitation at the cell surface should, then, be possible (if the bio-surface remains constant) using geochemical surface complexation models similar to those used for inorganic solids. Using thermodynamic equilibria, these models can predict when the reactions are most likely to occur at a solid surface between the active sites and metal ions. Mullen et al. [37] indicated that an equilibrium model of adsorption is useful in describing bacterial–metal interactions, although this approach may not be accurate when metal precipitation occurs. Using such an approach Warren and Ferris [36] have successfully demonstrated that there is a continuum between cation sorption and precipitation on microbial surfaces. In Figure 5, three distinct surface reaction steps are evident from isotherms relating the solid phase concentration of Fe(III) (Fe_s) to the equilibrium proton condition and the Fe(III) concentration in solution ($[Fe_D]/[H^+]^3$). These steps are: (1) an initial sorption stage with Lang-muir type behavior, (2) surface site saturation up to supersaturation and (3) eventual precipitation where there is an increase in solid phase Fe concentration (Fe_s). These three steps were modeled with the use of a simple mass action expression using a generalized bacterial surface binding site (BH):

$$BH + Fe^{3+} + 2H_2O \equiv BFe(OH)_2{}^0 + 3H^+$$

Unfortunately, bacterial surfaces are chemically and structurally heterogeneous as well as dynamic, ever-changing surfaces and these initial steps at modeling may have to become more complex to satisfy all bacterial conditions.

3.3 INDIRECT AND DIRECT INITIATION OF MINERAL FORMATION

The mechanisms by which bacteria initiate the formation of minerals in bulk solution vary widely between species. There may be a combination of biochem-ical and surface-mediated reactions during the process. As described earlier, bacterial surface layers may passively adsorb and indirectly serve as a nucle-ation template. However, bacteria can more directly initiate mineral precipita-tion by producing reactive compounds which also bind metals or catalyze metal reactions (enzymes, metallothioneins, siderophores, etc.). Bacteria can also instigate the spontaneous precipitation of metals by altering the geochemistry of their microenvironment [23,35].

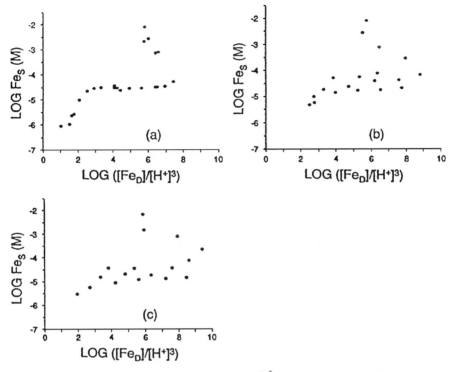

Figure 5. Log Fe$_s$ as a function of log $([Fe_D]/[H^+]^3)$ for the three bacterial strains: *B. subtilis* (a), *B. licheniformis* (b), and *P. aeruginosa* (c). Fes and Fe$_D$ mean the solid phase concentration of Fe(III) and the Fe(III) concentration in solution, respectively. These results clearly show that there is continuum between three stages, i.e., Fe sorption, nucleation and precipitation to these microbial surfaces. The linear portion of the curves at low $([Fe_D]/[H^+]^3)$ values indicates that sorption is directly proportional to $([Fe_D]/[H^+]^3)$ with slopes close to unity. Surface site saturation and onset of supersaturation is shown where the curves plateau as $([Fe_D]/[H^+]^3)$ values increase. The precipitation stage is shown by a sharp increase in solid-phase Fe concentration. Data are more variable for the two encapsulated species (*B. licheniformis* and *P. aeruginosa*), probably due to some breakage of the capsules during analyses. Reprinted with permission from [36]. Copyright (1998) American Chemical Society

Metal complexing ligands such as metallothioneins and siderophores [38] serve to sequester metals from the environment for incorporation into cellular components [39] including (possibly) Fe in magnetosomes [40]. Reactive inorganic ligands may also be produced as a result of cellular metabolic by-products including the production of H_2S and HPO_4^{2-}. The former reacts with metal ions to form metal sulfides (Figure 6), a common reaction in anoxic sediments with sulfate-reducing bacteria (SRBs) [41,42]. Phosphate pumped out of cells can also react with these same cations to form cell-associated and extracellular

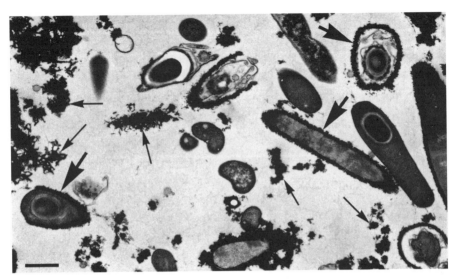

Figure 6. Thin section of a number of *Desulfomaculum* cells which have produced hydrogen sulfide as a metabolic endproduct so that metallic sulfides are produced on cell surfaces (large arrows) and in the external milieu (small arrows). In this specific case, Cr(VI) is also being reduced to Cr(III) and the major precipitates are Cr sulfides. Precipitated mineral phases were identified by EDS. Scale bar = 800 nm

precipitates [43,44]. Microbially mediated formation of phosphorites and sulfides has been documented in a variety of modern and ancient sediments [18].

Several genera couple the reduction and oxidation of metal species with the generation of energy in the form of ATP [45–47]. Under anaerobic conditions, oxidized metal species (including solid phases) can be used as terminal electron acceptors during bacterial respiration in a process called dissimilatory metal reduction [45] which often leads to the reduction of preformed minerals (e.g., Fe oxides; see Chapter 11). Reduction and oxidation of metal species often work hand-in-hand in natural environments since oxidation of a metal may provide a metal-reducing bacterium with a usable source of oxidized metal thus cycling metals [48].

The formation of solid mineral phases induced by bacterial processes may result in the deposition of mineral deposits over geological time scales through diagenesis of sediments [49]. Simulations of low temperature diagenesis using cells that had adsorbed metals indicated that the process of complete biomineralization starts at the cell wall surface and continues to form crystals, eventually filling the entire cell [19]. The mineral phases produced under laboratory conditions are compositionally and structurally identical to those found in natural settings. Observations made *in situ* in a variety of natural sediments and soils indicate that mineralized cells are highly resistant to degradation and gradually accumulate to form mineral deposits [50]. However, unless cells are so completely mineralized that their cell shape is retained or signature compounds are not degraded (e.g., isoprenoids), there may be little evidence to suggest biogenic origins of the deposit. Elemental deposits of Cu [51] and Au [52,53] contain cells that have been extensively coated with precipitates enabling researchers to implicate microbial origins to the deposits [23].

Biomineralization on the surface of an S-layered cyanobacterium, *Synechococcus* strain GL24, is responsible for the formation of large, natural, carbonate bioherms (modern, active stromatolites or thrombolites) in Fayetteville, Green Lake, New York [54]. This groundwater lake is rich in Ca^{2+}, HCO_3^- and SO_4^{2-}. The mechanism responsible for the precipitation of calcitic minerals is again a two-step process where the initial step involves the binding of Ca^{2+} to the cell followed by the binding of the anion to calcium [55]. The geochemistry of the lake (pH $= 7.9$) lies within the stability field for gypsum ($CaSO_4 \cdot 2H_2O$) which readily precipitates on the S-layer during periods when cells are not photosynthetically active. When the cells are photosynthesizing, they discharge OH^- as a metabolic end-product and increase the pH around the cell surfaces to pH 8.3 which promotes the formation of calcite ($CaCO_3$ [54]). Planktonic cells produce a 'whiting event' in the lake water and eventually settle to the lake bottom contributing to its marl sediment. When cells exist as part of a biofilm, on the bioherm, the mineralized cells add to bioherm development. Thompson and Ferris [54] and Schultze-Lam and Beveridge [55] conducted laboratory simulations using Sr^{2+}, Mg^{2+} and Ca^{2+} which resulted in the formation of strontianite, celestite, hydromagnesite and mixed calcium–strontium carbonates, showing that the process was not entirely dependent on Ca. Carbonate mineral formation mediated by bacteria in a number of other environments, mostly marine, is generally the end product of the direct chemical precipitation of cell-bound Ca or Mg resulting in aragonite and high magnesium calcite [50]. The ability of *Synechococcus* GL24 to alter the geochemistry of its environment, and therefore to influence the mineral phase that precipitates on its S-layer, is a remarkable example of the degree to which a microorganism can contribute to the geochemical cycling of metals and the formation of mineral deposits over geological time.

In biofilms, the natural microbial communities of mixed populations with differing metabolic functions can cycle, transform and accumulate metals and initiate the formation of authigenic mineral phases [56]. For example, Brown *et al.* [48] were able to study the dynamics of biofilms growing on a granitic pluton in the deep subsurface at an underground research facility in the Canadian Shield. This particular site was excavated for the purpose of disposing of nuclear fuel waste. Biofilms developed on the rock faces where groundwater was seeping from fractures in the granite. The outer oxic layer of the biofilm was actively oxidizing iron leached from the biotite of the granite into ferrihydrite or hematite, whereas the anaerobic inner face of the biofilm was reducing the iron back to Fe(II) which then reacted with HCO_3^- to form a black siderite layer. These two processes occurred with micrometers of each other (exemplifying the microenvironment concept) and may be a modern analog of the development of Archean banded-iron formations where reduced and oxidized forms of iron can be found in close proximity to each other. As well, biomineralization within biofilms is thought to play a major role in the corrosion of solid materials [29]. Mineral deposition by bacteria on a metal surface can shift the corrosion potential of the interface. Manganese oxide deposition seems to produce a positive shift whereas bioprecipitated sulfides, a negative shift; both processes result in accelerated corrosion.

4 METAL REDOX TRANSFORMATIONS AND BIOMINERALIZATION

There are two main mechanisms of metal redox transformation that are employed by bacteria. For instance, bacterial reduction or oxidation of metals can be indirectly catalyzed by the production of inorganic/organic reductants or extracellular catalysts produced as part of an organism's cellular metabolism, or metals can be transformed by the action of enzymes directly associated with cellular systems designed to derive energy from the reaction (e.g., dissimilatory iron reduction). Through biotransformation and subsequent biomineralization of metals in soil environments, bacteria can defend themselves against toxic metals in their proximate surroundings or gain energy for growth and metabolism. In addition to energy and detoxification, however, there may be a number of benefits bacteria gain by this sequence of reactions. For example, as biominerals are formed on the surfaces of bacteria, the cell can be protected from toxic oxygen species and ultraviolet radiation, as well as from physical and biochemical attack from other microorganisms [57,58]. Perhaps this has helped bacteria survive during their long existence since the dawn of life. When looking at the different strategies employed by aerobic and anaerobic bacteria using metals in their metabolism, there are some general trends. Typically anaerobic or facultative anaerobic bacteria use metals as terminal electron acceptors to generate energy, and the accumulation of the reduced or oxidized metals on the surface of the cell leads to mineral formation under favorable conditions. On the other hand, aerobic bacteria generally reduce or oxidize metal species as more of a detoxification mechanism. By catalyzing redox transformations, cells can convert the metal to a less toxic form such as an insoluble precipitate.

Biomineralization related to metal redox transformations of iron and manganese will be discussed in the following sections, since species transformations of those two metals are dependent on microbially mediated redox reactions especially under most anaerobic conditions in soil environments. In addition, the high capacity of toxic metals to adsorb onto the surface of iron and manganese oxides or hydroxides as well as their high abundance in soils makes these metals important in the chemistry of soil systems.

4.1 IRON

Iron can be assimilated, reduced by dissimilatory iron-reducing bacteria (DIRB) or be oxidized by a variety of microorganisms in acidic to basic environments through direct (enzymatic) or indirect (non-enzymatic) action [35,45,59–62]. Evidence of iron-interactive bacterial activity can be found in almost any environment through geological history from ancient sediments such as the banded-iron formations in the deep subsurface [48] to more recent settings such as in acidic mine tailings [63] and on corroded materials [29].

4.1.1 Iron Oxidation

In soils where the pH is circumneutral and the level of oxygen is low, there are several microorganisms which oxidize iron, such as *Sphaerotilus* and *Gallionella*. These Fe oxidizers thrive at anoxic–oxic transition zones where the oxygen concentration is low and the influx of Fe(II) is high. This is because Fe-oxidizing prokaryotes cannot compete with chemical oxidation under aerated conditions [64]. In low pH environments the activity of *Thiobacillus ferrooxidans*, an Fe(II)-oxidizing bacterium often growing in pyritic mine tailings, promotes the precipitation of iron hydroxides and hydroxysulfates in conjunction with acidification of the local environment on a massive scale. *T. ferrooxidans* was found to play a major role in the genesis of acid sulfate soils and waters (e.g., acid mine drainage) by oxidation of pyritic minerals to form H_2SO_4 [65]. These bacteria derive energy from the solubilization of iron pyrites and other sulfide minerals [66]. However, there is little energy derived from this oxidation (Fe(II) → Fe(III)); large quantities of iron must be oxidized in order to support growth.

This bacterium has been exploited by the mining industry for years to leach metals from sulfidic ores. In a study by Mann and Fyfe [67], a variety of iron oxide mineral phases associated with bacteria were identified in a thin specimen collected from a mine sediment core in Elliot Lake, Ontario. These included goethite (α-FeOOH), ferrihydrite ($5Fe_2O_3 \cdot 9H_2O$), akagaeneite (β-FeOOH), hematite (Fe_2O_3) and maghemite (γ-Fe_2O_3). The most common mineral forms associated with acidic mine tailings, are usually poorly ordered Fe precipitates with various amounts of trace metal impurities [63]. Recently, in a banded-iron ore from Gunma, Japan, small goethite aggregates were attributed to iron biomineralization by bacteria and the precipitates varied from amorphous Fe–P–(S) to Fe–P–S precipitates and to Fe–S–(P) (schwertmannite-like) forms [68].

Highly reactive iron oxides and hydroxides readily adsorb aqueous metal ions including toxic trace metals found in mining or industrial environments. One of the benefits of microbially mediated formation of Fe(III) oxides and hydroxides at field sites is the thermodynamic stability of the iron compounds as opposed to sulfide minerals which can easily release toxic metals if oxidized. Biomineralization of Fe(III) may serve to immobilize toxic metals and reduce the transport of these species through the environment.

4.1.2 Iron Reduction

The bioavailability, toxicity and mobility of aqueous trace metals are highly controlled by their association with a solid surface such as iron or manganese oxides and hydroxides. Whether the dissolution of these metal-adsorbed solid phases is due to chemical or biological processes, their reduction may release potentially toxic metals into natural environments. Consequently, dissimilatory iron reduction by bacteria is being actively studied to understand the fundamental impact of this process (see also Chapter 11).

The bioavailable form of Fe(III) for reduction in aquatic sediments by DIRB was found to range from amorphous and poorly crystalline oxides [69,70] to crystalline oxides (e.g., goethite and hematite [71,72]), and even structural Fe in clays [73]. The mechanism behind this process and the factors controlling the rate and extent of these reactions are still poorly understood and are under intense investigation.

It is believed that reduction of iron mineral phases may be either non-enzymatically or enzymatically mediated, though it is a matter of much debate. In the former case, solid hydrous ferric oxide (HFO) is reduced by changes in pH or redox potential induced by bacteria, or by products of bacterial metabolism (e.g., nitrite, sulfide), and is not related with any direct enzymatic activity. In the latter case, however, some enzymes (reductases) from bacteria are involved in the reduction of HFO and during these processes bacteria can gain energy for maintenance and growth. The reduction of Fe(III) is coupled to the oxidation of an organic substrate and electrons are shuttled to the HFO to drive the synthesis of ATP (anaerobic respiration; for a detailed discussion see Chapter 11). Most of the Fe(III) reduction in subsurface environments is due to enzymatic Fe(III) reduction by microorganisms rather than non-enzymatic processes (e.g., [60]). Obligate anaerobes such as *Geobacter metallireducens* and *Desulfovibrio* spp. as well as facultative anaerobes such as *Shewanella putrefaciens* and *S. alga* were found to couple anaerobic respiration-linked metal reduction to anaerobic organic carbon oxidation [59,74–78]. Under anoxic conditions, DIRB can also utilize a variety of oxidized metal species including Cr(VI), U(VI), and Se(VI) as electron acceptors coupled to oxidation of organic matter or H_2. The reduction of HFO by *S. putrefaciens*, for instance, may require the direct contact between the cells and solid substrate for the transfer of electrons [74]. In order to derive energy from this transfer, cytochromes may be localized in the outer membrane in a novel energy deriving process [79], since it was previously thought that electron transport chains were limited to the plasma membrane. During the reductive process, the Fe(II) liberated from the HFO under anaerobic conditions may bind back onto the HFO inhibiting further reduction and forming secondary mineral phases. Fredrickson *et al.* [80] observed the formation of several secondary iron minerals in the presence of different inorganic ligands. In a bicarbonate-containing medium, siderite was formed and, in a medium containing HCO_3–phosphate, vivianite was observed along with fine-grained magnetite. In a medium containing 2,6-anthraquinone disulfonate (AQDS) as a humic acid analog, which can serve as an electron shuttle for the reductive dissolution of Fe oxides [81], the rate of Fe(III) reduction increased and crystalline magnetite was formed.

4.1.3 Magnetic Iron Minerals

Bacteria also mediate the precipitation of magnetic iron minerals including magnetite (Fe_3O_4), greigite (Fe_3S_4) and pyrrhotite (Fe_7S_8). Magnetic precipitates external to the cell (with no direct control by bacteria of the resulting mineral character) are found to be indistinguishable from those produced by abiotic processes [62]. The bacteria responsible for the formation of these types of minerals (such as poorly crystallized magnetite [82]) are the same dissimilatory iron reducers mentioned above. Intracytoplasmic magnetic minerals are more crystalline and are produced by magnetotactic bacteria which can actually control the size, location, shape and crystallographic orientation of the magnetic mineral particles [62,83]. These bacteria synthesize intracellular, membrane-bounded [84] Fe_3O_4 or FeS particles called magnetosomes (Figure 7) [62]. Arrangements of magnetosomes within cells can impart a permanent magnetic dipole moment to the cell, which effectively makes a compass needle of each cell. Magnetotactic bacteria can exhibit magnetotaxis which aligns the cell along geomagnetic field lines while it swims [40]. This enables these microaerophilic bacteria to swim downwards towards more anoxic regions in the water column. When magnetotactic bacteria die, their magnetosomes can escape the cell and are preserved in marine and freshwater sediments [85].

4.2 MANGANESE

Manganese can exist in a number of oxidation states in nature, the most abundant being (+II), (+III) and (+IV). Mn(II) is soluble between pH 6 and 9, whereas the oxidized species (Mn(III) and Mn(IV)) form highly insoluble oxides and hydroxides [86]. Under neutral pH and oxic conditions, the

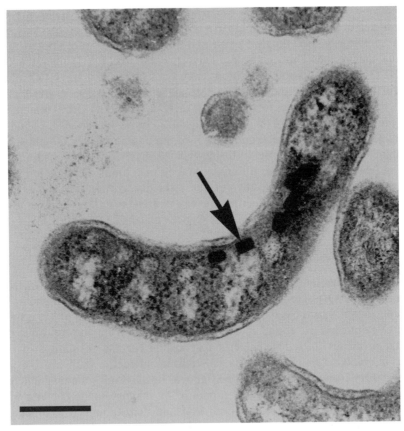

Figure 7. Thin section of the magnetotactic spirillum, *Magnetospirillum magnetotacticum*, showing the magnetosomes arranged in a line along the longitudinal axis of the cell. The arrow points to one magnetosome and the electron density is due to the magnetite which has a single magnetic domain. Scale bar = 400 nm

oxidation of Mn(II) to Mn(III) or Mn(IV) is thermodynamically favored, but the rate is generally slow because of the high activation energy for Mn(II) oxidation. However, redox transformations between reduced and oxidized forms are strongly catalyzed by bacteria though the mechanism of this Mn oxidation and reduction is still unclear (e.g., [87–89]; see Chapter 11).

Microbial Mn(II) oxidation is a major process that can produce Mn oxide coatings on soil particles 10^5 times faster than abiotic oxidation [58], and can also occur in marine sediments and near hydrothermal vents [90]. Bacteria such as *Leptothrix*, *Pedomicrobium*, *Hyphomicrobium*, *Caulobacter*, *Arthrobacter*, *Micrococcus*, *Bacillus*, *Chromobacterium*, *Pseudomonas*, *Vibrio*, and *Oceanospirillum* have all been implicated [91] (Figure 8). Like Fe oxides, Mn oxides are

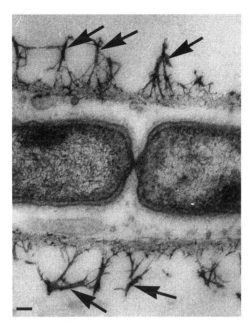

Figure 8. Thin section of *Leptothrix* sp. which is precipitating manganese oxide on its outermost structure called a sheath. The arrows point to the manganese mineral phase identified by EDS. Scale bar = 150 nm

highly reactive mineral phases, and help control the mobility of metals in sediments and soils through adsorption on their surfaces. Biogenic Mn oxides formed by *L. discophora* SS-1 have significantly higher Pb adsorption capacity and larger surface area than abiotically precipitated and commercial Mn oxides [92].

Several pathways have been proposed for Mn oxidation; they can be operationally classified as indirect oxidation or direct oxidation. Indirect pathways cause changes in the pH or redox potential of the environment, and are thought to result from the metabolic production of an oxidant (e.g., hydrogen peroxide) [91]. Direct oxidation results from enzymatic conversion utilizing single proteins or protein–polysaccharide complexes. Though the functional significance is still unclear, there must be some physiological advantage. For example, oxidation may be protective against harmful oxygen species [75] and Mn toxicity [57], or an inorganic energy source for growth [87]. Diverse genera have the ability to oxidize Mn but common genetic motifs have not been found among them. Some enzymes responsible for Mn oxidation have been identified, purified and characterized [93]. These two facts may indicate that cellular systems have evolved independently over the course of time serving to give some advantage over other species [75].

Both indirect and direct Mn oxidation leads to extracellular precipitation on or within the surface layers depending on the bacteria (Figure 8 and [75]). The amorphous structure of biologically precipitated Mn oxides and hydroxides makes them difficult to be identified by X-ray diffraction techniques although crystalline todorokite was identified in a sample from a microbial mat taken from a hot spring in Yellowstone [94]. Vernadite (δ-MnO_2) is a typically identified amorphous phase in nature [87,95], whereas manganite (γ-$MnOOH$) [96], hausmannite (Mn_3O_4) [97], and Mn-oxide birnessite [98] are some forms identified in laboratory studies.

5 BIOMINERALIZATION AND BIOREMEDIATION

Bioremediation is a proven technology for many environmental contaminants such as polycyclic aromatic hydrocarbons, explosives, pesticides, polychlorinated biphenyls and other organic toxicants, but its use for inorganics and metal remediation is just becoming recognized as an efficient and effective strategy. The problem with remediation of metals, unlike organic agents, is that they cannot be broken down into harmless metabolites and they tend to have an inhibitory effect on microbial activity which severely limits remediation strategies. In addition, field site-specific approaches are required to remove metal contaminants depending on their speciation (i.e., cationic or anionic). Highly toxic and mobile oxyanions of uranium, chromium, arsenic, and selenium are of particular concern in many soils and groundwaters throughout the world. Their speciation in the environment and hence their reactivity, bioavailability, toxicity and mobility are significantly controlled by microbe-mediated reactions. Cadmium, lead, zinc, nickel, cobalt, and other base metals generally exist as cationic forms in aqueous phase and adsorption onto bacterial cell walls can remove these toxicants very effectively [13,37]. Biosorption techniques do not actually require the enzymatic processes of living cells. The electronegative sites on the surface layers of bacteria described earlier account for most of the complexation capacity of this biomass. Although metal accumulation can be rapid, the maximum adsorptive capacity is a function of the amount of biomass and the geochemistry of the system. In comparison, metal oxyanions such as chromate, arsenate and selenate do not bind to any significant extent on the electropositive sites on the bacterial surface or through metal–ligand bridging. Bioremediation of these anions is predominantly based on microbially catalyzed redox conversions to insoluble forms. A majority of the metal redox transformations are mediated through enzymatic catalysis and, therefore, the rate of reduction or oxidation may be dependant on the enzyme activity and biomass concentration. However, since nucleation and precipitation of mineral phases may proceed spontaneously under favorable conditions once the activation energy barrier is overcome [36], enhanced metal immobilization can be achieved. The following sections will review the redox transformation and biomineralization of several toxic oxyanions, the mineral phases formed, and the mechanisms behind these processes.

5.1 ARSENIC

Arsenic can enter the aqueous environment through the dissolution of arsenic containing minerals such as arsenopyrite ($FeAsS$), realgar (AsS) and orpiment (As_2S_3). However, many anthropogenic sources exceed the natural input by nearly four times the natural input [99]. Arsenic can exist as arsenate (As(V)), arsenite (As(III)), arsenic metal (As^0) and arsine (As(-III)) according to the oxidation state, but little attention has been paid to the behavior of As^0 and As(-III) due to their unstable electrochemistry and rare occurrence in nature. Soluble species of arsenic, As(V) and As(III) are each highly toxic but As(III) is about 200 times more toxic than As(V) [100]. In addition, under field conditions As(III) is observed to have more mobility than As(V) [101] which adsorbs onto or coprecipitates with iron oxides at circumneutral pH range [102].

Over the past two decades, research on the geochemical cycling of arsenic has revealed the important role bacteria play in determining its speciation and contribution to arsenic cycling in the environment. This microbial cycling of arsenic stems from detoxification (reduction followed by As(III) expulsion [103]), energy generation (oxidation) and respiration (reduction) with both soluble forms of arsenic [104].

Several microorganisms (e.g., *Alcaligenes faecalis*, *P. arsenitoxidans*, *T. ferro-oxidans*) can oxidize aqueous As(III) or the arsenic in arsenopyrite, orpiment or enargite (Cu_5AsS_4) and use them as energy sources [105–108]. Arsenite oxidase-producing bacteria oxidize As(III) to As(V) and can be used to help certain aqueous arsenic removal strategies [100]. However, As(V) adsorbed on iron oxides can be released into water under anoxic conditions when reduction of $Fe(III)O_x$ also begins to release Fe(II) [109]. This process of As(V) release at low redox potential might be catalyzed by dissimilatory iron-reducing bacteria [45].

In addition to As(V) reduction by detoxification (e.g., As(III) has been observed to be produced from As(V) by certain bacteria and As(III) is then expelled [103]), some microorganisms have been found to use As(V) as an electron acceptor during respiration. Up to 1997 four arsenate-respiring bacteria have been isolated and characterized, i.e., *Chrysiogenes arsenatis*, *Sulfurospirillum arsenophilus*, *Sulfurospirillum barnesii*, and *Desulfotomaculum auripigmentum* [104]. However, despite the similar reductive abilities of these As (V)-respiring bacteria, they appear to be distinct from one another in their physiological and phylogenetical characteristics.

The reduction of As(V) to As(III) is thought to increase the mobility and toxicity of the metal in solution. However, a newly discovered As(V)-respiring bacterium, *D. auripigmentum*, has the unusual ability of reducing As(V) and S(VI) to As(III) and S(−II) under anaerobic conditions which precipitates as arsenic trisulfide (As_2S_3) or orpiment [110,111]. In natural environments, the formation of this mineral had been assumed to be abiotic and had only been observed in extreme environments such as geothermal reservoir fluids and hot springs. By TEM, this solid is found both intra- and extracellularly as particulate clusters, which could be surrounded by a membrane indicating that the particles are membrane mediated and not surface mediated. The ability of this novel sulphate-reducing bacterium to use both SO_4^{2-} and As(V) as terminal electron acceptors is remarkable since arsenic is highly toxic and, especially, since it is thermodynamically more favorable to reduce As(V) than SO_4^{2-} [111]. The formation of As_2S_3 in natural environments is thought to be a missing sink for arsenic cycling in various aquatic and sedimentary environments [112].

5.2 SELENIUM

Selenium, which has similar geochemical behavior to sulfur, is often concentrated in sulfide minerals. Anthropogenic activities such as fossil fuel combustion, mining, smelting, petroleum refining and fly ash production contaminate the environment with selenium along with agricultural fertilizers and sewage sludge [113–115]. Selenium is an essential trace nutrient in a wide range of organisms from bacteria to humans; in higher amounts, however, selenium can be extremely toxic [116,117].

Although selenium has four oxidation states (−II, 0, IV and VI) as does sulfur, selenium usually occurs in one of two soluble Se-oxyanion forms that are subject to bacterial redox transformations; selenite (SeO_3^{2-}, [Se(IV)]) and selenate (SeO_4^{2-}, [Se(VI)]). The geochemical cycling of selenium parallels that of sulfur so that selenium is predominantly cycled through biological pathways [118]. These transformations include reduction (assimilatory and dissimilatory), oxidation, methylation (followed by volatilization), and demethylation. During assimilatory reduction, SeO_4^{2-} or SeO_3^{2-} are transported into the cell and reduced to Se^{2-} to be incorporated into essential cellular proteins [119]. However, with excess amounts of selenium, the cells begin to incorporate selenium rather than their macronutrient, sulfur, leading to a breakdown in cellular metabolism and eventual death.

Reduction of soluble selenate and selenite to elemental selenium (Se^0) has been reported with Se^0 being deposited outside of the cell [120], in the periplasmic space [121] or in the cytoplasm of cells [122]. Aerobic and anaerobic reduction of selenium oxyanions is thought to be mediated by both detoxification and dissimilatory mechanisms. Under aerobic conditions, *P. stutzeri* can non-enzymatically reduce both soluble species to a solid precipitate (6.3 mmol L^{-1} of SeO_4^{2-} in 24 h) [123]. Recoveries of elemental selenium from these experiments were 79 % and 68 % for SeO_3^{2-} and SeO_4^{2-}, respectively. A strain of *Wolinella succinogenes* was isolated by enrichment which had adapted to selenium toxicity and could detoxify by reduction both SeO_3^{2-} and SeO_4^{2-} at concentrations of 1 and 10 mmol L^{-1}. The reduction product was a red amorphous Se^0 precipitate which was concentrated as a single non-crystalline cytoplasmic granule in each cell [124]. Tomei *et al.* [125] also found cytoplasmic granules of reduced selenium within *Desulfovibrio desulfuricans* implicating SRBs in selenium cycling. Despite the chemical proximity between sulfur and selenium, strain SES-1 which is a dissimilatory selenate reducer was sulfate-independent during its respiration in anoxic sediments which contain 310 mmol L^{-1} sulfate [126].

The potential for utilizing selenium-precipiting bacteria in a bioremediation application has been investigated. A pilot-scale operation to remove selenium from contaminated agricultural drainage water using a facultative anaerobic selenium-reducing bacteria was successful [127]. Using *Thauera selenatis* which reduces NO_3^- and SeO_4^{2-}, 98 % of the selenium was removed as Se^0. The mechanism of reduction involved the use of NO_3^- and SeO_4^{2-} as terminal electron acceptors under anaerobic conditions forming NO_2^- and SeO_3^{2-} [128]. However, in order to reduce SeO_3^{2-} to elemental selenium, the nitrite reductase used by *T. selenatis* during denitrification is required [129]. This means that SeO_4^{2-} is completely reduced to Se^0 only when NO_3^- is present [130]. Losi and Frankenberger, Jr. [131] assessed the potential of *Enterobacter cloacae* strain SLD1a-1 for use in a bioremediation application. This strain can grow under both of aerobic and anaerobic conditions, although SeO_4^{2-} was removed from the solution only after dissolved O_2 decreased so as to form a microaerophilic environment. The reduced selenium precipitates were found in the bulk solution and as discrete surface elemental Se^0 particles (< 0.1 μm) on the bacterium. The authors hypothesize that the enzymatic reduction is membrane associated where Se^0 is produced near the cell membrane and rapidly expelled.

5.3 CHROMIUM

Chromium oxidation states range from $(-II)$ to $(+VI)$, but only $(+III)$ and $(+VI)$ states are normally dominant within the environment. Chromate (CrO_4^{2-}), bichromate $(HCrO_4^-$ and dichromate $(Cr_2O_7^{2-})$ are soluble forms of Cr(VI) that are not strongly sorbed to soil components under slightly acidic to alkaline conditions and are thus highly mobile [132]. Owing to a wide variety of uses, industrial discharge of chromium into the environment often occurs. Hexavalent chromium, Cr(VI), is carcinogenic and mutanogenic to living organisms and also teratogenic to mammals including humans [133]. Therefore, in recent years there has been increased public concern about possible health hazards associated with the soluble and particulate forms of Cr(VI). Conversely, Cr(III) readily forms oxides and hydroxides such as $Cr(OH)_3$ that are very insoluble above pH 5.5 [134]. Removal techniques have been mainly dependent on the reduction of Cr(VI) to Cr(III) using chemical treatment processes followed by a precipitation step (e.g., [135,136]). However, certain bacteria can reduce CrO_4^{2-} (e.g., [137–139b]). Once reduced by microbial processes, the Cr(III) can then bind to electronegative cell layers and form stable nucleation sites for further metal precipitation (Figure 1; 139a). Cells are often observed with amorphous Cr(III) hydroxides or are uniformly coated by Cr(III), the latter possibly representing the initial stages of nucleation and precipitation. A recent report has shown that pseudomonad isolated from a chronium-contaminated site was capable of a two-pronged attack on Cr(VI); a soluble reductase precipitated the metal outside the cell and a reactive surface immobilized Cr on the outer membrane and capsule [139b]. The capacity of microbial biomass to remove chromium from solution represents a possible effective and more economic alternative to conventional remediation strategies.

Since the first report on chromium-reducing *Pseudomonas* strains isolated from chromate-contaminated sewage sludge in the 1970s [140], several additional chromate-reducing bacteria have been reported. These include additional strains of *Pseudomonas*, as well as strains of *Micrococcus*, *Escherichia*, *Enterobacter*, *Bacillus*, *Aeromonas* and *Achromobacter* [137, 141–145]. The mechanisms by which these microorganisms reduce Cr(VI) are variable and species dependent. Cr(VI)-reducing bacteria generally grow under circumneutral pH conditions and some species are capable of both aerobic and anaerobic reduction of chromate. Several species use Cr(VI) as a terminal electron acceptor in their respiratory chains and these species include *P. aeruginosa* [143], *B. subtilis* [143], *P. fluorescens* [144], *Enterobacter cloacae* [141] and some SRBs [146,147]. On the other hand, in some gram-negative strains, chromium reduction could be due to detoxification where outer membrane proteins scavenge the metal or plasma membrane proteins participate in its efflux leading to decreased uptake [148].

Soluble forms of Cr(VI) are generally used to model chromium reduction; however, reduction of Cr(VI) (in the form of the mineral crocoite ($PbCrO_4$)) to Cr(III) (as $Cr(OH)_3$) has been reported with a *P. chromatophila* strain by Lebedeva and Lyalikova [149]. Some laboratory studies describe the reduced Cr precipitates as $Cr(OH)_3$, $CrO(OH)$ or amorphous chromium hydroxides [141,146,150]; however, formation of mixed Fe–Cr hydroxides has also been observed [139a, 151]. Intracellular accumulation of reduced chromium is also found [137,152] but there are few data on its nature. The likelihood of finding Cr precipitates associated with bacterial cells in the natural environment is assured by the relative stability of $Cr(OH)_3$ under a wide range of redox potential and pH although there are few reports linking Cr precipitates to microbial reduction in the field. Losi *et al.* [136] attributed the reduction and precipitation of Cr(VI) from contaminated groundwater in an alfalfa field soil to the activity of the indigenous microbial population stimulated by amendments of manure.

Bench to pilot scale bioremediation applications using chromate-reducing bacteria have been employed to remove Cr(VI) from industrial effluents and soil solutions. Using *E. cloacae*, chromate was reduced in an industrial wastewater but only 40% of the precipitated Cr(III) could be removed by centrifugation [153]. Fine grained Cr(III) precipitates also remained in solution after using indigenous bacteria to reduce chromate from a industrial leachate in a pilot scale study [154]. Turick *et al.* [155] found they could stimulate indigenous anaerobic Cr(VI)-reducing bacteria by the addition of nutrients to a Cr(VI)-contaminated soil (200 mg kg^{-1}). With only the addition of glucose, 64% of the Cr(VI) was reduced. Chirwa and Wang [156,157] utilized *Bacillus* sp. and *P. fluorescens* LB300 in a fixed film continuous flow bioreactor where bacterial cells were immobilized on an inert matrix under aerobic conditions. As Cr(VI) was reduced to Cr(III), it adsorbed to the biofilm and was effectively removed from solution.

5.4 BIOREMEDIATION APPROACHES

Laboratory studies have confirmed the potential of microbes to transform and remove metals from solution and are currently being employed for metal remediation in engineered bioremediation strategies. *In situ* immobilization of mobile toxic metal contaminants from groundwater can inhibit further migration into sensitive areas as well as limit their toxicity and bioavailability (see reviews [46,158]). A recent technique for *in situ* bioremediation of groundwater is the construction of a biobarrier. This is becoming an effective process whereby the contaminated plume flows through a 'reactive wall' which contains inorganic and/or biological constituents to transform or immobilize metals. For instance, the use of SRBs in biobarriers will immobilize metals as insoluble sulfides [159]. Some of the newest strategies involve the introduction of dissimilatory iron-reducing bacteria into biobarriers to provide a source of reduced iron to react with metals such as Cr(VI) and organics in the contaminant plume [160]. On-site bioremediation systems such as pump-and-treat techniques are used to facilitate the biological precipitation of metals from solution and stirred batch reactors to leach metals from soils. As well, on-site continuous flow bioreactor systems employ the use of the increased surface area and electronegative binding sites created by bacterial biofilms to remove metals from solution.

A popular approach to successful site-specific bioremediation may begin by enriching metal tolerant microbial populations from a contaminated site and selecting possible isolates that can transform the metal into a form more easily removed or immobilized. The use of indigenous strains of bacteria versus genetically engineered microorganisms (GEMs) (e.g., [161]) in remediating contaminated sites is a debatable issue. GEMs may have all the necessary genes for immobilizing a toxic metal; however, they rarely survive once reintroduced into the field since they are apt to be outcompeted by the natural microbiota. Using indigenous strains isolated from contaminated sites is another approach that has advantages including tolerance to the metals at the site and the obvious adaptations they have to the environment at the site. The moral, ethical and political issues involved with using GEMs are not a problem if indigenous strains are used. All contaminated sites that we have sampled contain tolerant metal-immobilizing species.

Bioremediation for the removal of inorganics may soon compete with chemical methods in efficiency and cost effectiveness. Many new, potentially effective strategies in bioremediation are being developed which are employing newly discovered bacterial species. In addition, fundamental research into bacterial mechanisms of metal transformation and precipitation of toxic metals is allowing the optimization of such bioremediation processes. It is plausible that in most contaminated sites one can isolate microorganisms that have adapted ways to detoxify their surroundings or to utilize the particular toxic contaminant to their advantage. It is estimated that 99.9% of soil microbes are uncultureable and, as of yet, undiscovered [162,163]. In many instances while investigating the potential of isolated bacteria to convert metals for bioremediation application, insights into the global biogeochemical cycling of metals are also discovered.

ACKNOWLEDGEMENTS

The authors thank Bob Harris, Dianne Moyles and Anu Saxena of our laboratory for their assistance with some aspects of the work reported in this chapter.

The work reported from the authors' laboratory has been funded by a Natural Science and Engineering Council of Canada (NSERC) grant to TJB. JSM was supported by a NSERC Industrial Postgraduate Scholarship during his tenure as a M.Sc. student. The electron microscopy was performed in the NSERC Guelph Regional STEM Facility which is partially funded by a NSERC-Major Facilities Access grant to TJB.

REFERENCES

1. Mann, H. (1990). Biosorption of heavy metals by bacterial biomass. In *Biosorption of Heavy Metals*, ed. Volesky, B., CRC Press, Boca Raton, FL, p. 93.
2. Beveridge, T. J. (1981). Ultrastructure, chemistry and function of the bacterial wall, *Int. Rev. Cytol.*, **72**, 229.
3. Langley, S. and Beveridge, T. J. (1999). Metal binding by *Pseudomonas aeruginosa* PAO1 is influenced by growth of the cells as a biofilm, *Can. J. Microbiol.*, **45**, 1.
4. Barns, S. M. and Nierzwicki-Bauer, S. (1997). Microbial diversity in ocean, surface and subsurface environments. In *Geomicrobiology: Interactions between Microbes and Minerals*, Reviews in Mineralogy, Vol. 35, ed. Banfield, J. F. and Nealson, K. H., Mineralogical Society of America, Washington, DC, p. 35.

5. Banfield, J. F. and Hamers, R. J. (1997). Processes at minerals and surfaces with relevance to microorganisms and prebiotic synthesis. In *Geomicrobiology: Interactions between Microbes and Minerals*, Reviews in Mineralogy, Vol. 35, ed. Banfield, J. F. and Nealson, K. H., Mineralogical Society of America, Washington, DC, p. 81.
6. Beveridge, T. J. and Murray, R. G. E. (1976). Uptake and retention of metals by cell walls of *Bacillus subtilis*. *J. Bacteriol.*, **127**, 1502.
7. Doyle, R. J., Matthews, T. H. and Streips, U. N. (1980). Chemical basis for selectivity of metal ions by the *Bacillus subtilis* cell wall, *J. Bacteriol.*, **143**, 471.
8. Beveridge, T. J. and Murray, R. G. E. (1980). Sites of metal deposition in the cell wall of *Bacillus subtilis*, *J. Bacteriol.*, **141**, 876.
9. Beveridge, T. J., Forsberg, C. W. and Doyle, R. J. (1982). Major sites of metal binding in *Bacillus licheniformis* walls, *J. Bacteriol.*, **150**, 1438.
10. Graham, L. L. and Beveridge, T. J. (1994). Structural differentiation of the *Bacillus subtilis* cell wall, *J. Bacteriol.*, **176**, 1413.
11. Beveridge, T. J. (1988). The bacterial surface: general considerations towards design and function, *Can. J. Microbiol.*, **34**, 363.
12. Beveridge, T. J. (1984). Bioconversion of inorganic materials: mechanism of the binding of metallic ions to bacterial walls and the possible impact on microbial ecology. In *Current Perspectives in Microbial Ecology*, ed. Klug, M. J. and Reddy, A., American Society for Microbiology, Washington, DC, p. 601.
13. Volesky, B. (1990). Removal and recovery of heavy metals by biosorption. In *Biosorption of Heavy Metals*, ed. Volesky, B., CRC Press, Boca Raton, FL, p. 7.
14. Beveridge, T. J. (1995). The periplasmic space and the concept of the periplasm in gram-positive and gram-negative bacteria, *ASM News*, **61**, 125.
15. Graham, L. L., Harris, R., Villiger, W. and Beveridge, T. J. (1991). Freeze-substitution of gram-negative eubacteria: general cell morphology and envelope profiles, *J. Bacteriol.*, **172**, 1623.
16. Beveridge, T. J. (1989). Role of cellular design in bacterial metal accumulation and mineralization, *Annu. Rev. Microbiol.*, **43**, 147.
17. Ferris, F. G. and Beveridge, T. J. (1986). Site specificity of metallic ion binding in *Escherichia coli* K-12 lipopolysaccharide, *Can. J. Microbiol.*, **32**, 52.
18. Ferris, F. G. (1989). Metallic ion interactions with the outer membrane of gram-negative bacteria. In *Metal Ions and Bacteria*, ed. Beveridge, T. J. and Doyle, R. J., John Wiley and Sons, New York, p. 295.
19. Beveridge, T. J. (1989). Metal ions and bacteria. In *Metal Ions and Bacteria*, ed. Beveridge, T. J. and Doyle, R. J., John Wiley and Sons, New York, p. 1.
20. Geesey, G. G. and Jang, L. (1989). Interactions between metal ions and capsular polymers. In *Metal Ions and Bacteria*, ed. Beveridge, T. J. and Doyle, R. J., John Wiley and Sons, New York, p. 325.
21. Sleytr, U. B. and Beveridge, T. J. (1999). Bacterial S-layers, *Trends Microbiol.*, **7**, 253.
22. Schultze-Lam, S., Harauz, G. and Beveridge, T. J. (1992). Participation of a cyanobacterial S layer in fine-grain mineral formation, *J. Bacteriol.*, **174**, 7971.
23. Douglas, S. and Beveridge, T. J. (1998). Mineral formation by bacteria in natural microbial communities, *FEMS Microbiol. Ecol.*, **26**, 79.
24. Costerton, J. W., Lewandowski, Z., Caldwell, D. E., Korber, D. R. and Lappin-Scott, H. (1995). Microbial biofilms, *Annu. Rev. Microbiol.*, **49**, 711.
25. Beveridge, T. J., Makin, S. A., Kadurugamuwa, J. L. and Li, Z. (1997). Interactions between biofilms and the environment, *FEMS Microbiol. Rev.*, **20**, 291.
26. Costerton, J. W., Lewandowski, Z., DeBeer, D., Caldwell, D., Korber, D. and James, G. (1994). Biofilms, the customized microniche, *J. Bacteriol.*, **176**, 2137.

27. Revsbech, N. P. and Jorgensen, B. B. (1988). Microelectrodes: their use in microbial ecology, *Adv. Microb. Ecol.*, **9**, 293.
28. Beveridge, T. J., Hughes, M. N., Lee, H., Leung, K. T., Poole, R. K., Savvaidis, I., Silver, S. and Trevors, J. T. (1997). Metal-microbe interactions: contemporary approaches, *Adv. Microb. Physiol.*, **38**, 177.
29. Little, B. J., Wagner, P. A. and Lewandowski, Z. (1997). Spatial relationships between bacteria and mineral surfaces. In *Geomicrobiology: Interactions between Microbes and Minerals*, Reviews in Mineralogy, Vol. 35, ed. Banfield, J. F. and Nealson, K. H., Mineralogical Society of America, Washington, DC, p. 123.
30. Brown, D. A., Beveridge, T. J., Keevil, W. C. and Sherriff, B. L. (1998). Evaluation of microscopic techniques to observe iron precipitation in a natural microbial biofilm, *FEMS Microbiol. Ecol.*, **26**, 297.
31. Urrutia, M. M. and Beveridge, T. J. (1993). Remobilization of heavy metals retained as oxyhydroxides or silicates by *Bacillus subtilis* cells, *Appl. Environ. Microbiol.*, **59**, 4323.
32. Fein, J. B., Daughney, C. J., Yee, N. and Davis, T. (1997). A chemical equilibrium model for metal adsorption onto bacterial surfaces, *Geochim. Cosmochim. Acta*, **61**, 3319.
33. Daughney, C. J., Fein, J. B. and Yee, N. (1998). A comparison of the thermodynamics of metal adsorption onto two common bacteria, *Chem. Geol.*, **144**, 161.
34. Stumm, W. (1992). *Chemistry of the Solid Water Interface*, John Wiley & Sons, New York.
35. Fortin, D., Ferris, F. G. and Beveridge, T. J. (1997). Surface-mediated mineral development by bacteria. In *Geomicrobiology: Interactions between Microbes and Minerals*, Reviews in Mineralogy, Vol. 35, ed. Banfield, J. F. and Nealson, K. H., Mineralogical Society of America, Washington, DC, p. 161.
36. Warren, L. A. and Ferris, F. G. (1998). Continuum between sorption and precipitation of Fe(III) on microbial surfaces, *Environ. Sci. Technol.*, **32**, 2331.
37. Mullen, M. D., Wolf, D. C., Ferris, F. G., Beveridge, T. J., Flemming, C. A. and Bailey, G. W. (1989). Bacterial sorption of heavy metals, *Appl. Environ. Microbiol.*, **55**, 3143.
38. Macaskie, L. E. and Dean, A. C. R., (1990). Metal-sequestering biochemicals. In *Biosorption of Heavy Metals*, ed. Volesky, B., CRC Press, Boca Raton, FL, p. 199.
39. Wood, J. M. (1984). Evolutionary aspects of metal ion transport through cell membranes. In *Metal Ions in Biological Systems*, ed. Sigel, H., Marcel Dekker, New York, p. 223.
40. Blakemore, R. (1975). Magnetotactic bacteria, *Science*, **190**, 377.
41. Fortin, D., Southam, G. and Beveridge, T. J. (1994). Nickel sulfide, iron-nickel sulfide and iron sulfide precipitation by a newly isolated *Desulfotomaculum* species and its relation to nickel resistance, *FEMS Microbiol. Ecol.*, **14**, 121.
42. Lovley, D. R. (1995). Bioremediation of organic and metal contaminants with dissimilatory metal reduction, *J. Indust. Microbiol.*, **14**, 85.
43. Macaskie, L. E., Blackmore, J. D. and Empson, R. M. (1988). Phosphatase overproduction and enhanced uranium accumulation by a stable mutant of a *Citrobacter* sp. isolated by a novel method, *FEMS Microbiol. Lett.*, **55**, 157.
44. Macaskie, L. E, Jeong, B. C. and Tolley, M. R. (1994). Enzymatically accelerated biomineralization of heavy metals: application to the removal of americium and plutonium from aqueous flows, *FEMS Microbiol. Rev.*, **14**, 351.
45. Lovley, D. R. (1993). Dissimilatory metal reduction, *Annu. Rev. Microbiol.*, **47**, 263.
46. Lovley, D. R. and Coates, J. D. (1997). Bioremediation of metal contamination, *Curr. Opin. Biotechnol.*, **8**, 285.

47. Ehrlich, H. L. (1997). Microbes and metals, *Appl. Microbiol. Biotechnol.*, **48**, 687.
48. Brown, D. A., Kamineni, D. C., Sawicki, J. A. and Beveridge, T. J. (1994). Minerals associated with biofilms occurring on exposed rock in a granitic underground research laboratory, *Appl. Environ. Microbiol.*, **60**, 3182.
49. Beveridge, T. J., Meloche, J. D., Fyfe, W. S. and Murray, R. G. E. (1983). Diagenesis of metals chemically complexed to bacteria: laboratory formation of metal phosphates, sulfides and organic condensates in artificial sediments, *Appl. Environ. Microbiol.*, **45**, 1094.
50. Ferris, F. G., Shotyk, S. and Fyfe, W. S. (1989). Mineral formation and decomposition by microorganisms. In *Metal Ions and Bacteria*, ed. Beveridge, T. J. and Doyle, R. J., John Wiley & Sons, New York, p. 413.
51. Sillitoe, R. H., Folk, R. L. and Saric, N. (1996). Bacteria as mediators of copper sulphide enrichment during weathering, *Science*, **272**, 1153.
52. Lyalikova, N. N. and Mokeicheva, L. Y. (1969). The role of bacteria in gold migration deposits, *Geokhimiya*, **38**, 905.
53. Southam, G. and Beveridge, T. J. (1994). The in vitro formation of placer gold by bacteria, *Geochim. Cosmochim. Acta*, **58**, 4527.
54. Thompson, J. B. and Ferris, F. G. (1990). Cyanobacterial precipitation of gypsum, calcite and magnesite from natural alkaline lake water, *Geology*, **18**, 995.
55. Schultze-Lam, S. and Beveridge, T. J. (1994). Nucleation of celestite and strontianite on a cyanobacterial S-layer, *Appl. Environ. Microbiol.*, **60**, 447.
56. Ferris, F. G., Schultze, S., Witten, T. C., Fyfe, W. S. and Beveridge, T. J. (1989). Metal interactions with microbial biofilms in acidic and neutral pH environments, *App. Environ. Microbiol.*, **55**, 1249.
57. Ghiorse, W. C. (1984). Biology of iron and manganese-depositing bacteria, *Annu. Rev. Microbiol.*, **38**, 515.
58. Tebo, B. M., Ghiorse, W. C., van Waasbergen, L. G., Siering, P. L. and Caspi, R. (1997). Bacterially mediated mineral formation: insights into manganese(II) oxidation from molecular genetic and biochemical studies. In *Geomicrobiology: Interactions between Microbes and Minerals*, Reviews in Mineralogy, Vol. 35, ed. Banfield, J. F. and Nealson, K. H., Mineralogical Society of America, Washington, DC, p. 225.
59. Lovley, D. R. (1991). Dissimilatory Fe(III) and Mn(IV) reduction, *Microbiol. Rev.*, **55**, 259.
60. Lovley, D. R. (1997). Microbial Fe(III) reduction in subsurface environments, *FEMS Microbiol. Rev.*, **20**, 305.
61. Konhauser, K. O. (1998). Diversity of bacterial iron mineralization, *Earth-Sci. Rev.*, **43**, 91.
62. Bazylinski, D. A. and Moskowitz, B. M. (1997). Microbial biomineralization of magnetic iron minerals: microbiology, magnetism and environmental significance. In *Geomicrobiology: Interactions between Microbes and Minerals*, Reviews in Mineralogy, Vol. 35, ed. Banfield, J. F. and Nealson, K. H., Mineralogical Society of America, Washington, DC, p. 181.
63. Fortin, D., Davis, B. and Beveridge, T. J. (1996). Role of *Thiobacillus* and sulfate-reducing bacteria in iron biocycling in oxic and acidic mine tailings, *FEMS Microbiol. Ecol.*, **21**, 11.
64. Emerson, D. and Moyer, C. (1997). Isolation and characterization of novel iron-oxidizing bacteria that grow at circumneutral pH, *Appl. Environ. Microbiol.*, **63**, 4784.
65. Ivarson, K. C. and Sojak, M. (1978). Microorganisms and ochre deposits in field drains of Ontario, *Can. J. Soil Sci.*, **58**, 1.
66. Nordstrom, D. K. and Southam, G. (1997). Geomicrobiology of sulfide mineral oxidation. In *Geomicrobiology: Interactions between Microbes and Minerals*, Reviews

in Mineralogy, Vol. 35, ed. Banfield, J. F. and Nealson, K. H., Mineralogical Society of America, Washington, DC, p. 361.

67. Mann, H. and Fyfe, W. S. (1989). Metal uptake and Fe-, Ti-oxide biomineralization by acidophilic microorganisms in mine waste environments, Elliot Lake, Canada, *Can. J. Earth Sci.*, **26**, 2731.

68. Akai, J., Akai, K., Ito, M., Nakano, S., Maki, Y. and Sasagawa, I. (1999). Biologically induced iron ore at Gunma iron mine, Japan, *Am. Miner.*, **84**, 171.

69. Lovley, D. R. and Phillips, E. J. P. (1986). Availability of ferric iron for microbial reduction in bottom sediments of the freshwater tidal Potomac River, *Appl. Environ. Microbiol.*, **52**, 751.

70. Urrutia, M. M., Roden, E. E., Fredrickson, J. K. and Zachara, J. M. (1998). Microbial and surface chemistry controls on reduction of synthetic Fe(III) oxide minerals by the dissimilatory iron-reducing bacterium *Shewanella alga*, *Geomicrobiol. J.*, **15**, 269.

71. Roden, E. E. and Zachara, J. M. (1996) Microbial reduction of crystalline Fe(III) oxides: influence of oxide surface area and potential for cell growth, *Environ. Sci. Technol.*, **30**, 1618.

72. Zachara, J. M., Fredrickson, J. K., Li, S. M., Kennedy, D. W., Smith, S. C. and Gassman, P. L. (1998). Bacterial reduction of crystalline Fe^{3+} oxides in single phase suspensions and subsurface materials, *Am. Mineral.*, **83**, 1426.

73. Ernsten, V., Gates, W. P. and Stucki, J. W. (1996). Microbial reduction of structural iron in clays – a renewable source of reduction capacity, *J. Environ. Qual.*, **27**, 761.

74. Lovley, D. R. and Phillips, E. J. P. (1988). Novel mode of microbial energy metabolism: organic carbon oxidation coupled to dissimilatory reduction of iron or manganese, *Appl. Environ. Microbiol.*, **54**, 1472.

75. Nealson, K. H., Rosson, R. A. and Myers, C. R. (1989). Mechanisms of oxidation and reduction of manganese. In *Metal Ions and Bacteria*, ed. Beveridge, T. J. and Doyle, R. J., John Wiley & Sons, New York, p. 295.

76. Nealson, K. H. and Myers, C. R. (1992). Microbial reduction of manganese and iron: new approaches to carbon cycling, *Appl. Environ. Microbiol.*, **58**, 439.

77. Lovley, D. R., Giovannoni, S. J., White, D. C., Champine, J. E., Phillips, E. J. P., Gorby, Y. A. and Goodwin, S. (1993). *Geobacter metallireducens* gen. nov. sp. nov., a microorganism capable of coupling the complete oxidation of organic compounds to the reduction of iron and other metals, *Arch. Microbiol.*, **159**, 336.

78. Lovley, D. R., Coates, J. D., Saffarini, D. A. and Lonergan, D. J. (1997). Dissimilatory iron reduction. In *Transition Metals in Microbial Metabolism*, ed. Winkelmann, G. and Carrano, C., Harwood Academic Publishers, Amsterdam, The Netherlands, p. 187.

79. Myers, C. R. and Myers, J. M. (1992). Localization of cytochromes to the outer membrane of anaerobically grown *S. putrefaciens* MR-1, *J. Bacteriol.*, **174**, 3429.

80. Fredrickson, J. K., Zachara, J.M, Kennedy, D. W., Dong, H. L., Onstott, T. C., Hinman, N. W. and Li, S. M. (1998). Biogenic iron mineralization accompanying the dissimilatory reduction of hydrous ferric oxide by a groundwater bacterium, *Geochim. Cosmochim. Acta*, **62**, 3239.

81. Lovley, D. R., Fraga, J. L., Blunt-Harris, E. L., Hayes, L. A., Phillips, E. J. P. and Coates, J. D. (1998). Humic substances as a mediator for microbially catalyzed metal reduction, *Acta Hydrochim. Hydrobiol.*, **26**, 152.

82. Sparks, N. H. C., Mann, S., Bazylinski, D. A., Lovley, D. R., Jannasch, H. W. and Frankel, R. B. (1990). Structure and morphology of magnetite anaerobically-produced by a marine magnetotactic bacterium and a dissimilatory iron-reducing bacterium, *Earth Planet. Sci. Lett.*, **98**, 14.

83. Frankel, R. B., Zhang, J. P. and Bazylinski, D. A. (1998). Single magnetic domains in magnetotactic bacteria, *J. Geophys. Res.-Solid Earth*, **103**, 30601.
84. Gorby, Y. A., Beveridge, T. J. and Blakemore, R. P. (1988). Characterization of the bacterial magnetosome membrane, *J. Bacteriol.*, **170**, 834.
85. Vali, H. and Kirschvink, J. L. (1990). Observations of magnetosome organization, surface structure, and iron biomineralization of undescribed magnetic bacteria: evolutionary speculations. In *Iron Biominerals*, ed. Frankel, R. B. and Blakemore, R. P., Plenum Press, New York, p. 97.
86. Stumm, W. and Morgan, J. J. (1996). *Aquatic Chemistry*, 3rd edn, John Wiley & Sons, New York.
87. Nealson, K. H., Tebo, B. M. and Rosson, R. A. (1988). Occurrence and mechanisms of microbial oxidation of manganese, *Adv. Appl. Microbiol.*, **33**, 299.
88. Ghiorse, W. C. (1988). Microbial reduction of manganese and iron. In *Biology of Anaerobes*, ed. Zehnder, A. J. B., John Wiley & Sons, New York, p. 305.
89. Ehrlich, H. L. (1996). *Geomicrobiology*, Marcel Dekker, New York.
90. Fortin, D., Ferris, F. G. and Scott, S. D. (1998). Formation of Fe-silicates and Fe-oxides on bacterial surfaces in samples collected near hydrothermal vents on the Southern Explorer Ridge in the northeast Pacific Ocean, *Am. Miner.*, **83**, 1399.
91. Gounot, A. M. (1994). Microbial oxidation and reduction of manganese: consequences in groundwater and applications, *FEMS Microbiol. Rev.*, **14**, 339.
92. Nelson, Y. M., Lion, L. W., Ghiorse, W. C. and Shuler, M. L. (1999). Production of biogenic Mn oxides by *Leptothrix discophora* SS-1 in a chemically defined growth medium and evaluation of their Pb adsorption characteristics, *Appl. Environ. Microbiol.*, **65**, 175.
93. Adams, L. F. and Ghiorse, W. C. (1987). Characterization of extracellular Mn [2+] oxidizing activity and isolation of Mn [2+] oxidizing protein from *Leptothrix discophora* SS-1, *J. Bacteriol.*, **169**, 1279.
94. Ferris, F. G., Fyfe, W. S. and Beveridge T. J. (1987). Manganese oxide deposition in a hot spring microbial mat, *Geomicrobiol. J.*, **5**, 33.
95. Tipping, E., Thompson, D. W. and Davidson, W. (1984). Oxidation products of Mn(II) in lake waters, *Chem. Geol.*, **44**, 359.
96. Greene, A. C. and Madgwick, J. C. (1991). Microbial formation of manganese oxides, *Appl. Environ. Microbiol.*, **57**, 1114.
97. Mann, S., Sparks, N. H. C., Scott, G. H. E. and de Vrind-de Jong, E. W. (1988). Oxidation of manganese and formation of Mn_3O_4 (hausmannite) by spore coats of a marine *Bacillus* sp, *Appl. Environ. Microbiol.*, **54**, 2140.
98. Carlson, L. and Schwertmann, V. (1987). Iron and manganese oxides in Finnish ground water treatment plants, *Wat. Res.*, **21**, 165.
99. Nriagu, J. O. and Pacyna, J. M. (1988). Quantitative assessment of worldwide contamination of air, water and soils by trace metals, *Nature* (London), **333**, 134.
100. Williams, J. W. and Silver, S. (1984). Bacterial resistance and detoxification of heavy metals, *Enz. Microb. Technol.*, **6**, 530.
101. Gulens, J. and Champ, D. R. (1979). Influence of redox environments on the mobility of arsenic in groundwater. In *Chemical Modeling in Aqueous Systems*, ed. Jenne, E. A., American Chemical Society, Washington, DC, p. 81.
102. Mok, W. M. and Wai, C. M. (1994). Mobilization of arsenic in contaminated river waters. In *Arsenic in the Environment*, ed. Nriagu, J. O., John Wiley & Sons, New York, p. 99.
103. Rosen, B. P., Silver, S., Gladysheva, T. B., Ji, G., Oden, K. L., Jagannathan, S., Shi, W., Chen, Y. and Wu, J. (1994). The arsenite oxyanion-translocating ATPase:

bioenergetics, functions and regulation. In *Phosphate in Microorganisms*, ed. Torriani-Gorini, A., Yagil, E. and Silver, S., American Society for Microbiology, Washington, DC, p. 97.

104. Newman, D. K., Ahmann, D. and Morel, F. M. M. (1998). A brief review of microbial arsenate respiration, *Geomicrobiol.*, **15**, 255.

105. Osborn, F. H. and Ehrlich, H. L. (1976). Oxidation of arsenite by a soil isolate of *Alcaligenes*, *J. Appl. Bacteriol.*, **41**, 295.

106. Ilyaletdinov, A. N. and Abdrashitova, S. A. (1981). Autotrophic oxidation of arsenic by a culture of *Pseudomonas arsenitoxidans*, *Mikrobiologiya*, **50**, 197.

107. Ehrlich, H. L. (1963). Bacterial action on orpiment, *Econ. Geol.*, **58**, 991.

108. Ehrlich, H. L. (1964). Bacterial oxidation of arsenopyrite and enargite, *Econ. Geol.*, **59**, 1306.

109. Aggett, J. and O'Brien, G. A. (1985). Detailed model for the mobility of arsenic in lacustrine sediments based on measurements in Lake Ohakuri, *Environ. Sci. Technol.*, **19**, 231.

110. Newman, D. K., Beveridge, T. J. and Morel, F. M. M. (1997). Precipitation of arsenic trisulfide by *Desulfotomaculum auripigmentum*, *Appl. Environ. Microbiol.*, **63**, 2022.

111. Newman D. K., Kennedy, E. K., Coates, J. D., Ahmann, D., Ellis, D. J., Lovley, D. R. and Morel, F. M. M. (1997). Dissimilatory arsenate and sulfate reduction in *Desulfotomaculum auripigmentum*, *Arch. Microbiol.*, **168**, 380.

112. Moore, J. N., Ficklin, W. H. and Johns, C. (1988). Partitioning of arsenic and metals in reducing sulfidic sediments, *Environ. Sci. Technol.*, **22**, 432.

113. McCready, R. G. L., Campbell, J. N. and Payne, J. I. (1966). Selenite reduction by *Salmonella heidelberg*, *Can. J. Microbiol.*, **12**, 703.

114. Doran, J. W. (1982). Microorganisms and the biological cycling of selenium. In *Advances in Microbial Ecology*, ed. Marshall, K. L., Plenum Press, New York, p. 1.

115. Oremland, R. S., Steinberg, N. A., Presser, T. S. and Miller, L. G. (1991). In situ bacterial selenate reduction in the agricultural drainage systems of western Nevada, *Appl. Environ. Microbiol.*, **57**, 615.

116. Yang, G., Wang, S., Zhou, R. and Sun, S. (1983). Endemic selenium intoxication of humans in China, *Am. J. Clin. Nutr.*, **37**, 872.

117. Ohlendorf, H. M. (1989). Bioaccumulation and effects of selenium in wildlife. In *Selenium in Agriculture and the Environment*, ed. Jacobs, L. W., Soil Science Society of America, Madison, WI, p. 133.

118. Shrift, A. (1973). Metabolism of selenium by plants and microorganisms. In *Organic Selenium Compounds: their Chemistry and Biology*, ed. Klayman, D. L. and Gunter, W. H. H., John Wiley & Sons, New York, p. 763.

119. Karle, J. A. and Schrift, A. (1986). Use of selenate, selenide and selenecysteine for the synthesis of formate dehydrogenase by a cysteine requiring mutant of *Escherichia coli* K-12. *Biol. Trace Element. Res.*, **11**, 27.

120. Harrison, G. I., Laishley, E. J. and Krouse, H. R. (1980). Stable isotope fractionation by *Clostridium pasteurianum*. 3. Effect of SeO_3^{2-} on the physiology and associated sulfur isotope fractionation during SO_3^{2-} and SO_4^{2-} reductions, *Can. J. Microbiol.*, **26**, 952.

121. Gerrard, T. L., Telford, J. N. and Williams, H. H. (1974). Detection of selenium deposits in *Escherichia coli* by electron microscopy, *J. Bacteriol.*, **119**, 1057.

122. Silverberg, B. A., Wong, P. T. S. and Chau, Y. K. (1976). Localization of selenium in bacterial cells using TEM and energy dispersive X-ray analysis, *Arch. Microbiol.*, **107**, 1.

123. Lortie, L., Gould, W. D., Rajan, S., McCready, R. G. L. and Cheng, K.-J. (1992). Reduction of selenate and selenite to elemental selenium by a *Pseudomonas stutzeri* isolate, *Appl. Environ. Microbiol.*, **58**, 4042.
124. Tomei, F. A., Barton, L. L., Lemanski, C. L. and Zocco, T. G. (1992). Reduction of selenate and selenite to elemental selenium by *Wolinella succinogenes*, *Can. J. Microbiol.*, **38**, *1328.*
125. Tomei, F. A., Barton, L. L., Lemanski, C. L., Zocco, T. G., Fink, N. H. and Sillerud, L. O. (1995). Transformation of selenate and selenite to elemental selenium by *Desulfovibrio desulfuricans*, *J. Indust. Microbiol.*, **14**, 329.
126. Oremland, R. S., Hollibaugh, J. T., Maest, A. S., Presser, T. S., Miller, L. G., and Culbertson, C. W. (1989). Selenate reduction to elemental selenium by anaerobic bacteria in sediments and culture: biogeochemical significance of a novel, sulfate-independent respiration, *Appl. Environ. Microbiol.*, **55**, 2333.
127. Cantafio, A. W., Hagen, K. D., Lewis, G. E., Bledsoe, T. L., Nunan, K. M. and Macy, J. M. (1996). Pilot-scale selenium bioremediation of San Joaquin drainage water with *Thauera selenatis*, *Appl. Environ. Microbiol.*, **62**, 3298.
128. Macy, J. M., Lawson, S. and DeMoll-Decker, H. (1993). Bioremediation of selenium oxyanions in San Joaquin drainage water using *Thauera selenatis* in a biological reactor system, *Appl. Microbiol. Biotechnol.*, **40**, 588.
129. Macy, J. M., Rech, S., Auling, G., Dorsch, M., Stackebrandt, E. and Sly, L. (1993). *Thauera selenatis* gen. nov. sp. nov., a member of the beta-subclass of *Proteobacteria* with a novel type of anaerobic respiration, *Int. J. Syst. Bacteriol.*, **43**, 135.
130. DeMoll-Decker, H. and Macy, J. M. (1993). The periplasmic nitrite reductase of *Thauera selenatis* may catalyze the reduction of selenite to elemental selenium, *Arch. Microbiol.*, **160**, 241.
131. Losi, M. E. and Frankenberger, Jr., W. T. (1997). Reduction of selenium oxyanions by *Enterobacter cloacae* strain SLD1a-1: isolation and growth of the bacterium and its expulsion of selenium particles, *Appl. Environ. Microbiol.*, **63**, 3079.
132. Zachara, J. M., Ainsworth, C. C., Cowan, C. E. and Resch, C. T. (1989). Adsorption of chromate by subsurface soil horizons, *Soil. Sci. Soc. Am. J.*, **53**, 418.
133. Yassi, A. and Nieboer, E. (1988). Carcinogenicity of chromium compounds. In *Chromium in the Natural and Human Environments*, ed. Nriagu, J. O. and Nieboer, E. John Wiley & Sons, New York, p. 443.
134. Rai, D., Sass, B. M. and Moore, D. A. (1987). Chromium(III) hydrolysis constants and solubility of chromium(III) hydroxide, *Inorg. Chem.*, **26**, 345 .
135. Eary, L. E. and Rai, D. (1988). Chromate removal from aqueous waste by reduction with ferrous ion, *Environ. Sci. Technol.*, **22**, 972.
136. Losi, M. E., Amrhein, C. and Frankenberger, Jr., W. T. (1994). Bioremediation of chromate-contaminated groundwater by reduction and precipitation in surface soils, *J. Environ. Qual.*, **23**, 1141.
137. Horitsu, H., Futo, S., Miyazawa, Y., Ogai, S. and Kawai, K. (1987). Enzymatic reduction of hexavalent chromium by hexavalent chromium tolerant *Pseudomonas ambigua* G-1, *Agric. Biol. Chem.*, **51**, 2417.
138. Cervantes, C. (1991). Bacterial interactions with chromate, *Antonie van Leeuwenhoek*, **59**, 229.
139. Turick, C. E., Apel, W. A. and Carmiol, N. S. (1996). Isolation of hexavalent chromium-reducing anaerobes from hexavalent-chromium-contaminated and non-contaminated environments, *Appl. Microbiol. Biotechnol.*, **44**, 683.
139a. McLean, J. S. and Beveridge, T. J. (2000). Isolation and characterization of a chromium-reducing bacterium from a chromated-copper arsenate site, *Environ. Microbiol.* **2**, 611.

139b. McLean, J. S. and Beveridge, T. J. (2001). Chromate reduction by a pseudomonad isolated from a site contaminated with chromate copper arsenate. *Appl. Environ. Microbiol.* **67**, 1076.

140. Romanenko, V. I. and Koren'kov, V. N. (1977). A pure culture of bacteria utilizing chromates and bichromates as hydrogen acceptors in growth under anaerobic conditions. Translated from *Mikrobiologiya*, **46**, 414.

141. Wang, P. C., Mori, T., Komori, K., Sasatsu, M., Toda, K. and Ohtake, H. (1989). Isolation and characterization of an *Enterobacter cloacae* strain that reduces hexavalent chromium under anaerobic conditions, *Appl. Microbiol. Biotechnol.*, **55**, 1665.

142. Kavasnikov, E. I., Stepanyuk, V. V., Klyushnikova, T. M., Serpokrylov, N. S., Simonova, G. A., Kasatkina, T. P. and Panchenko, L. P. (1985). A new chromium reducing gram variable bacterium with mixed type of flagellation, *Microbiol.*, **54**, 69.

143. Gvozdyak, P. I., Mogilevich, N. F., Ryl'skii, A. F. and Grishchenko, N. I. (1986). Reduction of hexavalent chromium by collection strains of bacteria, *Mikrobiologiya*, **55**, 962.

144. Bopp, L. H. and Ehrlich, H. L. (1988). Chromate resistance and reduction in *Pseudomonas fluorescence* strain LB300, *Arch. Microbiol.*, **150**, 426.

145. Ishibashi, Y., Cervantes, C. and Silver, S. (1990). Chromium reduction in *Pseudomonas putida*, *App. Environ. Microbiol.*, **56**, 2268.

146. Fude, L., Harris, B., Urrutia, M. M. and Beveridge, T. J. (1994). Reduction of Cr(VI) by a consortium of sulfate-reducing bacteria (SRBIII), *Appl. Environ. Microbiol.*, **60**, 1525.

147. Tebo, B. M. and Obraztsova, A. Y. (1998). Sulfate-reducing bacterium grows with Cr(VI), U(VI), Mn(IV) and Fe(III) as electron acceptors, *FEMS Microbiol. Lett.*, **162**, 193.

148. Mondaca, M. A., Gonzalez, C. L. and Zaror, C. A. (1998). Isolation, characterization and expression of a plasmid encoding chromate resistance in *Pseudomonas putida* KT2441, *Lett. Appl. Microbiol.*, **26**, 367.

149. Lebedeva, E. V. and Lyalikova, N. N. (1979). Reduction of crocoite by *Pseudomonas chromatophila* species nova, *Mikrobiologiya*, **48**, 517.

150. Baillet, F., Magnin, J. P., Cheruy, S. and Ozil, P. (1998). Chromium precipitation by the acidophilic bacterium *Thiobacillus ferrooxidans, Biotechnol. Lett.,* **20**, 95.

151. McLean, J., Beveridge, T. J. and Phipps, D. (1999). Chromate removal from contaminated groundwater using indigenous bacteria. In *Proceedings of the In Situ and On Site Bioremediation Fifth International Symposium*, ed. Alleman, B. C. and Leeson, A., Battelle Press.

152. Kong, S., Johnstone, D. L., Yonge, D. R., Petersen, J. N. and Brouns, T. M. (1994). Long-term intracellular chromium partitioning with subsurface bacteria, *Appl. Microbiol. Biotechnol.*, **42**, 403.

153. Ohtake, H., Fujii, E. and Toda, K. (1990). Reduction of toxic chromate in an industrial effluent by use of a chromate reducing strain of *Enterobacter cloacae*, *Environ. Sci. Technol.*, **11**, 663.

154. Andersson, K. and Rosen, M.(1997). Biological reduction of hexavalent chromium in a leachate from Vargon Alloys AB, *Vatten.*, **53**, 245.

155. Turick, C. E., Graves, C. and Apel, W. A. (1998). Bioremediation potential of Cr(VI)-contaminated soil using indigenous microorganisms, *Bioremediation J.*, **2**, 1.

156. Chirwa, E. M. N. and Wang, Y. T. (1997). Hexavalent chromium reduction by *Bacillus* sp. in a packed bed bioreactor, *Environ. Sci. Technol.*, **31**, 1446.

157. Chirwa, E. M. N. and Wang, Y. T. (1997). Chromium reduction by *Pseudomonas fluorescens* LB300 in a fixed film bioreactor, *J. Environ. Eng.*, **123**, 760.
158. Frankenberger, Jr., W. T. and Losi, M. E. (1995). Applications of bioremediation in the cleanup of heavy metals and metalloids. In *Bioremediation: Science and Applications*, ed. Skipper, H. D. and Turco, R. F., SSSA Special Publication 43, Soil Science Society of America, Madison, WI, p. 173.
159. McGregor, R. G., Ludwig, R., Blowes, A. D. and Pringle, E. (1999). Remediation of a heavy metal plume using a reactive wall. In *Proceedings of the In Situ and On Site Bioremediation Fifth International Symposium*, ed. Alleman, B. C. and Leeson, A., Battelle Press.
160. Gerlach, R., Cunningham, A. and Caccavo, Jr., F. C. (1999). Chromium elimination with microbially reduced iron-redox-reactive biobarriers. In *Proceedings of the In Situ and On Site Bioremediation Fifth International Symposium*, ed. Alleman, B. C. and Leeson, A., Battelle Press.
161. Pazirandeh, M., Wells, B. M. and Ryan, R. L. (1998). Development of bacterium-based heavy metal biosorbents: enhanced uptake of cadmium and mercury by *Escherichia coli* expressing a metal binding motif, *Appl. Environ. Microbiol.*, **64**, 4068.
162. Amann, R., Snaidr, J. and Wagner, M. (1996). In situ visualization of high genetic diversity in a natural microbial community, *J. Bacteriol.*, **178**, 3496.
163. Barker, W. W., Welch, S. A. and Banfield, J. F. (1997). Biogeochemical weathering of silicate minerals. In *Geomicrobiology: Interactions between Microbes and Minerals*, ed. Banfield, J. F. and Nealson, K. H., Reviews in Mineralogy, Vol. 35, Mineralogical Society of America, Washington, DC, p. 391.

Part II Impact of Soil Particle–Microorganism Interactions on the Terrestrial Environment

(A) *Ion Cycling and Organic Pollutant Transformation*

7 Effect of Mineral Colloids on Biogeochemical Cycling of C, N, P, and S in Soil

G. GUGGENBERGER

University of Bayreuth, Germany

K. M. HAIDER

Kastanienallee 4, Deisenhofen, Germany

Interactions between Soil Particles and Microorganisms
Edited by P. M. Huang, J.-M. Bollag and N. Senesi. © 2002 John Wiley & Sons, Ltd

1 INTRODUCTION

Inorganic soil colloids exert a profound influence on almost all soil functions. Of particular importance is the role of the mineral colloids in the stabilization and degradation of soil organic matter (SOM) and its associated nutrients. Thereby, they directly affect the cycling of C, N, P, and S in soil.

Generally more than 95% of N and S and between 20 and 90% of the P in surface soils are present in SOM. Biochemically controlled processes of transformation of SOM are closely connected with the mobilization of N, P, and S into the soil solution or their immobilization from the solution [1,2] and to the release or fixation of trace gases such as CH_4, CO_2, OCS (carboxide sulfide), H_2, N_2O, and NO [3,4]. These microbially mediated processes are the basis for the interrelation of the C, N, P, and S cycling within the soil–plant system as shown in Figure 1. The close relationship between the organic forms of C, N, P, and S is well established and discussed in many reviews [e.g., 5–8]. Depending upon climate and management conditions, the C/N ratios range between 10 and 15, the C/P ratios between 50 and 100, and the C/S ratios between 60 and 100. The release of mineral nutrients is generally closely connected with the mineralization of C, causing the close relation among these elements. However, some element-specific pathways can occur (e.g., denitrification of N, source-specific mineralization of phosphate and sulfate esters [5]).

Numerous biotic and abiotic factors affect the cycling of organic C, N, P, and S. Of particular importantce are the temperature and water regimes, because they control the amount of carbon fixed by photosynthesis and the rate of decomposition and mineralization of organic matter [9]. Among the numerous

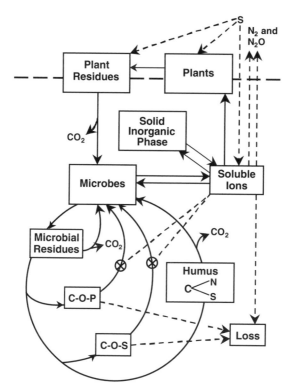

Figure 1. Schematic illustration of interrelations of C, N, S, and P cycling in soil–plant systems. Reprinted from [5], Copyright (1981), with permission from Elsevier Science

other factors that influence the ability of the decomposer community to utilize organic compounds as a source of C and nutrients, the textural composition of soil and soil structure are probably the second most important ones. Specific interactions between mineral soil colloids, microbial activity and metabolic products and SOM components have been reviewed in detail (e.g., by Oades [10], and Robert and Chenu [11]) and are also the subject of the chapter by Chenu and Stotzky (Chapter 1, this book). In the last decade, studies on the interactions of organic materials and microorganisms with minerals in the stabilization of soil aggregates have gained increasing importance, and the reviews of van Veen and Kuikman [12], Oades [13], Ladd *et al.* [14], and Baldock (chapter 3, this book) are referred.

In this chapter, the role of the mineral phase in soil on the cycling of C, N, P, and S will be addressed. This will be achieved first by examinations of the relation between contents of C, N, P, and S and soil parameters. Subsequently, stabilization and turnover of organic matter and associated nutrients will be discussed with respect to interactions of substrates and metabolic products with primary particles (i.e., phyllosilicate clay minerals and Fe- and Al-oxyhydroxides) and as related to soil aggregation. Hence, in our review particles larger than colloids are also included.

2 CONTENTS OF SOIL ORGANIC CARBON AND ITS RELATION TO N, P, AND S

2.1 INFLUENCE OF SOIL TEXTURE AND MINERALOGY

Soils contain organo-mineral complexes of varying sizes, structures, and organization. As a result of this heterogeneity, which occurs on scales ranging from nanometers to centimeters, SOM may be exposed or protected from microbial decomposition. Similarly, concentrations of oxidants, reductants, and nutrients vary among these heterogeneous microsites in soils, as does pH and water content.

Clay and organic matter interact chemically or by formation of microaggregates that render organic substances less susceptible to biodegradation [15,16]. When supplied with similar input of organic material, fine-textured soils usually contain more organic matter than coarse-textured soils [17,18]. Likewise, the clay content is positively correlated with SOM content when other factors such as vegetation, mean annual temperature, and drainage class are similar [19,20]. The protective effects of clay on soil C, N, P, and S also have been included in many computer models that describe the turnover of soil organic C and associated nutrients; e.g., Jenkinson and Rayner [21], van Veen *et al.* [22], Parton *et al.* [23]. With increasing clay content, a greater proportion of the newly formed microbial products becomes and remains physically protected [24].

Hassink [18] tested whether a general relationship between soil texture and the capacity to preserve organic C and N can be found. He reported highly significant positive correlations between the clay plus silt content of a soil and the amounts of C and N associated with this fraction (Figure 2a,b). From the observed relationship between C in the clay- and silt-size fraction and the percentage of soil particles $< 20 \, \mu m$, Hassink [25] calculated the protective capacity of a soil for C as

$$g \, C \, kg^{-1} \, soil = 4.09 + 0.39 \% \text{ of particles} < 20 \, \mu m$$

and further defined the saturation deficit as the difference between the protective capacity and the actual amount of clay- plus silt-associated C. After incubation with ^{14}C- labeled ryegrass, the amount of ^{14}C respired showed a significant

a)   C in size fraction < 20 μm (g kg⁻¹soil)

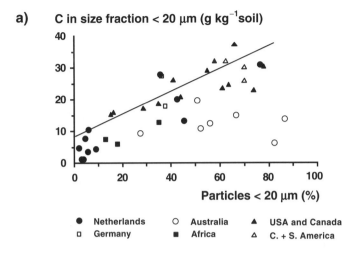

b)   N in size fraction < 20 μm (g kg⁻¹soil)

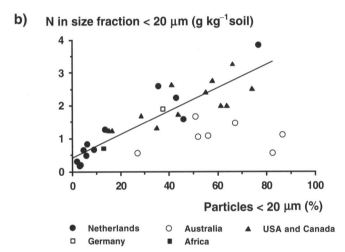

Figure 2. Relationship between C **(a)** and N **(b)** in the particle-size fraction $< 20\,\mu m$ (clay plus silt in g kg⁻¹ soil) and the percentage of the particle-size fraction $< 20\,\mu m$ in uncultivated and grassland soils of temperate and tropical regions; Australian sites are excluded from regression. From [18], with kind permission from Kluwer Academic Publishers

correlation with the saturation deficit, whereas the correlation between $^{14}CO_2$ production and soil texture was weak [25]. It may thus be concluded that the degree of saturation of the protective capacity of a soil predicts the decomposition rate of residue C (and likewise probably also that of residue N, P, and S) better than does soil texture alone.

2.1.1 Surface area

A close relationship between mineral-associated organic C and N with the surface area measured by the Brunauer–Emmett–Teller adsorption isotherm using N_2 adsorption (N_2–BET) has been found for marine sediments with surface loadings of 0.6–1.5 mg organic cm^{-2} [26,27]. These loadings were considered to represent the 'monolayer equivalent' range for organic matter associated with mineral particles [27]. Mayer [28] found that in soils, the high-density soil organic C ($d > 1.9\,g\,cm^{-3}$) showed a close relationship with the surface area as well, with a good fit into the monolayer equivalent range for about half of the soils. Soils with high carbonate content, low pH, or poor drainage showed organic C concentrations above the monolayer equivalent range, whereas concentrations below this range were found for arid soils as result of low primary production. In contrast, Kaiser and Guggenberger [29] found only a poor correlation between both parameters for a range of topsoil and illuvial horizons from European soils.

The N_2-BET method measures only the external area of clays and oxides, and laboratory experiments have shown that clay minerals may sequester certain organic species in the interlayer area [30,31]. If the negative charge of the organic species is reduced by acid soil reaction, expandable smectite-type clay minerals can intercalate SOM [30]. Further, Wang and Huang [31] showed that humic macromolecules can be intercalated in nontronite through interaction with phenol monomers and *in situ* polymerization at a pH near neutrality. However, there is little field evidence that intercalation of organic substances in interlayers of phyllosilicate clay minerals is an important process in the stabilization of SOM [32,33]. Many organic molecules are too large to enter the internal surfaces of minerals. Theng and Orchard [34] have compiled literature data on external and total surfaces of soil minerals (Table 1). It is obvious that the external surfaces of crystalline Fe and Al oxides are comparable with those of nonexpanding layer silicates (kaolinites, illite), whereas those of short-range ordered minerals such as imogolite, allophane, and ferrihydrite, and of non-crystalline $Al(OH)_3$ are at least an order of magnitude larger because of their very small dimensions of the unit particle.

The N_2-BET surface area of organic matter is in the range of the minerals but largely depends on sample pre-treatment, i.e., whether it is oven-dried, air-dried, or lyophilized [36,37]. It should be also noted that N_2 is subject to molecular sieving at 77 K due to activated diffusion in micropores $<\sim 1$ nm. It therefore severely underestimates the total surface area of organic matter. The surface area of organic materials analyzed by CO_2 adsorption at 273 K is about two orders of magnitude larger than that measured by N_2 adsorption and about 95–98 % of the SOM surface is formed by micropores [38]. Laboratory experiments have shown that removal of organic matter increases the N_2-BET surface area of soil [39,40], while coating of soils with organic matter led to a

Table 1. Surface area of some layer silicates, Fe and Al oxides, short-range ordered clays, and humic acid. Compiled by Theng and Orchard. Reprinted with permission from [34]. Copyright CRC Press, Boca Raton, Florida

Mineral	Specific surface area ($m^2\,g^{-1}$)	
	External (BET)[a]	Total
Kaolinite	11–26	same
Halloysite (varying particle shape)	30–109	330–720 (EG)[b]
Illite		
< 0.2 μm e.s.d.	51	same
0.2–0.5 μm e.s.d.	24	same
Montmorillonite (Wyoming)	14	800 (CPB)[b]
Goethite	17–81	same
Gibbsite	26–58	same
Imogolite[c]	303–397	822–1031 (EGME)[b]
Allophane[c]	292–582	638–897 (EG, EGME)[b]
Non-crystalline $Al(OH)_3^d$	285	not analyzed
Ferrihydrite	225–340	not available[e]
Humic acid[c]	0.7; 18[f]	536[g]

[a] These values were derived from adsorption of N_2 at $-195\,°C$, applying the Brunauer–Emmett–Teller (BET) adsorption isotherm.
[b] Abbreviations: EG, ethylene glycol; EGME, ethylene glycol monoethyl ether; CPB, cetylpyridinium bromide.
[c] The discrepancy between external and total surface area may be ascribed to inaccessibility of some interparticle pores to nitrogen as the particles 'coalesce' during the outgassing treatment. Assumptions as to molecular coverage by the adsorbate ($0.333\,mg\,m^{-2}$ for EG and $0.286\,mg\,m^{-2}$ for EGME) are also open to question.
[d] Data from Kaiser et al. [35].
[e] The total area is expected to be similar to that for allophane if the same value for molecular coverage by EG or EGME is assumed.
[f] The lower and higher values refer to oven-dried and freeze-dried material, respectively.
[g] This value is calculated on the basis of a particle diameter of 7 nm and a density of $1.6\,g\,cm^{-3}$.

smaller surface area [29]. Thus it can be concluded from N_2-BET data that, if only pores $> \sim 1\,nm$ are considered, sorbed organic matter masks mineral surfaces by reducing the surface roughness of the minerals [38,39]. Because the actual surface area in soil, based on the N_2-BET method, is a function of the mineral surface area and of its alteration by adsorbed organics, it may not be a good indicator of the capacity of minerals to store organic C, N, P, and S.

2.1.2 Mineralogy

Iron and Al oxides have been assumed to be major sorbents of organic C, because they are positively charged in the pH range of most soils (4–7), whereas layer silicates contain a permanent negative charge (Table 2). However, edge sites of layer silicates are also amphoteric, and therefore, the pH-dependent charge is associated with exposed Al (O, OH) and Fe (O, OH) groups [41,42].

Table 2. Point of zero charge (PZC) for some soil oxides and short-range ordered minerals. Reprinted with permission from [34]. Copyright CRC Press, Boca Raton, Florida

Mineral	Formula; molar ratio	PZC
Goethite	α – FeOOH	8.1
Hematite	α – Fe_2O_3	8.5
Gibbsite	γ – $Al(OH)_3$	9.4
Silica	$SiO_2 \cdot 4H_2O$	2–3
Imogolite (natural)	$(OH)_3Al_2O_3SiOH$; Al/Si = 2.0	6.8
Allophane (synthetic)	Al/Si = 1.74	6.9
	Al/Si = 1.43	6.5
	Al/Si = 1.15	5.5
Ferrihydrite (natural)	5 $Fe_2O_3 \cdot 9H_2O^a$	5.3–7.5

^{a} The presence or incorporation of Si in the structure depresses the PZC; the PZC of freshly formed, Si-free ferrihydrite is 7.6.

According to Theng and Orchard [34], the crystal edge represents between 1% (montmorillonite) and 20% (kaolinite) of the surface area of the layer silicates. Organic matter is also a variable charge soil component and can be regarded as a negatively charged organic polyelectrolyte containing carboxyls with $pK_a < 5$ as major functional groups [43]. Hence, Fe and Al oxides and edge sites of clay minerals can attract and bind the negatively charged organic substances by electrostatic interactions.

However, the strong pH dependence of organic matter sorption, the release of OH^- during the sorption, and the competition of organic polyanions with inorganic anions that form inner-sphere complexes with the mineral sorbate, i.e. phosphate, suggest that ligand exchange reactions between carboxyl groups of organic molecules at the mineral surface are most important in the sorption of organic matter on mineral phases [44–46]. Investigations with organic model compounds showed that, especially, the formation of bidentate complexes between two organic ligands in the *ortho* position of an aromatic ring and a metal at the surface of oxides and hydroxides causes a strong chemisorptive binding [45–47]. Sorption experiments with forest floor leachates on soils and soil minerals provided evidence that this type of binding occurs also in the sorption of natural organic matter [35]. According to solution ^{13}C NMR spectroscopy, sorption of dissolved organic matter (DOM) to soil material leads to a preferential depletion of carbonyl and aromatic C in solution (i.e., preferential adsorption) whereas alkyl C accumulates in solution; the same trends being observed after sorption of DOM to goethite and amorphous $Al(OH)_3$ (Table 3).

The strong affinity of organic carbon to Fe and Al oxides explains why good correlations are frequently found between concentrations of organic carbon and indicators of Al and Fe oxides, e.g., oxalate-extractable Al or dithionite–citrate–bicarbonate-extractable Fe for whole soils, clays, and density fractions

Table 3. Distribution of C species of dissolved organic matter in the soil solution before and after sorption experiments with mineral subsoil horizons and with oxyhydroxide phases according to liquid-state ^{13}C NMR spectroscopy. From [35]. Reproduced by permission of Blackwell Science Ltd

Sample	C species (%)			
	Carbonyl C 160–210 ppm	Aromatic C 110–160 ppm	O-alkyl C 50–110 ppm	Alkyl-C 0–50 ppm
Solution before the sorption experiment				
Original	21	31	28	20
Solution after sorption to mineral soil				
2Bw horizon	18	27	29	26
Bw horizon	19	28	28	25
Solution after sorption to oxyhydroxide phases				
Non-crystalline Al(OH)$_3$	16	24	29	31
Goethite	17	25	28	30

[29,48–50]. Shang and Tiessen [50] have shown that the size and crystallinity of the oxides exert a pronounced influence on concentrations of organic C, N, and P. Both factors strongly affect the surface area and thus the reactivity of the minerals. In particular, non-crystalline Al(OH)$_3$ seems important in interacting with organic matter, probably because of the weak tendency of Al to form crystalline oxides compared with Fe [10]. Stabilization of SOM as complexes of organic matter with non-crystalline Al(OH)$_3$ or short-range ordered Al-silicates was found in volcanic soils [51], in acid ranker soils [52], in acid alpine podzols [53], and cryptopodzolic soils [54]. In addition to the ability of short-range ordered positively charged minerals to interact with organic matter through ligand exchange reactions, their geometry may be well suited for physically protecting organic matter [55]. Microporosity may also play a role for protecting organic matter in non-crystalline minerals. Liu and Huang [56] showed that the microporosity of mineral colloids increases with the decrease of the crystallinity.

Minerals not only contribute to stabilization of SOM by sorptive stabilization and abiotic polymerization of low molar mass organic compounds, but they may also contribute to the degradation of organic compounds. In particular Mn(IV) oxides such as birnessite have been shown to enhance mineralization of low molar mass organic substances [57] or of humic materials [58] in suspension. In an incubation experiment with an artificial soil, Miltner and Zech [59] have shown that birnessite also enhances the mineralization of beech litter. The presence of birnessite increased the decomposition constant of the more stable C pool, k_2, by a factor of 1.5 compared with the control. Because microbial biomass was lower in the birnessite treatment than in the control, it was concluded that abiotic oxidation of the organic matter by the Mn oxide enhanced the decomposition. According to Sunda and Kieber [58], the oxidative decomposition processes caused by birnessite are not complete but produce low molar mass organic substances that are easily available substrates for the decomposer community and are rapidly mineralized.

Compared with the associations of soil organic C on minerals, little is known about associations of organic N, P, and S on oxides and layer silicates. However, the few studies dealing with mineral-bound organic C, N, P, and S indicate that N, P, and S are in general enriched with the same minerals as is C [50,60,61] and that sorption of dissolved organic N (DON) on minerals and soil follows the same patterns as does dissolved organic C (DOC) [62,63]. For N, this may be explained by the fact that much proteinaceous material is an integrated part of organic matter macromolecules [64,65]. This may also hold true for some of the organic P and S, in particular those forms with C-P and C-S bondings.

Distinct differences, however, need to be recognized for the sorption of amino acid N [7,43]. Some amino acids carry a positive charge which makes them a possible sorbate to negatively charged surfaces such as layer silicates. In a laboratory experiment, Aufdenkampe et al. [66] found that organic N was preferentially sorbed from DOM on organic-free kaolinite indicating preferential sorption of basic amino acids with positively charged side chains. Henrichs and Sugai [67] reported preferential sorption of the basic amino acid, lysine, compared with neutral and acid amino acids on sediments. However, they assumed that sorption of lysine occurred on negatively charged organic material coating the mineral phase. In contrast, field and laboratory data on sorption of organic N to soil and mineral phases indicated that DON is sorbed passively with DOM [62,63]. Kaiser and Zech [63] found a preferential sorption of acidic N compounds (i.e., muramic acid) to oxides and illite compared with neutral compounds. Thus acidic functional groups also appear to be important in sorption of organic N sorption.

Most soil organic P is present as mono- and diesters of orthophosphoric acid, and a minor proportion can be assigned to phosphonates with a direct C-P bond [68,69]. Phosphate esters of inositols in their hexa- and pentaphosphate forms are the most common identifiable compounds [70]. Due to their high charge density, inositol phosphates occupy the same sorption sites as orthophosphate (i.e., non-crystalline and short-range ordered Fe and Al oxides). These organic esters of P are even more strongly adsorbed than inorganic P [71,72]. Consequently, Abekoe and Tiessen [73] found a close positive relationship between organic P and dithionite–citrate–bicarbonate-extractable Fe. Phosphodiesters have lower charge densities and their phosphate groups are considerably shielded from ionic interactions and are thus sorbed to a lesser extent than phosphomonoesters [72].

Less is known about sorption of organic S species on minerals. Because organic S in soils is composed primarily of sulfur-containing amino acids and of sulfate esters [7,74], preferential sorption of organic S on Fe and Al oxides can be also assumed. In summary, it appears that the same mineral phases control the sorption of organic C, N, P, and S on the mineral soil matrix.

2.2 INFLUENCE OF SOIL GENESIS

Soil genesis is intimately connected to weathering of primary minerals and formation of pedogenic minerals. In accordance with the state factors of soil formation [75], the types of mineral formed in a particular soil or soil horizon depend primarily on the parent material, the age of soil, the abiotic conditions of weathering (precipitation and temperature), and the soil biota including its vegetation. Storage and turnover of C, N, P, and S in soil vary with respect to soil age, climatic zones, and soil types [6,60,76–78].

2.2.1 Formation and Aging of Pedogenic Minerals and their Role in Soil Organic Matter Storage and Turnover

A promising tool to study the influence of aging of pedogenic minerals on SOM storage and turnover are chronosequence studies. Torn *et al.* [78] investigated mineral formation and the relationship with storage and turnover of soil organic C along a 4100-kiloyear (kyear) chronosequence at Hawaii. With weathering of the volcanic ash-lava mixture, short-range ordered minerals (allophane, imogolite and ferrihydrite) accumulated during the first 150 kyears as primary weathering products. This is reflected by a transition from Inceptisol to Andisol soils. As the minerals are metastable, they dehydrate with time to crystalline clays, including halloysite, kaolinite, gibbsite, goethite, and hematite [55,79]. From 150 to 4100 kyears of the soil chronosequence, the amount of non-crystalline minerals therefore decreased and that of crystalline minerals strongly increased, as tracked by the transition from Andisols to Ultisols to Oxisols [78]. The authors found that the abundance of non-crystalline minerals accounted for $> 40\%$ of the variation in organic C content across all investigated mineral horizons, substrate ages, and soil orders (Figure 3a). Organic-matter $\Delta^{14}C$ was highly and negatively correlated with the abundance of non-crystalline minerals (Figure 3b), suggesting long turnover times with high contents of short-range ordered minerals.

Such a sequence of reactivity is a general pattern found in soil development in humid environments [80,81]. Formation of non-crystalline minerals at intermediate stages of soil development is most important in weathering of Fe- and/or Al-rich acidic parent materials, such as volcanic ashes, basalt, gneiss [53,54,78]. Short-range ordered $Al(OH)_3$ polymers and ferrihydrite are also enriched in spodic horizons, together with organic C [82,83]. Mean radiocarbon ages of the organic matter of up to 4000 years have been reported for spodic horizons [82], indicating effective stabilization of organic matter by the mineral phase.

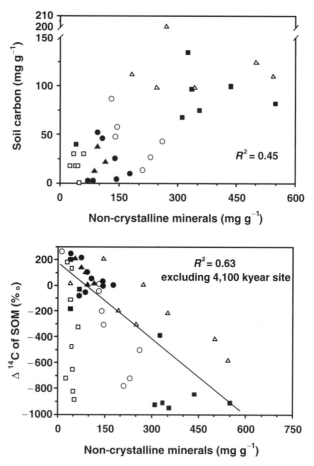

Figure 3. Soil C versus content of non crystalline minerals of mineral soil horizons from six chronosequence sites ($r^2 = 0.45$; $n = 35$) **(a)**, and $\Delta^{14}C$ versus content of non-crystalline minerals in all mineral horizons from the sites **(b)**; the solid line shows the regression between mineral abundance and $\Delta^{14}C$, excluding 4100-kyearr site ($r^2 = 0.63$; $n = 39$). Substrate age: △, 0.3 kyears; ●, 2.1 kyears; ▲, 20 kyears; ■, 150 kyears; ○1400 kyears; □, 4100 kyears. From [78]. Reproduced by permission of Nature

Vice versa, the restrained aging of non-crystalline minerals in spodic horizons should be seen in strong connection with the inhibitory effects of organic compounds on crystallization [84–86]. Hence, in a mutual interaction, non-crystalline mineral phases can prevent the organic substances from being bio-degraded, which, in turn, prevents the non-crystalline mineral phases from being transformed into forms of better crystallinity [82].

2.2.1.1 Biological influences on mineral weathering It has been postulated that low molar mass organic acids such as oxalic, citric, and salicylic acids, and lignin-derived phenolic compounds rather than humic compounds are effective agents for mineral degradation in soil [87,88]. On the other hand, the possibility that humic substances interact with and to dissolve minerals was hypothesized long ago [89] and is still being considered nowadays [90].

Humic substances constitute approximately 70% of the weight of SOM, but the concentrations of low molar mass organic acids are more difficult to assess. These natural organic acids are derived from plant residues, as well as from microbial metabolism. Their concentrations depend largely on the extent of microbial activity, which, however, will enhance both their formation and their degradation. Phenolic and aliphatic acids are continuously produced through the activity of microorganisms in soil [91,92]. Whitehead [93] detected a number of aromatic acids including p-hydroxybenzoic, p-coumaric, vanillic, and ferulic acids. Baker [94] compared the metal-dissolving capacities of various acids at several minerals and found that their capacities are of the same order as that of humic acids (Table 4).

Microbially degraded lignin is also important in complexing and dissolving metal oxides and ions. Lignin degradation in soil occurs by mechanisms including free radicals leading to cleavage reactions in both the side chains and rings (Figure 4). The resulting partly degraded lignin has more hydroxyl and carboxyl groups than the mother polymer. The former compact structure becomes gradually more dispersed and more hydrophilic. Larger cleavage products of lignin with hydroxyl and carboxyl groups that are sometimes in vicinal positions can complex with surfaces of metal oxides or with metal ions in solutions. Thereby partly degraded lignin becomes stabilized against further decay, but still contains larger fragments with typical structural features of lignin [95,96]. The resulting organo-mineral complexes contribute to mineral weathering and to mobilization of trace elements.

Table 4. Action of various organic compounds on minerals and metals. Table based on data published in [94]

Sapmle	Ion determined	Extractant µg of metal ions extracted per 1 h by 0.1% w/v)			
		Salicylic acid	Oxalic acid	Pyrogallol	Humic acid
Galena	Pb	130	95	35	200
Bornite	Cu	260	650	55	190
Chalcocite	Cu	4450	9750	920	3800
Hematite	Fe	< 3	80	10	470
Pyrolusite	Mn	4200	15500	5510	1000
Calcite	Ca	11900	980	2040	10500
Copper	Cu	5500	2620	1190	5700
Lead	Pb	41800	660	1470	41800

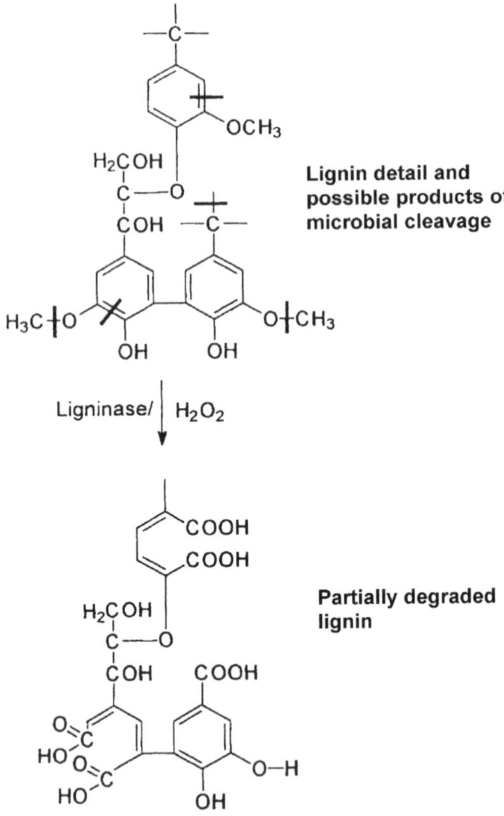

Figure 4. Structural features of a microbially altered lignin. From [95]. Reproduced by permission of Wiley–VCH

Chemolithotrophic bacteria involved in the N, S, and Fe cycles (nitrifiers, sulfur- and iron-oxidizing bacteria) are involved in the weathering of minerals [97,98], but a major part of the weathering processes in soils occurs mainly in the rhizosphere of plants with the participation of organotrophic microorganisms and root exudates [99]. Such weathering processes directly or indirectly involve the mobilization of major and trace elements and can have favorable or unfavorable effects on the production of plants and their quality.

Several publications have dealt with the effects of root systems on silicate weathering and found a weight loss of biotite and a production of kaolinite [100]. The first step of this silicate weathering process involves the mobilization of the interlayer K by ion exchange reactions that can lead to the formation of new minerals by modification of the crystallochemical structure under the effect of acid and/or complexing organic agents. The effect of plant roots and of microorganisms associated with the roots cannot be distinguished, but micro-organisms themselves are efficient in mineral weathering. In the rhizosphere, they can act directly on the minerals by solubilization and insolubilization processes and promote the availability of K, P, Fe, Mg, or adsorbed NH_4^+ [101,102]. Marschner et al. [103] have shown that the plant uptake of K, P, Fe, and Mg is significantly enhanced by rhizospheric weathering and pro-moted, in particular, by microbial rhizospheric weathering processes. The rhizospheric weathering processes highly influence the bioavailability of trace elements not only to plants but also to animals and humans through the food chain.

A considerable portion of the P in soils is linked in either inorganic or organic forms to surfaces of solid soil minerals. Exuded organic acids, phenols, or hydroxymates solubilize P either from solid surfaces or by dissolving P-con-taining minerals [104]. Furthermore, phenolic compounds formed during com-posting of plant residues or water-soluble humic compounds can be active in releasing sorbed P from solid mineral particles by solubilization or complex-ation with metal ions [105]. Humic or fulvic acids, as well as low molar mass aliphatic acids, can block sites on inorganic soil materials and thus reduce P adsorption. The efficiency of organic ligands in suppressing P adsorption is expressed as [106]:

$$\text{Efficiency in} \% = \left(1 - \frac{\text{phosphate sorbed in the presence of ligands}}{\text{phosphate sorbed when applied alone}}\right) \times 100$$

By using this formula, Klees [107] measured the inhibiting effect of various organic ligands on P adsorption by goethite (α-FeOOH). Table 5 shows that the adsorbed ligands effectively blocked the adsorption of P, and P was not able to displace the adsorbed ligands from the goethite surface.

Table 5. Inhibition of P adsorption by preceding adsorption of different organic ligands on the surface of goethite (α-FeOOH). From [107]. Reproduced by permission of the copyright holder

Organic compound	Efficiency (%) for the following concentrations			
	$0.2\,\text{mmol L}^{-1}$	$2\,\text{mmol L}^{-1}$	$10\,\text{mmol L}^{-1}$	$20\,\text{mmol L}^{-1}$
Oxalic acid	2.1	3.2	3.6	3.9
Citric acid	9.6	16.4	19.1	21.3
Catechol	4.2	9.4	11.6	13.6
Protocatechuic acid	11.2	19.9	22.6	25.0
Pyrogallol	45.4	58.3	60.1	63.1
Gallic acid	42.8	54.6	57.5	58.8

Conditions: 500 mg goethite suspended in 50 mL acetate buffer at pH 5.5 and shaken for 24 h with or without the ligand solution. Then P was added to a concentration of 60 μ mol L^{-1} and shaken for another 24 h. After centrifugation, the residual P concentration was determined.

However, if a low molar mass organic ligand is present during the formation of the oxides, adsorption of P by the oxides may be enhanced. In a recent study using atomic force microscopy, Liu and Huang [56] showed that $1\,mmol\,L^{-1}$ citrate, through structural perturbation greatly increased the surface roughness or irregularity, specific surface area, and reaction sites of Fe oxides formed. The concurrent increase in specific reaction sites for P adsorption apparently more than compensated for the blocking of the P adsorption sites by citric acid. A similar mechanism of this citrate effect on enhancing P adsorption by Al hydroxides through fundamental structural perturbation has been reported by Kwong and Huang [108]. More specific interactions of metal oxides with natural organic compounds are presented by Violante *et al.* (Chapter 4, this book).

2.2.2 Influence of climate

Besides textural and mineralogical control, the amounts of C and N in soil, as well as those of S and P to a minor extent, are related to production and degradation of organic matter at a given site. The balance between both processes depends on climatic conditions [23,109]. As a result, soil organic C, N, P, and S are strongly correlated with climatic variables, such as temperature and moisture [75].

Analyses of grassland soils of the Great Plains showed that with increasing mean annual temperature (MAT) at a particular mean annual precipitation (MAP) and texture, there is a decrease in the soil organic C content [19] (Figure 5). The same soils also show a strong response of soil organic C to MAP, with organic C increasing to about 800 mm and then leveling off. In these soils, on which photosynthetic activity is often limited by moisture, this is primarily a result of increases in plant productivity and, hence, inputs of C [109]. There is considerable interaction between MAP and the soil's content in clay and silt with respect to the C concentration in the soil investigated [19]. The clay effect is assumed in slowing turnover and the higher silt contents in increasing water holding capacity and thus the plant productivity. Generally, soil organic N is closely related to soil organic C. These results correspond to those predicted by Parton *et al.* [23] using the CENTURY SOM model.

Amelung *et al.* [110] investigated the response of soil organic C, N, and S to varying temperature and precipitation in the distribution of these elements among different particle-size separates. Figure 6 shows that soil organic C in the clay-size pools correlates positively with the logarithm of the MAP/MAT ratio ($r = 0.72***$), as does N ($r = 0.81***$). More SOM appears to be preferentially stabilized by clay at higher temperature. In contrast, unprotected SOM (partly degraded particulate organic matter) in the fine sand fractions is rapidly degraded as MAT increases and MAP decreases [110,111]. It is suggested that a

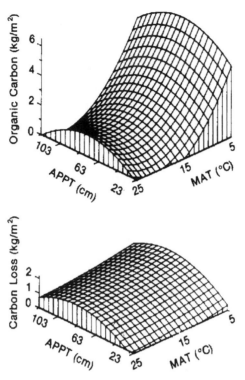

Figure 5. Response of soil organic C to continuous variation in mean annual tempera-ture (MAT) and annual precipitation (APPT) for a surface soil with loam texture (20 % silt, 40 % clay), as deduced from regression analysis of 500 rangeland soils. From [19]. Reproduced by permission of Soil Science Society of America

higher MAT promotes both the decomposition of unprotected SOM and the transfer of corresponding degradation products and metabolic products of microbial activity to the clay fraction. Temperature is most important for the enrichment of SOM in clay in a moderate range of textures (17–45 % of clay). Principal component analysis showed that clay and fine sand fractions contain sensitive soil organic C and N pools related to climate, whereas S seems to be controlled by factors other than those regulating the dynamics of soil organic C and N.

2.2.3 Organic and inorganic forms of N, P and S in different soil orders

Weathering of primary minerals supplies P and S to the plant- and microbially available pools in soil and, therefore, initiates the biological cycle with formation of organic forms of P and S. Phosphorus and S solubilized by mineral weathering

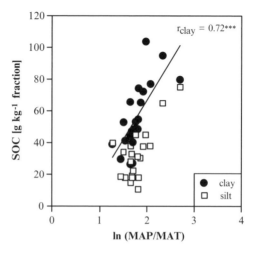

Figure 6. Soil organic C in the clay ($< 2\,\mu m$) and silt (2–$20\,\mu m$) fractions as related to ln [mean annual precipitation (MAP)/mean annual temperature (MAT)]. From [110]. Reproduced, by permission of Soil Science Society of America

or mineralization can again be sorbed to secondary mineral surfaces [112]. Sorbed P, in particular, can eventually become surrounded or occluded by Fe and Al oxides, rendering the P largely unavailable to the biota [113]. Organic forms of P can also be strongly stabilized by reactions with pedogenic minerals [50].

A well recognized procedure to isolate different pools of soil P is the Hedley fractionation [114] by which phosphate ions and other forms of labile, non-labile, non occluded, and occluded inorganic and organic P are removed sequentially with a series of successively stronger reagents. Cross and Schlesinger [115] reviewed literature data on Hedley fractionations of soil to test whether the contribution of geochemical versus biological processes to soil P availability in soil varies with pedogenesis in accordance with the Walker and Syers [6] paradigm. Cross and Schlesinger [115] showed that total soil P decreases as a function of soil development from an average of $864\,mg\,P\,kg^{-1}$ soil in Entisols to between 200 and $430\,mg\,kg^{-1}$ soil in Ultisols and Oxisols. As a percentage of total P, HCl-extractable P (largely P in primary minerals) ranges from a maximum of 66% in the Entisols to less than 1% in the highly weathered Oxisols (Figure 7). An inverse pattern was observed for the NaOH and sonicated-NaOH inorganic P (largely P adsorbed to surfaces of and occluded within Al and Fe oxides). These fractions were least prominent in less weathered soil and most prominent in Ultisols and Oxisols. The residual pool which comprises occluded inorganic P and recalcitrant organic P did not vary across the soil orders. The percentage of organic P increased from 5% in

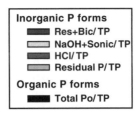

Phosphorus fractions (% of total P)

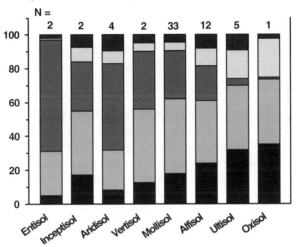

Figure 7. Soil P fractions for each soil order expressed as a percentage of total P. Reprinted from [115], Copyright (1981), with permission from Elsevier Science

Entisols to 35% in Oxisols (Fig. 7). Such an increase is not only observed for total organic P but also for the labile organic P pool (bicarbonate organic P related to resin- and bicarbonate-extractable total P). This indicates an increasing importance of organic P as a source of plant-available P with soil age.

Similar results on variations of different P fractions were reported by Crews *et al.* [116] across a chronosequence in Hawaii. However, they showed that organic P had a maximum at intermediate soil development and then declined in the older sites. This may be explained with the maximum storage of SOM in soils rich in short-range ordered minerals [78]. Further, Crews *et al.* [116] provided evidence that P is not permanently occluded by secondary Fe and Al minerals.

Comprehensive studies on forms of S in different soil orders are lacking. Organic N closely follows soil organic C, and the highest organic N reserves are to be expected in soils with intermediate weathering, such as Andisols, Alfisols, and Mollisols. Most of the inorganic N forms do not accumulate in soil and are susceptible to losses as dissolved NO_3^- or gaseous N_2O and/or NH_3. An exception is NH_4^+, which can be reversibly adsorbed to clay minerals in neutral to weakly acid soils or fixed in a nonexchangeable form between mineral lattices [117]. NH_4^+ fixation is most common in vermiculite and montmorillonite clays, where NH_4^+ ions fit in similar locations between clay layers as does K^+ in natural hydrous mica (illite). Hence, the highest proportions of inorganic N are to be expected in soil orders that represent little to intermediately weathered soil.

3 ROLE OF MINERAL COLLOIDS AND PRIMARY SOIL PARTICLES IN STABILIZATION AND CYCLING OF C, N, P, AND S

Models developed to describe turnover of SOM usually divide organic compounds into several compartments with specific ranges of degradation rates and residence times. In reality, however, no distinct pool boundaries exist, and there is a continuous transformation of organic C, N, P, and S with overlapping residence times. Furthermore, even incompletely humified plant residues can be stabilized by sorption, by entrapment, or by complexation with metal oxides. Various size fractions of soil material can be used as a criterion for the residence times of C in chemically stabilized pools of SOM [21,118] and in SOM pools stabilized by adsorption on soil colloids or entrapped in microcompartments of pores or aggregates [13,119]. Both mechanisms can equally and simultaneously contribute to the stabilization of SOM and associated nutrients.

Separation of soil into primary organomineral complexes is based on the concept that SOM associated with mineral particles of different size differs in structure and function [118]. Primary organomineral complexes are usually isolated after complete dispersion of soils. Christensen [61,118] achieved this by the use of probe-type sonication. However, the clay-size ($< 2\,\mu m$) primary organomineral complexes are a mixture of 'true' primary particles and microaggregates of fine clay mineral particles held together by organic and inorganic 'cements' and electrostatic forces [120]. As much as a complete dispersion of soil by high energy sonication of soil is desirable, a major constraint of high-energy sonication is the possible redistribution of particulate SOM among size separates [121]. Free particulate organic matter (POM) may be destroyed by sonication and be present in the silt- and clay-size separates after ultrasonic treatment. Another drawback of sonication concerns the redistribution of microbial C. Even limited ultrasonic treatments cause release of microbial cell contents [122] which are probably adsorbed by silt- and clay-size separates. However, as microbial biomass usually does not exceed 5% of SOM [123], the redistribution of microbial cell components is considered to be limited to a small proportion of the total SOM.

To minimize redistribution of POM during ultrasonic dispersion, a two-step sonication of soil is proposed [110,124]. At the first step, macroaggregates (250–2000 μm) are dispersed by low ultrasonic energy, and the coarse POM released is separated by sieving. At the final ultrasonic dispersion, the < 250 μm suspension is completely dispersed by higher energy input. Schmidt *et al.* [125] evaluated several ultrasonic dispersion and particle-size fractionation methods for isolating primary organomineral complexes from soil and recommended calorimetric calibration of ultrasonic instruments as a simple and effective way to ensure reproducibility and comparison among laboratories. They also showed that dispersion with energy in the range 30–590 J mL^{-1} evidently caused no detachment and redistribution of organic matter in contrast to higher energies.

Beside the different procedures for sample dispersion, the use of different size classes in different countries complicates generalization of data from the literature to assess the distribution of C, N, P, and S across primary soil particles with respect to element cycling. Most studies follow the ISSS or the USDA classification schemes, where clay is < 2 μm, silt is 2–20 μm (ISSS) or 2–50 μm (USDA), and sand is 20–2000 μm (ISSS) or 50–2000 μm (USDA). However, some studies also use other ranges.

3.1 CONTENTS OF SOIL ORGANIC MATTER IN DIFFERENT SIZE SEPARATES

To compare concentrations of SOM in different size separates independently from different carbon levels in soil, Christensen [61] defined the enrichment factor, E. It relates the content of an organic matter component in a particular size fraction to the content of that component in the whole soil, e.g.:

$$E_{C,N,P,S} = (\text{mg C, N, P, or S g}^{-1} \text{ separate})/(\text{mg C, N, P, or S g}^{-1}\text{whole soil})$$

Generally, the highest enrichment of C was found in fine silt and clay-size separates, with E_C ranging from 1.3 to 3.9 for silt and to > 10 for clay [61,118]. Excluding soils high in POM, sand-size separates usually show E_C factors < 0.1. The C-enrichment of clay and silt was found to be inversely related to the proportion of these size separates in whole soil, and the enrichments decline following a logarithmic function as clay and silt yields increase [61]. Clay-size separates generally account for more than 50 % of the C in whole soil [126–128]. However, if redistribution of POM-derived organic matter is avoided by the two-step ultrasonic dispersion, C-enrichment in coarse sand-size separates may sometimes be at a maximum [110] (Table 6).

Table 6. Enrichment factors (E) for organic C , total N, organic P, and total S in particle-size separates of topsoils

Reference	USDA Soil Taxonomy	Site	Particle size	E_{OC}[a]	E_N[b]	E_{OP}[c]	ES[d]
Anderson et al. [129][e]	Mollisol	arable field Saskatchewan, Canada	< 0.2 μm	1.97	2.60	n.d.	4.35
			0.2–2 μm	2.57	2.55	n.d.	2.34
			2–5 μm	3.85	3.35	n.d.	2.45
			5–50 μm	0.68	0.56	n.d.	0.49
			50–2000 μm	0.10	0.06	n.d.	0.07
Amelung et al. [110]	primarily Mollisols (incl. Alfisol, Vertisol, Aridisol)	native grassland North America	< 2 μm	1.89	2.21	n.d.	3.51
			2–20 μm	1.05	0.91	n.d.	1.29
			20–250 μm	0.32	0.24	n.d.	0.37
			250–2000 μm	3.54	1.65	n.d	1.33
Tiessen et al. [60]	Mollisols	native grassland Saskatchewan, Canada	< 0.2 μm	1.26	2.00	2.00	n.d.
			0.2–2 μm	1.64	2.18	2.87	n.d.
			2–5 μm	1.70	1.98	2.01	n.d.
			5–50 μm	0.82	0.90	0.81	n.d.
			50–2000 μm	0.54	0.48	0.32	n.d.
Christensen [127] and Rubæk et al. [130]	Alfisol	arable field Askov, Denmark	< 2μm	5.04	6.18	5.38	n.d.
			2–20 μm	2.90	2.50	2.05	n.d.
			20–2000 μm	0.10	<0.10	<0.20	n.d.
Guggenberger et al. [131] and Rubæk et al. [130]	Inceptisol	arable field Bavaria, Germany	< 2μm	2.51	2.39	2.06	n.d.
			2–20 μm	1.00	0.83	0.66	n.d.
			20–2000 μm	0.26	0.16	0.13	n.d.

[a] E_{OC} refers to the enrichment factor for organic C; [b] E_N refers to the enrichment factor for total N; [c] E_{OP} refers to the enrichment factor for organic P; [d] E_S refers to the enrichment factor for total S; [e] in this paper, EN refers to the enrichment factor for organic N.

Table 6 also shows that the trends of enrichment factors for organic C (E_{OC}), total N (E_N), organic P (E_{OP}), and total S (E_{OS}) among the size fractions are similar for a given soil. This suggests that particle-size fractionation yields characteristic pools not only for C accumulation but also for the accumulation of N, P, and S. However, the degree of enrichment is different for each of the elements. In the clay-size fraction, the enrichment factors increase in the order $E_{OC} < E_N = E_{OP} < E_S$. This may be partly explained by accumulation of mineral N and S in the finer fractions. Jensen et al. [132] reported that, depending on the clay content of the soils investigated, mineral-fixed $NH_4^+ - N$ in the clay and silt fractions varies between 3 and 16% of total N. A similar percentage was found by Kowalenko and Ross [133] for a clay loam. However, a preferential accumulation in the clay fraction was found not only for total N and S but also for organic N [129]. Therefore, the enrichment of N, P, and S in the smaller size separates is, to a large part, a result of the enrichment of these elements within the organic matter in the smaller fraction. This enrichment pattern is reversed in the (coarse) sand fraction, where N, P, and S are more depleted than C.

3.2 QUALITY AND ORIGIN OF SOIL ORGANIC MATTER IN DIFFERENT SIZE SEPARATES

3.2.1 Ratios of C/N, C/OP, and C/S in Soil Organic Matter

In essentially any case where soils have been texturally separated, a trend toward lower C/N ratios in the finer size particles has been observed [61,129,134]; C/N ratios in the clay-size separates are often < 10. This trend holds also true for the C/OP and the C/S ratios (Figure 8). Although sand-size separates show C/OP ratios > 100 and C/S ratios > 200, the corresponding ratios in the clay-size fractions are around 50. This narrowing with decreasing particle size

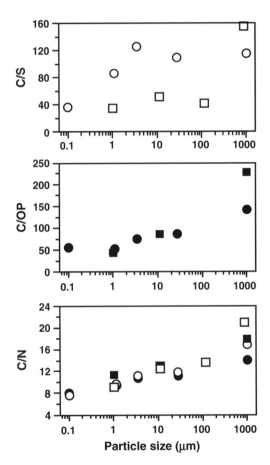

Figure 8. Plots of C/N, C/OP, and C/S *versus* mean diameter (μm) for particle-size fractions of some North American and European soils (Figure complied from data given in : ●, Tiessen *et al.* [60]; ■, Rubæk *et al.* [130] and Guggenberger *et al.* [131]; ○, Anderson *et al.* [129] (organic N was measured); □, Amelung *et al.* [110])

suggests that not only are nitrogenous organic substances selectively concentrated in the finer fractions, but also organic substances containing P and S. The ratios of these soil organic matter constituents provide a clue to the origin and the degree of decomposition of the organic matter associated with differently sized primary particles [118,134]. C/N, C/OP, and C/S ratios found for sand-size separates are in the same range of values reported for plant litter [7]. Those of the clay-size fraction, in contrast, are within or close to the range of ratios found in microbial biomass. The weakest relationship with soil texture were found for the C/S ratios. Microorganisms have C/N ratios of 7–12 [7], C/OP ratios of 14–36 [135], and C/S ratios of 57–85 [7].

A more detailed assessment of processes related to the distribution of C, N, P, and S in different primary particle-size separates is obtained by analysis of the structural composition of SOM, either by microscopy, spectroscopic methods, or analysis of biomarkers.

3.2.2 Chemical Composition and Origin of Soil Organic Matter

Electron microscopy gives direct access to the origin and the degree of decomposition of SOM within the different fractions. Golchin et al. [136] observed mostly large, undecomposed root and plant fragments in the 500–2000 μm fractions and more decomposed materials in the 10–100 μm fractions. According to Amelung et al. [110], organic matter in the coarse sand fraction is composed mainly of slightly decomposed plant residues, whereas SOM associated with fine sand consists of fairly altered organic debris and fine particles of roots. Microbial cells in primary particle-size fractions have not been detected by microscopy, probably as a result of their destruction by the high-energy sonication used for complete soil dispersion [137].

Cross polarization/magic angle spinning (CP/MAS) ^{13}C NMR spectroscopy is, at present, probably the most favored tool for detecting individual forms of C. Although this method is problematic with respect to quantification resulting from the application of the cross polarization technique [138–140], it provides important insight into the chemical structure of SOM associated with different primary particle fractions (Figure 9). In general, O-alkyl resonances mainly derived from carbohydrates or from ether bonds in degraded lignin predominate across the entire textural range, but they are minimized in the silt fraction [141–144]. The strongest O-alkyl signals are always observed in the sand fraction, and spectra of the organic matter associated with sand strongly resemble those of fresh plant litter. Aromatic C, and in particular the aromatic C with other carbon substituents, is maximal in silt but is lowest in the clay fraction(s). Alkyl C steadily increases with decreasing particle size and represents the major [142] or the second major C species in the clay fraction [141]. Oades et al. [142] identified long-chain material as the dominating form of alkyl C and considered this to be a contribution of natural waxes. Baldock et al. [145] showed that

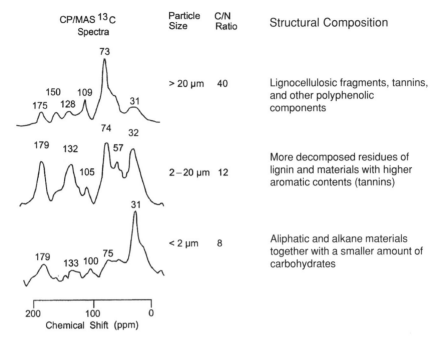

Figure 9. Composition of the organic matter associated with different particle-size fractions as investigated by CP/MAS ^{13}C NMR spectrometry. From [141], with kind permission from Kluwer Academic Publishers

microbial utilization of ^{13}C-labeled glucose resulted in the synthesis of O-alkyl C and alkyl C, mainly in the clay fraction. This suggests that clay-bound SOM primarily consists of microbial residues.

These observations are generally corroborated and extended by wet-chemical analysis of specific compounds of SOM in particle-size separates. Individual sugar monomers released from SOM upon acid hydrolysis culminate in sand and clay separates and have their minimum in the silt-size fraction [134,142]. Ratios of (galactose + mannose)/(arabinose + xylose) and of (rhamnose + fucose)/(arabinose + xylose) increase from sand to silt to clay, indicating primarily plant-derived polysaccharides in the sand-size separates and larger fractions of microbial polysaccharides in the clay.

Christensen and Bech-Andersen [146] reported that amino acids, in particular diaminopimelic acid which is confined to the peptidoglycans of cell wall of the prokaryotes, are recovered in elevated yields from clay. The clay-to-silt ratio for diaminopimelic acid was 2.6–3.0 in a loamy sand and 4.7–4.8 in a sandy loam soil. Besides amino acids, the amino sugars are an important form of organic N in soil and are primarily a result of microbial resynthesis [147]. Because muramic acid is largely derived from the peptidoglycans of the bacterial cell wall and glucosamine is found in both bacteria and fungi [148], the ratio of glucosamine to muramic acid can be used as an indicator of bacterial and fungal contribution to composition of SOM [149,150]. Zhang *et al.* [151] showed that the average concentration of total amino sugars in a range of North American native grassland soils increased markedly as the particle size decreased from coarse sand to clay (Figure 10). The major proportion of the three hexosamines glucosamine, galactosamine, and mannosamine (69%), and of muramic acid (79%) was attached to clay. The ratio of glucosamine to muramic acid varied markedly among the particle-size fractions (Figure 10), indicating that either the microbial origin or the stability of amino sugars was different between the separates. Due to its negative charge, muramic acid may be preferentially bound to positively charged Fe and Al oxides in the clay fraction [63]. Because most of the bacterial biomass has been found in the vicinity of clay [152,153], high concentrations of bacterial-derived compounds (diaminopimelic acid, muramic acid) in the clay-size separates may be caused by a high production of microbial-derived substances. Hence, the enrichment of nitrogenous compounds in the clay separates can be explained, to a large part, by accumulation of microbial- and, in particular, bacterial-derived amino acids and amino sugars.

The distribution of different organic forms of P in different particle size separates has been examined by means of solution ^{31}P NMR spectroscopy of alkaline extracts. Compared with bulk soils, diester-P structures were enriched in the clay fractions of a range of North American soils [154]. In contrast, monoester-P structures were enriched in particle size classes coarser than clay.

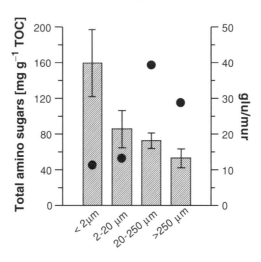

Figure 10. Mean concentration of total amino sugars (▨) and ratios of glucosamine to muramic acid (glu/mur) (●) in particle-size separates of 18 surface soils of the North American Prairie. Figure based data published in [151]

Similarly, Rubæk *et al.* [130] reported that the dialyzed NaOH extracts of clay were enriched in teichoic acid-P and other diester-P forms compared with the extracts of silt and clay. Because the major organic P structure in microorganisms is diester-P, including teichoic acid-P and plants contribute primarily to monoester-P in soil [155], this indicates that organic P associated with clay is also primarily of microbial origin. There is evidence that diester-P produced by microorganisms can be stabilized on mineral surfaces, with Fe oxides being most effective [156].

Concurrent with the increase of microbial compounds with decreasing particles sizes, yields of lignin phenols released upon CuO oxidation decrease [131,157]. This drop is accompanied by increasing side-chain oxidation, as indicated by the acid-to-aldehyde ratios of the vanillyl units, ranging from values in sand near 0.2 to ratios in clay of up to 1.0. The increasing oxidative degradation of lignin with decreasing particle size is consistent with decreasing proportions of O-substituted aromatic C in ^{13}C NMR spectra [144]. Hence, lignin in the sand-size fraction appears to be in slightly modified vascular plant debris, whereas lignin residues in the clay-size fractions are highly degraded.

3.2.3 Role of Dissolved Organic Matter in the cycling of C, N, P, and S

DOM represents an intermediate decomposition product of SOM that is rich in functional polar groups [158]. In particular in forest soils, appreciable amounts of DOM are translocated from the forest floor into the illuvial mineral soil horizons (e.g., mixed forest, 207 kg DOC ha^{-1} year^{-1} [159]; mixed deciduous forest, 405 kg DOC ha^{-1} year^{-1}, [160]) where DOM is effectively chemisorbed to Fe and Al oxides. Comparative analyses of the chemical composition of the forest floor leachate and solid-phase organic C in spodic horizons using ^{13}C NMR spectroscopy revealed a strong resemblance of the organic C accumulated in the mineral soil horizon to the organic C in the leachate DOM [29,161]. Leaching of organic matter affects not only C cycling in soil but also that of N, P, and S. Schoenau and Bettany [1] reported that the fulvic acids in the B and C horizons of forest and grassland soils are enriched in N, P, and S relative to humic acids and fulvic acids in the respective surface horizons. This indicates that the organic matter being translocated downwards is richer in N, P, and S than fractions not susceptible to leaching. Qualls *et al.* [160], quantifying the fluxes of dissolved organic C, N, and P in a deciduous forest, found an average annual output from the forest floor in the form of dissolved organic matter of 18, 28, and 14% of the input in solid litterfall for C, N, and P, respectively. 94% of the N and 64% of the P in the solution percolating from the forest floor into the mineral soil was organic.

Thus the translocation of nutrient-rich organic compounds to deeper horizons and the effective immobilization on hydroxide surfaces affects turnover of the nutrients in surface horizons by removing labile products that would normally be available for microbial growth and uptake. Kaiser and Guggenberger [29] provided evidence that mobilization of DOM in surface layers and subsequent sorption to minerals in deeper horizons is an important process in the stabilization of organic C and organically bound nutrients not only in spodic horizons but generally in soils having a considerable sorption capacity for DOM (i.e., Fe and Al oxides). This pathway in the stabilization of SOM and associated nutrients may be particularly relevant for clay-size separates of soil.

3.3 DECOMPOSITION AND STABILIZATION OF C, N, P AND S RELATED TO PRIMARY PARTICLES

Depending on the time scale, the turnover of organic matter and its associated nutrients can be studied by (i) short-term studies where mineralization and stabilization of decomposition products are investigated during laboratory incubation with or without the use of ^{14}C-, ^{13}C- or ^{15}N-labeled materials and by (ii) long-term field experiments using naturally or artificially ^{14}C- or ^{13}C-labeled organic matter.

Christensen [61] reviewed the literature on C, N, and S mineralization during laboratory incubation experiments. In general, the fine particle size separates are the most responsible for the mineralization of organic C, N, and S. Clay usually accounts for more than 60% of the mineralization, whereas some 30% is made up of mineralization from silt-size separates, and sand generally contributes less than 10%. Catroux and Schnitzer [162] reported that the percentage of mineralizable N after 28 days of incubation was particularly large for the fine clay fraction. Lowe and Hinds [163] examined the N and S mineralization from silt (2–50 μm) and three clay-size separates (1–2 μm, 0.2–1 μm, < 0.2 μm). They found that the mineralizations of N and S were highly correlated, and the proportion of S mineralized from individual size separates was greater than that of N. Christensen [164] measured CO_2 evolution from clay, silt, and sand-size separates and found highest C mineralization rates for sand. The higher mineralization rate of C in sand does not conflict with results demonstrating a low mineralization rate of N and S in sand. The high CO_2 production from sand indicates a high decomposability of POM, which may be coupled with immobilization of N [61].

These data showing a rapid turnover of parts of the clay-associated SOM fit well with investigations dealing with substrate incorporation into the different size fractions. Chichester [165] observed a preferential enrichment of added inorganic ^{15}N in newly formed SOM in clay. Cheshire and Mundie [166] compared the distribution of native and labeled carbohydrates among size separates from soils incubated with ^{14}C-labeled glucose, ryegrass, and straw. They found that, depending on the substrate source, the highest activity was located in either the clay- (soil incubated with glucose) or the sand-sized fractions (soil incubated with plant materials). However, microbial sugars (mannose, rhamnose, and fucose) produced during incubation of labeled plant materials showed the largest concentrations in clay. This indicates that after microbial resynthesis, utilizable substrate C and N ends up as microbial residues primarily in the clay-size separates [146]. In contrast, more recalcitrant plant remains are concentrated in the sand-size separates.

Amelung et al. [167] used natural ^{13}C-labeling of animal manure collected from steers fed on maize (C4 plant) to study incorporation of substrate C in particle-size fractions of soil cultivated with a pasture of C3 plants. During the 70 days of the experiment, the authors observed a continuous accumulation of particulate C in the coarse sand-size separates. The proportion of dung-derived C in the fine clay-fraction peaked 42 days after dung application, coinciding with the maximum leaching rate measured in lysimeters. The authors suggested that this resulted from an increasing sorption of microbially released DOM from the dung patches. Hence, the results provided evidence that the coarse sand- and fine clay-fractions may reflect short-term dynamics of SOM. In contrast, the contribution of dung-derived C to total C hardly changed in the coarse clay-, the fine sand-, and plus silt-size fractions during the time scale of the experiments, indicating a longer-term SOM dynamics

For longer-term studies, the natural abundance of ^{13}C can be used to estimate the turnover of soil organic C when the original photosynthetic pathway of organic inputs is naturally or artificially changed. Balesdent et al. [168] investigated the natural ^{13}C abundance in size separates of a temperate Hapludalf topsoil planted with maize for 23 years after clearing of a pine forest (C3 vegetation). Confirming the short-term experiments, they found that within the coarse sand, 61% of the C was from maize, indicating rapid turnover. Smaller incorporation of maize-derived C were found for silt (12% C4–C), coarse clay (18% C4–C), and fine clay (26% C4–C). Similar results were reported for a North American prairie (C4) soil (Ochraqualf) that was cultivated with timothy or wheat (C3) for 97 years [169]. However, in this soil, after 27 years of cultivation, the authors observed that the δ^{13}C values remained fairly constant for fine clay and higher than those of the larger size separates, suggesting little further turnover of prairie-derived SOM in this fraction. Studies in tropical soils, employing natural ^{13}C abundance, also suggested that fine particle-size separates contain stabilized SOM. Martin et al. [170] reported that after 25 years of tree (C3) invasion on grass savanna (C4) < 10, 28, 56, and 70% of the total soil organic C in sand, coarse silt, fine silt, and clay, respectively, was savanna derived. Similarly, Vitorello et al. [171] calculated turnover times of 59, 6, and 4 years for SOM in clay, silt, and sand-size separates, respectively, obtained from an Oxisol. These data provide evidence that clay also contains SOM components with very slow turnover rates.

Compared with the δ^{13}C approach, the use of ^{14}C-dating to estimate turnover in particle-size separates is scarce. In addition, calculated mean residence times are susceptible to various problems [172]. Anderson and Paul [173] applied ^{14}C-dating to Ap horizons and found that coarse clay and fine silt contained the oldest and thus the most stable SOM. In contrast, SOM associated with fine clay was the most enriched in ^{14}C derived from atomic bomb tests, indicating rapid turnover. However, as suggested by Christensen [61], the rapid turnover of clay-bound SOM in this experiment may have been caused by an artifact resulting from contributions from recently added substrates and from cell contents of microbial biomass killed during sample pretreatments and soil dispersion. In contrast to the former study, Scharpenseel et al. [174] described SOM associated with fine clay as being the oldest fraction. However, high levels of recent (atomic-bomb-derived) ^{14}C in organo-mineral associations (humins) indicated that fine particle-size separates also can act as a sink for recent inputs of organic matter [175,176].

In summary, a large portion of C in sand-size separates is rapidly mineralized, whereas N, P, and S appear to be preferentially immobilized by microorganisms and redistributed after cell death to finer separates. SOM associated with silt has slow turnover rates. This may be attributed to accumulation of recalcitrant organic materials, such as highly aromatic and insoluble remnants of lignin degradation. For clay, however, the results of the different methods to assess turnover of soil organic C and associated nutrients suggest that this is a heterogeneous pool. Clay appears to contain SOM pools of rapid and of slow turnover. Dissolved organic substrates appear to pass rapidly via porewater to clay-size fractions where microbial biomass is concentrated [152,153] and that may effectively utilize the substrates. Until complete mineralization of C, N, P, and S, several cycles of synthesis and decomposition of microbial biomass can elapse (microbial loop). Moreover large portions of microbial metabolic products (i.e., cell contents) are water soluble and labile and are thus a major reservoir of readily available organic N, P, and S. Rubæk et al. [130] found that the highly active and easily mineralizable organic phosphorus as determined with a sensitive macroporous resin method was mainly associated with clay and that this fraction was primarily of microbial origin (teichoic acid-P and other forms of diester-P).

The microbial utilization of water-soluble substrates is in direct competition with sorption to Fe and Al oxides located in the clay fraction. Nelson et al. [177] and others have reported that sorption of biologically labile compounds to minerals resulted in a significant reduction of their bioavailability. Sorptive stabilization of SOM by clay either separates requires a prior solubilization, as is the case for many microbial metabolites or for highly oxidatively-degraded lignocellulose-derived constituents (cf. Figure 4), or an attachment of clay particles to extracellular polysaccharides produced by bacteria and fungi [14,178]. Protective sorption of SOM by establishment of strong chemisorptive bonding between charged organic molecules and mineral surfaces may be responsible for the highest weight percentages of organic C, N, P, and S observed in the clay-size fraction. It may also explain the apparent inconsistency between essentially complete degradation of chemically recognizable lignin in the same soils where seemingly labile organic matter rich in N, P, and polysaccharides accumulates [134].

4 ROLE OF SECONDARY SOIL PARTICLES IN STABILIZATION AND CYCLING OF C, N, P, AND S

In the soil environment, primary particles are usually clustered to larger entities—the secondary soil particles or aggregates. The arrangement of primary and secondary particles with associated pores ranges in size from micrometers to centimeters and is defined as soil structure. Aggregation and soil structure determine the pore-size distribution and, hence, the accessibility of substrates to exoenzymes, microorganisms and faunal grazers. However, with respect to the definition of different pore-size classes there is quite a confusion in the literature. According to IUPAC, classification of pores is micropores (< 2 nm), mesopores (2–50 nm), and macropores (50–2000 nm). In soil science, the same terms are used for much larger pores. For Example, according to the Glossary of Soil Science Terms Committee of the SSSA, micropores, mesopores, and macropores of aggregates have a mean pore diameter of 5–$30\,\mu$m, 30–$75\,\mu$m, and 75–$>5000\,\mu$m respectively, while pores of 0.1–$5\,\mu$m are defined as multimicropores and those of $< 0.1\,\mu$m as cryptopores [179]. In order to avoid this confusion, throughout this chapter we replaced micro-, meso-, and macropores as used by soil scientists in the original literature by small, intermediate, and large pores.

Van Veen and Kuikman [12] calculated that 95% of the pore space in a silt loam soil is not accessible to bacteria. The limited accessibility of substrates to the decomposer community is of great importance for the protection of SOM from decomposition [95,180]. Decomposition rates may also be limited if the tortuosity of pores within the aggregates results in a low diffusion of reactants and extracellular compounds such as enzymes, to the site of the substrates [181]. Hence, the physical protection of SOM from mineralization by soil organisms and their enzymes is determined by the size and organization of pores between the particles, which themselves are a function of the size of the particles. In addition, aggregation also affects factors other than accessibility. For example, Sexstone et al. [182] reported low oxygen concentrations in the center of wet aggregates, which may limit respiration. Compared with primary organomineral complexes, secondary organomineral complexes are much more dynamic, as they are directly affected by formation and degradation of aggregates.

4.1 AGGREGATION OF MINERAL COLLOIDS

Soil texture has a major influence on the form, stability, and resilience of soil aggregates, and, consequently, soil structure [183]. In sandy soils, there is only little cementation by organic materials of fine clays and other non-crystalline and crystalline inorganic materials. In loams, the cohesive nature of clays and their shrink–swell capacity creates aggregates during drying and wetting cycles [13]. The greater the clay content, the greater the shrink–swell capacity and the more vigorous the cycles of structural formation during dry–wet cycles. Maximum development of aggregation occurs in smectite-rich clays, e.g., in Vertisols.

4.1.1 Importance of the Soil Biota

The influence of the soil biota in aggregate formation is most important in soils with a low shrink–swell capacity [13]. However, the major role of the soil biota and associated SOM lies in the *stabilization* of the aggregates by living and dead organic binding agents. Conceptualized models suggests that the building blocks are stable microaggregates (< 250μm), which are bound together to form macro-aggregates (> 250μm) [184,185]. A more detailed discussion about the concept of aggregate hierarchy is provided by Baldock (chapter 3, this book).

4.1.1.1 Associations of secondary organominerals with plant roots and fungal hyphae   The macrofauna (e.g., earthworms) are important in the formation of macroaggregates, whereas plant roots, fungal hyphae, and bacteria, as well as their decomposition products, are relevant for the biotic stabilization of aggregates. This exerts a pronounced influence on the physical stabilization of SOM within macroaggregates. Plant roots and fungal hyphae are larger in size and are located at different sites within the soil micro-environment compared with bacteria. Because of their larger size, they can bridge the pores (~ 50 μm) between the microaggregates that compose the macroaggregates. In particular, in soils under grassland or where annual organic inputs are large, formation and stabilization of macroaggregates are directly or indirectly related to the growth and decomposition of plant roots and hyphae of mycorrhizal fungi [186,187]. Roots form a mesh with strands of > 10 μm, grow in coarser pores, and are thus distributed around larger aggregates. Hyphae of mycorrhizal fungi can be regarded as an extension of the root system, and the developing mycelia with hyphal diameters up to 10 μm grow throughout finer pore systems and around smaller aggregates. Together, the roots and hyphae form a network of materials with extracellular mucilages on their surfaces and act as a 'sticky string bag' that stabilizes macroaggregates.

4.1.1.2 Associations of secondary organominerals with bacteria   Many soil bac-teria produce extracellular polysaccharides. Foster [188] has shown that indi-vidual bacterial cells and colonies isolated in small pores are often surrounded by microbial extracellular polysaccharides. The extracellular polysaccharides can become coated with clay particles so that bacterial colonies often form the core of microaggregates [11,189] (see also Chenu and Stotzky, chapter 1, this book). Hattori and co-workers [190,191] found an increasing number of bacteria with increasing aggregate size, although the largest increase was observed for the small microaggregates (10–20 μm). These particles represent the inner part of large microaggregates, and the bacteria were located within pores of 2–6 μm diameter. Their distribution was clumpy which can be interpreted to be caused by the discrete occurrence of organic matter residues within pores [34]. With respect to cycling of SOM it should be noted that bacteria within these pores are inaccessible to protozoa and thus cannot be preyed upon. Bacteria in the interior of microaggregates are often small [192] and many exist in a nongrowing or dormant state [34]. In small pores that are no longer accessible to recolonizing bacteria, the extracellular polysaccharides may persist after death of the bacteria that secreted them, continuing to link various parts of the microaggregates [193,194].

Bacteria living at the outer surface of microaggregates occupy pores of > 6 μm and are not physically protected from protozoan predators. However, compared with starving bacteria within microaggregates, bacteria located in intermediate pores near the surface of microaggregates are large [192]. Killham et al. [195] suggested that bacterial cells near the surface of microaggregates are able to intercept most soluble organic substrates that diffuse within the soil matrix. Hence, they efficiently prevent readily decomposable substrates from reaching microorganisms in the interior of microaggregates.

4.1.1.3 Role of the composition of the microbial community in SOM cycling Beare et al. [196] suggested that the composition of the microbial community is important in determining the rates of SOM decomposition and of nutrient turnover and availability. They showed that conventional tillage (CT) agroecosystems contain a bacteria-dominated microbial community with faster residue decomposition and nutrient mineralization, whereas no-tillage (NT) agroecosystems contain a community richer in fungi with slower residue decomposition and greater nutrient retention.

Frey et al. [197] examined the factors controlling fungal and bacterial biomass in NT and CT agroecosystems and assessed the relation of the structure of the microbial community with soil aggregation and the accrual of SOM. Organic C, total N, and the mean weighted diameter of water-stable aggregates (MWD) were significantly higher in NT relative to CT at three of six investigated sites. Fungal hyphal length was 1.9 to 2.5 times higher in surface soil with NT compared with CT at all sites sampled, whereas there was no consistent tillage effect on bacterial abundance. Bacterial and fungal biomass in surface soils with NT or CT were not strongly related to soil texture, MWD, and pH, but they were positively correlated with total organic C and total N (Table 7).

Table 7. Correlation coefficients among soil physical and chemical characteristics, bacterial biomass, fungal biomass, and the proportion of total biomass composed of fungi for the 0–5 cm depth increment of six soils under conventional tillage and tillage. Adapted from [197], Copyright (1999), with permission from Elsevier Science, with additional material from S. D. Frey, personal communication (1998)

Soil characteristics	Bacterial biomass C ($\mu g\,g^{-1}$ soil)	Fungal biomass C ($\mu g\,g^{-1}$ soil)	Fungal contribution (% of total biomass)
Sand, $g\,kg^{-1}$	NS	NS	NS
Silt, $g\,kg^{-1}$	0.42*	NS	NS
Clay, $g\,kg^{-1}$	−0.39*	NS	NS
Mean weighted diameter, mm	NS	NS	NS
pH	NS	NS	NS
Water content, $g\,g^{-1}$ dry soil	0.67***	0.85***	0.67***
Total organic C, $mg\,g^{-1}$ soil	0.44**	0.45**	NS
Total N, $mg\,g^{-1}$ soil	0.40**	0.51*	0.42*
Particulate organic matter C, $mg\,g^{-1}$ soil	0.65***	0.49**	NS

*, **, *** Significant at the 0.05, 0.01, and 0.001 probability levels, respectively; NS = not significant.

Soil water content, measured at the time of sampling explained the highest proportion of the variation in bacterial and fungal biomass. However, fungal biomass was linearly related to soil moisture while bacterial biomass was relatively constant at higher water contents. The authors suggested that bacteria are substrate limited at higher water contents required for maximal growth rates. Whereas fungi could utilize more recalcitrant SOM, bacteria require more labile sources of SOM.

4.2 PHYSICALLY PROTECTED SOIL ORGANIC MATTER

The concept of physical protection of SOM is based on the differentiation between free and intraaggregate SOM pools. Intraaggregate SOM is incorporated and physically stabilized within aggregates, whereas free organic matter is found between aggregates. Intermicroaggregate SOM may act as a cementing agent that glues microaggregates together to form macroaggregates and is, in this case, intramacroaggregate SOM. One important SOM component that can occur in free and in occluded form is particulate organic matter (POM), which is derived from surface and root litter.

4.2.1 Particulate Organic Matter

Alperin *et al.* [198] combined the concepts of chemical stability and physical protection into a unified classification system, which assumes that there are two categories of natural organic matter: labile (readily degradable) and resistant (slowly degradable). Furthermore, the system assumes that there are three categories of microenvironments: free particulate (unprotected), occluded (physically entrapped in a microenvironment), and adsorbed (attached to surfaces by physical or chemical mechanisms). These interactions between inorganic and organic soil colloids give rise to six different categories of SOM:

(1) free particulate: intrinsically labile,
(2) free particulate: intrinsically resistant,
(3) occluded particulate: potentially labile but currently protected as a result of inaccessibility,
(4) occluded particulate: intrinsically resistant and further protected by its inaccessibility,
(5) adsorbed: potentially labile, but currently protected by adsorption,
(6) adsorbed: intrinsically resistant and further protected by adsorption.

Free POM can be directly separated from whole soil or aggregate fractions as light fraction by density fractionation using, e.g., sodium polytungstate [199,200], whereas occluded POM requires destruction of aggregates either by slaking [201] or by dispersion in, e.g., sodium hexametaphosphate with [199] or without [200] the aid of limited sonication.

Gregorich et al. [202] isolated free POM by density fractionation of soil suspensions and POM enclosed in aggregates by further ultrasonic dispersion. Both fractions together contained between 6 and 15 % of total soil organic C in cultivated soils and more than 50 % in a soil under native sod. Puget et al. [201] found that 26 % of the total POM-associated C of a cultivated soil occurred in free form and was located in the interaggregate pore space, but 74 % of total POM-associated C was occluded within the aggregates.

Golchin et al. [178] showed that occluded POM has lower C/N ratios than free POM. The higher alkyl C and lower O-alkyl C proportions in CP/MAS ^{13}C NMR spectra indicated a more decomposed state of occluded POM than of free POM. In a subsequent study, Golchin et al. [203] suggested that macroaggregates are stabilized mainly by carbohydrate-rich root or plant debris occluded within aggregates. Jastrow [199] also showed that POM within macroaggregates is an important agent that facilitates the binding of microaggregates into macroaggregates. She also found that 80 % of the organic C accumulated in macroaggregates after 10 years of restored tallgrass prairie on an agricultural soil occurred in the mineral-associated fraction of macroaggregates, suggesting that inputs of organic debris occurred relatively rapidly into particles that are associated with mineral matter. This indicates that plant debris is involved in the formation of aggregates and is, itself, stabilized by occlusion within the aggregates but that the main driving factors in the stabilization of aggregates and aggregate hierarchy are the microbial decomposition processes [185].

4.2.2 Microbial Metabolites and Residues

Fresh plant debris contains readily available C sources and is favored for use by the soil microorganisms and are hot spots of microbial activity. During the assimilation of the substrate by microorganisms, extracellular polysaccharides and other substances are released [204,205]. This leads to an encrustment of plant fragments by mineral particles and formation and stabilization of aggregates [178,200,204,205]. Particulate plant debris thus becomes the center of water-stable aggregates and is protected from rapid decomposition. As decomposition proceeds within aggregates, the more labile portions of the organic cores are consumed by the decomposers and more resistant materials, being poor in O-alkyl C and rich in alkyl C and aromatic C, are concentrated as occluded particles within the aggregates. As a result, intraaggregate POM becomes less available to microorganisms, the microbial production of binding agents diminishes as the organic cores become less digestible. Eventually, aggregates become increasingly unstable and relatively recalcitrant POM is released from intimate association with mineral particles. Consequently, free POM is composed not only of recently deposited plant residue but contains a mixture of fresh plant debris and older stable plant-derived material [200,202]. This conceptual model originally developed for microaggregate formation and accumulation of intramicroaggregate SOM [178] has been confirmed and extended to macroaggregate formation and stabilization of intramacroaggregate SOM [200]. Baldock (chapter 3, this book) discusses in detail how these organic binding agents relate to aggregate turnover.

Although this model relies on the gluing effects of labile organic substances, such as polysaccharides, there are indications that more stable products of microbial resynthesis are also involved in the stabilization of SOM. One substance of importance is glomalin, a glycoprotein with N-linked oligosaccharides, that is produced by hyphae of arbuscular mycorrhizal (AM) fungi [206]. Wright and Upadhyaya [207] discovered that glomalin is abundant in soils and provided evidence that it is stable in soil. Later, they found a good relation between aggregate stability and glomalin contents for a range of soils [208], see Figure 11.

Chantigny et al. [149] observed a strong positive correlation between the mean weighted diameter of water-stable aggregates (MWD) and fungal glucosamine ($r^2 = 0.68***$) and bacterial muramic acid ($r^2 = 0.48***$) and proposed that soil aggregation may be more closely related to total microbial cell-wall mass (living and residual) than to the viable fraction. Larger contents of fungal glucosamine in the surface soil samples from NT than CT agroecosystems coincided with larger values for MWD, POM, and SOM, and the additional

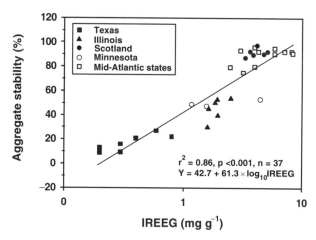

Figure 11. Relationship between stability of 1–2 mm-size aggregates and easily extractable glomalin (IREEG) shown for $\log_{10}$ scale. From [208], with kind permission from Kluwer Academic Publishers

amino sugars stored were distributed only into the larger aggregate-size classes [209]. The enrichment of SOM by fungal glucosamine suggests that the accrual of hyphal cell-wall residues is an important process in NT agroecosystems leading to a higher storage of soil organic C and N.

4.3 DECOMPOSITION AND STABILIZATION OF C, N, P, AND S RELATED TO SECONDARY PARTICLES

The nature and stability of soil aggregates thus depends on the types of the organic stabilizing agents at each hierarchical level of organization [13]. Concurrently, the spatial location of the organic substances within the hierarchically structured aggregates determines the degree to which SOM is physically protected from decomposition. Hence, feedbacks between cycling of SOM and associated nutrients and aggregate cycling can occur [185,200].

A significant proportion of the physically protected SOM and their associated nutrients in soil is believed to have an intermediate mean residence time of about 10–50 years [181,210]. They may decompose much faster upon disturbances of the aggregates [200,210]. This is the case when aggregates are disrupted in the field by cultivation of an undisturbed soil. When slaked in the laboratory (disruption of dry soil aggregates upon rapid addition of water), water-stable macroaggregates of native and cultivated soil contain more C, N, and P than microaggregates, resulting from the organic matter located in the intermicroaggregate pore space [181] (Table 8). After crushing the macroaggregates to the size of microaggregates, the percentage of mineralizable C and N was larger than for intact macroaggregates resulting from the release of this physically protected SOM [210,213,214]. The microaggregates had the least percentage of mineralizable C and N. In addition, the C/N and C/OP ratios of microaggregates were narrower than those of macroaggregates (Table 8). Hence, it appears that the macroaggregates contain SOM that is qualitatively more labile and less highly processed than that associated with microaggregates. SOM located in the intermicroaggregate pores of macroaggregates are, therefore, believed to be responsible for much of the SOM losses observed in degrading agroecosystems [210,215,216].

Organic matter contained in the macroaggregates also has a more rapid turnover than that in microaggregates [217]. Studies on different soil types such as Oxisol, Alfisol, and Inceptisol, where C3 vegetation was replaced by C4 vegetation showed that macroaggregates contained younger C than microaggregates (Table 9). These observations agree with the data of Buyanovsky et al. [119] who investigated the amounts of ^{14}C derived from labeled soybean residues in the physically protected fractions of soil. Partially processed residues

Table 8. Total organic carbon (TOC) concentrations and C/N and C/P ratios of aggregate size fractions obtained by slaking

Reference	USDA Soil Taxonomy	Site	Aggregate size	TOC (mg g^{-1} aggregate)	C/N	C/P
Elliott [210][a]	Mollisol	native sod Nebraska	2000–8000 μm	34.2	11.7	n.a.[b]
			500–2000 μm	33.9	10.5	113
			300–500 μm	29.8	10.1	99
			208–300 μm	24.6	9.8	84
			90–208 μm	23.7	9.6	84
			53–90 m	19.5	9.2	74
			< 53 μm	21.3	9.5	75
Buyanovsky et al. [119]	Alfisol	soybean crop Missouri	1000–2000 μm	22.3	10.2	n.a.
			500–1000 μm	17.5	9.3	n.a.
			250–500 μm	17.3	9.0	n.a.
			100–250 μm	22.3	8.5	n.a.
			< 100 μm	12.0	7.6	n.a.
Monreal et al. [211][c]	Mollisol	wheat-fallow Saskatchewan	> 250 μm	23.0	10.5	n.a.
			50–250	17.4	10.2	n.a.
			< 50 μm	16.5	9.7	n.a.
Angers et al. [212]	Alfisol	arable field France	1000–4000 μm	23.7	16.1	n.a.
			250–1000 μm	13.7	11.1	n.a.
			50–250 μm	10.8	11.5	n.a.
			< 50 μm	7.1	8.7	n.a.

[a] Normalized to a sand-free basis to account for the different content of primary sand particles in the particle size separates.
[b] Not analyzed.
[c] 'Moist' soil.

Table 9. Percentage of young C (C4-derived C) in aggregates

Reference	USDA Soil Taxonomy	Site	Experiment	Aggregate size	C4-C (%)
Skjemstad et al. [218][a]	Oxisol	Australia	C4 grass for 35 years on former C3 rainforest	200–2000 μm	49
				< 200 μm	41
			C4 grass for 83 years on former C3 rainforest	200–2000 μm	75
				< 200 μm	67
Puget et al. [219][b]	Alfisol	France	corn (C4) for 6 years on former C3 arable crops	200–6000 μm	15
				< 200 μm	6
			corn (C4) for 23 years on former C3 arable crops	200–6000 μm	46
				< 200 μm	38
Angers and Giroux [220][b]	Inceptisol	Québec	corn (C4) for 15 years on former C3 hay system	2000–6000 μm	21
				1000–2000 μm	18
				500–1000 μm	11
				200–500 μm	8
				50–200 μm	3
				< 50 μm	1

[a] Aggregates obtained by hand-grinding.
[b] Water-stable aggregates of slaked soils.

within macroaggregates showed turnover rates from 1–3 years, whereas the more highly humified residues in microaggregates had longer residence times of ~ 7 years.

Jastrow [199] studied gradients of water-stable macroaggregate formation and SOM accumulation in a chronosequence of prairie restoration. An important finding was that the rate constant (k) for change in aggregation was more than 35 times the k for total organic C accumulation. This relatively rapid turnover of macroaggregates may suggest that physical protection of intermicroaggregate binding agents (such as POM) inside macroaggregates may not be a major mechanism of stabilization of SOM [185]. However, Jastrow and Miller [185] also suggested that new macroaggregates are being rapidly formed, and the majority of the intermicroaggregate binding agents are probably not unprotected for long time.

At the same sites, Jastrow et al. [221] used the photosynthetic switch of C4 to C3 vegetation to calculate average turnover times of aggregate-associated organic matter. The average turnover time for old C4-derived C was 412 years for microaggregates compared with an average turnover time of 140 years for macroaggregates. Net inputs rates of new C3-derived C increased with aggregate size from 0.73 to 1.13 g kg^{-1} fraction year^{-1}. Equal net input rates were reported for microaggregates and small (< 1000 μm) macroaggregates, suggesting that the formation and degradation of microaggregates may be more dynamic than is predicted from their stability in cultivated soils or by the observed turnover times for old C. This contradicts the original concept of aggregate hierarchy that includes extremely stable microaggregates that, once formed, turn over very slowly [16] and indicates that at least two different pools of organic C characterized with different turnover rates exist within microaggregates.

As already discussed, large microaggregates are disrupted after microbial mineralization of available components of intraaggregate POM. At the same time, organomineral complexes and SOM within small microaggregates remain strongly associated with mineral particles that encrust new cores of organic microaggregates [178,221]. Thus the turnover of mineral-associated SOM and SOM within small aggregates is probably slower than the degradation of either large microaggregates or their intraaggregate POM. The nature of SOM protected in small clay- and silt-sized microaggregates can be characterized by high-energy UV photo-oxidation. This procedure destroys complex organic materials exposed to UV radiation through oxidation, even in the presence of clay, but SOM within small microaggregates is protected. Photo-oxidation of clay- and silt-sized aggregates of an Australian Hapludoll for 4 h resulted in an enrichment of components of SOM with a higher mean residence time than SOM of bulk soil [222]. The chemical structure of this UV-resistant, old SOM was very similar to that of the original SOM.

5 SOIL COLLOIDS AND THEIR INTERACTIONS IN SORPTION, COMPLEXATION, AND BINDING OF LOW MOLAR MASS ORGANIC COMPOUNDS

Polycondensation of low molar mass organic compounds by abiotic catalysis may contribute to formation of SOM [223]. A variety of organic compounds, including phenols, carbohydrates, nitrogenous substances and xenobiotics, including pesticides, can interact with soil colloids to catalyze their sorption, polymerization, entrapment, and therefore, their availability and deactivation. It was observed that 50–70 % of single phenols such as vanillic, ferulic, protocatechuic, caffeic acids, and several others, when incubated in soils (at concentrations $> 100 \, \mathrm{mg \, kg^{-1}}$), became degraded to CO_2. At low amendments, however $(1–100 \, \mathrm{mg \, kg^{-1}}$ of phenolic compound in soil), about one third or more became incorporated into soil organic matter and into the humic matrix [224,225]. Wang et al. [226] and Shindo and Huang [227,228] showed that birnessite, Fe and Al oxides, and clay and silt separated from soils affect the oxidative polymerization of polyphenols in suspension. In the case of Mn oxides, this is effected by their action as Lewis acids to oxidize the polyphenols to semiquinones. $^{13}\mathrm{CNMR}$ spectra of the polymeric reaction products of polyphenols with birnessite strongly resemble that of natural organic substances [229].

Such reactions of soil minerals with natural and anthropogenic organics also catalyze chemical breakdown of organic molecules. Birnessite greatly accelerates the abiotic deamination and decarboxylation of glycine, especially in the presence of pyrogallol and other phenolic compounds [227]. Kaolinite, nontronite, and natural soil clays also can catalyze similar reactions and, therefore, have an important role in N transformations in soils [230,231].

Soil colloids, especially clays and colloidal metal oxides, also participate in the hydrolysis, oxidation, and decarboxylation of various pesticides and are, therefore, active in inactivation and degradation of these and similar xenobiotic compounds [232]. A large number of low molar mass toxic organic compounds are produced in industry and unintentionally released into the environment or intentionally used in agriculture or households. In soils, they are either degraded by microbes or by abiotic reactions or remain in soil as such or as their metabolites. Numerous such compounds are retained by organic or inorganic soil colloids by various types of interactions, including van der Waal's forces, hydrogen bonds, formation of charge transfer complexes, and covalent bonding [233]. Hydrophobic sorption of nonpolar residues or trapping of molecules in a molecular sieve formed by humic materials have also been suggested [234–237]. Adsorption of xenobiotics to soil particles results from attraction by solid surfaces, and soil colloids are mainly responsible for the adsorption phenomenon, because of their small particle size and large surface area. The humic matter in soils is generally more active in these sorptive processes, but clay minerals and clay–humus complexes also have an important role. It is the ability of soils to retain xenobiotics or other organic molecules and to prevent them from moving within or outside the soil matrix [238].

Covalent binding was observed for several of the low molar mass organic compounds but it may not always occur. It is becoming more apparent that humic compounds and clays (or both together) can entrap or sequester low molar mass compounds in structural voids, in the hydrophobic interior of micelle-like humic aggregates, or in micropores of organic and inorganic colloids [239–241]. Consequently, retention and resultant unavailability of otherwise toxic chemicals in soils is a key factor in avoiding their interactions with organisms and/or pollution of other compartments, such as air or ground and surface waters [242].

6 CONCLUSIONS

With regard to the effects of mineral colloids on biogeochemical cycling of C, N, P, and S in soils, short-term incubation experiments as well as long-term field studies with stable and radioactive isotope studies showed that chemical and physical interactions of minerals with soil organic matter results predominantly in the stabilization of the organic substances. However, minerals can also contribute to a more rapid cycling of SOM and of organically bound nutrients. Laboratory experiments provide evidence that some soil minerals, such as birnessite, may enhance the cycling of organically bound nutrients by catalytic degradation of the organic matter. With respect to biological processes, in particular phyllosilicate clay minerals may contribute to an enhanced cycling of SOM by creating a favorable physical environment for microorganisms. There is evidence from incubation experiments and isotope studies that in the clay fraction SOM of slow turnover as well as SOM of rapid turnover occur. The latter is probably related to microbial residues which are often loosely attached to phyllosilicates and are available to posterior microbial generations (microbial loop). Stable SOM, in contrast, is tightly bonded to Fe and Al oxyhydroxides of the clay fraction by formation of innersphere complexes. The release of the stored organic nutrients depends on the mean residence time of SOM associated with the mineral phase.

The mean residence time of SOM varies widely depending on the type of the various organomineral associations and, in addition to that, on the spatial location within the aggregate structure of soil. In Table 10, some approaches to pool SOM fractions with different turnover rates are compared. Pools I and II described in this table are generally related to plant fragments, divided into easily available cell constituents and lignocelluloses at various degrees of degradation, respectively. As these fragments become more degraded and reduced in size, the microbially available components become exhausted and resistance

Table 10. Comparison of estimated mean residence times of soil organic matter in soil physical fractions

Pool	Mean residence time (years)			
	Jenkinson and Rayner [21]	Parton et al. [23]	Buyanovsky et al. [119]	Carter [243]
I	Decomposable plant material, 0.24	Metabolic plant residues, 0.1–1	Vegetative fragments 2–0.2 mm, 0.5–1	Litter, 1–3
II	Resistant plant material, 3.33	Structural plant residues, 1–5	Vegetative fragments 0.05–0.025 mm, 1–3	Free POM (light fraction), 1–15
III	Soil biomass, 2.44	Active SOM pool, 1–5	OM in aggregates 2–1 mm, 1–4	Microbial biomass, 0.1–0.4
IV	Physically stabilized OM, 72	Slow SOM pool, 25–50	OM in aggregates 1–0.1 mm, 2–10	Intermicroaggregate OM[a], 5–50
V	Chemically stabilized OM, 2857	Passive SOM pool, 1000–1500	OM in fine silt, ~400 OM in fine clay, ~1000	Intramicroaggregate OM[b] physically sequestered, 50–1000 chemically sequestered, 1000–3000

[a] Organic matter stored within macroaggregates but external to microaggregates; includes coarse occluded particulate organic matter and microbial organic matter.
[b] Organic matter stored within microaggregates; includes fine occluded particulate organic matter and microbial derived organic matter.

to degradation increases. Pool III also includes material with a relatively short mean residence time and consists of soil biomass as well as processed but readily available organic matter within large aggregates. On average, about 20–30% of the total C in SOM is involved in the pools I–III. To maintain a relatively constant nutrient level and release by mineralization, the pools have to be renewed continuously by fresh plant residues. The balance between decay and renewal processes controls nutrient availability and is sensitive to management practices [244]. Pool IV also is affected by cultivation, as physical disturbance (caused e.g. by ploughing) destroys macroaggregates and large microaggregates and thus releases physically protected SOM [181]. Pool V contains the most stable SOM which is protected by chemical and physical mechanisms. It represents 50–70% of the total C in SOM and is seldom affected by management practices.

A strict distinction between physical and chemical effects of soil minerals on the stabilization of SOM cannot often be made, as SOM exists in a continuum of decay that is heterogeneously distributed within the soil matrix [95,245]. There are distinct feedback mechanisms between SOM and primary and secondary colloids with direct effects on nutrient cycling. Organic materials may destroy primary minerals, alter them, or impede crystallization of secondary minerals. Less crystallized minerals favor SOM accumulation and reduce turnover of C, N, P, and S. Organic substances also act as binding agents for aggregates. In turn, SOM becomes occluded by minerals, which reduces turnover of SOM.

Although many studies in the last decade have greatly improved knowledge of the types and directions of interrelationships between transformation of SOM and changes in the mineral phase at different scales and its influence on turnover of soil C, N, P, and S, future research should be directed more to a detailed quantification of these processes. One such example is the role of minerals in the catalytic degradation of organic matter and the concurrent release of organically bound nutrients. The role of this process in enhancing the cycling of C, N, P, and S in soil environments *in situ* still needs to be verified and quantified. Another topic that deserves further attention concerns the quantification of the diffusion of organic substrates into micro- and mesopores of mineral clusters and their accessibility to degrading exoenzymes. Likewise, the quantification of fluxes of organic substrates within the different pore size classes of aggregates are of importance with regard to the accessibility of microorganisms themselves.

REFERENCES

1. Schoenau, J. J and Bettany, J. R. (1987). Organic matter leaching as a component of carbon, nitrogen, phosphorus, and sulfur cycles in a forest, grassland, and gleyed soil, *Soil. Sci. Soc. Am. J.,* **51,** 646.
2. Duxbury, J. M., Harper, L. A. and Mosier, A. R. (1993). Contribution of agroecosystems to global climate change. In *Agricultural Ecosystem Effects on Trace Gases*

and Global Climate Change, ed. Harper, L. A. *et al.*, ASA Special Publ. 55, American Soil Association, Madison, WI, p. 1.

3. Cicerone, R. J. and Oremland, R. S. (1988). Biogeochemical aspects of atmospheric methane, *Global Biogeochem. Cycles*, **2**, 299.

4. Conrad, R. (1995). Soil microbial processes involved in production and consumption of atmospheric trace gases. In *Advances in Soil Ecology*, Vol. 1, ed. Jones J. G., Plenum Press, New York, p. 207.

5. McGill, W. B. and Cole, C. V. (1981). Comparative aspects of cycling of organic C, N, S and P through soil organic matter, *Geoderma*, **26**, 267.

6. Walker, T. W. and Syers, J. K. (1976). The fate of phosphorus during pedogenesis. *Geoderma*, **15**, 1.

7. Stevenson, F. J. (1986). *Cycles of Soil*, John Wiley & Sons, New York.

8. Johnson, D. S. (1995). The role of carbon in the cycling of other nutrients. In *Carbon Forms and Functions in Forest Soils*, ed. McFee, W. W. and Kelly, W. J., Soil Science Society of America, Madison, WI, p. 299.

9. Oades, J. M. (1995). An overview of processes affecting the cycling of organic carbon in soils. In *The Role of Nonliving Organic Matter in the Earth's Carbon Cycle*, ed. Zepp, R. G. and Sonntag, Ch. John Wiley and Sons, Chichester, p. 293.

10. Oades, J. M. (1989). An introduction to organic matter in mineral soils. In *Minerals In Soil Environments*, 2nd edn, ed. Dixon, J. B. and Weed, S. B., SSSA Book Series No 1. Soil Science Society of America, Madison, WI, p. 89.

11. Robert, M. and Chenu, C. (1992). Interactions between soil minerals and microorganisms. In *Soil Biochemistry*, Vol 7, ed. Stotzky, G. and Bollag, J.-M., Marcel Dekker, New York, p. 307.

12. van Veen, J. A. and Kuikman, P. J. (1990). Soil structural aspects of decomposition of organic matter by micro-organisms, *Biogeochemistry*, **11**, 213.

13. Oades, J. M. (1993). The role of biology in the formation, stabilization and degradation of soil structure, *Geoderma*, **56**, 377.

14. Ladd, J. N., Foster, R. C., Nannipieri, P. and Oades, J. M. (1996). Soil structure and biological activity. In *Soil Biochemistry*, Vol 9, ed. Stotzky, G. and Bollag, J.-M., Marcel Dekker, New York, p. 23.

15. Martin, J. P. and Haider, K. (1986). Influence of mineral colloids on turnover rates of soil organic carbon. In *Interactions of Soil Minerals with Natural Organics and Microbes*, ed. Huang, P. M. and Schnitzer, M., SSSA Special Publication 17, Soil Science Society of America, Madison, WI., p. 283.

16. Tisdall, J. M. and Oades, J. M. (1982). Soil organic matter and structural stability: mechanisms and implications for management, *Plant Soil*, **76**, 319.

17. Jenkinson, D. S. (1988). Soil organic matter and its dynamics. In *Rusell's Soil Conditions and Plant Growth*, 11th edn, ed. Wild, A., Longman, New York, p. 564.

18. Hassink, J. (1997) The capacity of soils to preserve organic C and N by their association with clay and silt particles, *Plant Soil*, **191**, 77.

19. Burke, I. C., Yonker, C. M., Parton, W. J., Cole, C. V., Flach, K. and Schimel, D. S. (1989). Texture, climatic and cultivation effects on soil organic matter content in U.S. grassland soils, *Soil Sci. Soc. Am. J.*, **53**, 800.

20. Parton, W. J. and Rasmusssen, P. E. (1994). Long-term effects of crop management in wheat-fallow: II.CENTURY model simulation, *Soil Sci. Soc. Am. J.*, **58**, 530.

21. Jenkinson, D. S. and Rayner, J. H. (1977). The turnover of soil organic matter in some of the Rothamsted classical experiments, *Soil Sci.*, **123**, 298.

22. van Veen, J. A., Ladd, J. H. and Frissel, M. J. (1984). Modelling C and N turnover through the microbial biomass in soil. *Plant Soil*, **76**, 257.

23. Parton, W. J., Schimel, D. S., Cole, C. V. and Ojima, D. S. (1987). Analysis of factors controlling soil organic matter levels in Great Plains grasslands, *Soil Sci. Soc. Am. J.*, **51**, 1173.

24. van Veen, J. A., Ladd, J. N. and Amato, M. (1985). Turnover of carbon and nitrogen through the microbial biomass in a sandy loam and a clay soil incubated with [^{14}C(U)]glucose and [^{15}N]$(NH_4)_2SO_4$ under different moisture regimes, *Soil Biol. Biochem.*, **17**, 257.

25. Hassink, J. (1996). Preservation of plant residues in soils differing in unsaturated protective capacity, *Soil Sci. Soc. Am. J.*, **60**, 487.

26. Mayer, L. M. (1994). Surface area control of organic carbon accumulation in continental shelf sediments. *Geochim. Cosmochim, Acta*, **58**, 1271.

27. Keil, R. G., Montlucon, D. B., Prahl, F. R. and Hedges, J. I. (1994). Sorptive preservation of labile organic matter in marine sediments, *Nature*, **370**, 549.

28. Mayer, L. M. (1994b). Relationships between mineral surfaces and organic carbon contentrations in soils and sediments, *Chem. Geol.*, **114**, 347.

29. Kaiser, K. and Guggenberger, G. (2000) The role of DOM sorption to mineral surfaces in the preservation of organic matter in soils, *Org. Geochem.*, **31**, 711.

30. Churchman, G. J. and Theng, B. K. G. (1984). Interactions of halloysites with amides: mineralogical factors affecting complex formation, *Clay Minerals.*, **19**, 161.

31. Wang, M. C. and Huang, P. M. (1986). Humic macromolecule interlayering in nontronite through interaction with phenol monomers, *Nature* (London), **323**, 529.

32. Theng, B. K. G., Churchman, G. J. and Newman, R. H. (1986). The occurrence of interlayer clay–organic complexes in two New Zealand soils. *Soil Sci.*, **142**, 262.

33. Theng, B. K. G., Tate, K. R. and Becker-Heidmann, P. (1992). Towards establishing the age, location, and identity of the inert soil organic matter of a Spodosol. *Z. Pflanzenernähr, Bodenk.*, **155**, 181.

34. Theng, B. K. G. and Orchard, V. A. (1995). Interactions of clays with microorganisms and bacterial survival in soil: a physicochemical perspective. In *Environmental Impact of Soil Componenent Interactions, Vol II. Metals, Other Inorganics, and Microbial Activities*, ed. Huang, P. M., Berthelin, J., Bollag, J.-M., McGill, W. B. and Page, A. L., CRC Lewis, Boca Raton, FL, p. 123.

35. Kaiser, K., Guggenberger, G., Haumaier, L. and Zech, W. (1997). Dissolved organic matter sorption on subsoils and minerals studied by ^{13}C-NMR and DRIFT spectroscopy, *Eur. J. Soil. Sci.*, **48**, 301.

36. Chiou, C. T., Lee, J.-F. and Boyd, S. A. (1990). The surface area of soil organic matter, *Environ. Sci. Technol.*, **24**, 1164.

37. Pennell, K. D., Boyd, S. A. and Abriola, L. M. (1995). Surface area of soil organic matter reexamined, *Soil Sci. Soc. Am. J.*, **59**, 1012.

38. De Jonge, H. and Mittelmeijer-Hazeleger, M. C. (1996). Adsorption of CO_2 and N_2 on soil organic matter: Nature of porosity, surface area, and diffusion mechanisms, *Environ. Sci. Technol.*, **30**, 408.

39. Feller, C., Shouller, E., Thomas, F., Rouiller, J. and Herbillon, A. J. (1992). N_2–BET specific surface areas and their relationships with secondary constituents and organic matter contents, *Soil Sci.*, **153**, 293.

40. Theng, B. K. G., Ristori, G. G., Santi, C. A. and Percival, H. J. (1999). An improved method for determining the specific surface areas of topsoils with varied organic matter content, texture and clay mineral composition, *Eur. J. Soil Sci.*, **50**, 309.

41. Schulthess, C. P. and Huang, P. M. (1991). Humic and fulvic acid adsorption by silicon and aluminum oxide surfaces on clay minerals, *Soil Sci. Soc. Am. J.*, **55**, 34.

42. Kubicki, J. D., Itoh, M. J., Schroeter, L. M. and Apitz, S. E. (1997). Bonding mechanisms of salicylic acid adsorbed onto illite clay. An ART-FTIR and molecular orbital study, *Environ. Sci. Technol.*, **31**, 1151.
43. Stevenson, F. J. (1994) *Humus Chemistry: Genesis, Composition, Reactions*, 2nd edn, John Wiley & Sons, New York.
44. Tipping, E. (1981) The adsorption of aquatic humic substances by iron oxides, *Geochim. Cosmochim. Acta*, **45**, 191.
45. Gu, B., Schmitt, J., Chen, Z., Liang, L. and McCarthy, J. F. (1994). Adsorption and desorption of natural organic matter on iron oxide: mechanisms and models, *Environ. Sci. Technol.*, **28**, 38.
46. Gu, B., Schmitt, J., Chen, Z., Liang, L. and McCarthy, J. F. (1995). Adsorption and desorption of different organic matter fractions on iron oxide, *Geochim. Cosmochim. Acta*, **59**, 219.
47. Jekel, M. R. (1986). Interactions of humic acids and aluminum salts in the flocculation process, *Wat. Res.*, **20**, 1535.
48. Adams, W. A. and Kassim, J. K. (1984). Iron oxyhydroxides in soils developed from Lower Palaeozoic sedimentary rocks in mid-Wales and implications for some pedogenetic processes, *J. Soil Sci.*, **35**, 117.
49. Evans, L. J. and Wilson, W. G. (1985). Extractable Fe, Al, Si, and C in B horizons of podzolic and brunisolic soils from Ontario, *Can. J. Soil Sci.*, **65**, 489.
50. Shang, C. and Tiessen, H. (1998). Organic matter stabilization in two semiarid tropical soils: size, density, and magnetic separations, *Soil Sci. Soc. Am. J.*, **62**, 1247.
51. Shoji, S. and Fujiwara, Y. (1984). Active aluminum and iron in the humus horizons of Andisols from northeastern Japan: Their forms, properties, and significance in clay weathering, *Soil Sci.*, **137**, 216.
52. Condron, L. M., Tiessen, H., Trasar-Cepeda, M. C., Moir, J. O. and Stewart, J. W. B. (1993). Effects of liming on organic matter decomposition and phosphorus extractability in an acid humic Ranker soil from north-western Spain, *Biol. Fertil. Soils*, **15**, 279.
53. Bäumler, R. and Zech, W. (1994). Characterization of andisols developed from nonvolcanic material in Eastern Nepal, *Soil Sci.*, **158**, 211.
54. Blaser, P., Kernebeek P., Tebbens, L., van Breemen, N., Luster, J. (1997). Cryopodzolic soils in Switzerland, *Eur. J. Soil Sci.*, **48**, 411.
55. Wada, K. (1989). Allophane and imogolite. In *Minerals In Soil Environments*, 2nd edn., ed. Dixon J. B. and Weed, S. B., SSSA Book Series No 1. Soil Science Society of America, Madison, WI, p. 1051.
56. Liu, C. and Huang, P. M. (1999). Atomic force microscopy and surface characteristics of iron oxide formed in citrate solutions, *Soil Sci. Soc. Am. J.*, **63**, 65.
57. Wang, M. C. and Lin, C.-H. (1993). Enhanced mineralization of amino acids by birnessite as influenced by pyrogallol, *Soil Sci. Soc. Am. J.*, **57**, 88.
58. Sunda, W. G. and Kieber, D. J. (1994). Oxidation of humic substances by manganese oxides yields low molecular weight organic substrates, *Nature*, **367**, 62.
59. Miltner, A. and Zech, W. (1998). Oxides change the degradation kinetics of C pools in litter material, *Z. Pflanzenernähr. Bodenk.*, **161**, 93.
60. Tiessen, H., Stewart, J. W. B. and Hunt, H. W. (1984). Concepts of soil organic matter transformations in relation to organo-mineral particle size fractions, *Plant Soil*, **76**, 287.
61. Christensen, B. T. (1992). Physical fractionation of soil and organic matter in primary particle size and density separates, *Adv. Soil Sci.*, **20**, 1.
62. Qualls, R. G. and Haines, B. L. (1991). Geochemsitry of dissolved organic nutrients in water percolating through a forest ecosystem, *Soil Sci. Soc. Am. J.*, **55**, 1112.

63. Kaiser, K. and Zech, W. (2000) Sorption of dissolved organic nitrogen by soil minerals and subsoils, *Eur. J. Soil Sci.*, **51**, 403.
64. Knicker, H. and Hatcher, P. G. (1997). Survival of protein in an organic-rich sediment: Possible protection by encapsulation in organic matter, *Naturwissenschaften*, **84**, 231.
65. Knicker, H. and Kögel-Knabner, I. (1998). Soil organic nitrogen formation examined by means of NMR spectroscopy. In *Nitrogen-Containing Macromolecules in the Bio- and Geosphere*, ed. Stankiewicz, B. A. and van Bergen, P. F., ACS Symp. Ser. 707, American Chemical Society, Washington, DC, p. 333.
66. Aufdenkampe, A. K., Hedges, J. I., and Richey, J. E. (2001). Sorptive fractionation of dissolved organic nitrogen and amino acids onto fine sediments within the Amazon Basin, *Limnol. Ocean.* (in press).
67. Henrichs, S. M. and Sugai, S. F. (1993). Adsorption of amino acids and glucose by sediments of Resurrection Bay, Alaska, USA: Functional group effects, *Geochim. Cosmochim. Acta*, **57**, 823.
68. Tate, K. R. and Newman, R. H. (1982). Phosphorus fractions of a climosequence of soils in New Zealand tussock grassland, *Soil Biol. Biochem.*, **14**, 191.
69. Condron, L. M., Frossard, E., Tiessen, H., Newman, R. H., Stewart, J. W. B. (1990). Chemical nature of organic phosphorus in cultivted and uncultivated soils under different environmental conditions, *J. Soil Sci.*, **41**, 41.
70. Duxbury, J. M., Smith, M. S. and Doran, J. W. (1989). Soil organic matter as a source and a sink of plant nutrients. In *Dynamics of Soil Organic Matter in Tropical Ecosystems.*, ed. Coleman, D. C., Oades, J. M. and Uehara, G., NifTAL Project, University of Hawaii Press, Honolulu, HA, p. 33.
71. Anderson, G., Williams, E. G. and Moir, J. O. (1974). A comparison of the sorption of inorganic orthophosphate and inositol hexaphosphate by six acid soils, *J. Soil Sci.*, **25**, 51.
72. Stewart, J. W. B. and Tiessen, H. (1987). Dynamics of soil organic phosphorus, *Biogeochemistry*, **4**, 41.
73. Abekoe, M. K. and Tiessen, H. (1998). Phosphorus forms, lateritic nodules and soil properties along a hillslope in northern Ghana, *Catena*, **33**, 1.
74. Freney, J. R. (1986). Forms and reactions of organic sulfur compounds in soils. In *Sulfur in Agriculture*, ed. Tabatabi, M. A., American Soil Association, Madison, WI, p. 207.
75. Jenny, H. (1980), *The Soil Resource, Origin and Behavior*, Springer, New York.
76. Post, W. M., Emanuel, W. R., Zinke, P. J. and Stangenberger, A. G. (1982). Soil carbon pools and world life zones, *Nature*, **298**, 156.
77. Post, W. M., Pastor, J., Zinke, P. J. and Stangenberger, A. G. (1985). Global patterns of soil nitrogen storage, *Nature*, **317**, 613.
78. Torn, M. S., Trumbore, S. E., Chadwick, O. A., Vitousek, P. M. and Hendricks, D. M. (1997). Mineral control of soil organic carbon storage and turnover, *Nature*, **389**, 170.
79. Schwertmann, U. and Taylor, R. M. (1989). Iron oxides. In *Minerals In Soil Environments*, 2nd edn, ed. Dixon J. B. and Weed, S. B., SSSA Book Series No 1. Soil Science Society of America, Madison, WI, p. 379.
80. Alexander, E. B. (1985). Arduino, E, *et al.*: Estimating relative ages from iron-oxide / total-iron ratios of soils in the western Po valley, Italy. A Discussion, *Geoderma*, **35**, 257.
81. Arduino, E., Barberis, E., Carraro, F. and Forno, M. G. (1984). Estimating relative ages from iron-oxide/total-iron ratios of soils in the western Po valley, Italy, *Geoderma*, **33**, 39.

82. Guillet, B. (1990) Le vieillissement des matières organique et des associations organo-minérales des andosols et des podzols, *Sci. Sol*, **28**, 285.
83. McKeague, J. A., Cheshire, M. V., Andreux, F. and Berthelin, J. (1986). Organo-mineral complexes in relation to pedogenesis. In *Interactions of Soil Minerals with Natural Organics and Microbes*, ed. Huang, P. M. and Schnitzer, M., SSSA Special Publication 17, Soil Science Society of America, Madison, WI, p. 549.
84. Schwertmann, U. (1966). Inhibitory effect of soil organic matter on the crystal-lisation of amorphous ferric hydroxide, *Nature* (London), **212**, 645.
85. Violante, A. and Huang, P. M. (1985). Influence of inorganic and organic ligands on precipitation products of aluminum, *Clays Clay Miner.*, **33**, 181.
86. Huang, P. M. (1991). Kinetics of redox reactions on manganese oxides and its impact on environmental quality. In *Rate of Chemical Processes in Soils*, ed. Sparks, D. L. and Suarez, D. L., SSSA Special Publication 27, Soil Science Society of America, Madison, WI, p. 191.
87. Schalscha, E. B., Appelt, H. and Schatz, A. (1967). Chelation as a weathering mechanism. 1. Effect of complexing agents on the solubilization of iron from minerals and granodiorit, *Geochim. Cosmochim. Acta*, **31**, 587.
88. Huang, W. H. and Keller, W. D. (1970). Dissolution of rock-forming silicate minerals in organic acids: simulated first-stage weathering of fresh mineral surfaces, *Am. Mineral.*, **55**, 2076.
89. Sprengel, C. (1826). Über Pflanzenhumus, Humussäure und Humussäure-Salze, *Kastner's Arch. Ges. Naturlehre.*, **8**, 145.
90. Schnitzer, M. (1992). Bedeutung der organischen Bodensubstanz für die Bodenbil-dung und Bodenprozesse, Transportprozesse in Böden und die Bodenstruktur, *Ber. Landwirtsch.*, **206**, 63.
91. Lewis, J. A. and Starkey, R. L. (1969). Decomposition of plant tannins by some soil microorganisms, *Soil Sci.*, **107**, 235.
92. Huang, P. M. and Violante, A. (1986). Influence of organic acids on crystallization and surface properties of precipitation products of aluminum. In *Interactions of Soil Minerals with Natural Organics and Microbes*, ed. Huang, P. M. and Schnitzer, M., SSSA Special Publication 17, Soil Science Society of America, Madison, WI, p. 159.
93. Whitehead, D. C. (1963). Identification of p-hydroxybenzoic, vanillic, p-coumaric and ferulic acids in soils, *Nature* (London), **202**, 417.
94. Baker, W. E. (1973). The role of humic acids from Tasmanian podzol soils in mineral degradation and metal mobilization, *Geochim. Cosmochim. Acta*, **37**, 269.
95. Haider, K. (1999). Von der toten organischen Substanz zum Humus, *Z. Pflanze-nernähr. Bodenk.*, **162**, 363.
96. Haider, K. (1992). Problems related to the humification process in soils of tempe-rate climates. In *Soil Biochemistry*, Vol 7, ed. Stotzky, G. and Bollag, J.-M., Marcel Dekker, New York, p. 55.
97. Berthelin, J. (1988). Microbial weathering processes. In *Physical and Chemical Weathering Cycles*, ed. Lerman, A. and Meyerbeck, M, Reidel, Dordrecht, The Netherlands. p. 339.
98. Ehrlich, H.L. (1990), *Geomicrobiology*, 2nd ed., Marcel Dekker, New York.
99. Berthelin, J. , Leyval, C., Laheurte, F. and de Guidici, P. (1990). Involvement of roots and rhizosphere microflora in the chemical weathering of soil minerals. In *Plant Root Growth—An Ecological Perspective*, ed. Atkinson, D., Spec. Publ. 10, British Ecology. Society., Blackwell, Oxford, p. 187.
100. Spyridakis, D. E., Chester, S. G. and Wilde, S. A. (1967). Kaolinization of biotite as a result of coniferous and deciduous seedling growth, *Soil Sci. Soc. Am. Proc.*, **31**, 203.

101. Azcon, R., Barea, J. M. and Hayman, D. S. (1976). Utilization of rock phosphate in alkaline soils by plants inoculated with mycorhizal fungi and phosphate solubilizing bacteria, *Soil Biol. Biochem.*, **8**, 135.
102. Marschner, H., Römheld, V., Horst, W. J. and Martin, P. (1986). Root -induced changes in the rhizosphere: Importance for the mineral nutrition of plant, *Z. Pflanzenernähr. Bodenk.*, **149**, 441.
103. Marschner, H., Römheld, V. and Kissel, M. (1986) Localization of phytosiderophore release and of boron uptake along intact barley roots, *Physiol. Planarum*, **71**, 157.
104. Mengel, K. (1985). Dynamics and availability of major nutrients in soils, *Adv. Soil Sci.*, **2**, 96.
105. Violante, A., Colombo, C. and Buondonno, A. (1991). Competitive adsorption of phosphate and oxalate by aluminum oxides, *Soil Sci. Soc. Am. J.*, **55**, 65
106. Deb, D. L. and Datta, N. P. (1967). Effects of associating anions on phosphorus retention in soil: I Under variable phosphorus concentrations, *Plant Soil*, **26**, 303.
107. Klees, K. (1993). Untersuchungen über den Einfluss organischer Säuren auf das Adsorption- und Desorptionsverhalten von Phosphat an Goethit (α- FeOOH) und die Auflösung eines Eisen (III)-Phosphates. Dissertation TU-Braunschweig.
108. Kwong, K. F. Ng Kee and Huang, P. M. (1978). Sorption of phosphate by hydrolytic reaction products of aluminum, *Nature* (London), **271**, 336.
109. Sala, O., Parton, W. J., Joyce, L. and Lauenroth, W. K. (1988). Primary production of the central grassland region of the United States, *Ecology*, **69**, 40.
110. Amelung, W., Zech, W., Zhang, X., Follett, R. F., Tiessen, H., Knox, E. and Flach, K.-W. (1998). Carbon, nitrogen, and sulfur pools in particle-size fractions as influenced by climate, *Soil Sci. Soc. Am. J.*, **62**, 172.
111. Trumbore, S. E., Chadwick, O. A. and Amundson, R. (1996). Rapid exchange between soil carbon and atmospheric carbon dioxide driven by temperature change, *Science*, **272**, 393.
112. Tiessen, H. and Stewart, J. W. B. (1985). The biogeochemistry of soil phosphorus. In *Planetary Ecology*, ed. Caldwell, D. E., Brierley; J. A. and Brierly, C. L., Van Nostrand Reinhold, New York, p. 463.
113. Uehara, G. and Gillman, G. P. (1981), *The Mineralogy, Chemistry and Physics of Tropical Soils with Variable Charge Clays*, Westview, Boulder, CO.
114. Hedley, M. J., Stewart, J. W. B., and Chauhan, B. S. (1982). Changes in inorganic and organic soil phosphorus fraction induced by cultivation practices and by laboratory incubations, *Soil Sci. Soc. Am. J.,* **46**, 970.
115. Cross, A. F. and Schlesinger, W. H. (1995). A literature review and evaluation of the Hedley fractionation: Applications to the biogeochemical cycle of phosphorus in natural ecosystems, *Geoderma*, **64**, 197.
116. Crews, T. E., Kitayama, K., Fownes, J. H., Riley, R. H., Herbert, D. A., Mueller-Dombois, D., Vitousek, P. M. (1995). Changes in soil phosphorus fractions and ecosystem dynamics across a long chronosequence in Hawaii, *Ecology*, **76**, 1407.
117. Miller, R. W. and Donahue, R. L. (1990), *Soils. An Introduction to Soils and Plant Growth*, 6th edn., Prentice Hall, Englewood Cliffs, NJ.
118. Christensen, B. T. (1996). Carbon in primary and secondary organomineral complexes. In *Structure and Organic Matter Storage in Agricultural Soils, Advances in Soil Science*, ed. Carter, M. R. and Stewart, B. A., CRC Press, Boca Raton, FL, p. 97.
119. Buyanovsky, G. A., Aslam, M. and Wagner, G. H. (1994). Carbon turnover in soil physical fractions, *Soil Sci. Soc. Am. J.*, **58**, 1167.
120. Emerson, W. W., Forster, R. C. and Oades, J. M. (1986). Organo-mineral complexes in relation to soil aggregation and structure. In *Interactions of Soil Minerals*

with Natural Organics and Microbes, ed. Huang, P. M. and Schnitzer, M., SSSA Special Publication 17, Soil Science Society of America, Madison, WI, p. 521.

121. Elliott, E. T. and Cambardella, C. A. (1991). Physical separation of soil organic matter, *Agric. Ecosyst. Environ.*, **34**, 443.

122. Ahmed, M. and Oades, J. M. (1984). Distribution of organic matter and adenosine triphosphate after fractionation of soils by physical procedures, *Soil Biol. Biochem.*, **16**, 465.

123. Jenkinson, D. S. and Powlson, D. S. (1976). The effect of biocidal treatments on metabolsm in soil—V. A method for measuring soil microbial biomass, *Soil Biol. Biochem.*, **8**, 209.

124. Amelung, W. and Zech, W. (1999). Minimisation of organic matter disruption during particle-size fractionation of grassland epipedons, *Geoderma*, **92**, 73.

125. Schmidt, M. W. I., Rumpel, C. and Kögel-Knabner, I. (1999). Evaluation of an ultrasonic dispersion procedure to isolate primary organomineral complexes from soil, *Eur. J. Soil Sci.*, **50**, 87.

126. Tiessen, H. and Stewart, J. W. B. (1983). Particle-size fractions and their use in studies of soil organic matter: II. Cultivation effects on organic matter composition in size fractions, *Soil Sci. Soc. Am. J.*, **47**, 509.

127. Christensen, B. T. (1985). Carbon and nitrogen in particle size fractions isolated from Danish arable soils by ultrasonic dispersion and gravity-sedimentation, *Acta Agric. Scand.*, **35**, 175.

128. Gregorich, E. G., Kachanoski, R. G. and Voroney, R. P. (1988). Ultrasonic dispersion of aggregates: Distribution of organic matter in size fractions, *Can. J. Soil. Sci.*, **68**, 395.

129. Anderson, D. W., Saggar, S., Bettany, J. R. and Stewart, J. W. B. (1981) Particle size fractions and their use in studies of soil organic matter: I. The nature and distribution of forms of carbon, nitrogen and sulfur, *Soil Sci. Soc. Am. J.*, **45**, 767.

130. Rubæk, G. H., Guggenberger, G., Zech, W. and Christensen, B. T. (1999). Organic phsophorus in soil size-separates characterized by phosphorus-31 NMR and resin extraction, *Soil Sci. Soc. Am. J.*, **63**, 1123.

131. Guggenberger, G., Christensen, B. T. and Zech, W. (1994). Land-use effects on the composition of organic matter in particle-size seaprates of soil: I. Lignin and carbohydrate signature, *Eur. J. Soil Sci.*, **45**, 449.

132. Jensen, E. S., Christensen, B. T. and Sorensen, S. (1989). Mineral-fixed ammonium in clay- and silt-size fractions of soils incubated with ^{15}N ammonium sulphate for five years, *Biol. Fertil. Soils*, **8**, 298.

133. Kowalenko, C. G. and Ross, G. J. (1980). Studies on the dynamics of 'recently' clay-fixed NH_4^+ using ^{15}N, *Can. J. Soil Sci.*, **60**, 61.

134. Hedges, J. I. and Oades, J. M. (1997). Comparative organic geochemistries of soils and marine sediments, *Org. Geochem.*, **27**, 319.

135. Brookes, P. C., Powlson, D. S. and Jenkinson, D. S. (1984). Phosphorus in soil microbial biomass, *Soil Biol. Biochem.*, **16**, 169.

136. Golchin, A., Oades, J. M., Skjemstad, J. O. and Clarke, P. (1994). Study of free and occluded particulate organic matter in soils by solid state ^{13}C CP/MAS NMR spectroscopy and scanning electron microscopy, *Aust. J. Soil Res.*, **32**, 285.

137. Stemmer, M., Gerzabek, M. H. and Kandeler, E. (1998). Organic matter and enzyme activity in particle-size fractions of soils obtained after low-energy sonication, *Soil Biol. Biochem.*, **30**, 9.

138. Fründ, R. and Lüdemann, H.-D. (1989). The quantitative analysis of solution and CPMAS-C-13 NMR spectra of humic material, *Sci. Total. Environ.*, **81/82**, 157.

139. Snape, C. O., Axelson, D. E., Botto, R. E., Delpuecht, J. J., Tekely, P., Gerstein, B. C., Pruski, M., Maciel, G. E. and Wilson, M. A. (1989). Quantitative reliability of aromaticity and related measurements on coals by ^{13}C NMR: A debate, *Fuel*, **68**, 547.
140. Skjemstad, J. O., Clarke, P., Taylor, J. A., Oades, J. M. and McClure, S. G. (1996). The chemistry and nature of protected carbon in soil, *Aust. J. Soil Res.*, **34**, 251.
141. Baldock, J. A., Oades, J. M., Waters, A. G., Peng, X., Vassallo, A. M. and Wilson, M. A. (1992). Aspects of the chemical structure of soil organic materials as revealed by solid-state ^{13}C NMR spectroscopy, *Biogeochemistry*, **16**, 1.
142. Oades, J. M., Vassallo, A. M., Waters, A. G. and Wilson, M. A. (1987). Characterization of organic matter in particle size and density fractions from a red-brown earth by solid-state ^{13}C N.M.R, *Aust. J. Soil Res.*, **25**, 71.
143. Preston, C. M., Newman, R. H. and Rother, P. (1994). Using ^{13}C CPMAS NMR to assess effects of cultivation on the organic matter of particle size fractions in a grassland soil, *Soil Sci.*, **157**, 26.
144. Guggenberger, G., Zech, W., Haumaier, L. and Christensen, B. T. (1995). Land-use effects on the composition of organic matter in particle-size seaprates of soil: II. CPMAS and solution ^{13}C NMR analysis, *Eur. J. Soil Sci.*, **46**, 147.
145. Baldock, J. A., Oades, J. M., Vassallo, A. M. and Wilson, M. A. (1989). Incorporation of uniformly labelled ^{13}C-glucose carbon into the organic fraction of a soil. Carbon balance and CP/MAS ^{13}C NMR measurements, *Aust. J. Soil Res.*, **27**, 725.
146. Christensen, B. T. and Bech-Andersen, S. (1989). Influence of straw disposal on distribution of amino acids in soil particle size fractions, *Soil Biol. Biochem.*, **21**, 35.
147. Parsons, J. W. (1981). Chemistry and distribution of amino sugars in *Soil Biochemistry*, Vol. 5, ed. Paul, E. A. and Ladd, J. N., Marcel Dekker, New York, p. 197.
148. Brock, T. D. and Madigan, M. T. (1988), *Biology of Microorganisms*, 5th edn, Prentice Hall, Englewood Cliffs, NJ.
149. Chantigny, M. H., Angers, D. A., Prévost, D., Vézina, L.-P. and Chalifour, F.-P. (1997). Soil aggregation and fungal and bacterial biomass under annual and perennial cropping systems, *Soil Sci. Soc. Am. J.*, **61**, 262.
150. Amelung, W., Zhang, X., Flach, K. W. and Zech, W. (1999) Amino sugars in native grassland soils along a climosequence in North America, *Soil Sci. Soc. Am. J.*, **63**, 86.
151. Zhang, X., Amelung, W., Yuan, Y. and Zech, W. (1998). Amino sugar signature of particle-size fractions in soils of the native prairie as affected by climate, *Soil Sci.*, **163**, 220.
152. Filip, Z. (1977). Einfluβ von Tonmineralen auf die mikrobielle Ausnutzung der kohlenstoffhaltigen Substanzen und Bildung der Biomasse, *Ecol. Bull.*, **25**, 173.
153. Ladd, J. N., Foster, R. C. and Skjemstad, J. O. (1993). Soil structure: Carbon and nitrogen metabolism, *Geoderma*, **56**, 401.
154. Sumann, M., Amelung, W., Haumaier, L. and Zech, W. (1998). Climatic effects on soil organic phosphorus in the North American Great Plains identified by phosphorus-31 nuclear magnetic resonance, *Soil Sci. Soc. Am. J.*, **62**, 1580.
155. Tate, K. R. (1984). The biological transformation of P in soil, *Plant Soil*, **76**, 245.
156. Miltner, A., Haumaier, L. and Zech, W. (1998). Transformations of phosphorus during incubation of beech leaf litter in the presence of oxides, *Eur. J. Soil Sci.*, **49**, 471.
157. Amelung, W., Flach, K. W. and Zech, W. (1999). Lignin in particle-size fractions of native grassland soils as influenced by climate, *Soil Sci. Soc. Am. J.*, **63**, 1222.
158. Guggenberger, G., Schulten, H.-R. and Zech, W. (1994). Formation and mobilization pathways of dissolved organic matter: evidence from chemical structural studies of organic matter fractions in acid forest floor solutions, *Org. Geochem.*, **21**, 51.

159. McDowell, W. H. and Likens, G. E. (1988). Origin, composition, and flux of dissolved organic carbon in the Hubbard Brook Valley, *Ecol. Monogr.*, **58**, 177.
160. Qualls, R. G., Haines, B. L. and Swank, W. T. (1991). Fluxes of dissolved organic nutrients and humic substances in a deciduous forest, *Ecology*, **72**, 254.
161. David, M. B., Vance, G. F. and Krzyszowska, A. J. (1995). Carbon controls on spodosol nitrogen, sulfur, and phosphorus cycling. In *Carbon Forms and Functions in Forest Soils*, ed. McFee, W. W. and Kelly, J. M., Soil Science Society of America, Madison, WI, p. 329.
162. Catroux, G. and Schnitzer, M. (1987). Chemical, spectroscopic, and biological characteristics of the organic matter in particle size fractions separated from an Aquoll, *Soil Sci. Soc. Am. J.*, **51**, 1200.
163. Lowe, L. E. and Hinds, A. A. (1983). The mineralization of nitrogen and sulphur from particle size separates of gleysolic soils, *Can. J. Soil Sci.*, **63**, 761.
164. Christensen, B. T. (1987). Decomposability of organic matter in particle size fractions from field soils with straw incorporation, *Soil. Biol Biochem.*, **19**, 429.
165. Chichester, F. W. (1970). Transformations of fertilizer nitrogen in soil. II. Total and N^{15}-labelled nitrogen of soil organo-mineral sedimentation fractions, *Plant Soil*, **33**, 437.
166. Cheshire, M. V. and Mundie, C. M. (1981). The distribution of labelled sugars in soil particle size fractions as a means of distinguishing plant and microbial carbohydrate residues, *J. Soil Sci.*, **32**, 605.
167. Amelung, W., Bol, R. and Friedrich, C. (1999). Natural ^{13}C abundance: A tool to trace the incorporation of dung-derived carbon into soil particle-size fractions, *Rapid Commun. Mass Spectrom.*, **13**, 1291.
168. Balesdent, J., Mariotti, A. and Guillet, B. (1987). Natural ^{13}C abundance as a tracer for studies of soil organic matter dynamics, *Soil Biol. Biochem.*, **19**, 25.
169. Balesdent, J., Wagner, G. H. and Mariotti, A. (1988). Soil organic matter turnover in long-term field experiments as revealed by carbon-13 natural abundance, *Soil Sci. Soc. Am. J.*, **52**, 118.
170. Martin, A., Mariotti, A., Balesdent, J., Lavelle, P. and Vuattoux, R. (1990). Estimation of organic matter turnover rate in a savanna soil by ^{13}C natural abundance measurements, *Soil Biol. Biochem.*, **22**, 517.
171. Vitorello, V. A., Cerri, C. C., Andreux, F., Feller, C. and Victoria, R. L. (1989). Organic matter and natural carbon-13 distribution in forested and cultivated oxisols, *Soil Sci. Soc. Am. J.*, **53**, 773.
172. Trumbore, S. E. (1993). Comparison of carbon dynamics in tropical and temperate soils using radiocarbon measurements, *Global Biogeochem. Cycles*, **7**, 275.
173. Anderson, D. W. and Paul, E. A. (1984). Organo-mineral complexes and their study by radiocarbon dating, *Soil Sci. Soc. Am. J.*, **48**, 298.
174. Scharpenseel, H. W., Tsutsuki, K., Becker-Heidmann, P. and Freytag, J. (1986). Untersuchungen zur Kohlenstoffdynamik und Bioturbation von Mollisolen, *Z. Pflanzenernähr. Bodenk.*, **149**, 582.
175. Goh, K. M., Stout, J. D. and Rafter, T. A. (1977). Radiocarbon enrichment of soil organic matter fractions in New Zealand soils, *Soil Sci.* **123**, 385.
176. Hsieh, Y. P. (1996). Soil organic carbon pools of two tropical soils inferred by carbon signatures, *Soil Sci. Soc. Am. J.*, **60**, 1117.
177. Nelson, P. N., Dictor, M.-C. and Soulas, G. (1994). Availability of organic carbon in soluble and particle-size fractions from a soil profile, *Soil Biol. Biochem.*, **26**, 1549.
178. Golchin, A., Oades, J. M., Skjemstad, J. O. and Clarke, P. (1994). Soil structure and carbon cycling, *Aust. J. Soil Res.*, **32**, 1043.

179. Glossary of Soil Science Terms Committee (1997) Glossary of Soil Science Terms 1996, Soil Science Society of America, Madison, WI.
180. Elliott, E. T. and Coleman, D. C. (1988). Let the soil work for us, *Ecol. Bull.*, **39**, 23.
181. Elliott, E. T., Paustian, K. and Frey, S. D. (1996). Modeling the measurable or measuring the modelable: a hierarchical approach to isolating meaningful soil organic matter fractionations. In *Evaluation of Soil Organic Matter Models Using Existing Long-Term Datasets*, ed. Powlson, D. S., Smith, P. and Smith, J. U., Springer, Berlin, p. 161.
182. Sexstone, A. J., Revsbech, N. P., Perkin, T. B. and Tiedje, J. M. (1985). Direct measurement of oxygen profiles and denitrification rates in soil aggregates, *Soil Sci. Soc. Am. J.*, **49**, 645.
183. Kay, B. D. (1997). Soil structure and organic carbon: a review. In *Soil Processes and the Carbon Cycle*, ed. Lal; R., Kimble, J. M., Follett, R. F. and Stewart, B. A., CRC Press, Boca Raton, FL, p. 169.
184. Oades, J. M. and Waters, A. G. (1991). Aggregate hierarchy in soils, *Aust. J. Soil Res.*, **29**, 815.
185. Jastrow, J. D. and Miller, R. M. (1998). Soil aggregate stabilization and carbon sequestration: feedbacks through organomineral associations. In *Soil Processes and the Carbon Cycle*, ed. Lal, R., Kimble, J. M., Follett, R. F. and Stewart, B. A., CRC Press, Boca Raton, FL,, p. 207.
186. Tisdall, J. M. and Oades, J. M. (1979). Stabilization of soil aggregates by the root system of ryegrass, *Aust. J. Soil Res.*, **17**, 429.
187. Tisdall, J. M. (1996). Formation of soil aggregates and accumulation of soil organic matter. In *Structure and Organic Matter Storage in Agricultural Soils*, ed. Carter, M. R. and Stewart, B. A., CRC Press, Boca Raton, FL, p. 57.
188. Foster, R. C. (1978). Ultramicromorphology of some South Australian soils. In *Modification of Soil Structure*, ed. Emerson, W. W., Boden, R. D. and Dexter, R. A., John Wiley & Sons, Chichester, p. 103.
189. Chenu, C. (1995). Extracellular polysaccharides: an interface between microorganisms and soil constituents. In *Environmental Impact of Soil Component Interactions. Natural and Anthropogenic Organics,* ed. Huang, P. M., Berthelin, J., Bollag, J-.M., McGill, W. B. and Page, A. L., CRC Lewis, Boca Raton, FL, p. 217.
190. Hattori, T. and Hattori, R. (1976). The physical environment in soil microbiology: an attempt to extend principles of microbiology to soil microorganisms, *CRC Crit. Rev. Microbiol.*, **4**, 423.
191. Vargas, R. and Hattori, T. (1986). Protozoan predation of bacterial cells in soil aggregates, *FEMS Microbiol. Ecol.*, **38**, 233.
192. Foster, R. C. (1988). Microenvironments of soil microorganisms, *Biol. Fertil. Soils*, **6**, 189.
193. Foster, R. C. and Martin, J. K. (1981), *In situ* analysis of soil components of biological origin. In *Soil Biochemistry*, Vol 5, ed. Paul, E. A. and Ladd, J. N., Marcel Dekker, New York, p. 75.
194. Haider, K. (1999). Microbe-soil contaminant interactions. In: *Bioremediation of Contaminated Soils*. In *Agronomy Monograph, Vol 37, ed. Kral, D. M.*, ASA, CSSA, Soil Science Society of America, Madison, WI, p. 33.
195. Killham, K., Amato, M. and Ladd, J. N. (1993). Effect of substrate location in soil and soil pre-water regime on carbon turnover, *Soil Biol. Biochem.*, **25**, 57.
196. Beare, M. H., Parmelee, R. W., Hendrix, P. F., Cheng, W., Coleman, D. C. and Crossley, C. A. Jr. (1992). Microbial and faunal interactions and effects on litter nitrogen decomposition in agroecosystems, *Ecol. Monogr.*, **62**, 569.

197. Frey, S. D., Elliott, E. T., Paustian, K. (1999). Bacterial and fungal abundance and biomass in conventional and no-tillage agroecosystem along two climatic gradients, *Soil Biol. Biochem.*, **31**, 573.

198. Alperin, M. J., Balesdent, J, Benner, R. H., Blough, N. V., Christman, R. F., Druffel, e. R. M., Frimmel, F. H., Guggenberger, G., Repeta, D. J., Richnow, H. H., Swift, R. S. (1994). How can we best characterize and quantify pools and fluxes of nonliving organic matter: Group report. In *The Role of Nonliving Organic Matter in the Earth's Carbon Cycle*, ed. Zepp, R. G. and Sonntag, C., Dahlem Workshop Report ES 16, John Wiley & Sons, Chichester, p. 67.

199. Jastrow, J. D. (1996). Soil aggregate formation and the accrual of particulate and mineral-associated organic matter, *Soil Biol. Biochem.*, **28**, 665.

200. Six, J., Elliott, E. T., Paustian, K. and Doran, J. W. (1998). Aggregation and soil organic matter accumulation in cultivated and native grassland soils, *Soil Sci. Soc. Am. J.*, **62**, 1367.

201. Puget, P., Besnard, E. and Chenu, C. (1996). Une méthode de fractionnement des matières organique particulaires des sols en fonction de leur localisation dans les agrégats, *C. R. Acad. Sci. Sér. II*, **322**, 965.

202. Gregorich, E. G., Drury, C. F., Ellert, B. H. and Liang, B. C. (1997). Fertilization effects on physically protected light fraction organic matter, *Soil Sci. Soc. Am. J.*, **61**, 482.

203. Golchin, A., Oades, J. M., Skjemstad, J. O. and Clarke, P. (1995). Structural and dynamic properties of soil organic matter as reflected by ^{13}C natural abundance, pyrolysis mass spectrometry and solid-state ^{13}C NMR spectroscopy in density fractions of an oxisol under forest and pasture, *Aust. J. Soil Res.*, **33**, 59.

204. Haynes, R. J. and Francis, G. S. (1993). Changes in microbial biomass C, soil carbohydrate composition and aggregate stability induced by growth of selected crop and forage species under field conditions. *J. Soil Sci.*, **44**, 665.

205. Haynes, R. J. and Beare, M. H. (1996). Aggregation and organic matter storage in meso-thermal, humid soils. In *Structure and Organic Matter Storage in Agricultural Soils*, ed. Carter, M. R. and Stewart, B. A., CRC Press, Boca Raton, FL, p. 213.

206. Wright, S. F., Franke-Snyder, M., Morton, J. B. and Upadhyaya, A. (1996) Time-course study and partial characterization of a protein on hyphae of arbuscular mycorrhizal fungi during active colonization of roots, *Plant Soil*, **181**, 193.

207. Wright, S. F. and Upadhyaya, A. (1996) Extraction of an abundant and unusual protein from soil and comparison with hyphal protein of arbuscular mycorrhizal fungi, *Soil Sci.*, **161**, 575.

208. Wright, S. F. and Upadhyaya, A. (1998). A survey of soils for aggregate stability and glomalin, a glycoprotein produced by hyphae of arbuscular mycorrhizal fungi, *Plant Soil*, **198**, 97.

209. Guggenberger, G., Frey, S. D., Six, J., Paustian, K. and Elliott, E. T. (1999). Bacterial and fungal cell-wall residues in conventional and no-tillage agroecosystems, *Soil Sci. Soc. Am. J.*, **63**, 1188.

210. Elliott, E. T. (1986). Aggregate structure and carbon, nitrogen, and phosphorus in native and cultivated soils, *Soil Sci. Soc. Am. J.*, **50**, 627.

211. Monreal, C. M., Schnitzer, M., Schulten, H.-R., Campbell, C. A. and Anderson, D. W. (1995). Soil organic structures in macro- and microaggregates of a cultivated brown chernozem, *Soil Biol. Biochem.*, **27**, 845.

212. Angers, D. A., Recous, S. and Aita, C. (1997) Fate of carbon and nitrogen in water-stable aggregates during decomposition of ^{13}C^{15}N-labelled wheat straw *in situ*, *Eur. J. Soil Sci.*, **48**, 295.

213. Gupta, V. V. S. R. and Germida, J. J. (1988). Distribution of microbial biomass and its activity in different soil aggregate size classes as affected by cultivation, *Soil. Biol Biochem.*, **20**, 777.

214. Beare, M. H., Cabrera, M. L., Hendrix, P. F. and Coleman, D. C. (1994). Aggregate-protected and unprotected organic matter pools in conventional- and no-tillage soils, *Soil Sci. Soc. Am. J.*, **59**, 787.

215. Cambardella, C. A. and Elliott, E. T. (1993). Carbon and nitrogen distribution in aggregates from cultivated and native grassland soils, *Soil Sci. Soc. Am. J.*, **57**, 1071.

216. Cambardella, C. A. and Elliott, E. T. (1994). Carbon and nitrogen dynamics of soil organic matter fractions from cultivated grassland soils, *Soil Sci. Soc. Am. J.*, **58**, 123.

217. Angers, D. A. and Carter, M. R. (1996). Aggregation and organic matter storage in cool, humid agricultural soils. In *Structure and Organic Matter Storage in Agricultural Soils*, ed. Carter, M. R. and Stewart, B. A., Adv.Soil Sci., CRC Press, Boca Raton, FL, p. 193.

218. Skjemstad, J. O., Le Feuvre, R. P. and Prebble, R. E. (1990). Turnover of soil organic matter under pasture as determiend by ^{13}C natural abundance, *Aust. J. Soil Res.* **28**, 267.

219. Puget, P., Chenu, C. and Balesdent, J. (1995). Total and young organic matter distributions in aggregates of silty cultivated soils, *Eur. J. Soil Sci.*, **46**, 449.

220. Angers, D. A. and Giroux, M. (1996). Recently deposited organic matter in soil water-stable aggregates, *Soil Sci. Soc. Am. J.*, **60**, 1547.

221. Jastrow, J. D., Boutton, T. W. and Miller, R. M. (1996). Carbon dynamics of aggregate-associated organic matter estimated by carbon-13 natural abundance, *Soil Sci. Soc. Am. J.*, **60**, 801.

222. Skjemstad, J. O., Janik, L. J., Head, M. J. and McClure, S. G. (1993). High energy ultraviolet photo-oxidation: a novel technique for studying physically protected organic matter in clay- and silt-sized aggregates, *J. Soil Sci.*, **44**, 485.

223. Huang, P. M. (1990). Role of soil minerals in transformations of natural organics and xenobiotics in soil. In *Soil Biochemistry*, Vol. 6, ed. Bollag, J.-M. and Stotzky, G., Marcel Dekker, New York, p. 29.

224. Kassim, G., Stott, D. E., Martin, J. P. and Haider, K. (1982). Stabilization and incorporation into biomass of phenolic and benzenoid carbons during biodegradation in soil, *Soil Sci. Soc. Amer. J.*, **46**, 305.

225. Martin, J. P. and Haider, K. (1986). Influence of mineral colloids on turnover rates of soil organic carbon. In *Interactions of Soil Minerals with Natural Organics and Microbes*, ed. Huang, P. M. and Schnitzer, M., SSSA Special Publication 17, Soil Science Society of America, Madison, WI, p. 283.

226. Wang, T. S. C., Wang, M. C., Ferng, Y. L. and Huang, P. M. (1983). Catalytic synthesis of humic substances by natural clays, silts, and soils, *Soil Sci.*, **135**, 350.

227. Shindo, H. and Huang, P. M. (1984) Significance of Mn(IV) oxide in abiotic formation of organic nitrogen complexes in natural environments, *Nature*, **308**, 57.

228. Shindo, H. and Huang, P. M. (1984) Catalytic effects of manganese(IV), iron(III), aluminum, and silicon oxides on the formation of phenolic polymers, *Soil Sci. Soc. Am. J.*, **48**, 927.

229. Wang, M. C. and Huang, P. M. (1992). Significance of Mn(IV) oxide in the abiotic ring cleavage of pyrogallol in natural in natural environments, *Sci. Total Environ.*, **113**, 147.

230. Haider, K. (1995). Sorption phenomena between inorganic and organic compounds in soils: Impacts on transformation processes. In *Environmental Impact of*

   Soil Componenent Interactions, Vol I, ed. Huang, P. M., Berthelin, J., Bollag, J.-M., McGill, W. B. and Page, A. L., CRC Lewis, Boca Raton, FL, p. 21.
231. Bollag, J.-M., Myers, C., Pal, S. and Huang, P. M. (1995). The role of abiotic and biotic catalysts in the transformation of phenolic compounds. In *Environmental Impact of Soil Componenent Interactions*, Vol I, ed. Huang, P. M., Berthelin, J., Bollag, J.-M., McGill, W. B. and Page, A. L., CRC Lewis, Boca Raton, FL, p. 299.
232. Schwarzenbach, R. P., Gschwend, P. M. and Imboden, D. M. (1993), *Environmental Organic Chemistry*, John Wiley & Sons, New York.
233. Hassett, J. J. and Banwart, W. L. (1989). The sorption of nonpolar organics by soils and sediments. In *Reactions and Movement of Organic Chemicals in Soil*, ed. Sawhney, B. L. and Brown, K., Soil Science Society of America, Madison, WI, p. 31.
234. Chiou, C. T. (1989) Theoretical considerations of the partition uptake of nonionic organic compounds by soil organic matter. In *Reactions and Movement of Organic Chemicals in Soil*, ed. Sawhney, B. L. and Brown, K., Soil Science Society of America, Madison, WI, p. 1.
235. Engebretson, R. R., Amos, T. and von Wandruzka, R. (1996). Quantitative approach to humic acid associations, *Environ. Sci. Technol.*, **30**, 990.
236. Dec, J. and Bollag, J.-M. (1998). Determination of covalent and non-covalent binding between xenobiotic chemicals and soil, *Soil Sci.*, **162**, 858.
237. Haider, K. M., Spiteller M., Dec, J., and A. Schäffer (2000). Silylation of soil organic matter, extraction of humic compounds and soil-bound residues. In *Soil Biochemistry* Vol. 19, ed. J.-M. Bollag and G. Stotzky, Marcel Dekker, New York, p. 139.
238. Luthy, R. G., Aiken, G. R., Brusseau, M. L., Cunningham, S. D., Gschwend, P. M., Pignatello, J. J., Reinhard, M., Traina, S. J., Weber, W. J. Jr. And Westall, J. C. (1997). Sequestration of hydrophobic organic contaminants by geosorbents, *Environ. Sci. Technol.*, **31**, 3341.
239. Haider, K., Spiteller, M., Wais, A. and Fild, M. (1993). Evaluation of the binding mechanism of anilazine and its metabolites in soil organic matter, *Int. J. Environ. Anal. Chem.*, **51**, 125.
240. Dec, J., Haider, K., Schäffer, A., Fernandes, E., and Bollag, J.-M. (1997). Use of a silylation procedure and ^{13}C-NMR spectroscopy to characterize bound and sequestered residues of cyprodinil in soil, *Environ. Sci. Technol.*, **31**, 2991.
241. Xing, B. and Pignatello, J. J. (1997). Dual model of low-polarity compounds in glassy poly (vinyl chloride) and soil organic matter, *Environ. Sci. Technol.*, **31**, 792.
242. Alexander, M. (1995). How toxic are toxic chemicals in soil?, *Environ. Sci. Technol.*, **29**, 2713.
243. Carter, M. R. (1996). Analysis of soil organic matter storage in agroecosystems. In *Structure and Organic Matter Storage in Agricultural Soils*, ed. Carter, M. R. and Stewart, B. A., Adv. Soil Sci., CRC Press, Boca Raton, FL, p. 3.
244. Wander, M. M., Traina, S. J., Stinner, B. R. and Peters, S. E. (1994) Organic and conventional management effects on biologically active soil organic matter pools, *Soil Sci. Soc. Am. J.*, **58**, 1130.
245. Haider, K. (2000). Reply to the comments on 'Von der toten organischen Substanz zum Humus', *J. Plant Nutr. Soil Sci.*, **163**, 123.

8 Impact of Interactions between Microorganisms and Soil Colloids on the Transformation of Organic Pollutants

J. DEC AND J.-M. BOLLAG
The Pennsylvania State University, USA

P. M. HUANG
University of Saskatchewan, Canada

N. SENESI
Università di Bari, Italy

Interactions between Soil Particles and Microorganisms
Edited by P. M. Huang, J.-M. Bollag and N. Senesi. © 2002 John Wiley & Sons, Ltd

1 INTRODUCTION

Microorganisms are considered to be one of the primary factors that determine the fate of organic xenobiotics in soil [1]. Environmentally, the most desired result of microbial activity is the degradation of toxic chemicals of both natural and synthetic origin. Some anthropogenic compounds may be used by microorganisms as a source of energy and nutrients for growth. The prevalent form of microbial transformation, however, is cometabolism, in which microorganisms transform the xenobiotic molecules, but are unable to use these molecules as a source of carbon or energy [2].

The ultimate cause of microbial transformation is the activity of the intracellular and extracellular enzymes produced by microorganisms [3,4]. Xenobiotics, whose chemical structures are similar to those of organic compounds occurring in nature are usually more susceptible to biodegradation than those having little structural resemblance to natural products [5]. This is because microbial enzymes that are specific to toxic chemicals of natural origin may also show specificity for the structurally analogous xenobiotics. Frequently, microorganisms need an acclimation period to undergo a genetic mutation during a prolonged exposure to xenobiotics that can induce the production of novel degrading enzymes [6,7].

Microbial enzymes are involved in complex relationships with other components of the soil system, which may considerably modify their ability to mediate the transformation of anthropogenic compounds. The efficiency of intracellular enzymes depends on everything that happens to microorganisms. As a result of this dependence, the activity of cellular enzymes is frequently identified with that of microorganisms, so the latter tend to be viewed as the subject matter of the investigatory insight, whereas the role of enzymes is likely to be underrated. In the case of extracellular enzymes, however, the perspective changes. Once released from their producers, extracellular enzymes start functioning as a sole cause of xenobiotic biodegradation and automatically become the center of attention, driving out microorganisms to the background.

Mineral colloids that are omnipresent in soil environments contribute greatly to the complexity of biodegradation processes [8]. The influence of minerals relies partly on the fact that they themselves may possess catalytic properties and be able to mediate xenobiotic transformation. Frequently it is difficult to determine whether a pollutant is transformed in soil abiotically or biotically [9]. In many cases, significant biological and abiotic degradations take place simultaneously. Biodegradation processes can be also complicated by the adsorption phenomena occurring on mineral surfaces. Both biological agents (i.e., microorganisms or extracellular enzymes) and xenobiotics can be subject to adsorption. Adsorption of microorganisms reduces their mobility and may also alter their physiological activity. Extracellular enzymes adsorbed on mineral colloids are usually less active than free enzymes; on the other hand, they may show increased stability. Microorganisms and enzymes can transform degradable xenobiotics as long as the latter are present in the soil solution. However, when xenobiotics are removed from the aqueous phase through adsorption to mineral particles, their bioavailability drastically decreases and so does their biodegradation [10].

Humic substances (HS) that are associated with mineral colloids add another dimension to the problem of xenobiotic biodegradation in soil environments. Organic matter covers a considerable portion of mineral surfaces, thus reducing the mineral's ability to adsorb microorganisms, enzymes, or xenobiotics [11,12]. On the other hand, HS are strong adsorbents themselves, and the adsorption of microbes, enzymes and xenobiotics on the organic fraction may be difficult to distinguish from that occurring on the mineral surfaces [13]. This makes it difficult to determine the effect of each of these fractions on biodegradation.

To date, considerable work has been done to explain the mechanisms of microbial degradation [1]. Also, extensive studies have been carried out to determine the effect of minerals [8] and soil organic matter (SOM) [14,15] on the fate of xenobiotics. Relatively little has been done, however, to integrate the information about these important aspects of environmental pollution. The purpose of this paper is to review literature data that add up to the present knowledge of how the effects of minerals, organic matter and microorganisms are interrelated.

2 BIODEGRADATION OF XENOBIOTICS

2.1 MICROBIAL DEGRADATION

Present knowledge on microbial degradation of xenobiotics has been accumulated through countless *in vitro* experiments in which xenobiotic compounds were exposed to specific microorganisms isolated from soil or other natural environments. Monitoring the degradative activity of microorganisms under field conditions may generate ambiguous results, as it is difficult to distinguish between physico-chemical and microbial transformation reactions. The ability of microorganisms to transform xenobiotics relies largely on the four major processes—oxidation, reduction, hydrolysis, and synthetic reactions—that are at work in the general metabolism of all biological systems.

If a microorganism fails to degrade a given pollutant under optimal conditions, it may be because it lacks the enzyme that can mediate the transformation. In many cases, transformation can be achieved with another microorganism that happens to be equipped with the necessary enzyme. Certain xenobiotic compounds, however, especially those whose chemical structures differ considerably from the chemical structures of organic compounds naturally occurring in soil, cannot be biodegraded before the microorganism populations manage to develop novel specific enzymes through genetic mutation and natural selection. The molecular mechanisms of genetic adaptation to xenobiotic compounds and induction of specific enzymes have been reviewed by van der Meer *et al.* [7] and Singh *et al.* [16].

Among the major microbial processes that may involve xenobiotics are oxidation reactions, such as oxidative coupling, hydroxylation, β-oxidation, epoxidation, N-dealkylation, decarboxylation, ether cleavage, aromatic ring cleavage, heterocyclic ring cleavage, or sulfoxidation.

Oxidative coupling reactions are important in the synthesis of humus from the monomer products of microbial metabolism [11,17–20], and in the incorporation of certain xenobiotics (e.g., phenols and anilines) into the polymeric structure of SOM [21]. They are mediated by various phenoloxidases produced by many fungi (e.g., *Aspergillus*, *Geotrichum*, *Penicillium*, or *Trichoderma*), actinomycetes, and bacteria (*Bacillus*, *Arthrobacter*, or *Pseudomonas*) [22].

Hydroxylation of the xenobiotic molecule is frequently a first step in its microbial transformation. The introduction of the hydroxyl group can occur at aliphatic moieties as well as at aromatic rings, often making the compound more water soluble so it becomes more available to microorganisms. Enzymes catalyzing this reaction are hydroxylases, monooxygenases, or mixed function oxidases. Hydroxylation requires molecular oxygen from air, but under anaerobic conditions oxygen can be derived from water [23]. In the case of aromatic compounds, the addition of polar hydroxy groups may facilitate ring cleavage. For instance, 2,4-dichlorophenol is hydroxylated by an oxygenase, and the aromatic ring of the resulting dichlorocatechol is cleaved by a dioxygenase [24]. Aliphatic hydroxylation often leads to the degradation of the side chain (e.g., in decarboxylation, deamination, and dealkylation). In a study with eight actinomycetes of the genera *Amycolatopsis* and *Streptomycetes* [25], various benzoates were hydroxylated to catechol derivatives that were then transformed by ring cleavage. The dinitroaniline herbicide butraline was hydroxylated in the side chain by a soil fungus of *Paecilomyces* sp. [26].

Xenobiotics that possess fatty acid side chains can be transformed by β-oxidation, (sequential cleavage of two-carbon fragments until the chain is only four or two carbons long). The reaction is the same in bacteria, actinomycetes, and fungi. In the case of 2,4-dichlorophenoxyalkanoic acid, which has an even number of carbon atoms in the side chain, β-oxidation results in the formation of 2,4-dichlorophenoxyacetate [16].

Microbial epoxidation (the insertion of an oxygen atom into a carbon–carbon double bond) is especially common with chlorinated cyclodiene insecticides, such as aldrin, isodrin, or heptachlor, which were incubated with *Aspergillus niger, A. flavus, Penicillium chrysogenum,* or *P. notatum* [27]. Epoxides also occur in the degradation pathways for unsaturated aliphatic compounds [28,29]. Microbial oxidation of polyaromatic hydrocarbons (PAHs) involves an epoxidation step followed by epoxide hydrolysis [30]. Epoxides are often more toxic and stable than their parent compounds.

N-alkylated herbicides, such as phenylureas, acylanilides, carbamates, *s*-triazines, dinitroanilines and bipyridyls may undergo dealkylation in the initial stages of the microbial catabolism. In soil, atrazine is primarily *N*-dealkylated forming deethylatrazine, deisopropylatrazine, or a combination of these metabolites [31]. The insecticide mexacarbate incubated with the fungus *Trichoderma viride* was transformed to 4-methylamino-3,5-xylyl methylcarbamate as a major dealkylation product [32].

Decarboxylation by soil microorganisms is common to a variety of compounds, such as decarboxylate benzoic acids, bipyridyls, and chlorinated hydrocarbons [33]. *Pseudomonas putida* P111 metabolized various *ortho*-chlorobenzoates to corresponding catechols by dioxygenation and spontaneous liberation of CO_2 via an unstable chlorocarboxy-diol [34]. A dioxygenase in *Alcaligenes eutrophus* converted 2,4-dichlorophenoxyacetic acid to 2,4-dichlorophenol with the release of CO_2 [35]. During incubation of 4,4'-dichlorobenzylic acid with yeast *Rhodotorula gracilis*, the compound was subject to a combined decarboxylation–dehydrogenation reaction, resulting in the formation of 4,4'-dichlorobenzophenone and CO_2. The reaction was stimulated by the presence of citric acid, but α-ketoglutarate had an inhibitive effect [36].

Ether cleavage, or *O*-dealkylation (the removal of one of the hydrocarbons linked by the oxygen atoms) is probably catalyzed by mixed function oxidases in the presence of reduced pyridine nucleotides and molecular oxygen. Ether linkages or alkoxy groups are present in many pesticides, such as benzoic acids, organophosphates, carbamates, methoxy-*s*-triazines, phenylureas, or phenoxyalkanoates. With cell-free extracts from *Arthrobacter* sp. and *Pseudomonas* sp., the ether cleavage of phenoxyalkanoic herbicides, such as 2-methyl-4-chlorophenoxyacetic acid (MCPA), produced phenols as intermediates [37]. Dimethyl ether was converted by ammonia monooxygenase in *Nitrosomonas europaea* to a mixture of formaldehyde and methanol [38]. The bacterium *Sphingomonas* sp. SS33 utilized 4,4'-dichlorodiphenyl ether for growth [39]. On the other hand, 2,4-dichlorodiphenyl ether was not used as a carbon source; it was cleaved, however, with the release of corresponding chlorophenols and chlorocatechols [39].

Aromatic rings of xenobiotics are cleaved by many microorganisms, especially bacteria. The benzene ring must undergo dihydroxylation prior to cleavage and the hydroxyl groups must be at *ortho* or *para* positions to each other. Phenol, for instance, is known to be transformed to catechol by the enzyme phenol hydroxylase, which is present in many *Pseudomonas* spp. and other bacteria [40–42]. The aromatic ring of catechol is cleaved between the two hydroxyl groups by a catechol 1,2-dioxygenase (*ortho* pathway), and the resulting muconic acid is quickly mineralized via Krebs cycle intermediates [43]. In an alternative pathway involving a catechol 2,3-dioxygenase, ring cleavage takes place adjacent to the two hydroxyl groups of catechol (*meta* pathway) [44] followed by the mineralization of the resulting 2-hydroxymuconic semialdehyde.

Most common heterocyclic ring xenobiotics that are degraded (with ring cleavage) by microorganisms contain one six-membered ring (e.g., pyridines, triazines, and pyrimidines). For instance, the degradation pathway for atrazine involves the dehalogenation and hydroxylation of the herbicide by a hydrolase of *Pseudomonas* (atrazine chlorohydrolase) [45]. Further steps, mediated by other hydrolases, can lead to mineralization (evolution of CO_2) via gradual hydroxylation to cyanuric acid [46] and aromatic ring cleavage [47] resulting in the release of biuret and then urea. The degradation of chlorpyrifos (a pyridine-based herbicide) in soil has been reported to involve both chemical and microbiological processes [48]. The major products were 3,5,6-trichloro-2-pyridinol and carbon dioxide. In soils in which the pyridinol did not accumulate, large quantities of carbon dioxide were produced, suggesting that chlorpyrifos metabolism involved cleavage of the heterocyclic aromatic ring. A *Nocardia* sp. isolated from soil transformed pyridine, pyridine *N*-oxide, and 2-hydroxypyridine with ring cleavage between carbons 2 and 3 [49].

Sulfoxidation or oxidation of organic sulfides and sulfites to sulfoxides and sulfates can occur through both biological and chemical reactions, but microbial reactions dominate the process [50]. For example, *Cunninghamella elegans* metabolized chlorpromazine and methdilazine within 72 h [51]. As determined by ^{1}H NMR, the major metabolites of chlormazine included chlormazine sulfoxide (36%) and 7-hydroxychlorpromazine sulfoxide (5%). In experiments with methdilazine, methdilazine sulfoxide was one of the major products (30%). Another example is the oxidation of *s*-ethyl *N*,*N*-dipropylcarbamothioate (EPTC) to EPTC-sulfoxide by *Rhodococcus* sp. isolated from soil [52].

Reduction, a separate category of the major microbial processes, affects the nitroaromatics, organic sulfoxides, and halogenated hydrocarbons. Nitro groups are reduced to amines via the formation of nitroso and hydroxyamino groups as was the case for organophosphorus pesticides such as parathion, paraoxon, ethyl *p*-nitrophenyl benzenethiophosphonate (EPN), fenitrothion, or hexahydro-1,3,5-trinitro-1,3,5-triazine (RDX) [53–55]. Sulfoxides can be reduced to sulfides. *Klebsiella pneumoniae*, for instance, reduced the organophosphorus pesticide fensulfothion to fensulfothion sulfide [56]. Halogenated hydrocarbons, such as dichlorodiphenyltrichloroethane (DDT), heptachlor, lindane, or methoxychlor may undergo dehydrodehalogenation. DDT is reductively dechlorinated to 1,1-dichloro-2,2-bis(p-chlorophenyl)ethane (DDD) and dehydrochlorinated to 1,1-dichloro-2,2-bis(p-chlorophenyl)ethylene (DDE) [57] by *Aerobacter aerogenes* containing reduced Fe(II) cytochrome oxidase. During incubation with municipal anaerobic sludge hexahydro-1,3,5-trinitro-1,3,5-triazine (RDX) was reduced to 1-nitroso or 1,3-dinitroso derivatives [55]. Reduction of nitroaromatics has been reported to occur aerobically or anaerobically [58]. Incubation of methoxychlor with *Aerobacter aerogenes* resulted in the reductive dechlorination of the substrate to a methoxydichloro compound [59]. The initial step of lindane degradation by cell-free extracts of *Clostridium rectum* was also shown to be reductive dechlorination [60]. Anaerobic microorganisms are involved in reductive dehalogenation of chlorinated aromatic hydrocarbons [61]. Among other reductive reactions, one can distinguish saturation of double or triple bonds, reduction of aldehydes to alcohols, reduction of ketones to secondary alcohols, and reduction of certain metals [62].

Hydrolytic reactions, another major category of microbial processes, involve many enzymes, such as esterase, acrylamidase, phosphatase, hydrolase, and lyase. Xenobiotics that have ester or amide linkages are known to be readily hydrolyzed. In the case of hydrolytic dehalogenation, a halogen is exchanged with a hydroxyl group originating from water, so the process can also be considered a hydrolytic reaction. Many fungi excrete hydrolytic enzymes into the soil environment. For instance, various esterases, which are produced by *Pseudomonas* sp., *Flavobacterium* sp., and a recombinant *Streptomyces*, have been shown to hydrolyze a number of organophosphate pesticides, such as methyl and ethyl parathion, diazinon, fensulfothion, dursban and coumaphos [63,64]. Enzymatic hydrolysis of organophosphate, phenylcarbamate and dithioate pesticides may lead to a great decrease in their toxicity [65]. The herbicide metamitron was degraded by *Arthrobacter* sp. through hydrolytic cleavage of the amide bond in the triazine ring [66]. Hydrolysis was an initial stage of the transformation of the insecticide phosalone by *Acinetobacter calcoaceticus* [67]. A cytosolic enzyme isolated from *Pseudomonas* sp. strain CRL-OK hydrolyzed the carbamate linkage of the insecticide carbaryl (1-naphthyl N-methylcarbamate) [68].

Synthetic reactions, the last of the major categories of microbial processes, lead to the formation of linkages between two molecules of the same xenobiotic, or between molecules of a xenobiotic and another compound. Conjugation reactions, such as methylation and acetylation commonly occur during microbial metabolism of xenobiotics. The herbicide metolachlor was metabolized by soil microorganisms via conjugation pathway with glutathione [69]. A similar transformation pathway was determined for alachlor [70]. During aerobic and anoxic incubation with *Pseudomonas aeruginosa*, 2,4-dinitrophenol was reduced to 4-amino-2-nitrotoluene, 2-amino-4-nitrotoluene, and 2,4-diaminotoluene [71]. The aryloamines were then involved in microbial acetylation to acetamide derivatives.

2.2 TRANSFORMATION BY EXTRACELLULAR ENZYMES

One of the aspects of microbial activity is the contribution of extracellular enzymes to the overall degradation effect. The involvement of soil enzymes in xenobiotic transformation was first inferred based on degradative processes occurring in sterile soils [72]. Microorganisms are the major source of extracellular enzymes in soil. However, considerable amounts of soil enzymes can also be derived from plant debris, plant root exudates, and soil fauna. The degradation of xenobiotics by extracellular enzymes was demonstrated in many studies involving organophosphorus and acylanilide herbicides [4,72,73]. Malathion, for instance, was hydrolyzed to a mono-carboxylic acid (via loss of an ethyl group) by a carboxyl esterase that was isolated from soil [74]; the enzyme exhibited typical Michaelis–Menton kinetics. Three parathion hydrolases were purified and characterized after their isolation from gram-negative *Flavobacterium* sp. strain ATCC 27551 and two other bacteria derived from the soil microbial community [75]. Depending on the bacteria, the hydrolases differed in their chemical structure and molecular weight and they exhibited different affinities for ethyl parathion and EPN. A parathion-methyl hydrolase detected in soil seemed to occur in more than one form [76]. The herbicide derivative 3, 4-dichloroaniline was transformed to 3,3',4,4'-tetrachloroazobenzene by a peroxidase extracted from soil [77]. Also, a propanil-degrading acylamidase was extracted from soil by three different methods [78].

When parathion hydrolase from a *Pseudomonas* sp. isolated from soil was added to soil polluted with diazinone, the enzyme degraded more than 98% of the chemical during incubation for 24 h [79]. Partially purified enzyme preparations (acylamidases) from soil microorganisms, such as *Pseudomonas striata*, *Fusarium solani*, and *Bacillus sphaericus* degraded many acylanilide, phenylcarbamic acid, and phenylurea herbicides [80]. An amidase from *Pseudomonas alcaligenes* isolated from soil was found to degrade various phenylcarbamate herbicides, such as CIPC (isopropyl-3-chlorophenylcarbamate), IPC (isopropyl-*N*-phenylcarbamate), BIPC (1-methyl-prop-2-ynyl-3-chlorophenylcarbamate), and swep (methyl-3,4-dichlorophenyl) [81].

Enzymes commonly occurring in soil, such as esterases, amidases, phosphatases, and proteases catalyze the hydrolysis of the respective chemical bonds in both xenobiotic molecules and natural products. Among enzymes extracted from soil by Tabatabai and Fu [82] were several oxidoreductases, many hydrolases and a lyase. The reactions catalyzed by these enzymes involved xenobiotic substrates, such as chloroanilines (peroxidase) or malathion (carboxylesterase).

Extracellular phenoloxidases were found to catalyze the transformation of phenolic or anilinic compounds to their polymerized products [83]. For example, 2,4-dichlorophenol, an intermediate of 2,4-D, was coupled to itself by a laccase of *Rhizoctonia praticola* to yield dimeric to pentameric products. In the presence of this laccase, 2,4-dichlorophenol reacted with various humic constituents, such as syringic acid, oricinol, or vanillin to form a variety of hybrid oligomers. Various phenols and anilines were also enzymatically incorporated to natural humic acids (HA) [84]. The above outlined oxidative coupling reactions were carried out only *in vitro*, but there are indications they may also occur in soil. Many soil microorganisms were found capable of producing and excreting enzymes, including phenoloxidases [82]. Phenoloxidases were obtained from several fungi (*Geotrichum candidum*, *Stachybotrys chartarum*, *Epicoccum nigrum*) and bacteria (*Bacillus* spp., *Pseudomonas* spp., *Arthrobacter* spp., *Chaetomium thermophilium*) that were isolated from soil or composting solid waste [85–87]. Extracellular phenoloxidases are believed to participate in humus formation by mediating the polymerization of the monomer products of microbial metabolism. They are also implicated in binding of xenobiotics to soil by catalyzing the oxidation of pollutants to free radicals, followed by chemical coupling of the oxidation products to organic matter. There are indications that participation of enzymes in humification and binding may be triggered by the very presence of toxic substrates that induce the production of extracellular phenoloxidases [21].

3 ABIOTIC CATALYTIC TRANSFORMATION

3.1 THE INVOLVEMENT OF MINERAL SURFACES IN XENOBIOTIC TRANSFORMATION

Transformation of xenobiotics in terrestrial systems is greatly influenced by mineral components of soil [88]. Most organic chemicals, including xenobiotics, exhibit a strong affinity to HS. However, clay minerals, with their high concentration in soil, large surface area and relatively high charge density, contribute to the overall xenobiotic transformation at least as much as does the organic matter.

Transformation processes occurring in soil are the function of the multicomponent association between HS, clays, metal oxides, $CaCO_3$ and other minerals [9]. According to various estimates, as much as 90% of SOM may be associated with soil minerals [89,90]. In soils of organic matter content greater than a few percent, mineral surfaces are believed to be completely blocked by HS and do not participate in the adsorption of xenobiotics [91]. This percentage of organic matter may be underestimated, because in the case of atrazine, mineral soil surfaces were involved in adsorption when SOM content was as high as 6–8 % [11]. On the other hand, there are indications that the surfaces of smectites in the interlayer spaces are largely uncovered by organic matter [11]. Interactions between different solid components reduce the surface for abiotic transformations. The external surfaces of clay minerals may be covered by organic matter and mineral oxides that bind or replace exchangeable cations responsible for many transformation reactions.

Both the spatial distribution of ions and the charge distribution within polarizable species are strongly influenced by the electric field created by charged surfaces. Consequently, the tendency of organic xenobiotics to undergo abiotic transformation is often much higher near the surface than in the bulk solution. Molecules in direct contact with the surface can undergo many transformations catalyzed by the adsorption sites [9]. On the other hand, some reactions are inhibited by proximity of the surface. Organic pollutants (OP) in the adsorbed state are in general more persistent than OP in the free state [92–94].

The significance of soil mineral-catalyzed abiotic transformation of pesticides in the environment has become widely recognized only recently [8,95,96]. Even in soil environments where biological activity is intense, abiotic transformation is of importance. Abiotic and biotic transformations of many pesticides take place simultaneously [96].

Soil organic matter can induce surface-catalyzed reactions of adsorbed pesticides, but theoretically it could also hinder the degradation of some pesticides by decreasing both their availability to microbial attack and their concentration in the soil solution [96]. The processes of adsorption and abiotic degradation of xenobiotics through the action of the surfaces of soil minerals vary with the structural and surface properties of the minerals, saturating cations and their hydration status, molecular structures of xenobiotics, and associated environmental factors [96].

In addition to clay minerals with well ordered crystalline structures, soils contain mineral colloids, such as metal oxides, aluminosilicates and carbonates, that exist in poorly ordered forms [88]. The contribution of these short-range ordered (SRO) colloids to the transformation of organic chemicals on mineral surfaces is difficult to estimate. The poorly ordered colloids appear to play a dual role in humification and binding interactions between xenobiotics and soil. First, they may be directly involved in the oxidation of organic chemicals of organic and anthropogenic origin with the formation of the desorption-resistant organic residues [97,98]. Second, they may influence the efficiency of enzymes that contribute to the same reactions [99]. The SRO mineral colloids may also have an effect on abiotic transformation of xenobiotics through the formation of their coatings on crystalline minerals and organic matter and the subsequent alteration of surface chemistry of soil components [88].

3.2 OXIDATIVE TRANSFORMATION OF ORGANIC COMPOUNDS

Soil minerals play an important role in catalyzing the abiotic polymerization of the natural phenolic compounds and the subsequent formation of HS [8,88,100–102]. Manganese oxides, which occur in soils and sediments [103], are very reactive in promoting the humification [104]. They act as Lewis acids to accept electrons from diphenols, resulting in their oxidative polymerization. The rate-determining step in the formation of HA from phenols is apparently the formation of a semiquinone radical involving a single electron transfer reaction [105]. Semiquinones couple with each other to form a stable HA polymer. The coupling of free radicals requires little activation energy, in contrast to electron transfer reactions [106]. Coupling of semiquinones, rather than the formation of quinones, should thus be kinetically the reactive pathway in the transformation of phenolic compounds to HS.

The catalytic ability of Fe oxides in the rapid oxidative polymerization of phenolic compounds increases in the following order: ferrihydrite > goethite > maghemite > lepidocrocite > hematite [88]. Ferrihydrite, which is an SRO Fe oxide with high specific surface, is most reactive in catalyzing the oxidative polymerization reaction. Besides the nature of Fe oxides, the catalytic ability of Fe oxides is related to the structure and functionality of phenolic compounds [107].

The oxidative polymerization of phenolic compounds is substantially influenced by the catalysis of Al hydroxides [108]. The Al^{3+} ions apparently promote delocalization of electrons from phenolic oxygen atoms into the π orbital system through displacement of protons from the phenolic groups. This catalyzes the oxidation of phenolic compounds.

Manganese(IV) oxides are most reactive among metal oxides in catalyzing abiotic oxidation of phenolic compounds [104,107]. The comparison of catalytic effects of Mn oxides and tyrosinase on the oxidative polymerization of diphenols at pHs common to soil environments indicated that the relative catalytic effects of Mn(IV) oxides and enzymes such as tyrosinase would vary with the type of phenolic compound [109].

Manganese oxides have been shown to oxidize rapidly phenolic acids to form a number of soluble products [110]. Mass spectrometric data show that some of the soluble products of the reaction have somewhat higher molecular weights than the parent compounds. The soluble products of the reaction of ferulic acid and MnO_2 do not contain any ferulic acid-derived polymers. The oxidized products of ferulic acid are apparently rapidly sorbed on the surfaces of MnO_2. FTIR data [111] show that in the birnessite–catechol system, some of the reaction products are adsorbed on the mineral surface, with the spectrum of the reaction products resembling that of synthetic HA [112] obtained from catechol by oxidation with sodium periodate [113]. Furthermore, mass spectrometric data indicate that in the abiotic catalysis by birnessite, adsorption of reaction products on the mineral surfaces limits the extent of polymerization, favoring the formation of components with lower degrees of aromatic ring condensation and lower molecular weights compared with those generated by catalysis of tyrosinase [111].

The redox reaction between birnessite and anilines may represent a pathway for transformation of aromatic amines in soil environments in the absence of oxygen and substantial microbial activity [114]. The reaction rate increases with decreasing pH and was first order with respect to birnessite and the substrate. The principal oxidation products of aniline and p-toluidine are azobenzene and 4,4'-dimethylazobenzene. Substituted anilines can be relatively resistant to microbial degradation, so the birnessite oxidation may be considered an important transformation pathway.

Clay-size layer silicates also have the ability to catalyze the oxidation of phenolic compounds [8,115]. For instance, aromatic radical cations form on the intracrystal surfaces of transition metal-saturated layer silicates [116]. When reacted under moderate conditions with Cu(II) or Fe(II) ions, aromatic molecules donate electrons to the metal cations, leading to the formation of polymers [117]. Layer silicates accelerate the formation of HA to varying degrees [118]. The promoting effect of 2:1 layer silicates is higher than that of 1:1 layer silicates. This is attributed to the larger specific surface and lattice imperfections of the former than the latter, which favor the adsorption of O_2 molecules or radicals.

Calcium-saturated nontronite, which is an Fe(III)-containing 2:1 layer silicate, can convert hydroquinone to humic macromolecules in the interlayers; most of the interlayer humic macromolecules are highly resistant to alkali extraction and may thus be humin type materials [119]. The ability of nontronite to promote the oxidation of phenolic compounds is related to the structure and functionality of these compounds [120]. Catechol with two o-OH groups is evidently more easily cleaved than hydroquinone with two p-OH groups, while pyrogallol with three hydroxyls in adjacent positions is even more easily cleaved than catechol.

Primary minerals common in soil environments also have the ability to catalyze the oxidative polymerization of phenolic compounds [121]. Among primary minerals, the degree of acceleration of the oxidative polymerization of hydroquinone is greatest in the tephroite system, which increases the total HA yields more than ninefold because: (1) tephroite (ideal chemical formula, $MnSiO_4$) is an Mn-bearing silicate, (2) part of the Mn in tephroite is present in the higher valence states, and (3) the oxidation of diphenols by Mn(III) and Mn(IV) is thermodynamically favorable. Therefore, the catalytic role of primary minerals in the oxidative polymerization of phenolic compounds should not be overlooked.

SRO aluminosilicates such as allophane have been shown to catalyze the oxidative degradation of phenolic compounds [88]. However, the catalytic ability of other SRO aluminosilicates remains obscure. The role of these SRO aluminosilicates in the transformation of other organic compounds in soils has yet to be uncovered.

Many other organic compounds such as salicylic acid and pyruvic acid are degraded by mineral colloids such as Mn oxides by electron transfer reactions [122–124]. Mn oxides can even oxidatively transform monophenolic compounds, particularly those containing electron donating substituent groups on the aromatic ring [125,126].

Organic compounds such as 2,4-D and ethyl ether can be degraded by the catalysis of birnessite [127]. These organic compounds can be adsorbed on birnessite and rapidly oxidized, both producing CO_2 as a major product (Figure 1), but by somewhat different mechanisms. In the case of 2,4-D only, methanol-extractable Mn^{2+} is detected. Therefore, the solid degradation of organochlorine herbicides can occur by the catalysis of birnessite, which is a common component in soil environments.

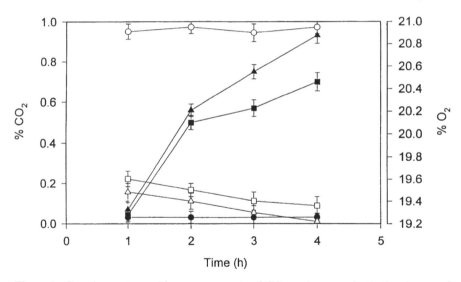

Figure 1. Gas chromatographic measurements of CO_2 and oxygen in the headspace of reaction vials containing birnessite and 2,4-D at 1, 2, 3 and 4 h, respectively. The values indicate the mole percent of CO_2 and oxygen in the headspace gas. •, CO_2 from air control; ■, CO_2 from ethyl ether control; ▲, CO_2 from 0.3 μmol 2,4-D plus ethyl ether sample: O, oxygen from air control; □, oxygen from the ethyl ether control; △, oxygen from 0.3 μmol 2,4-D plus ethyl ether. The error bars indicate the standard error for two measurements. Reprinted from [127], Copyright (1996), with permission from Elsevier Science

3.3 BRØNSTED AND LEWIS ACIDITY OF MINERAL SURFACES AND HYDROLYSIS OF ORGANIC COMPOUNDS

Mineral surfaces may detoxify adsorbed OP by catalysis through the ability of mineral colloids to behave as Brønsted acids and donate protons or to act as Lewis acids and accept electron pairs. Brønsted acidity derives essentially from the dissociation of water molecules coordinated to surface-bound cations. Therefore, this acidity is strongly influenced by the hydration status and polarizing power of surface-bound and structural cations on mineral colloids [128]. Lewis acidity arises from structural ions such as Al and Fe exposed at the edges of minerals [129].

The surfaces of mineral colloids have the ability to protonate many uncharged organic molecules. The Brønsted acid strength is related to the electronegativity and polarizing power of the exchangeable and structural metal cations: $H^+ > Al^{3+}$, $Fe^{3+} > Mg^{2+} > Ca^{2+} > Na^+ > K^+$ [129]. Highly polarizing cations (small, high charge) and low water contents promote Brønsted acidity.

Certain pesticides are degraded by catalysis of mineral colloids, which was first reported by formulation chemists who used clays as carriers and diluents [130]. Degradation of pesticides on clays by surface catalysis was also found for several organophosphate and s-triazine pesticides and attributed to the surface acidity of mineral colloids [131–133]. The nature of the mineral colloid, its saturating cation, and hydration status determine the rate and mechanism of degradation of parathion [134].

The hydrolysis of quinalphos (O,O-diethyl O-quioxalin-2-yl phosphorothioate) on homoionic Cu(II)-, Fe(III)-, Al-, Na-, K-, and Ca-bentonites in hydroalcoholic solution involved two different mechanisms, depending on the nature of the exchangeable cations [135]. With Cu-, Fe-, and Al-clays, the pesticide hydrolyzed to 2-hydroxyquinoxaline, whereas O-ethyl O-quinoxalin-2-yl thiophosphoric acid was the main product of hydrolysis on the Na-, K-, and Ca-clays.

The amino form of 3-aminotriazole can be converted to the imino form [136]. Triazine compounds can also be protonated on dry clay minerals. Soil catalytic degradation of atrazine through hydrolysis is an important pathway [137]. Surface acidity of minerals also catalyzes hydrolysis of the chloro-s-triazine herbicides to the non-phytotoxic 2-hydroxy-s-triazines [138]. This reaction deserves attention in a wide range of other organic compounds in soil environments [129,139].

Besides the Brønsted acidity, the Lewis acidity of metals is important in mineral-catalyzed hydrolysis reactions. Aluminum and Fe oxides in water and especially in the dry state have the ability to catalyze organic hydrolysis reactions, at least for those that are hydroxyl ion-catalyzed [140]. Dissolved metals and metal-containing surfaces play a significant role in the catalysis of the hydrolysis of OP. The ability of a metal ion to catalyze the hydrolysis varies with its ability to complex with reactant molecules and shift electron density and conformation in ways favorable to the reaction [141]. Metals vary in their complex formation constants, which reflect differences in metal–ligand bond strengths and solvation forces [142].

The reactivity of hydrolyzable OP arises from the presence of electrophilic (electron deficient) sites within the molecules [142]. The S_N2 mechanism (nucleophilic substitution) involves attack of the electrophilic sites by OH^- or H_2O, generation of a higher number intermediate, subsequent elimination of the leaving group, and the formation of a hydrolysis product (X and Y in this mechanism are elements or functional groups attached to C) [142]:

S_N2

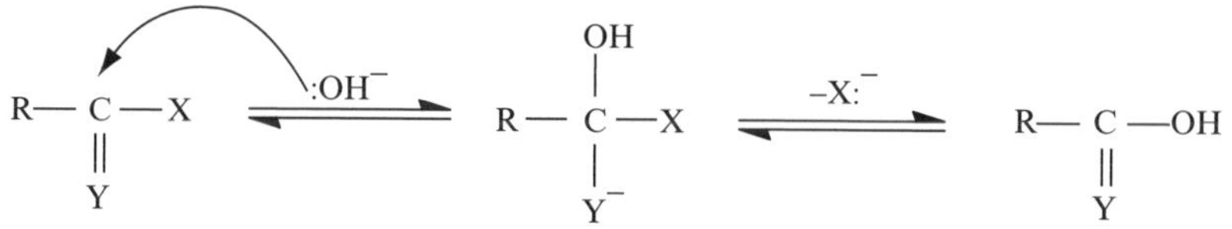

In the case of the S_N1 mechanism (nucleophilic substitution monomolecular), the reaction proceeds with the loss of the leaving group to generate a lower coordination number intermediate and then the generation of the hydrolysis product by nucleophilic addition (X, Y and Z in this mechanism are elements or functional groups attached to P) [142]:

S_N1

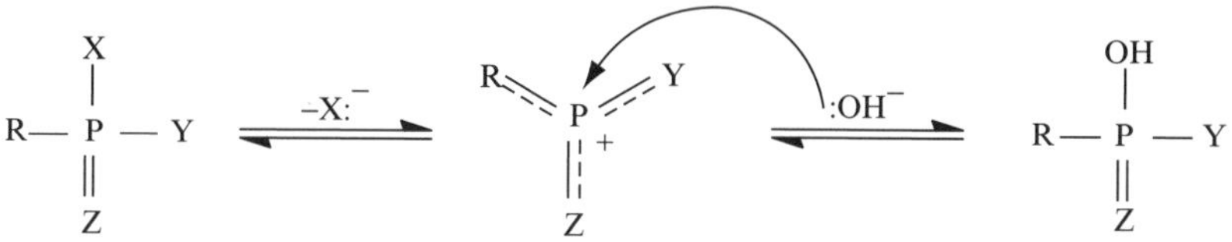

Metal ions can catalyze hydrolysis in a way similar to acid catalysis [142]. Metal ions and protons coordinate to OP so that the electron density is shifted away from the site of nucleophilic attack to facilitate the reaction. Protons have an extremely high charge density and great polarizing power; metal catalysis is, thus, insignificant in acidic conditions. By contrast, metal ions can readily coordinate two or more ligand donor sites on a molecule and can greatly outnumber protons in neutral and alkaline conditions [143].

The hydrolysis of organic compounds that possess good leaving groups is limited by the rate of nucleophilic attack. These organic compounds are susceptible to types 1, 3, 4, 6 and possibly 5 of metal catalysis (Table 1). The hydrolysis of organic compounds that possess poor leaving groups is limited by the breakdown of the tetragonal intermediate and, therefore, types 2, 7 and

Table 1. Mechanisms of metal catalysis. Reprinted with permission from [142]. Copyright CRC Press, Boca Raton, Florida

Type 1	The metal coordinates the electrophile, shifting the electronic distribution in the molecule in a way that enhances its reactivity.
Type 2	The metal coordinates to the leaving group, increasing its leaving ability.
Type 3	The metal acts as a center for simultaneous attachment of both the electrophile and attacking nucleophile (template effect).
Type 4	The metal coordinates the nucleophile and induces deprotonation (which increases the reactivity of the nucleophile).
Type 5	Coordination of the substrate with the metal induces conformation changes that facilitate reaction.
Type 6	Coordination of the substrate with the metal makes the molecule more positive, lessening unfavorable electrostatic interaction with the nucleophile.
Type 7	Coordination with the metal blocks inhibitory reverse reaction paths, such as (1) loss of OH^- from a tetrahedral intermediate instead of loss of the leaving group X^-, or (2) nucleophilic attack by X^-.

possibly 5 are important. A number of pesticides and other OP are in the intermediate region, where metal catalysis may shift the rate-limiting step from nucleophilic attack to breakdown of the tetragonal intermediate or vice versa. Many organic compounds are susceptible to metal catalysis. These include carboxylic acid ester, amides, anilides, phosphate-containing esters, and other hydrolyzable compounds [139,142].

Many metals that have catalytic ability form low-solubility inorganic solids in the pH range of soil environments. Surface-bound metals, thus, must be accessible to the reaction with solute species for a metal-catalyzed effect to be initiated. Research has established that metal oxides/hydroxides act as hydrolysis catalysts for phosphate esters [144]. Much has been subsequently reported about the reactivities of naturally occurring solids such as oxides, carbonates, sulfides and aluminosilicates [142,145].

The surfaces of metal oxides, such as TiO_2, α-FeOOH, or Al_2O_3, catalyzed the hydrolysis of the thionate (P=S) and oxonate (P=O) forms of chlorpyrifos-methyl [O,O-dimethyl O-(3,5,6-trichloro-2-pyridinyl)phosphorothioate and O,O-dimethyl-O,3,5,6-trichloro-2-pyridil phosphate] in buffer solutions (pH 2–5) [146]. Paraoxon (O,O-diethyl O-nitrophenyl phosphate) was also subject to surface-catalyzed hydrolysis [146]. Metal hydrolysis is believed to occur through (1) metal complexation of the thionate-S or oxonate-O, which increases the electrophilicity of the P atom; (2) induced deprotonation of metal-coordinated water, generating metal-hydroxo species that can serve as nucleophiles, and (3) metal coordination of the leaving group, facilitating its exit [146].

Aluminosilicate minerals have been observed to catalyze a number of hydrolysis reactions. Surface-catalyzed parathion hydrolysis, for example, has been observed following addition of kaolinite and soil samples [132,134,146,147].

Bidentate complex formation by interlamellar layer exchangeable cations is believed to be responsible for montmorillonite-catalyzed phosmet [*O,O*-dimethyl phosphorodithioate *S*-ester with *N*-(mercaptomethyl) phtalamide] hydrolysis [146,148].

The type of exchangeable cations (e.g., Fe(III), Al, Ca, and Na) within montmorillonite has been shown to substantially affect pathways and rates of fenoxyprop-ethyl [(±)-ethyl 2-[4-(6-chloro-2-benzoxazolyloxy)-phenoxy]propanoate] hydrolysis [145,149]. The hydrolysis of 1-(4-methoxyphenyl)-2,3-epoxypropane in the presence of montmorillonite and kaolinite was influenced by clay water content, the type of the exchangeable cations within the clay, and the extent of ester adsorption [150]. In studies involving minerals composed of a single ion, monophenyl terephthalate hydrolysis was found to be catalyzed by Al_2O_3 [151]. The observed decrease in the rate of surface-catalyzed reaction with increasing ionic strength can be only partially explained by a decrease in adsorption of the anionic monophenyl terephthalate substrate; a lowering in OH^- concentration near the surface with increasing ionic strength may also be responsible.

The neutral substrate phenyl picolinate is susceptible to surface-catalyzed hydrolysis by TiO_2 and α-FeOOH, but not to catalysis by Al_2O_3 and Fe_2O_3 [146,152]. The intensity of the TiO_2-catalyzed reaction to changes in pH and ionic strength suggests that hydrolysis by OH^- attack, which is predominant in catalyst-free solution, is superseded by hydrolysis via H_2O attack in the presence of a catalyst.

Phenyl phosphate, a monoester dianion, has been found to adsorb onto mineral surfaces composed of Ti(IV), Mn(IV), and Fe(III) and is subject to surface-catalyzed hydrolysis [153].

Surface catalysis by metals on minerals is observed when all participating reactants are adsorbed to a significant extent and when rate constants for the reactions at the mineral–water interface are comparable with or exceed rate constants for the reaction in homogeneous solution. However, relatively little is known about the conformation and stoichiometry of adsorbed species. This hampers our fundamental understanding of surface catalysis. Nevertheless, based on the current state of the art, two generalizations can be made: (1) auxiliary donor groups that facilitate metal catalysis by metal ions in solution should also facilitate surface catalysis, and (2) phenomena of only secondary importance in reactions of dissolved complexes, e.g., electrostatic and hydrophobic interactions, may play a much greater role in surface catalysis [142].

3.4 CATALYTIC EFFECTS OF HUMIC SUBSTANCES ON THE TRANSFORMATION OF ORGANIC POLLUTANTS

Soil organic matter (SOM), and especially HS are known to exert catalytic or inhibitory effects on the abiotic hydrolysis of several OP [14]. In aqueous media, HA and fulvic acids (FA) were able to enhance the acid hydrolysis of phenoxy acetic acids and esters [154], and chloro-s-triazines [155,156] and to retard the alkaline hydrolysis of the n-octyl ester of 2,4-D (2,4-DOE) [157]. The rates of pH-independent, non-acid-base-catalyzed, abiotic hydrolysis of organophosphorothioate esters, such as chlorpyrifos and diazinon, halogenated alkenes, aziridine derivatives, and others, in aqueous systems was apparently not affected by the presence of natural organic matter (NOM) [158].

The mechanism proposed to explain the catalytic effects of HA and FA on the dechlorohydroxylation of the chloro-s-triazines, simazine, atrazine, and propazine consists of the interaction through H-bonding between the surface carboxylic groups of the HA and FA molecules and the heterocyclic nitrogen atoms of the triazine ring (Figure 2) [14,155,156]. This interaction would strengthen the electron-withdrawing effect of the electron-deficient carbon bearing the chlorine atom, thus reducing the activation energy barrier for the hydrolytic cleavage of the C-Cl bond and facilitating the replacement of chlorine by the weakly nucleophilic water (Figure 2). This mechanism would also explain the correlation found between soil-catalyzed chemical hydrolysis of chloro-s-triazines and the amount adsorbed by SOM [159]. No evidence of catalytic action was observed for the weakly acidic functional groups of FA (high pK_a values), such as phenolic hydroxyl groups, and carboxylate groups [160,161]. Thus, undissociated carboxyl groups were suggested to be the only catalytically active sites of SOM for the abiotic hydrolysis of H-bonded chloro-s-triazines.

A first-order kinetics was determined for hydrolysis of atrazine in aqueous suspensions of HA and aqueous solution of FA at pH $\leq$ 7 [155,156]. An increase of FA concentration caused a higher hydrolysis rate constant and shortened the half-life of atrazine, but had no effect on the activation energy. The latter increased with an increase of the pH of the system, possibly as a

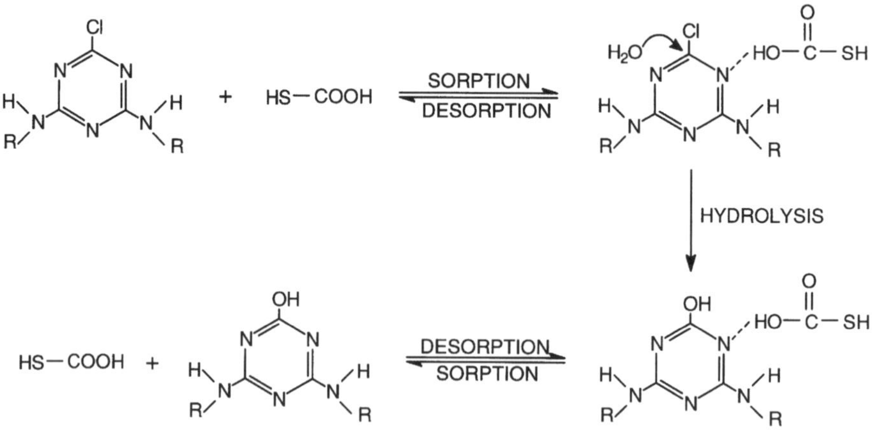

Figure 2. The mechanism of the catalytic effects of humic substances on the dechlorohydroxylation of the chloro-s-triazines. Reprinted with permission from [14]. Copyright CRC Press, Boca Raton, Florida

result of the decreased concentration of undissociated carboxylic groups [156]. Since no correlation was found between the rate of atrazine hydrolysis and the number of carboxylic groups of different HA, not only the number but also the arrangement of these groups in the HA molecule were believed to affect the catalytic effect of HA [155].

In the presence of HS, the rate of base-catalyzed hydrolysis of 2,4-DOE in aqueous systems at pH 9–10 was retarded by a factor equal to the fraction of the ester associated with the HS [157]. These observations were consistent with those for unreactive HS-bound 2,4-DOE in equilibrium with reactive aqueous-phase 2,4-DOE, and were interpreted in terms of changes in solution chemistry in the vicinity of the negatively charged surfaces of HS molecules at alkaline pHs [157].

An overall model for the catalytic effects of HS on the hydrolysis kinetics of hydrophobic OP was proposed by Perdue [162]. The model consisted of a combination of equations describing separately partitioning equilibria, micellar catalysis and general acid–base catalysis. Based on this model it was predicted that the overall catalytic effects of HS could be almost totally described in terms of micellar catalysis and partitioning equilibria, with only a minor contribution from general acid–base catalysis. The effect of general acid–base catalysis remained relatively unimportant even in the presence of rather high concentrations of HS. The model has been successfully tested with experimental data obtained for the base-catalyzed hydrolysis of 2,4-DOE [157] and the acid-catalyzed hydrolysis of atrazine [155]. Apparently, HS behave in a way similar to anionic surfactants. The catalytic effects were attributed to electrostatic stabilization of the transition state for the acid catalysis, in which the substrate becomes more positively charged, and to the destabilization of the transition state for base-catalyzed hydrolysis, in which the substrate becomes more negatively charged [162]. The model was also used to predict that HS may not affect significantly the base-catalyzed hydrolysis rate of OP that associate weakly with HS, such as parathion, whereas for OP that associate strongly with HS, such as DDT, the effect of HS is expected to be significant [162]. In organic soils, soil solutions, and sediments, where much higher concentrations of HS can be present, the impact of HS on hydrolysis kinetics of OP is, therefore, expected to be significant. In particular, hydrolysis reactions of DDT can be strongly retarded in these environments.

Inhibition by HA of hydrolytic enzymes in soil was indicated as an additional indirect effect that may operate in the hydrolysis reactions of specific OP [163].

3.5 PHOTOSENSITIZATION EFFECTS OF HUMIC SUBSTANCES ON ORGANIC POLLUTANTS

Photochemical transformations (photoreactions) induced by sunlight are important pathways in the abiotic degradation of OP in the top layer of soil and water [164]. Photoreactions may involve two distinct processes: direct and indirect photolysis. The former is initiated through direct absorption of sunlight by the OP molecule, which becomes energetically excited, and can then be transformed to photoproducts. Most OP, however, absorb radiation at less than 290 nm; sunlight at these wavelenghts is largely absorbed by the ozone layer and does not reach the soil. For that reason, indirect photolysis is more important than direct photolysis. In indirect photolysis, light is initially absorbed by a compound other than the OP molecule, which may then either transfer the energy to the OP molecule (photosensitization) or produce specific photoreactants (e.g., singlet oxygen) which may react with the target OP (photoinduction).

Indirect photochemical processes are particularly efficient on soil surfaces, where sunlight can be absorbed by both organic and inorganic chromophores present in soil particles and initiate or sensitize phototransformations of OP. Although the exact identity of these species is unknown, several clay minerals, metal oxides and oxyhydroxides, transition metal ions and various HS fractions have been suggested to be involved in indirect photolysis of OP in the environment [165].

Humic substances can strongly absorb sunlight and thus behave as potent and polyvalent photosensitizers and photoinitiators of several photoreactions involving nonabsorbing, photochemically stable OP. For example, photolysis of atrazine in irradiated aqueous solution was more extensive in the presence of 0.01 % dissolved FA than in its absence [166]. The reaction followed a first-order kinetics with respect to the herbicide concentration, and proceeded through the initial hydrolysis of atrazine to the 2-hydroxy analogue, followed by rapid partial N-dealkylation and then by complete N-dealkylation of the molecule (Figure 3). Photolysis of UV-irradiated ($\lambda = 254$ nm) aqueous solutions of prometryn also followed a first-order reaction kinetics in water and HA suspension at pH 3, whereas a second-order kinetic was measured in the presence of dissolved HA at pH 6 and 8 and FA at pH 3, 6, and 8 [167]. An additional, phyto-toxic dealkylated product, 4-amino-6-(isopropylamino)-1-triazine, was detected when the photolysis of prometryn was performed in aqueous solutions of HA or FA. Hydroxyl radicals photogenerated by UV irradiation of dissolved HA and FA were suggested as responsible for the dealkylation of the 2-hydroxy analogue of prometryn [167].

Humic substances are unanimously considered to be the primary source for the production of highly reactive, short-lived photoreactants such as the solvated electron, singlet oxygen, superoxide radical anion, peroxy radicals, hydroxy radicals, hydrogen peroxide, and various oxidoreductive species, including the photoexcited HS molecule and HS organic radicals. However, HS can also function as a scavenger of some phototransients such as hydroxyl radicals. Sunlight-induced formation of photoreactive species in the presence of HS and other fractions of dissolved and colloidal SOM and NOM has been reviewed recently [14,165,168–170].

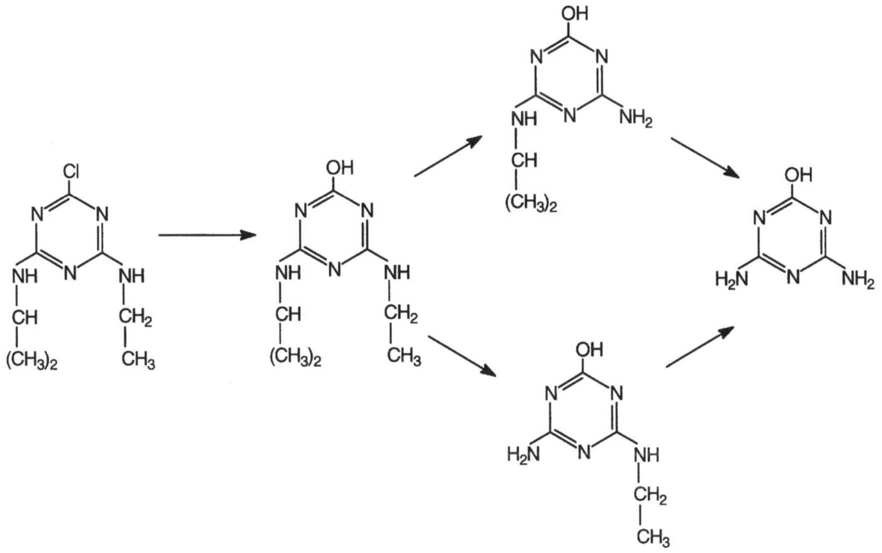

Figure 3. Photolysis of atrazine in irradiated aqueous solution of 0.01% fulvic acid leading to dechlorination, hydroxylation and N-dealkylation of the molecule. Reprinted from [166] by courtesy of Marcel Dekker Inc.

The solvated electron is a powerful reductant that may react rapidly with electronegative OP such as chlorinated organics, e.g., dioxins, that are dehalo-genated [168,169]. Singlet oxygen and superoxide radical anions are efficient (but selective) photooxidants of various OP including dissociated forms of phenolic compounds, e.g., chlorinated phenols, cyclic dienes and sulfur compounds [171]. For example, these species, produced in the presence of HS, are able to photooxidize (to sulfoxides) the sulfide groups of the thioether insecticides disulfoton, fenthion, methiocarb, and butocarboxim [164]. Peroxy radicals derived from HS are important photooxidants for alkylphenols [169].

The photoexcited HS parent molecule can also undergo direct photochemical reactions with some OP, through three major possible pathways that are: energy transfer (or photosensitization), charge transfer, and photoincorporation [168]. The first pathway is a mechanism of indirect photolysis, in which photoexcited HS molecules in the triplet state transfer energy to previously-bound OP molecules that cannot absorb sunlight themselves, thus acting as sensitizers of the OP photodegradation. This mechanism, referred to as 'static photosensitization' is similar to that followed by biological photosensitizers such as riboflavin. Up to half of the triplet-state HS molecules are estimated to have energies sufficiently high to photosensitize reactions of various OP, including DDT, polycyclic aromatic hydrocarbons (PAH), nitroaromatic compounds, polyenes, and diketones [172].

Photoinduced electron or hydrogen transfer (charge transfer) reactions, likely involving triplet-state HS intermediates [172], may occur between irradiated electron donor HS and polyaromatic electron acceptors such as PAH and paraquat [168]. Investigations conducted by electron spin resonance (ESR) spectroscopy have shown that visible and UV light irradiation of HS may enhance the concentrations of organic free radicals in HS [173], which are highly susceptible to free radical-mediated interactions with OP. The free radical increase measured by ESR in several donor–acceptor systems, such as HA–s-triazine and HA–urea herbicides, was ascribed to the unpairing of electrons originating from the formation of charge transfer complexes under the effect of light [174,175].

Finally, direct photoincorporation of OP, such as polychlorobenzenes, into the HS molecule may also occur through radical combination or cycloaddition [168].

4 BINDING PHENOMENA OCCURRING ON MINERAL AND HUMIC SURFACES AND THEIR EFFECT ON POLLUTANT BIOAVAILABILITY AND DEGRADATION

4.1 BINDING OF ORGANIC POLLUTANTS

Soil is a complex system; therefore, it is not an easy task to assign specific mechanism to sorption phenomena occurring in terrestrial environments. The current knowledge in this area has been established based on batch equilibrium experiments and the application of the mathematical equations developed to describe adsorption isotherms representing the amount of adsorbed chemical as a function of the concentration of that chemical in solution at equilibrium. Organic pollutants that were tested using the above outlined approach may be grouped into cationic, basic, acidic, and nonionic and nonpolar compounds. The testing was carried out using a variety of soil materials [95,176].

4.1.1 Organic Cations

There are few examples of organic cations that are used as pesticides or occur as other pollutants in soils. Two bipyridilium quarternary salts, diquat and paraquat, are the best known examples. These two herbicides are very soluble in water because of their ionic nature. Diquat and paraquat are adsorbed on montmorillonite and kaolinite in amounts approaching the exchange capacity of the respective layer silicates [177]. Although the cation exchange is the principal mechanism underlying the adsorption of these organic cations on clays and soils, other adsorption mechanisms such as H-bonding and ion–dipole interactions may be involved. Adsorption of cationic pesticides on clays is affected by their molecular weight, functional groups, and molecular configuration [178]. UV and IR spectroscopic data indicate that charge transfer between the organic cation and the anionic silicate framework may also be a reaction mechanism. Thermodynamic and spectroscopic studies indicate significant differences between the bonding energies of diquat and paraquat and their molecular configurations in the adsorbed state on mineral colloids [179]. The adsorption isotherms of these two organic cations on montmorillonite and kaolinite belong to the H or high affinity class [180].

Organic matter adsorbs diquat and paraquat less strongly than do layer silicates [129]. How strongly organic matter adsorbs these organic cations depends on the nature of the cation that initially occupies the exchange sites. The weakly held Na^+ is easily displaced by organic cations; Ca^{2+} is less easily exchanged from HS by organic cations, partially a result of stronger association of higher charge cations with carboxylate groups. Further, higher charge cations such as Ca^{2+} and Al^{3+} also electrostatically interconnect the macromolecules of soil HS, restricting diffusion of paraquat to exchange sites that become buried in the aggregated structure [181]. The conformation and dimensions of HS under the influence of ionic environments determine the availability of the adsorption sites.

Diquat and paraquat were found to become partly inactivated in highly organic soils [182]. Other cationic pesticides, such as the fungicide phenacridam chloride, the germicide thyamine, the plant growth regulator phosphon and chlordimeform, showed high levels of adsorption on SOM [183]. Adsorption of cationic compounds on SOM has been generally described by nonlinear, L-shaped Freundlich and Langmuir isotherms [183,184].

IR spectra [182–185] and potentiometric titration [186] indicated that ionic binding, via cation exchange, is the dominant mechanism for adsorption of diquat, paraquat and chlordimeform by HS. Divalent diquat and paraquat were suggested to react with two negatively charged sites on HS, e.g., two carboxylate groups or a carboxylate group and a phenolate group. However, HA and FA apparently retained paraquat and diquat at levels that were considerably lower than their exchange capacity [182]. Thus, not all negatively charged sites on the HS molecule appear to be positionally available to bind these large organic cations, probably because of steric hindrance effects. Adsorption of phosphon and phenacridane chloride through ionic bonding onto SOM is also reported [183].

Data from IR and ESR analyses of the interaction products of paraquat, diquat and chlordimeform with HS suggested the possible formation of charge-transfer complexes between the electron-acceptor, deactivated bipyridilium ring and electron-donor structural moieties of the HS molecule [182,185].

The potential for leaching and groundwater pollution of cationic organics should be extremely low in soils containing reasonable levels of layer silicate colloids and organic matter. However, strong adsorption in the interlayers of expandable layer silicate complexes and/or micropores of HS could protect the cationic organics from microbial attack and degradation.

4.1.2 Basic Compounds

Basic organic compounds become positively charged when they are protonated. One of the most important groups of basic compounds used in agriculture is the s-triazines, which include common herbicides such as atrazine, simazine, and prometone. These herbicides are strongly retained in soils.

Basic herbicides can be adsorbed by soil mineral colloids as cationic species through protonation of the nonionic molecule. Protonation is controlled by the pK_a of the compound (R), according to:

$$pK_a = \log[RH]/[R] + pH$$

Furthermore, the adsorption is also conditioned by the surface acidity of the clay [187]. Surface acidity can be about four pH units lower than the pH of the bulk solution [188]. The higher surface acidity of montmorillonite surfaces compared with the bulk solution can explain the strong adsorption of weak bases at a pH higher than their dissociation constants. The extent of protonation depends on the nature of the exchangeable cation on the mineral colloids. For example, the reaction is maximal for Al-saturated clay colloids and decreases in roughly the same order as the polarizing power of the cations (Table 2).

Table 2. Extent of protonation of 3-aminotriazole by montmorillonite saturated with different cations. Reprinted with permission from [136]. Copyright (1968) American Chemical Society

Amount of 3-aminotriazole added	Amount of 3-aminotriazolium ion formed (% of cation exchange capacity)*						
	Interlayer cation						
	Na^+	NH_4^+	Ca^{2+}	Mg^{2+}	Al^{3+}	Cu^{2+}	Ni^{2+}
25	6	17			23	3	22
50	6	28	18	29	47	12	46
100	6	48	23	45	100	32	42
200	6	49	21	42			

* Based on the optical density of the 1696 cm^{-1} band of 3-aminotriazolium ion, using montmorillonite saturated with the organic cation as standard. Film weights were standardized either by weighing or by measuring the optical density at 800 cm^{-1} due to the silicate structure.

Relatively little information is available on the interactions between herbicides and SRO metal oxides, hydroxides, and oxyhydroxides. Huang *et al.* [189] reported that, after the organic matter was destroyed, the removal of sesquioxides by the dithionite–citrate–bicarbonate (DCB) treatment caused further substantial reduction in the adsorption of atrazine by the soils (Figure 4). Their data indicate that, besides organic matter, sesquioxides provide adsorption sites for atrazine. However, the amount of extractable Al and Fe in different particle size fractions is not proportional to the extent of adsorption, suggesting that different forms of sesquioxides could have different reactivities for atrazine. The adsorption capacity of sesquioxidic components for atrazine is attributed to their high specific surface and proton donor functional groups.

The biological activity and transport in soil of *s*-triazine and triazole herbicides have been shown to be negatively correlated with the level of SOM, due to adsorption phenomena [183,190]. Adsorption of *s*-triazines by organic soil colloids was influenced by pH and generally showed nonlinear L-shaped isotherms [160]. Adsorption of the imidazolinones imazaquin and imazethapyr on soil HAs

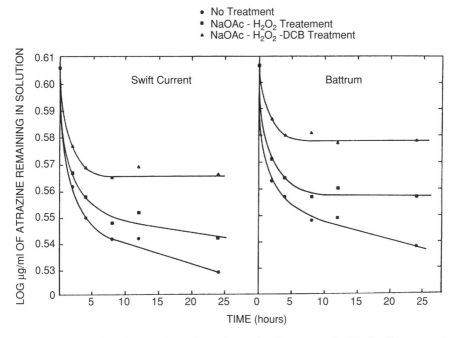

Figure 4. Dynamics of the adsorption of atrazine by two soils (Swift Current and Battrum) as influenced by selected removal of components. From [189]. Reproduced by permission of Lippincott, Williams & Wilkins

was described by linear [191] and nonlinear L-shaped isotherms [192,193], whereas that of rimsulfuron was better fitted by Langmuir-type isotherms [193].

Weakly basic pesticides such as s-triazines, amitrole and rimsulfuron, and amphoteric pesticides such as imidazolinones may become cationic through protonation depending on their basicity and the pH of the system that also controls the degree of ionization of acidic groups in HS molecules. Evidence that maximum adsorption of s-triazines on organic soils occurs at pH levels close to the pK_a of the herbicide is indicative of cation exchange [160]. However, the pH at the HS surface may be two orders of magnitude lower than that at the soil liquid phase [190]. Thus, surface protonation of s-triazines can occur even though the measured pH of the medium is greater than their pK_a. s-Triazine herbicides tend to be inactivated once they come into contact with acid soils. In nonacid soils, residual activity can persist due to relatively weak adsorption, controlled largely by physical adsorption rather than electrostatic attraction to soil particles.

Data from IR and FTIR analyses of the products resulting from interactions between s-triazines, imazethapyr or rimsulfuron and HA showed that ionic bonding can occur between protonated amino groups or heterocyclic nitrogen atoms of these compounds and a carboxylate group or, possibly, a phenolate group of the HA molecule [174,185,192–195]. Structural features of s-triazines, such as the type of the substituent in the 2-position and the nature of the alkyl groups at the 4- and 6-amino groups, are known to influence their basicity and steric hindrance and, hence, their reactivity with HS [196]. The weakly basic pesticides amitrole and dimefox could also be adsorbed by HA through ionic bonding after their protonation [14].

Heat of formation and data from IR and DTA suggested the occurrence of H-bonding between HA and s-triazines, involving multiple complementary sites available on both the HS and pesticide molecules [155,174,185,192,193,195].

The formation of charge-transfer complexes as a possible mechanism involved in the adsorption of s-triazines onto SOM was postulated by Hayes [190]. Subsequently, strong IR and/or ESR spectroscopic evidence was provided on the actual occurrence of this adsorption mechanism for s-triazines, amitrole, imazethapyr and rimsulfuron [174,185,192,193,195,197]. In particular, the increase of free radical concentration measured by ESR in the products of interaction between HA and the pesticides was ascribed to the formation of semiquinone free radical intermediates in the single-electron donor–acceptor transfers occurring between the electron-rich amino or heterocyclic nitrogens of the herbicide molecule and the electron-deficient, quinone-like structures of HA. A number of structural and chemical properties of both the HA and the pesticide molecule were shown to affect to various extent the efficiency to form the electron donor–acceptor system. In particular, the electron-acceptor nature of HA appeared to be related directly to their quinone content and inversely to their total acidity and content of carboxyls and phenolic hydroxyls

[174,195]. The presence of activated electron donors, such as a methoxyl group in the 2-position of the ring and isopropyl substituents in the amino groups, rendered prometone to be the most efficient electron donor among s-triazines [174,195].

s-Triazines are also believed to adsorb on HAs through a ligand exchange (cation bridge) mechanism involving the displacement of hydration water or other weak ligands partially holding a polyvalent metal ion associated to the HA molecule [14]. Hydrophobic adsorption on the surface or trapping within internal pores of the HA macromolecular sieve was also proposed as a possible nonspecific mechanism for the retention of s-triazines [198]. Finally, the formation of covalent bonds between amino groups of the s-triazine and carbonyl or quinone groups of HA, as previously suggested by Hayes [190], was successively confirmed by ESR analysis [174,195].

4.1.3 Acidic and Anionic Compounds

Acidic organic compounds possess acidic functional groups, such as carboxylic and phenolic, that dissociate to form anions. The anion forms of these acidic molecules generally adsorb little, if at all, on layer silicate colloids due to electrostatic repulsion between the anionic molecules and the negative charge of these soil components [129]. Therefore, acidic organic compounds do not adsorb on soils with permanent-charge minerals to the extent that cationic and basic molecules do.

The adsorption of 2,4-D by Mollisols has been reported to follow multiple first-order kinetics [199]. The rate constants of the fast and slow adsorption reactions of 2,4-D by the soils range from 4.8×10^{-3} to $122.7 \times 10^{-3}\,h^{-1}$ and 0 to $2.3 \times 10^{-3}\,h^{-1}$, respectively. The ability of the soils to bind 2,4-D is significantly correlated with their contents of organic matter and pyrophosphate-extractable Al and Fe. The metal oxides extracted by the pyrophosphate treatment are bound to organic matter [200] and are short-range in nature. These sesquioxidic components evidently possess reactive sites for 2,4-D. Precipitation products of Al and Fe are constantly interacting with natural organics in soils [201–204]. The hydrolytic reactions of Al and Fe are substantially perturbed by organic components of soils, resulting in the SRO metal oxides. Therefore, the retention of anionic pesticides and related organic pollutants by SRO metal oxides merits increasing attention.

Soils with variable-charge minerals (e.g., metal oxides and allophane) can strongly adsorb these acidic molecules by chemical adsorption through ligand exchange reaction [129]. Both carboxylic and phenolic groups can be bound to the surface of variable-charge minerals by direct coordination between the carboxylic or phenolic oxygen and the surface. The optimal pH for the adsorption of these acidic molecules happens to coincide approximately with the pK_a of the acidic molecules. The effect of surface charge of the mineral should, in principle, shift the pH of optimal adsorption away from the pK_a and toward the point of zero charge (PZC) of the mineral. Highly weathered soils, which contain large amounts of Al and Fe oxides, are bound to display their ability to adsorb and immobilize these acidic OP. Also, soils derived from volcanic ash contain microcrystalline and noncrystalline allophanes and imogolite or metal oxides. These minerals especially in noncrystalline state should contribute significantly to the ability of the soils to bind these acidic OP.

The bioactivity, residual toxicity, persistence and transport of acidic OP, and of their more common alkali salt (anionic) and ester formulations was found to be highest in soils rich in organic matter, and to correlate with the SOM content, even though the adsorption level of these compounds was much lower than that of cationic and basic OP [183,205]. Adsorption of 2,4-D and picloram on HA followed a Freundlich-type isotherm [206].

Acidic pesticides such as the chlorophenoxyalkanoic acids 2,4-D and 2,4,5-T, the pyridine derivatives picloram and asulam, and the halogenated benzoic acid dicamba were shown to be adsorbed by H-bonding onto HS mainly through carboxyl groups at pH values lower than the pollutant pK_a values [206–208]. Acidic pesticides in the anionic (salt) form, such as picloram, were suggested to bind to HS by a ligand exchange mechanism through polyvalent cations that can bridge the anionic OP to negatively charged groups of the HS molecule [209].

The marked quenching of free radical concentration measured by ESR spectroscopy in various soil HAs suggested the formation of covalent bonds between indigenous humic free radicals and phenoxy or aryloxy radical intermediates generated in the initial decarboxylation of the chlorophenoxyalkanoic molecule (Figure 5) [207]. The indigenous inorganic catalysts (e.g., cupric and ferric ions) and residual enzymatic activity (e.g., due to phenoloxidase) capable of mediating the oxidative pathway outlined in Figure 5, are known to be present in soil [21,88,97]. Phenoxy radicals may also be generated photochemically from phenoxyalkanoic compounds in aqueous solution and in the presence of light and air [210]. The ability of HA free radicals to couple with

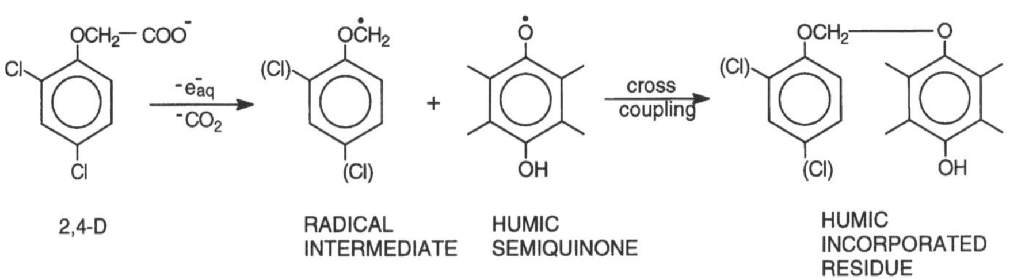

Figure 5. Incorporation of chlorophenoxyalkanoic intermediates into humic acid through the formation of covalent bonds between indigenous humic free radicals and phenoxy or aryloxy radical intermediates. Reprinted with permission from [14]. Copyright CRC Press, Boca Raton, Florida

chlorophenoxy radicals was found to be negatively correlated with the carboxyl content and the ratio of carboxyl to phenolic hydroxyl groups in HA, and with the number of chlorine atoms on the phenoxy ring of the herbicide [207]. Enzyme-catalyzed incorporation of chlorocatechol intermediates of 2,4-D and 2,4,5-T into HA polymers in the presence of peroxidase has also been shown [211].

Adsorption by van der Waals forces and hydrophobic or physical adsorption on the surface of HA has been also proposed as a possible nonspecific mechanism involved in the interaction with SOM of some acidic OP, such as picloram, dicamba and 2,4-D [206].

4.1.4 Nonionic, Nonpolar and Polar Organic Compounds

A huge group of organic molecules is classified as nonionic and nonpolar. In this context, nonpolar also refers to weakly polar, because few organic molecules are strictly nonpolar. Adsorption of this group of organic molecules on soil components is even weaker than that of the polar nonionic molecules. Adsorption mechanisms involving ion–dipole interaction and hydrogen bonding are not operative for nonpolar molecules. Adsorption of these molecules on silicate and oxide minerals in the presence of water is very limited. This group of molecules has low solubility in water and has a significant force of attraction to the organophilic phase in SOM. Retention of these molecules on soil materials is usually referred to as sorption rather than adsorption. The term sorption includes adsorption (a two-dimensional process) and a partitioning mechanism (a three-dimensional process). Two important variables that have been used to characterize and predict sorption of nonpolar organics by soils are: (1) the octanol/water partition coefficient, K_{ow}, (a property of the sorbate) and (2) the fraction of soil that is organic carbon, f_{oc}, a property of the sorbent. Sorption characteristics of nonpolar organics in most soils can be summarized as follows: (1) sorption isotherms are C-type or S-type, and low-affinity sorption is typical, (2) large soil particles reach equilibrium with a given organic sorbate more slowly than do small particles due to slow diffusion through micropores, (3) more hydrophobic sorbates have slower sorption rates than do sorbates of higher molecular weight, (4) temperature has little effect on sorption rates, and (5) sorption is reversible, but often very slow due to diffusion through micropores [129].

As is well known, soil HS are largely bound to mineral colloids [8,212]. Interactions with SOM have been shown to markedly affect the general behavior of a large number of nonionic OP that enter the soil, including their bioactivity, phytotoxicity, degradation, inactivation, persistence, mobility, transport, leaching and volatilization. Details on the effects of SOM, and especially of its humic fractions on various processes involving different classes of nonionic OP, such as substituted urea herbicides, chlorinated hydrocarbons, organophosphates, phenylcarbamates, substituted dinitroanilines, phenyl-amides, acetanilides, thiocarbamates and carbothioates, PAH, phthalic acid diesters (PAE), and single nonionic compounds, have been summarized in several recent reviews [14,213].

Linear C-type and nonlinear L-shaped Freundlich isotherms have been fre-quently obtained for the adsorption of nonionic pesticides to SOM and its humic fractions, whereas nonlinear S-shaped Freundlich isotherms and Lang-muir isotherms have been rarely observed [205]. For example, adsorption of the acetanilides metolachlor and butachlor on SOM fractions showed nonlinear L-shaped or S-shaped isotherms [214,215], whereas alachlor and triallate adsorp-tion on soil HA exhibited linear C-type isotherms [216,217]. Adsorption data of chlordane onto soil HAs were shown to be better fitted by Langmuir-type isotherms [218].

Several types of binding mechanisms have been found to be involved in the adsorption of nonionic OP on SOM and its humic fractions. Hydrogen bonding appeared to play an important role in the adsorption onto HS of several nonionic polar pesticides, including substituted ureas, phenylcarbamates, acet-anilides, cycloate, malathion, bromacil, and glyphosate [175,183,214,216,219], and PAE [205].

The formation of charge transfer bonds via single-electron donor–acceptor transfers has been suggested to occur between electron-donor substituted ureas and electron-acceptor moieties of HAs, on the basis of the increased free radical concentration measured in the formed complexes [175,195]. As a result of the absence of chlorine atoms on the phenyl ring, fenuron was more efficient as an electron donor than other ureas with one or more electron-withdrawing chlor-ine atoms that had a deactivating effect [195]. Other important nonionic OP that possess electron-accepting properties, such as DDT, alachlor, dioxins, and polychlorobiphenyls (PCB) were suggested to interact with complementary electron-donor units of HS to form charge-transfer bonds [216,220,221]. Spe-cific π–π (charge-transfer) interactions were also suggested to occur in the binding of PAH to HS [222,223].

A number of nonionic pesticides belonging to acylanilides, phenylcarba-mates, phenylamides, phenylureas, nitroanilines, and organophosphates (para-thion and methylparathion) were shown to be biodegraded in soil with the release of aromatic amines such as chloroanilines. Similarly, catechols and phenols were released when other classes of pesticides were subject to biodeg-radation. The released moieties were believed to couple covalently with several types of reactive functional groups present in SOM fractions and result in the irreversible incorporation into SOM [224,225]. The coupling processes may be mediated by chemical, photochemical, or enzymatic catalysts. The chemical attachment of chloroanilines to SOM appears to occur by at least two mechan-isms: (a) nucleophilic addition to carbonyl and quinone groups of SOM, leading to the formation of hydrolyzable bound molecules, probably anil, a

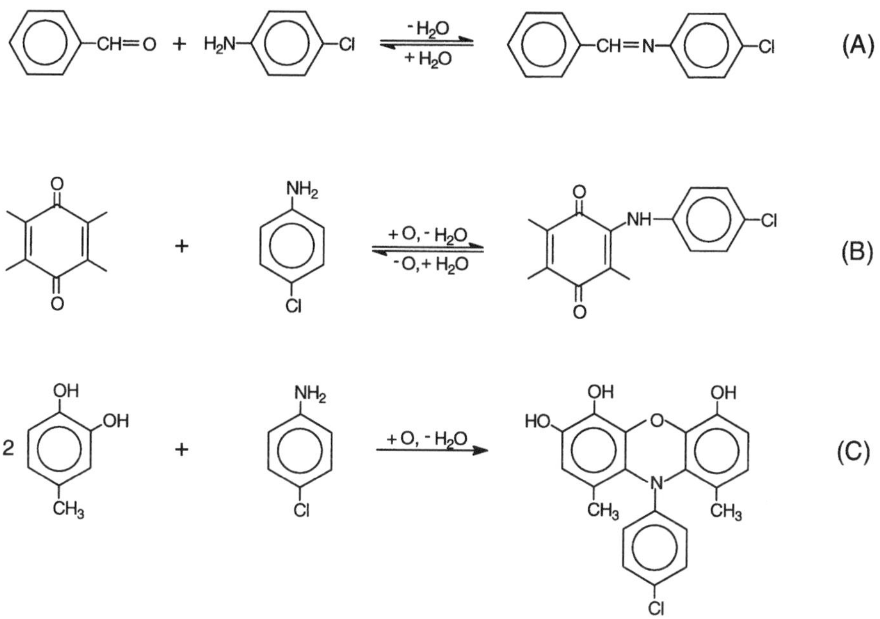

Figure 6. Reversible (A and B) and irreversible (C) incorporation of 4-chloroaniline, a degradation product of many pesticides, into soil organic matter through reaction with various functional groups present in humic acid. Reprinted with permission from [225]. Copyright (1976) American Chemical Society

Schiff base, or anilinoquinone (Figure 6A and B); and (b) oxidative-coupling reactions to phenolic SOM components, resulting in the formation of non-hydrolyzable heterocyclic rings and/or ethers (Figure 6C) [224–226]. Similar nucleophilic addition reactions were suggested to control the binding of benzidine, α-naphthylamine, and p-toluidine to HS [227].

The most important enzymes capable of catalyzing the covalent binding of substituted phenols (e.g., 2,4-dichlorophenol and chlorocatechol formed during the decomposition of 2,4-D, 2,4,5-T, PCP and 2,6-xylenol) and aromatic anilines to HS are the monophenol monooxygenases, laccases, and peroxidases [211, 228–233]. The enzymatic binding is mainly controlled by a free radical mechanism [22].

Adsorption to SOM through weak, additive short-range, van der Waals forces has been suggested for a number of nonionic and nonpolar OP, such as carbaryl and parathion [234], benzonitrile and DDT [235], alachlor and cycloate [216], and several thiocarbamates, carbothioates, and acetanilides [183].

Adsorption to hydrophobic sites on SOM surfaces and trapping within hydrophobic internal pores of the humic macromolecular sieve were suggested to be important nonspecific mechanisms for the retention of several nonionic, nonpolar OP that interact weakly with water. These include: phenylcarbamates [236], DDT and other chlorinated hydrocarbons [235], parathion [234], PAE [205,237], substituted anilines [183], leptophos, methazole, norflurazon, oxadiazinon, butralin, and profluralin [208], PCB [221], and metolachlor [214].

Hydrophobic retention of nonionic, nonpolar OP in SOM has also been regarded as a physical phenomenon of 'partitioning' between water and a nonspecific organic phase, in which the adsorbate dissolves in SOM [238]. Partitioning is modeled as an equilibrium process similar to partitioning of a chemical between two immiscible solvents, such as water and 1-octanol. In other words, SOM can be considered as a nonaqueous solvent into which the hydrophobic OP can partition from water [238].

4.2 BIODEGRADATION OF BOUND XENOBIOTICS

Binding interactions in soil are known to reduce the rates of biodegradation of naturally occurring chemicals, such as nucleic acids, proteins, or carbohydrates [239]. The degradation of organic pesticides may also be considerably reduced when they are retained by soil colloids. For instance, Ogram et al. [240] determined that a species of Flavobacterium could only degrade the herbicide 2,4-D when it was dissolved in water, and that sorption of 2,4-D on soil or clay completely protected the compound from biodegradation. Also, benzylamine mineralization by high cell densities was slower in the presence of montmorillonite (because of sorption) than in the absence of the clay [241].

Sorbed or immobilized xenobiotics seem to undergo biodegradation only after desorption has occurred [240,242–244]. The hydrolysis of parathion, at an initial concentration of $5\,mg\,kg^{-1}$, by a Xanthomonas sp. or its crude enzyme extract was suppressed as a result of adsorption of the herbicide on Na-montmorillonite [93]. Without clay, nearly all parathion degraded within 2 h, but more than $2.5\,mg\,kg^{-1}$ of parathion remained in the clay suspension. Similarly, quinoline was rapidly mineralized in the absence of clay by a bacterium isolated from subsurface sediments [245]. When Na-montmorillonite was present, however, the overall quinoline mineralization rate was considerably reduced. As observed by fluorescence spectroscopy, quinoline disappearance from the clay surface corresponded to the bioavailability of the surface-bound species. The uptake of the surface-bound quinoline was about 30 times slower than that of quinoline in solution. Apparently, the surface-bound molecules were protected from microbial attack and, hence, desorption was required prior to uptake.

The major reason for reduced biodegradation rates is the diminished bioavailability of chemicals involved in binding processes [10]. The availability of sorbed xenobiotics to microbial degraders varies with the chemical properties of the pollutant, the nature of the sorbent, the mechanism of sorption, and the properties of the degradative organisms [246]. For example, phenol (a nonionic compound at neutral pH) adsorbed to stream sediments was available to microorganisms, whereas cationic surfactants were unavailable [247]. The herbicide diquat sorbed to the external exchange sites of a nonexpanding clay mineral (kaolinite) was readily accessible to microorganisms. However, diquat molecules that were entrapped in the interlayers of the expanding clay mineral montmorillonite escaped biodegradation [248].

The mechanism of binding may change with time that xenobiotics reside in soil, leading to changes in the bioavailability of the retained chemicals [10]. For instance, a strain of *Pseudomonas* mineralized as much as 45% of phenanthrene during a 5-day incubation with the compound that was aged in soil for less than 1 day [249]. When phenanthrene was aged for 315 days, however, only 5% of the compound was subject to microbial mineralization. Based on these and other data, it was concluded that the initial fast adsorption of phenanthrene molecules on soil colloids did not constitute as big barrier to biodegradation as the slow diffusion of the chemical across organic matter and its sequestration in inaccessible microsites within the soil matrix [249].

Sorbed contaminants may be available to certain microorganisms and unavailable to others. Such was the case for soil-sorbed naphthalene that was degraded by *Pseudomonas putida* strain 17484, but it did not undergo degradation by a gram-negative soil isolate, designated NP-Alk [246,250]. A simple kinetic method was developed to examine the rates and extents of naphthalene biodegradation in soil-free and soil-containing systems [246]. For NP-Alk, sorption limited both the rate and extent of naphthalene mineralization. For *Pseudomonas putida*, both the rates and extents of naphthalene mineralization exceeded the predicted values and resulted in enhanced rates of naphthalene desorption from soil. The *Psueudomonas* strain attached reversibly to soil, whereas the attachment of NP-Alk was more extensive and irreversible. The mineralization rates of naphthalene indicated that the average *Pseudomonas putida* cell had access to naphthalene concentrated at the sorbent–water interface. With *Pseudomonas*, there was a sevenfold enhancement of naphthalene desorption from soil in comparison with NP-Alk. Apparently, by directly mineralizing surface-localized, labile sorbed naphthalene, the *Pseudomonas* strain established concentration gradients that promoted desorptive diffusion and mineralization of nonlabile naphthalene partitioned into SOM. With NP-Alk, on the other hand, naphthalene degradation appeared to rely mainly on the passive desorption of the surface-sorbed compound [246]. Using a model sorbent, hexadecyltrimethylammonium (HDTMA)-modified smectite, the differences in the efficiency of sorbed naphthalene utilization were still evident despite more rapid desorption of naphthalene from HDTMA–clay as opposed to natural soil [250]. The underlying basis of the observed differences remains to be determined.

The utilization of organics by microbes seems to be modified when organics are bound on metal oxide-coated clays, compared with clays of homoionic to nonpolymeric cations [251]. As the surface properties of these metal oxides vary with pH and ionic factors, the release of such bound organics may occur as the pH and ionic environments of natural microhabitats fluctuate. Once the organics have been released from the hydrous metal oxide–mineral complexes, they could either be degraded by soil microbes or be complexed by mineral colloids.

The nature of soil minerals may affect the degradability of pesticides adsorbed on their surfaces [96]. Addition of montmorillonite to diquat solutions containing mixed soil microbiota completely inhibits diquat mineralization when a sufficient amount of clay is present to allow total removal of diquat from the solution. On the other hand, the microbial degradation of diquat is not influenced by adsorption on kaolinite, since diquat is readily desorbed from the surface of kaolinite. Therefore, it is not possible to distinguish between microbial degradation in solution and on kaolinite surfaces.

Laboratory experiments on the microbial breakdown of environmental pollutants are mostly carried out in solution under conditions far different from those in nature, and abiotic factors in natural ecosystems may considerably influence the course of degradation [252]. For instance, C-P bonds in phosphate toxicants added to the culture of *Pseudomonas testosteroni* were readily cleaved by the bacteria despite the chemical stability of the compounds [253]. *Pseudomonas testosteroni* used either ionic methylphosphonate (MPn) or inorganic orthophosphate (Pi) as sole phosphorus sources and exhibited diauxic utilization of MPn and Pi. However, the utilization of MPn was suppressed in the presence of Pi that was the primary phosphorus source for bacterial growth. Such suppression did not occur in the presence of soil that had a high binding capacity for Pi thus making it inaccessible to bacteria. MPn, on the other hand, exhibited limited binding to this soil, therefore its biodegradation was only slightly reduced.

When sorption is at equilibrium, the concentration of degradable molecules in the soil solution is frequently sufficient to support the growth of microorganisms [1]. As these molecules are consumed, immobilized molecules can be desorbed from the solid phase and become available for biodegradation. Desorption may continue until the concentration of xenobiotics in the liquid phase is insufficient to maintain the microorganisms.

It has to be pointed out that some investigations provided evidence contradictory to the general principle that binding interactions should reduce the rates of xenobiotic transformation. For instance, the mineralization of phenanthrene by a strain of *Pseudomonas fluorescens* in the presence of humic acid or humic–clay complexes was considerably enhanced compared with that observed in the absence of soil components [254]. The enhanced degradation probably resulted from an increased concentration of phenanthrene in the vicinity of the bacterial cells attached to organic and mineral surfaces. Increased pollutant transformation in the presence of soil colloids was also observed for polycyclic aromatic hydrocarbons, phenol, and benzylamine [255–257].

It appears that, given sufficient time, the physically sorbed or sequestered xenobiotics may be almost completely degraded by soil microorganisms unless the soil is repeatedly contaminated. In fact, even the covalently bound chemicals undergo biodegradation. For instance, mixed cultures of soil bacteria were capable of partial mineralization (up to 10 % in 10–12 weeks) of chlorinated phenols that were chemically bound to humic acid during the enzyme-mediated oxidative coupling [258,259]. For reasons that remain to be determined, the mineralization rate for the humic acid-bound 2,4-dichlorophenol (3.0 % in 50 days) was greater than that for the free chemical (only 0.5 %). On the other hand, at longer incubation times (10–12 weeks), the mineralization of the former ceased to increase, indicating that only some 'surface' fraction of the bound chemical was degradable, while the remainder was bound to a 'core' that was totally inaccessible to microorganisms.

4.3 ADSORPTION OF MICROORGANISMS AND BIODEGRADATION OF XENOBIOTICS

The attachment of bacteria to solid surfaces is a common phenomenon in soil habitats. Microorganisms may use many different mechanisms to achieve and maintain the attachment. Adsorption is critical to the initial adhesion of a microbial cell when it encounters a surface. The next step may involve special attachment structures, such as pili or holdfasts. In most cases, however, microorganisms manage to strengthen the attachment by means of extracellular polymeric adhesives (polysaccharide or proteins), which are frequently products of their own metabolism. In a study on the attachment of a marine *Pseudomonas* sp. to a variety of surfaces, the number of bacteria that became attached depended on surface charge and surface hydrophobicity [260,261]. Hydrophobic plastics with little or no surface charge, such as teflon, polyethylene, polystyrene, or poly(ethylene terephthalate), retained large numbers of bacteria. Moderate numbers of bacteria were retained by hydrophobic metals with a positive (platinum) or neutral (germanium) surface charge. Materials with hydrophilic, negatively charged surfaces, such as glass, mica, or oxidized plastics retained very few bacteria. Based on these results it was concluded that both electrostatic and hydrophobic interactions are involved in bacterial attachment [260].

Soil microorganisms are vital factors in soil formation [96], contributing to mechanical and biochemical weathering of soil minerals. Biochemical weathering is usually considered to be the result of the secretory activity of living organisms (canopy drip, exudates, and microbial metabolic compounds). Microbial weathering processes include solubilization and insolubilization of elements that lead to the transformation and degradation of primary and secondary minerals, and the formation of new soil minerals and soil aggregates.

In surface soils, microorganisms are closely associated with mineral colloids [96]. This phenomenon can be observed by examining an ultra-thin section of clay soil by electron microscopy. Microorganisms can react with clay particles by producing extracellular polysaccharides (EPS) in the form of fibrils radiating from their surfaces. These clay–EPS particles persist even after the microbe dies. It is thought that microorganisms produce these fibrils in order to anchor themselves to mineral surfaces. It is frequently observed in ultra-thin sections that cell wall remnants are enclosed by a layer of mineral particles. One of the most intriguing properties of soil microenvironments is that due to the accumulation of H^+ at clay and organic surfaces, a surface pH may be two or three units more acidic than that of the aqueous bulk phase barely 100 nm away. The magnitude of the pH gradient will be accentuated by microbial metabolism and the associated proton release. This microenvironment property may have the following consequences: pH gradient, solubilization or precipitation of different nutrients, changes in enzyme–substrate interactions, and modification of microbial cell walls.

Sorption of microorganisms may be extremely extensive, especially in soils with high contents of clay and organic particles. In the study of Di Grazia et al. [262], for instance, naphthalene was metabolized in soil mainly by sorbed degraders whose population was twice as great as that of the free degraders. Some sorbed bacteria may be released back to the solution, but some may be sorbed irreversibly [263]. In an experiment with a periphytic marine bacterium, the attachment of microorganisms to solids appeared to depend on cell concentration [264]. When the concentration of the bacteria under study was relatively low (approximately $6 \times 10^5 \, mL^{-1}$), essentially all cells present in the medium were adsorbed on hydroxyapatite. However, only a small fraction was adsorbed when the bacterial population was 100 times greater (approximately 6×10^7).

The effect of sorption on the ability of microorganisms to degrade organic compounds is difficult to predict. Sometimes, the degradation may be enhanced, as was the case for a bacterium of Sphingomonas sorbed on porous Teflon that facilitated the release and degradation of 3-chlorodibenzofuran sorbed on the same material [265]. Hydroxyapatite particles, however, did not enhance bacterial activity, regardless of whether the bacteria or organic compounds (glucose or glutamic acid) were associated with the surface [264]. Microbial degradation may also be reduced as the cells undergo the adsorption to solid surfaces. Free bacteria contributed more to the uptake of glucose and amino acids from various surface waters than attached bacteria [266].

The utilization of sorbed organic compounds by microorganisms that adhere to the same surfaces is a common phenomenon [1]. Apparently, the adsorbed molecules are transported directly into the attached cells without entering the surrounding liquid. In a similar manner, microorganisms may consume the insoluble chemicals to which they adhere [267].

Knowledge of the effects of clay and other soil components on microbial activity under the field conditions is relatively limited [268]. These effects may vary with the type of microorganism. Microbial activity (growth, respiration, and the uptake of substrates) may be stimulated by clays to a greater extent in bacteria than in fungi [251]. Montmorillonite enhanced metabolic processes (uptake of glucose and melanin formation) in several fungi [269,270]. On the other hand, little correlation was found between the uptake of glucose and clay contents in the soils studied, but the output of CO_2 was reduced and the biomass was more stable in soils of high clay contents [271].

4.4 IMMOBILIZATION OF EXTRACELLULAR ENZYMES AND XENOBIOTIC TRANSFORMATION

Like microorganisms, enzymes may be immobilized on clay minerals and this immobilization may have a considerable impact on the rate of xenobiotic degradation. The adsorption of enzymes by clay minerals has been studied extensively in the past 40 years [272–274]. Because of their small size, clay particles possess a large surface area and significant cation exchange capacity. Anionic clays, such as smectites, tend to adsorb and inactivate enzymes [275]. Enzymes also may intercalate the interlayer spaces of expanding clay minerals. In the case of non-expending silicates, they are adsorbed on external sites, especially on the edges of clays. The adsorption is dependent on the type of clay mineral, the ratio of enzyme to clay, the pH of the clay suspension, the cation exchange capacity and specific surface area of the clay; the oxidation state of the cation saturating the clay; the isoelectric point of the enzyme, and the structure, shape, and molecular weight of the enzyme [276–278].

Various types of interactions, such as cation exchange [275], van der Waals forces [279], and hydrophobic binding [273] have been suggested to explain the adsorption mechanisms. Interactions between an enzyme and a clay rely largely on the formation of ionic bonds between opposite charges and other adsorption mechanisms, including van der Waals forces, hydrogen bonds, or hydrophobic interactions [278].

Enzymes are amphoteric in nature and their electric charge is positive at pH values below the isoelectric point (IEP) and negative when the pH is above the IEP. Most layer silicate clays show a permanent and pH-independent negative charge, except for a small fraction of the total charge, which is variable with pH as a result of the dissociation of Al-OH and Fe-OH groups at the clay edges. Electrostatic forces are important when the enzyme is in the cationic form that is at a pH below the isoelectric point. Immobilization on the clay surface may considerably change enzymatic activity. The activity of glucose oxidase was sharply reduced after adsorption on montmorillonite and kaolinite [280]. Similarly, the activities of clay-adsorbed arylsulfatase and bacterio-lytic endopeptitase were considerably lower than those of the unadsorbed enzymes. Adsorption of arylsulfatase on kaolinite was confined to the external surfaces, whereas that on montmorillonite also involved the interlayer surfaces. Phenoloxidases were strongly bound to homoionic bentonites with the loss of enzymatic activity, but less strongly bound to kaolinite [281]. The reduced activity of clay–enzyme complexes is attributed to steric and diffusional restrictions, direct involvement of the active site in binding to the support, and modified conformation of the immobilized enzyme [278,282].

Organic and inorganic polymers that coat clay surfaces may further influence enzyme activity. Glucose oxidase and urease, for instance, became strongly bound to smectite saturated with hexadecyltrimethylammonium cation (HDTMA$^+$), retaining 60–100% of activity they had in the free state [283,284]. Strong binding to HDTMA–smectite was also determined for peroxidase and arginase, but the enzymes lost all their activity [284]. The same effects were observed for arginase and urease after the saturation of smectite with bipyridil cations. Through the involvement of active sites on the enzyme molecule in binding interactions, the effect of minerals was in general unfavorable to enzymatic activity. However, when different clays were sequentially treated with 3-aminopropyltriethoxysilane and glutaraldehyde to prevent direct contact between enzyme molecules and mineral surfaces, immobilized enzymes retained most of their original activity [285,286].

Immobilization may have a considerable effect on the kinetic behavior of enzymes. The dependence of the reaction velocity on substrate concentration for a given enzyme is defined by the maximal initial velocity (V_{max}) and the Michaelis constant (K_m). V_{max} varies with the concentration of enzyme present, but K_m is independent of enzyme concentration and is indicative of the affinity of the enzyme for the substrate (the higher K_m, the lower affinity). K_m is equal to the substrate concentration at half-maximal velocity. The V_{max} values for the immobilized enzymes may be lower, and K_m constants may be higher than those for free enzymes. On the other hand, after immobilization, enzymes show generally increased stability toward physical, chemical, and biological denaturation [287].

Immobilization may cause conformational alterations of the enzyme molecule so the active site becomes less accessible to the substrate, causing a setback in substrate transformation [288]. The interactions with the polar or apolar sites on solid support may induce local differences in pH and the concentration of substrate and product in the vicinity of support and in the bulk solution, thus influencing the rate of transformation. Also, diffusion of the reaction components to active site on the enzyme molecule may be disturbed in the presence of mineral support and result in reduced transformation.

There are indications that enzymes can also bind to SOM. Partially purified fractions of enzymatically active HS were obtained by gel chromatography, ultrafiltration, and isoelectric focusing [289]. Gel chromatography was used to isolate humic–peroxidase and humic–catalase complexes [290]. Humic acid has an inhibitory effect on the activity of catalase and tyrosinase [291]. The enzymes are probably trapped within the macromolecular net of humic acid, and some molecules may also be immobilized at the surface by adsorption forces [290].

Enzymes can also function as electron donors while forming complexes with humic acid as an electron acceptor [292].

Immobilization of enzymes on humic polymers and clay minerals is the major factor that determines their performance in the soil environment. Because of immobilization, the optimal pH for enzymes in soil is usually higher than that of free enzymes. In other words, the concentration of hydrogen ions near a polyanionic enzyme associated with soil colloids is higher than that measured in the bulk solution. These changes in pH may influence enzyme activity, but the actual effect of pH may vary with ionic strength of the soil solution.

The association of enzymes with organo-mineral complexes is believed to shield the enzyme against proteolytic activity without hindering the diffusion of substrate molecules to the active sites [285]. However, co-adsorption of non-proteolytic enzymes and proteases in neighboring sites on the support material may enhance proteolysis of the complexed enzyme. Substrate affinity for the enzyme molecule may also differ in free and immobilized enzymes. According to many reports, the K_m values for enzymes in soil are higher than those for the same enzymes tested *in vitro*. Accordingly, decreases in enzyme activity were observed with increasing amounts of clays at different substrate concentrations [293]. Complexation of oxidoreductive enzymes, such as laccase or tyrosinase, which play a vital role in detoxifying xenobiotic phenols and anilines, with HA also had an inhibitory effect (enhanced K_m values) on the transformation of the toxic compounds [294].

Information on the involvement of immobilized enzymes in the transformation of xenobiotics in soil is sparse because the transformation reactions may also be catalyzed by intracellular enzymes or freshly released free enzymes, as well as by mineral surfaces. Burns and Edwards [72] listed 15 immobilized soil enzymes (from the class of oxidoreductases, transferases, and hydrolases) that were found to transform xenobiotics. For instance, a soil-bound carboxylesterase transformed malathion to a monocarboxylic acid derivative, and propanil was transformed to 3,4-dichloroaniline by a soil-bound aryl acetamidase. It was possible to recover from soil (by extraction with $0.05 \, \text{mol} \, \text{L}^{-1}$ phosphate buffer) the immobilized peroxidase involved in the transformation of 3,4-dichloroaniline to 3,3',4,4'-tetrachloroazobenzene [77].

Despite the observed transformation of xenobiotic compounds by soil enzymes it has to be stated that, in general, the intimate association of enzymes with soil colloids has a negative effect on pollutant degradation [287]. Enzyme inactivation, due to its adsorption by Na-saturated montmorillonite, was the dominant cause for the decrease in parathion hydrolysis by the crude enzyme extract from *Xanthomonas* sp. [93]. When different oxidoreductases (peroxidase, tyrosinase, laccase) were preincubated with HA or FA for 48 h, the transformation of 2,4-dichlorophenol by these enzymes was considerably diminished as compared with the transformation by non-preincubated enzymes [295]. Apparently, the enzymes were inhibited by their adsorption on humic molecules.

5 CONCLUSIONS

Literature data indicate that the transformation of xenobiotics in a multicomponent system, such as soil, is a result of the combined action of microorganisms, extracellular enzymes, mineral colloids and humic materials. Each of these four major components of the soil system not only participates in the transformation of xenobiotics but also modifies the participation of the three other factors. Microorganisms degrade xenobiotics in addition to degrading many natural products, including humic materials and extracellular enzymes. They also contribute to the weathering of minerals. Extracellular enzymes compete with the intracellular ones for both xenobiotic and natural substrates, so that it is difficult to distinguish between the extracellular and intracellular degradation. Humic acids are believed to catalyze certain transformation reactions, but their major impact lies in the ability to adsorb (microorganisms, enzymes and chemicals) or to be adsorbed (on soil minerals). Minerals exhibit mixed functions as well. They transform the naturally occurring and xenobiotic substrates abiotically; at the same time, they act as sorbents, thus altering the impact of microorganisms, enzymes, and chemicals. Adsorption and other binding interactions which occur on both mineral and humic surfaces are believed to reduce the bioavailability of xenobiotics.

Our knowledge of these complex interrelations is not yet sufficient, especially with regard to xenobiotic transformation. Extensive research has been done, for instance, on the activity of immobilized enzymes, but the substrates tested were mainly the products of natural metabolism, and the effect of immobilized enzymes on xenobiotic transformation, in most cases, can only be estimated based on the trends determined for non-xenobiotic substrates or for a few tested xenobiotics.

It is a common observation that certain recalcitrant compounds, which resist microbial degradation *in vitro*, may undergo transformation in soil environments. Is this because soil contains consortia of microorganisms capable of acting in concert to degrade a recalcitrant pollutant? Why then are mixed microbial populations isolated from soil not as effective? One can only speculate that microorganisms that control the degradation are immobilized in the soil matrix and may not be isolated by any of the established methods for an *in vitro* testing; many microorganisms are nonculturable. Immobilized, unextractable and nonculturable microorganisms constitute as much as 80–90 % of the total soil microbiota. Presently, however, there is no reliable method to find out the extent to which this 80–90 % majority may be accountable for biodegradation. It is also not clear how important the impact of genetic mutation and natural selection in natural environments may be, since most of the studies in this area have been carried out using pure cultures in liquid media.

As shown by many researchers cited in this paper, abiotic factors, such as catalytic activity of mineral and organic colloids, may be a major cause of the degradation of both degradable and recalcitrant compounds in soil. To date, however, the reports of abiotic transformation have been limited to a relatively narrow group of xenobiotics. Mixed biotic/abiotic mechanisms by which xenobiotic chemicals are degraded in soil environments will be an important area of future investigations.

REFERENCES

1. Alexander, M. (1999). *Biodegradation and Bioremediation*, 2nd edn, Academic Press, San Diego, CA.
2. Wackett, L. P. (1995) Bacterial co-metabolism of halogenated organic compounds. In *Microbial Transformation and Degradation of Toxic Organic Chemicals*, ed. Young, L. Y. and Cerniglia, C. E, Wiley-Liss, New York, p. 217.
3. Johnson, L. M. and Talbot, H. W. Jr. (1983). Detoxification of pesticides by microbial enzymes, *Experientia*, **39**, 1236.
4. Dick, W. A. and Tabatabai, M. A. (1999). Use of immobilized enzyms for bioremediation. In *Bioremediation of Contaminated Soils*, ed. Adriano, D. C., Bollag, J.-M., Frankenberger, W. T. Jr., and Sims, R. C., Agronomy Monograph Ser. 37, Madison, WI, p. 315.
5. Admassu, W. and Korus, R. A. (1996) Engineering of bioremediation processes: Needs and limitations. In *Bioremediation: Principles and Applications*, ed. Crawford, R. L. and Crawford, D. L., Cambridge University Press, Cambridge, p. 13.
6. Wiggins, B. A., Jones, S. H. and Alexander, M. (1987). Explanations for the acclimation period preceding the mineralization of organic chemicals in aquatic environments, *Appl. Environ. Microbiol.*, **53**, 791.
7. van der Meer, J. R., de Vos, W. W., Harayama, S. and Zehnder, A. J. B. (1992). Molecular mechanisms of genetic adaptation to xenobiotic compounds, *Microbiol. Rev.*, **56**, 677.
8. Huang, P. M. (1990). Role of soil minerals in transformations of natural organics and xenobiotics in the environment. In *Soil Biochemistry*, Vol. 6, ed. Bollag, J.-M. and Stotzky, G., Marcel Dekker, New York, p. 29.
9. Wolfe, N. L., Mingelgrin, U. and Miller, G. C. (1990). Abiotic transformations in water, sediments, and soil. In *Pesticides in the Soil Environment: Processes, Impacts, and Modeling*, ed. Cheng, H. H. Soil Science Society of America, Madison, WI, p. 103.
10. Alexander, M. (1995). How toxic are toxic chemicals in soil, *Env. Sci. Technol.*, **29**, 2713.
11. Stevenson, F. J. (1994). *Humus Chemistry: Genesis, Composition, Reactions*, John Wiley & Sons, New York.
12. Huang, W., Schlautman, M. A. and Weber, W. J. Jr. (1996). A distribution reactivity model for sorption by soils and sediments. 5. The influence of near-surface characteristics in mineral domains, *Environ. Sci. Technol.*, **30**, 2993.
13. Luthy, R. G., Aiken, G. R., Bruseau, M. L., Cunningham, S. D., Gschwend, P. M., Pignatello, J. J., Reinhard, M., Traina, S. J., Weber, W. J., Westall, J. C. (1997). Sequestration of hydrophobic organic contaminants by geosorbents, *Environ. Sci. Technol.*, **31**, 3341.
14. Senesi, N. and Miano, T. M. (1995). The role of abiotic interactions with humic substances on the environmental impact of organic pollutants. In *Environmental*

Impact of Soil Component Interactions. Natural and Anthropogenic Organics, Vol. I. ed. Huang, P. M., Berthelin, J., Bollag, J. M., McGill, W. B., and Page, A. L., CRC Press/Lewis, Boca Raton, FL, p. 311.

15. Haider, K. (1999) Microbe–soil–organic contaminant interactions. In *Bioremediation of Contaminated Soils*, ed. Adriano, D. C., Bollag, J.-M., Frankenberger, W. T. Jr., and Sims, R. C., Agronomy Monograph Ser. 37, Madison, WI, p. 33.

16. Singh, B. K., Kuhad, R. C., Singh, A., Lal, R. and Tripathi, K. K. (1999). Biochemical and molecular basis of pesticide degradation by microorganisms, *Crit. Rev. Biotechnol.*, **19**, 197.

17. Betts, W. B. and King, J. E. (1991). Oxidative coupling of 2,6-dimethoxyphenol by fungi and bacteria, *Mycol. Res. Cambridge*, **5**, 526.

18. Nutsubidze, N. N., Sarkanen, S., Schmidt, E. L., Shashikanth, S. (1998). Consecutive polymerization and depolymerization of kraft lignin by *Trametes cingulata*, *Phytochemistry*, **49**, 1203.

19. Clapp, C. E. and Hayes, M. H. B. (1999). Sizes and shapes of humic substances, *Soil Sci.*, **164**, 777.

20. Shevchenko, S. M., and Bailey, G. W. (1996). Life after death: Lignin–humic relationships reexamined, *Crit. Rev. Environ. Sci. Technol.*, **26**, 95.

21. Dec, J., and Bollag, J.-M. (2000). Phenoloxidase-mediated interactions of phenols and anilines with humic materials, *J. Environ. Qual.*, **29**, 665.

22. Sjoblad, R. D. and Bollag, J.-M. (1981). Oxidative coupling of aromatic compounds by enzymes from soil microorganisms. In *Soil Biochemistry*, Vol. 5, ed. Paul, E. A. and Ladd, J. N., Marcel Dekker, New York, p. 113.

23. Vogel, T. M. and Grbić-Galić, D. (1986). Incorporation of oxygen from water into toluene and benzene during anaerobic fermentative transformation, *Appl. Environ. Microbiol.*, **52**, 200.

24. Radnoti de Lipthay, J., Barkay, T., Vekova, J., Sorensen, S. J. (1999). Utilization of phenoxyacetic acid, by strains using either the *ortho* or *meta* cleavage of catechol during phenol degradation, after conjugal transfer of *tfd*A, the gene encoding a 2,4-dichlorophenoxyacetic acid/2-oxoglutarate, *Appl. Microbiol. Biotechnol.*, **51**, 207.

25. Grund, E., Knorr, C. and Eichenlaub, R. 1990. Catabolism of benzoate and monohydroxylated benzoates by *Amycolatopsis* and *Streptomyces* spp., *Appl. Environ. Microbiol.*, **56**, 1459.

26. Kearney, P. C., Plimmer, J. R., Williams, V. P., Klingebiel, U. I., Isensee, A. R., Laanio, T. L., Stolzenberg, G. E. and Zaylskie, R. G. (1974). Soil persistence and metabolism of *N-sec*-butyl-4-*tert*-butyl-2,6-dinitroaniline, *J. Agric. Food Chem.*, **22**, 865.

27. Korte, F., Ludwig, G. and Vogel, J. (1962). Umwandlung von Aldrin-[^{14}C] durch Mikroorganismen, Leberhomogenate und Moskito-Larven, *Liebigs Ann. Chem.*, **656**, 135.

28. Chiou, C. K. and Leak, D. J. (1996). Purification and characterization of two components of an epoxypropane isomerase/carboxylase of *Xanthobacter* Py2, *Biochem. J.*, **319**, 299.

29. van Hylckama Vlieg, J. E. T., Kingma, J., van den Wijngaard, A. J. and Janssen, D. B. (1998). A gluthatione *S*-transferase with activity towards *cis*-1,2-dichloroepoxyethane is involved in isoprene utilization by *Rhodococcus* sp. strain AD45, *Appl. Environ. Microbiol.*, **64**, 2800.

30. Anzenbacher, P., Niwa, T., Tolbert, L. M., Sirimanne, S. R. and Guengerich, F. P. (1996). Oxidation of 9-alkylanthracenes by cytochrome P450 2B1, horseradish peroxidase, and iron tetraphenylporphine/iodosylbenzene systems: Anaerobic and aerobic mechanisms, *Biochemistry*, **35**, 2512.

31. Boundy-Mills, K. L., de Souza, M. L., Mandelbaum, R. T., Wackett, L. P. and Sadowsky, M. J. (1997). The *atzB* gene of *Pseudomonas* sp. strain ADP encodes the second enzyme of a novel atrazine degradation pathway, *Appl. Environ. Microbiol.*, **63**, 916.

32. Benezet, H. J. and Matsumura, F. (1974). Factors influencing the metabolism of mexacarbate by microorganisms, *J. Agric. Food Chem.*, **22**, 427.

33. Havel, J. and Reineke, W. (1993). Microbial degradation of chlorinated aceto-phenones, *Appl. Environ. Microbiol.*, **59**, 2706.

34. Hernandez, B. S., Higson, F. K., Kondrat, R. and Focht, D. D. (1991). Metabolism of and inhibition by chlorobenzoates in *Pseudomonas putida* P111, *Appl. Environ. Microbiol.*, **57**, 3361.

35. Fukumori, F. and Hausinger, R. P. (1993). *Alcaligenes eutrophus* JMP134 '2,4-dichlorophenoxyacetic acid monooxygenase' is an α-ketoglutarate-dependent dioxy-genase, *J. Bacteriol.*, **175**, 2083.

36. Miyazaki, S., Boush, G. M. and Matsumura, F. (1969). Metabolism of ^{14}C-chloro-benzilate and ^{14}C-chloropylate by *Rhodotorula gracilis*, *Appl. Microbiol.*, **18**, 972.

37. Gamar, Y. and Gaunt, J. K. (1971). Bacterial metabolism of 4–chloro-2–methylphe-noxyacetate, formation of glyoxylate by side-chain cleavage, *Biochem. J.*, **122**, 527.

38. Hyman, M. R., Page C. L. and Arp, D. J. (1994) Oxidation of methyl fluoride and dimethyl ether by ammonia monooxygenase in *Nitrosomonas europaea*, *Appl. Environ. Microbiol.*, **60**, 3033.

39. Schmidt, S., Fortnagel, P. and Wittich, R.-M. (1993). Biodegradation and transfor-mation of 4,4'- and 2,4-dihalodiphenyl ethers by *Sphingomonas* sp. SS33, *Appl. Environ. Microbiol.*, **59**, 3931.

40. Molin, G., Nillson, I. (1985). Degradation of phenol by *Pseudomonas putida* ATCC 11172 in continuous culture at different ratios of biofilm surface to culture volume, *Appl. Environ. Microbiol.*, **50**, 946.

41. Powlowski, J.; Shingler, V. (1994). Genetics and biochemistry of phenol degradation by *Pseudomonas* sp. CF600, *Biodegradation*, **5**, 219.

42. van Schie, P. and Young, L. Y. (2000). Biodegradation of phenol: Mechanism and applications, *Bioremediation J.*, **4**, 1.

43. Harwoord, C. S. and Parales, R. E. (1996). The β-ketoadipate pathway and the biology of self-identity, *Annu. Rev. Microbiol.*, **50**, 553.

44. Bayly, R. C. and Barbour, M. G. (1984). The degradation of aromatic compounds by the *meta* and gentisate pathways. In *Microbial Degradation of Organic Com-pounds*, ed. Gibson, D. T., Marcel Dekker, New York. p. 253.

45. de Souza, M. L., Sadowsky, M. J. and Wackett, L. P. (1996). Atrazine chloro-hydrolase from *Pseudomonas* sp. strain ADP: Gene sequence, enzyme purification, and protein characterization, *J. Bacteriol.*, **178**, 4894.

46. Sadowsky, M. J., Tong, Z., de Souza, M., Wackett, L. P. (1998). AtzC is a new member of the amidohydrolase protein superfamily and is homologous to other atrazine-metabolizing enzymes, *J. Bacteriol.*, **180**, 152.

47. Cook, A. M., Beilstein, P., Grossenbacher, H., Hütter, R. (1985). Ring cleavage and degradative pathway of cyanuric acid in bacteria, *Biochem. J.*, **231**, 25.

48. Racke, K. D., Laskowski, D. A. and Schultz. (1990). Resistance of chlorpyrifos to enhanced biodegradation in soil, *J. Agric. Food Chem.*, **38**, 1430.

49. Shukla, O. P. and Kaul, S. M. (1986). Microbiological transformation of pyridine *N*-oxide and pyridine by *Nocardia* sp., *Can. J. Microbiol.*, **32**, 330.

50. Germida, J. J., Wainwright, M. and Gupta, V. V. S. R. (1992). Biochemistry of sulfur cycling in soil. In *Soil Biochemistry*, Vol. 7, ed. Stotzky, G. and Bollag, J.-M. Marcel Dekker, New York. p. 1.

51. Zhang, D., Freeman, J. P., Sutherland, J. B., Walker, A. E., Yang, Y. and Cerniglia, C. E. (1996). Biotransformation of chlorpromazine and methdilazine by *Cunninghamella elegans*, *Appl. Environ. Microbiol.*, **62**, 798.

52. Dick, W. A., Ankumah, R. O., McClung, G., Abou-Assaf, N. (1990). Enhanced degradation of *s*-ethyl *N,N*-dipropylcarbamothioate in soil and by an isolated microorganism. In *Enhanced Biodegradation of Pesticides in the Environment*, ed. Racke, K. D. and Coats, J. R., ACS Symp. Ser. 426, American Chemical Society, Washington, DC, p. 98.

53. Matsumura, F., and Benezet, H. J. (1978). Microbial degradation of insecticides. In *Pesticide Microbiology*, ed. Hill, I. R. and Wright, S. J. L., Academic Press, London, p. 648.

54. Singh, J., Comfort, S. D. and Shea, P. J. 1998. Remediation RDX-contaminated water and soil using zero-valent iron, *J. Environ. Qual.*, **27**, 1240.

55. Hawari, J., Halasz, A., Sheremata, T., Beaudet, S., Groom, C., Paquet, L., Rhofir, C., Ampleman, G. and Thiboutot, S. (2000). Characterization of metabolites during biodegradation of hexahydro-1,3,5-trinitro-1,3,5-triazine (RDX) with municipal anaerobic sludge, *Appl. Environ. Microbiol.*, **66**, 2652.

56. Timms, P. and MacRae, I. C. (1983). Reduction of fensulfothion and accumulation of the products fensulfothion sulfide by selected microbes, *Bull. Environ. Contam. Toxicol.*, **31**, 112.

57. Quensen III, J. F., Mueller, S. A., Jain, M. K. and Tiedje, J. M. (1998). Reductive dechlorination of DDE to DDMU in marine sediment microcosms, *Science*, **280**, 722.

58. Walker, J. E. and Kaplan, D. L. (1992). Biological degradation of explosives and chemical agents, *Biodegradation*, **3**, 369.

59. Baarschers, W. H., Bharath, A. I., Elvish, J. and Davies M. (1982). The biodegradation of methoxychlor by *Klebsiella pneumonia, Can. J. Microbiol.*, **28**, 176.

60. Kurihara, N., Ohisa, N., Nakajima, M., Kakutani, T. and M. Senda. (1981). Relationship between microbial degradability and polarographic half-wave potential of polychlorocyclohexenes and BHC isomers, *Agric. Biol. Chem.*, **45**, 1229.

61. Tiedje, J. M., Boyd, S. A. and Fathepure, B. Z. (1987). Anaerobic degradation of chlorinated aromatic hydrocarbons, *Dev. Ind. Microbiol.*, **27**, 117.

62. Bollag, J.-M., and Liu, S.-Y. (1990). Biological transformation processes of pesticides. In: *Pesticides in the Soil Environment: Processes, Impacts, and Modeling*, ed. Cheng, H. H. SSSA Book Series, no. 2, Soil Science Society of America, Madison, WI, p. 169.

63. Caldwell, S. R. and Raushel, F. M. (1991). Detoxification of organophosphate pesticides using an immobilized phosphotriesterase from *Pseudomonas diminuta*, *Biotechnol. Bioengin.*, **37**, 103.

64. Karam, J. and Nicell, J. A. (1997). Potential application of enzymes in waste treatment, *J. Chem. Tech. Biotechnol.*, **69**, 141.

65. Munnecke, D. M., Johnson, L. M., Talbot, H. W., and Barik, S. (1982). Microbial metabolism and enzymology of selected pesticides. In *Biodegradation and Detoxification of Environmental Pollutants*, ed. Chakrabarty, A. M., CRC Press, Boca Raton, FL, p. 1.

66. Engelhardt, G., Ziegler, W., Wallnöfer, P. R., Jarczyk, H. J., and Oehlmann, L. (1982). Degradation of the triazinone herbicide metamitron by *Arthrobacter* sp. DSM 20389, *J. Agric Food Chem.*, **30**, 278.

67. Golovleva, L. A., Baskunov, B. P., Finkelstein, Z. I. and Nefedova, M. Yu. (1983). Microbial degradation of the organophosphorus insecticide fosalone, *Biol. Bull. Acad. Sci. USSR*, **10**, 44.

68. Mulbry, W. W. and Eaton, R. W. (1991). Purification and characterization of the N-methylcarbamate hydrolase from *Pseudomonas* strain CRL-OK, *Appl. Environ. Microbiol.*, **57**, 3679.
69. Aga, D. S., Thurman, E. M., Yockel, M. E., Zimmerman, L. R. and Williams, T. D. (1996). Identification of a new sulfonic acid metabolite of metolachlor in soil, *Environ. Sci. Technol.*, **30**, 592.
70. Field, J. A. and Thurman, E. M. (1996). Gluthatione conjugation and contaminant transformation, *Environ. Sci. Technol.*, **30**, 1413.
71. Noguera, D. R. and Freedman, D. L. (1996). Reduction and acetylation of 2,4-dinitrotoluene by a *Pseudomonas aeruginosa* strain, *Appl. Environ. Microbiol.*, **62**, 2257.
72. Burns, R. G., and Edwards, J. A. (1980). Pesticide breakdown by soil enzymes, *Pest. Sci.*, **11**, 506.
73. Dick, W. A. and Tabatabai, M. A. (1992). Significance and potential use of enzymes. In *Soil Microbial Ecology*, ed. Metting, F. B., Marcel Dekker, New York. p. 315.
74. Getzin, L. W., and Rosefield, I. (1971). Partial purification and properties of a soil enzyme that degrades the insecticide malathion, *Biochim. Biophys. Acta*, **235**, 442.
75. Mulbry, W. W. and Karns. J. (1989). Purification and characterization of three parathion hydrolases from gram-negative bacterial strains, *Appl. Environ. Microbiol.*, **55**, 289.
76. Kishk, F. M., T. El-Essawi, S. Abdel-Ghafer, M. B. Abondonia. (1976). Hydrolysis of methylparathion in soils, *J. Agric. Food Chem.*, **24**, 305.
77. Bartha, R., Linke, H. A. B., and Pramer, D. (1968). Pesticide transformations: Production of chloroazobenzenes from chloroanilines, *Science*, **161**, 582.
78. Burge, W. D. (1972). Microbial populations hydrolyzing propanil and accumulation of 3,4-dichloroaniline and 3,3',4,4'-tetrachloroazobenzene in soils, *Soil Biol. Biochem.*, **4**, 379.
79. Munnecke, D. M. (1980). Enzymatic detoxification of waste organophosphate pesticides, *J. Agric. Food Chem.*, **28**, 105.
80. Villarreal, D. T., Turco, R. F. and Konopka, A. (1994). A structure-activity study with aryl acylamidases, *Appl. Environ. Microbiol.*, **60**, 3939.
81. Marty, J. L. and Vouges J. (1987). Purification and properties of a phenylcarbamate herbicide degrading enzyme of *Pseudomonas alcaligenes* isolated from soil, *Agric. Biol. Chem.*, **51**, 3287.
82. Tabatabai, M. A. and Fu, M. (1992). Extraction of enzymes from soil. In *Soil Biochemistry*, Vol. 7, ed. Stotzky, G. and Bollag, J.-M., Marcel Dekker, New York, p. 197.
83. Bollag, J.-M. (1992). Decontaminating soil with enzymes: an in situ method using phenolic and anilinic compounds, *Environ. Sci. Technol.*, **25**, 1876.
84. Bollag, J.-M., Dec, J. and Huang, P. M. (1998). Formation mechanisms of complex organic structures in soil habitats, *Adv. Agron.*, **63**, 237.
85. Bordeleau, L. M., and Bartha, R. (1972). Biochemical transformations of herbicide-derived anilines: purification and characterization of causative enzymes, *Can. J. Microbiol.*, **18**, 1865.
86. Martin, J. P., and Haider, K. (1969). Phenolic polymers of *Stachybotrys atra, Stachybotrys chartarum* and *Epiccocum nigrum* in relation to humic acid formation, *Soil Sci.*, **107**, 260.
87. Chefetz, B., Chen, Y. and Hadar, Y. (1998). Purification and characterization of laccase from *Chaetomium thermophilium* and its role in humification, *Appl. Environ. Microbiol.*, **64**, 3175.

88. Huang, P. M (1995). The role of short-range ordered mineral colloids in abiotic transformations of organic compounds in the environment. In *Environmental Impact of Soil Component Interactions:* Vol. 1, *Natural and Anthropogenic Organics*, ed. Huang, P. M., Berthelin, J., Bollag, J.-M., McGill, W. B., and Page, A. L., CRC Press/Lewis, Boca Raton, FL, p. 135.
89. Christiansen, B. T. (1992). Physical fractionation of soil and organic matter in primary particle size and density separates, *Adv. Soil Sci.*, **20**, 1.
90. Barriuso, E. and Koskinen, W. C. (1996). Incorporating nonextractable atrazine residues in soil size fractions as a function of time, *Soil Sci. Soc. Am. J.*, **60**, 150.
91. Menzer, R. E. and Nelson J. O. (1986). Water and soil pollutants. In *The basic Science of Poisons*, 3rd edn, ed. Doull, J. D., Klausen, C. D. and Amdur, M. O., Casarett and Doull's Toxicology, Macmillan, New York, p. 825.
92. Macalady, D. L. and Wolfe, N. L. (1985). Effects of sediment sorption and abiotic hydrolyses. I. Organophosphorothioate esters, *J. Agric. Food Chem.*, **33**, 167–173.
93. Masaphy, S., Fahima, T., Levanon, D., Henis, Y., and Mingelgrin, U. (1996). Parathion degradation by *Xanthomonas* sp. and its crude enzyme extract in clay suspensions, *J. Environ. Qual.*, **25**, 1248.
94. Guthrie, E. A. and Pfaender, F. K. (1998). Reduced pyrene bioavailability in microbially active soils, *Environ. Sci. Technol.*, **32**, 501.
95. Cheng, H. H. (1990). *Pesticides in the Soil Environment: Processes, Impact, and Modeling*, Soil Science Society of America, Madison, WI.
96. Huang, P. M. and Bollag, J.-M. (1998). Minerals–organics–microorganisms interactions in the soil environment. In *Structure and Surface Reactions*, ed. Huang, P. M., Senesi, N. and Buffle, J., John Wiley & Sons, New York, p. 3.
97. Chorover, J., Amistadi, M. K., Burgos, W. D. and Hatcher, P. G. (1999). Quinoline sorption on kaolinite–humic acid complexes, *Soil Sci. Soc. Am. J.*, **63**, 850.
98. Karthikeyan, K. G., Chorover, J., Bortiatynski, J. M. and Hatcher, P. G. (1999). Interaction of 1-naphthol and its oxidation products with aluminium hydroxide, *Environ. Sci. Technol.*, **33**, 4009.
99. Claus, H. and Filip, Z. (1990). Effects of clays and other solids on the activity of phenoloxidases produced by some fungi and actinomycetes, *Soil Biol. Biochem.*, **22**, 483.
100. Wang, T. S. C., Huang, P. M., Chou, C.-H, and Chen, J.-H. (1986). The role of soil minerals in the abiotic polymerization of phenolic compounds and formation of humic substances. In *Interactions of Soil Minerals with Natural Organics and Microbes*, ed. Huang, P. M. and Schnitzer, M., SSSA Special Publication No. 17, Soil Science Society of America, Madison, WI, p. 251.
101. Huang, P. M. (1991). Kinetics of redox reactions on surfaces of Mn oxides and its impact on environmental quality. In *Rate of Chemical Processes in Soils*, ed. Sparks, D. L. and Suarez, D. L., SSSA Special Publication No. 27, Soil Science Society of America, Madison, WI, pp. 191.
102. Pal, S., Bollag, J.-M., and Huang, P. M. (1994). Role of abiotic and biotic catalysts in the transformation of phenolic compounds through oxidative coupling reactions, *Soil Biol. Biochem.*, **26**, 813.
103. McKenzie, R. M. (1989). Manganese oxides and hydroxides. In *Minerals in Soil Environments*, 2nd edn, ed. Dixon, J. B. and Weed, S. B., Soil Science Society of America, Madison, WI, p. 439.
104. Shindo, H., and Huang, P. M. (1982). Role of Mn (IV) oxide in abiotic formation of humic substances in the environment, *Nature (London)*, **298**, 363.
105. Schnitzer, M. (1982). Quo vadis soil organic matter research. Trans. 12th Int. Congr., *Soil Sci.*, **6**, 67.

106. Chang, H. M. and Allan, G. G. (1971). Oxidation. In *Lignins*, ed. Sarkanen, K. V. and Ludwig, C. H., Wiley Interscience, New York, p. 433.
107. Shindo, H., and Huang, P. M. (1984). Catalytic effects of manganese(IV), iron(III), aluminum, and silicon oxides on the formation of phenolic polymers, *Soil Sci. Soc. Am. J.*, **48**, 927.
108. Wang, T. S. C., Wang, M. C., and Huang, P. M. (1983). Catalytic synthesis of humic substances by using aluminas as catalysts, *Soil Sci.*, **136**, 226.
109. Shindo, H., and Huang, P. M. (1992). Comparison of the influence of Mn(IV) oxide and tyrosinase on the formation of humic substances in the environment, *Sci. Total Environ.*, **117/118**, 103.
110. Lehmann, R. G. and Cheng, H. H. (1988). Reactivity of phenolic acids in soil and formation of oxidation products, *Soil Sci. Soc. Am. J.*, **51**, 1304.
111. Naidja, A., Huang, P. M. and Bollag, J.-M. (1998). Comparison of reaction products from the transformation of catechol catalyzed by birnessite or tyrosinase, *Soil Sci. Soc. Am. J.*, **62**, 188.
112. Niemeyer, J., Chen, Y. and Bollag, J.-M. (1992). Characterization of humic acids, composts and peats by diffuse reflectance Fourier-transform infrared spectroscopy, *Soil Sci. Soc. Am. J.*, **56**, 135.
113. Hannien, K. I., Klocking, R. and Helbig, B. (1987). Synthesis and characterization of humic acid-like polymers, *Sci. Total Environ.*, **62**, 201.
114. Laha, S and Luthy, R. G. (1990). Oxidation of aniline and other primary aromatic amines by manganese dioxide, *Environ. Sci. Technol.*, **24**, 363.
115. Solomon, D. H. (1968). Clay minerals as electron acceptors and/or electron donors in organic reactions, *Clays Clay Miner.*, **16**, 31.
116. Pinnavaia, T. J., Hall, P. L., Cady, S. S., and Mortland, M. M. (1974). Aromatic radical cation formation on intracrystal surfaces of transition metal layer lattice silicates, *J. Phys. Chem.*, **78**, 994.
117. Mortland, M. M. and Halloran, L. J. (1976). Polymerization of aromatic molecules on smectites, *Soil Sci. Soc. Am. J.*, **40**, 367.
118. Shindo, H. and Huang, P. M. (1985). The catalytic power of inorganic components in the abiotic synthesis of hydroquinone-derived humic polymers, *Appl. Clay Sci.*, **1**, 71.
119. Wang, M. C. and Huang, P. M. (1986). Humic macromolecule interlayering in Rontronite through interaction with phenol monomers, *Nature (London)*, **323**, 529.
120. Wang, M. C. and Huang, P. M. (1994). Structural role of polyphenols in influencing the ring cleavage and related chemical reactions as catalyzed by nontronite. In *Humic Substances in the Global Environment and Implications on Human Health*, ed. Senesi, N. and Miano T. M., Proc. 8th Int. Conf. Int. Humic Substances Society, Elsevier, Amsterdam, p. 173.
121. Shindo, H., and Huang, P. M. (1985). Catalytic polymerization of hydroquinone by primary minerals, *Soil Sci.*, **139**, 505.
122. Jaurequi, M. A. and Reisenauer, H. M. (1982). Dissolution of oxides of manganese and iron by root exudate components, *Soil Sci. Soc. Am. J.*, **46**, 314.
123. Stone, A. T. and Morgan, J. J. (1984). Reduction and dissolution of manganese (III) and manganese(IV) oxides by organics. I. Reaction with hydroquinone, *Environ. Sci. Technol.*, **18**, 450.
124. Stone, A. T. and Morgan, J. J. (1984). Reduction and dissolution of manganese(III) and manganese(IV) oxides by organics. 2. Survey of the reactivity of organics, *Environ. Sci. Technol.*, **18**, 617.
125. Lehmann, R. G., Cheng, H. H., and Harsh, J. B. (1987). Oxidation of phenolic acids by soil iron and manganese oxides, *Soil Sci. Soc. Am. J.*, **51**, 352.

126. Stone, A. T. (1987). Reductive dissolution of manganese (III)/(IV) oxides by substituted phenols, *Environ. Sci. Technol.*, **21**, 979.
127. Cheney, M. A., Sposito, G., McGrath, A. E., and Criddle, R. S. (1996). Abiotic degradation of 2,4-D (dichlorophenoxy acetic acid) on synthetic birnessite: a calorespirometric method, *Coll. Surface. A. Phy. Eng. Asp.*, **107**, 131.
128. Mortland, M. M. (1986). Mechanisms of adsorption of nonhumic organic species by clays. In *Interactions of Soil Minerals with Natural Organics and Microbes*, ed. Huang, P. M., and Schnitzer, M., SSSA Special Publication No. 17, Soil Science Society of America, Madison, WI, p. 59.
129. McBride, M. B. (1994). *Environmental Chemistry of Soils*, Oxford University Press, Oxford.
130. Fowker, F. M., Benesi, H. A., Ryland, R. B., Sawyer, W. M., Detling, K. D., Loeffler, E. S., Folckemer, F. B., Johnson, M. R. and Sun, Y. P. (1960). Clay-catalyzed decomposition of insecticides, *J. Agric. Food Chem.*, **8**, 203.
131. Brown, C. B. and White, J. L. (1969). Reactions of 12 *s*-triazines with soil clays, *Soil Sci. Soc. Am. Proc.*, **33**, 863.
132. Saltzman, S., Yaron, B. and Mingelgrin, U. (1974). The surface catalyzed hydrolysis of parathion on kaolinite, *Soil Sci. Soc. Am. Proc.*, **38**, 231.
133. Mingelgrin, U., Saltzman, S. and Yaron, B. (1977). A possible model for the surface-induced hydrolysis of organophosphorus pesticides on kaolinite clays, *Soil Sci. Soc. Am. J.*, **41**, 519.
134. Mingelgrin, U. and Saltzman, S. (1979). Surface reactions of parathion on clays, *Clays Clay Miner.*, **27**, 72.
135. Kozlowski, H. (1988). Catalytic hydrolysis of quinalphos on homoionic clays, *Pestic. Sci.*, **24**, 1.
136. Russell, J. D., Cruz, M. and White, J. L. (1968). The adsorption of 3-aminotriazol by montmorillonite, *J. Agric. Food Chem.*, **16**, 21.
137. Armstrong, D. E., Chesters, G., and Harris, R. F. (1967). Atrazine hydrolysis in soils, *Soil Sci. Soc. Am. Proc.*, **31**, 61.
138. Russell, J. D., Cruz, M., and White, J. L. (1968). Model of chemical degradation of *s*-triazines by montmorillonite, *Science*, **160**, 1340.
139. Huang, P. M. (2000). Abiotic catalysis. In *The Handbook of Soil Science*, ed. Sumner, M. E., CRC Press, Boca Raton, FL, p. 303.
140. Hoffmann, M. R. (1990). Catalysis in aquatic environments. In *Aquatic Chemical Kinetics*, ed. Stumm, W., John Wiley & Sons, New York, p. 71.
141. Hoffmann, M. R. (1980). Trace metal catalysis in aquatic environments, *Environ. Sci. Technol.*, **14**, 1061.
142. Stone, A. T. and Torrents, A. (1995). The role of dissolved metals and metal-containing surfaces in catalyzing the hydrolysis of organic pollutants. In *Environmental Impact of Soil Component Interactions*. Vol. 1, *Natural and Anthropogenic Organics*, ed. Huang, P. M., Berthelin, J., Bollag, J.-M. McGill, W. B., and Page, A. L., CRC Press/Lewis, Boca Raton, FL, p. 275.
143. Plastourgou, M. and Hoffmann, M. R. (1984). Transformation and fate of organic esters in layer-flow systems: The role of trace metal catalysis, *Environ. Sci. Technol.*, **18**, 756.
144. Wilkins, R. A. (1991). *Kinetics and Mechanisms of Reactions of Transition Metal Complexes*, 2nd edn, VCH, Weinheim.
145. Smolen, J. M. and Stone, A. T. (1998). Organophosphorus ester hydrolysis catalyzed by dissolved metals and metal-containing surfaces. In *Soil Chemistry and Ecosystem Health*, ed. Huang, P. M., Adriano, D. C., Logan, T. J. and Checkai, R. T., SSSA Special Publication No. 52, Soil Science Society of America, Madison, WI, p. 157.

146. Smolen, J. M. and Stone, A. T. (1998). Metal (hydr)oxide surface-catalyzed hydrolysis of chlorpyrifos-methyl, chlorpyrifos-methyl oxon, and paraoxon, *Soil Sci. Soc. Am. J.*, **62**, 636.
147. Saltzman, S., Mingelgrin, U. and Yaron, B. (1976). Role of water in the hydrolysis of parathion and methylparathion on koalinite, *J. Agric. Food Chem.*, **24**, 739.
148. Sanchez-Camazano, M. S., and Sanchez-Martin, M. J. S. (1983). Monmorillonite-catalyzed hydrolysis of phosmet, *Soil Sci.*, **136**, 89.
149. Pusino, A., Petretto, S. and Gessa, C. (1996). Montmorillonite surface-catalyzed hydrolysis of fenoxyaprop-ethyl, *J. Agric. Food. Chem.*, **44**, 1150.
150. El-Amamy, M. M. and Mill, T. (1984). Hydrolysis kinetics of organic chemicals on montmorillonite and kaolinite surfaces as related to moisture content, *Clays Clay Miner.*, **32**, 67.
151. Stone, A. T. (1989). Enhanced rates of monophenyl terephthalate hydrolysis in aluminum oxide suspensions, *J. Colloid Interface Sci.*, **127**, 429.
152. Torrents, A., and Stone, A. T. (1991). Hydrolysis of phenyl picolinate at the mineral/water interface, *Environ. Sci. Technol.*, **25**, 143.
153. Baldwin, D. S., Beattie, J. K., Coleman, L. M. and Jones, D. R. (1995). Phosphate ester hydrolysis facilitated by mineral phases, *Environ. Sci. Technol.*, **29**, 1706.
154. Struif, B., Weil, L., and Quentin, K. E. (1975). The behavior of herbicides, phenoxy acetic acids and their esters in water, *Wom. Wasser*, **45**, 53.
155. Li, G. C. and Felbeck, G. T., Jr. (1972). A study of the mechanism of atrazine adsorption by humic acid from muck soil, *Soil Sci.*, **113**, 140.
156. Khan, S. U. (1978). Kinetics of hydrolysis of atrazine in aqueous fulvic acid solution, *Pestic. Sci.*, **9**, 39.
157. Perdue, E. M. and Wolfe, N. L. (1982). Modification of pollutant hydrolysis kinetics in the presence of humic substances, *Environ. Sci. Technol.*, **16**, 847.
158. Macalady, D. L., Tratnyek, P. G., and Wolfe, N. L. (1989). Influences of natural organic matter on the abiotic hydrolysis of organic contaminants in aqueous systems. In *Aquatic Humic Substances. Influence on Fate and Treatment of Pollutants*, ed. Suffet, I. and MacCarthy, P., Adv. Chem. Ser. No. 219, American Chemical Society, Washington, DC, p. 323.
159. Burkhard, N. and Guth, J. A. (1981). Chemical hydrolysis of 2-chloro-4,6-bis (alkylamino)-1,3,5-triazine herbicides and their breakdown in soil under the influence of adsorption, *Pestic. Sci.*, **12**, 45.
160. Weber, J. B., Weed, S. B., and Ward, T. M. (1969). Adsorption of *s*-triazines by soil organic matter, *Weed Sci.*, **17**, 417.
161. Gamble, D. S. and Khan, S. U. (1985). Atrazine hydrolysis in soil: Catalysis by the acidic functional groups of fulvic acid, *Can. J. Soil Sci.*, **65**, 435.
162. Perdue, E. M. (1983). Association of organic pollutants with humic substances: Partitioning equilibria and hydrolysis kinetics. In *Aquatic and Terrestrial Humic Materials*, ed. Christman, R. F. and Gjessing, E. T., Ann Arbor Science, Ann Arbor, MI, p. 441.
163. Malini de Almeida, R., Pospisil F., Vockova, K., and Kutacek, M. (1980). Effect of humic acid on the inhibition of pea choline esterase and choline acyltransferase with malathion, *Biol. Plant.*, **22**, 167.
164. Miller, G. C., Hebert, V. R., and Miller, W. W. (1989). Effect of sunlight on organic contaminants at the atmosphere-soil interface. In *Reactions and Movement of Organic Chemicals in Soils*, ed. Sawhney, B. L. and Brown, K., SSSA Special Publication No. 22, Soil Science Society of America, Madison, WI, p. 99.
165. Senesi, N. and Loffredo, E. (1997). Minimizing environmental damage originating from pesticide utilization. Abiotic photochemical control and remediation. In

Modern Agriculture and the Environment, ed. Rosen, D., Hader, Tel-Or, and Chen, Y., Kluwer, Berlin, p. 47.

166. Khan, S. U. and Schnitzer, M. (1978). UV irradiation of atrazine in aqueous fulvic acid solution, *J. Environ. Sci. Health*, **3**, 299.

167. Khan, S. U. and Gamble, D. S. (1983). Ultraviolet irradiation of an aqueous solution of prometryn in the presence of humic materials, *J. Agric. Food Chem.*, **31**, 1099.

168. Cooper, W. J. Zika, R. G., Petasne, R. G., and Fischer, A. M. (1989). Sunlight-induced photochemistry of humic substances in natural waters: Major reactive species. In *Aquatic Humic Substances. Influence on Fate and Treatment of Pollutants*, ed. Suffet, I. and MacCarthy, P., Adv. Chem. Ser. No. 219, American Chemical Society, Washington, DC, p. 333.

169. Hoignè, J., Faust, B. C., Haag, W. R., Scully, F. E. Jr., and Zepp, R. G. (1989). Aquatic humic substances as sources and sinks of photochemically produced transient reactants. In *Aquatic Humic Substances. Influence on Fate and Treatment of Pollutants*, ed. Suffet, I. and MacCarthy, P., Adv. Chem. Ser. No. 219, American Chemical Society, Washington, DC, p. 363.

170. Frimmel, F. G. (1994). Photochemical aspects related to humic substances, *Environ. Int.*, **20**, 373.

171. Hoignè, J., Bader, H., and Nowell, L. H. (1987). Rate constants of OH radical scavenging by humic substances: Role in ozonation and in a few photochemical processes for the elimination of micropollutants, *Am. Chem. Soc. Div. Environ. Chem.*, **27**, 208.

172. Zepp, R. G., Schlotzhauer, P. F., and Sink, R. M. (1985). Photosensitized transformations involving electronic energy transfer in natural waters: Role of humic substances, *Environ. Sci. Technol.*, **19**, 74.

173. Senesi, N. and Schnitzer, M. (1977). Effect of pH, reaction time, chemical reduction and irradiation on ESR spectra of fulvic acid, *Soil Sci.*, **123**, 224.

174. Senesi, N. and Testini, C. (1982). Physico-chemical investigations of interaction mechanisms between s-triazine herbicides and soil humic acids, *Geoderma*, **28**, 129.

175. Senesi, N. and Testini, C. (1983). Spectroscopic investigations of electron donor-acceptor processes involving organic free radicals in the adsorption of substituted urea herbicides by humic acids, *Pestic. Sci.*, **14**, 79.

176. Huang, P. M., Berthelin, J., Bollag, J.-M., McGill, W. B., and Page, A. L. (1995). *Environmental Impact of Soil Component Interactions*, Vol. 1, *Natural and Anthropogenic Organics*, CRC Press/Lewis, Boca Raton, FL.

177. Weber, J. B., Perry, P. W., and Upchurch, R. P. (1965). The influence of temperature and time on the adsorption of paraquat, diquat, 2,4-D, and prometone by clays, charcoal, and an anion-exchange resin, *Soil Sci. Soc. Am. Proc.*, **29**, 678.

178. Mortland, M. M. (1970). Clay–organic complexes and interactions, *Adv. Agron.*, **22**, 75.

179. Burchill, S., Hayes, M. H. B., and Greenland, D. J. 1981. Adsorption. In *The Chemistry of Soil Processes*, ed. Greenland, D. J. and Hayes, M. H. B. John Wiley & Sons, New York, p. 221.

180. Lahav, N., and White, D. H. (1980). A possible role of fluctuating clay-water-systems in the production of ordered prebiotic oligomers, *J. Mol. Evol.*, **16**, 11.

181. Hayes, M. H. B. and Mingelgrin, U. (1991). Interactions between small organic chemicals and soil colloidal constituents. In *Interactions at the Soil Colloid–Soil Solution Interface*, ed. Bolt, J. *et al.*, Plenum Press, New York, p. 323.

182. Khan, S. U. (1974). Humic substances reactions involving bipyridilium herbicides in soil and aquatic environments, *Residue Rev.*, **52**, 1.

183. Weber, J. B. (1972). Interaction of organic pesticides with particulate matter in aquatic and soil systems, *Adv. Chem. Ser.*, **111**, 55.
184. Maqueda, C., Perez Rodriguez, J. L., Martin, F., and Hermosin, M. C. (1983). A study of the interaction between chlordimeform and humic acid from a typic chromoxevert soil, *Soil Sci.*, **136**, 75.
185. Senesi, N., D'Orazio, V. and Miano, T. M. (1995). Adsorption mechanisms of *s*-triazine and bipyridylium herbicides on humic acids from hops field soils, *Geoderma*, **66**, 273.
186. Narine, D. R. and Guy, R. D. (1982). Binding of diquat and paraquat to humic acid in aquatic environments, *Soil Sci.*, **133**, 356.
187. Bailey, G. W., White, J. L. and Rothberg, T. (1968). Adsorption of organic herbicides by montmorillonite: Role of pH and chemical character of adsorbate, *Soil Sci. Soc. Am. Proc.*, **32**, 222.
188. Bailey, G. W. and White, J. L. (1970). Factors influencing the adsorption, desorption, and movement of pesticides in soil, *Residue Rev.*, **32**, 29.
189. Huang, P. M., Grover, R., and McKercher, R. B. (1984). Components and particle size fractions involved in atrazine adsorption by soils, *Soil Sci.*, **138**, 20.
190. Hayes, M. H. B. (1970). Adsorption of triazine herbicides on soil organic matter, including a short review on soil organic matter chemistry, *Residue Rev.*, **32**, 131.
191. Che, M., Loux, M. M., Traina, S. J., Logan, T. J. (1992). Effect of pH on sorption and desorption of imazaquin and imazethapyr on clays and humic acid, *J. Environ. Qual.*, **12**, 698.
192. Senesi, N., La Cava, P., and Miano, T. M. (1997). Adsorption of imazethapyr to amended and non-amended soils and humic acids, *J. Environ. Qual.*, **26**, 1264.
193. Senesi, N., Loffredo, E., D'Orazio, V., Brunetti, G., Miano, T. M., and La Cava, P. (2001). Adsorption of pesticides by humic acids from organic amendments and soils. In *Humic Substances and Chemical Contaminants*, ed. Clapp, C. E., Hayes, M. H. B., Senesi, N., Bloom, P. R. and Jardine, P. M., ASA, CSSA, and SSSA, Madison, WI (in press).
194. Kalouskova, N. (1986). Kinetics and mechanisms of interaction of simazine with humic acids, *J. Environ. Sci. Health*, **B**, **21**, 251.
195. Senesi, N., Testini, C. and Miano, T. M. (1987). Interaction mechanisms between humic acids of different origin and nature and electron donor herbicides: A comparative IR and ESR study, *Org. Geochem.*, **11**, 25.
196. Weber, J. B. (1967). Spectrophotochemically determined ionisation constants of 1,3-alkylamino-*s*-triazines and the relationships of molecular structure and basicity, *Spectrochim. Acta*, **23A**, 458.
197. Müller-Wegener, U. (1987). Electron donor acceptor complexes between organic nitrogen heterocycles and humic acid, *Sci. Total Environ.*, **62**, 297.
198. Schulten, H.-R. (1996). Three-dimensional molecular structures of humic acids and their interactions with water and dissolved contaminants, *J. Environ. Anal. Chem.*, **64**, 147.
199. Huang, P. M. and Grover, R. (1990). Kinetics and components involved in the adsorption of 2,4-D by soils. In *Health and Safety in Agriculture*, ed. Dosman, J. A., and Cockcroft, D. W., CRC Press, Boca Raton, FL, p. 228.
200. Bascomb, C. L. (1968). Distribution of pyrophosphate extractable iron and organic carbon in soils of various groups, *J Soil Sci.*, **19**, 251.
201. Huang, P. M., and Violante, A. (1986). Influence of organic acids on crystallization and surface properties of precipitation products of aluminum. In *Interactions of Soil Minerals with Natural Organics and Microbes*, ed. Huang, P. M., and Schnitzer, M.,

SSSA Special Publication No. 17, Soil Science Society of America, Madison, WI, p. 159.

202. Schwertmann, U., Kodama, H. and Fischer, W. R. 1986. Mutual interactions between organics and iron oxides. In *Interactions of Soil Minerals with Natural Organics and Microbes*, ed. Huang, P. M. and Schnitzer, M., SSSA Special Publication No. 17, Soil Science Society of America, Madison, WI, p. 224.

203. Sposito, G. (1996). *The Environmental Chemistry of Aluminum*, CRC/Lewis, Boca Raton, FL, p. 464.

204. Huang, P. M. and Wang, M. K. (1997). Formation chemistry and selected surface properties of iron oxides. In *Advances in GeoEcology* 30: *Soil and Environment*, (ed.) Auerswald, K., Stanjek, H., and Bigham, J. M., Catena, Reikskirchen, Germany, p. 241.

205. Khan, S. U. (1980). Determining the role of humic substances in the fate of pesticides in the environment, *J. Environ. Sci. Health B*, **15**, 1071.

206. Khan, S. U. (1973). Equilibrium and kinetics studies of the adsorption of 2,4-D and picloram on humic acid, *Can. J. Soil Sci.*, **53**, 429.

207. Senesi, N., Miano, T. M. and Testini, C. (1986). Role of humic substances in the environmental chemistry of chlorinated phenoxyalkanoic acids and esters. In *Chemistry for Protection of the Environment 1985*, ed. Pawlowski, L., Alaerts, G., and Lacy, W. J. Stud. Environ. Sci. No. 29, Elsevier, Amsterdam, p. 183.

208. Carringer, R. D., Weber, J. B., and Monaco, T. J. (1975). Adsorption-desorption of selected pesticides by organic matter and montmorillonite, *J. Agric. Food Chem.*, **23**, 569.

209. Nearpass, D. C. (1976). Adsorption of picloram by humic acids and humin, *Soil Sci.*, **121**, 272.

210. Zepp, R. G., Baughman, G. L., and Schlotzhauer, P. F. (1981). Comparison of photochemical behavior of various humic substances in water: Sunlight-induced reactions of aquatic pollutants photosensitized by humic substances, *Chemosphere*, **10**, 109.

211. Stott, D. E., Martin, J. P., Focht, D. D., and Haider, K. (1983). Biodegradation, stabilization in humus and incorporation into soil biomass of 2,4-D and chlorocatechol carbons, *Soil Sci. Soc. Am. J.*, **47**, 66.

212. Schnitzer, M. and Kodama, H. (1977). Reactions of minerals with soil humic substances. In *Minerals in Soil Environments*, ed. Dixon, J. B. and Weed, S. B. Soil Science Society of America, Madison, WI, p. 741.

213. Senesi, N. (1993). Nature of interactions between organic chemicals and dissolved humic substances and the influence of environmental factors. In *Organic Substances in Soil and Water: Natural Constituents and their Influence on Contaminant Behaviour*, ed. Beck, A. J., Jones, K. C., Hayes, M. H. B., and Mingelgrin, U., Royal Society of Chemistry, London, p. 73.

214. Kozak, J., Weber, J. B., and Sheets, T. J. (1983). Adsorption of prometryn and metolachlor by selected soil organic matter fractions, *Soil Sci.*, **136**, 94.

215. Sato, T., Kohnosu, S., and Hartwig, J. F. (1987). Adsorption of butachlor to soils, *J. Agric. Food Chem.*, **35**, 397.

216. Senesi, N., Brunetti, G., La Cava, P., and Miano, T. M. (1994). Adsorption of alachlor by humic acids from sewage sludge and amended and non-amended soils, *Soil Science*, **157**, 176.

217. D'Orazio, V., Loffredo, E., Brunetti, G., and Senesi, N. (1999). Triallate adsorption onto humic acids of different origin and nature, *Chemosphere*, **39**, 183.

218. Loffredo, E., D'Orazio, V., Brunetti, G., and Senesi, N. (1999). Adsorption of chlordane onto humic acids from soils and pig slurry, *Org. Geochem.*, **30**, 443.

219. Miano, T. M., Piccolo, A., Celano, G., and Senesi, N. (1992). Infrared and fluorescence spectroscopy of glyphosate-humic acid complexes, *Sci. Total Environ.*, **123**, 83.

220. Miller, L. L. and Narang, R. S. (1970). Induced photolysis of DDT, *Science*, **169**, 368.

221. Strek, H. J. and Weber, J. B. (1982). Behavior of polychlorinated biphenyls (PCBs) in soils and plants, *Environ. Pollut. A*, **28**, 291.

222. Gauthier, T. D., Shane, E. C., Guerin, W. F., Seitz, W. R. and Grant, C. L. (1986). Fluorescence quenching method for determining equilibrium constants for polycyclic aromatic hydrocarbons binding to dissolved humic materials, *Environ. Sci. Technol.*, **20**, 1162.

223. Gauthier, T. D., Booth, K. A., Grant, C. L., and Seitz, W. R. (1987). Fluorescence quenching studies of interactions between polynuclear aromatic hydrocarbons and humic materials, *Am. Chem. Soc. Div. Environ. Chem.*, **27**, 243.

224. Hsu, T. S. and Bartha, R. (1974). Interaction of pesticides-derived choroaniline residues with soil organic matter, *Soil Sci.*, **116**, 444.

225. Hsu, T. S. and Bartha, R. (1976). Hydrolyzable and nonhydrolyzable 2,4-dichloroaniline-humus complexes and their respective rates of biodegradation, *J. Agric. Food Chem.*, **24**, 118.

226. Adrian, P., Lahaniatis, E. S., Andreux, F., Mansour, M., Scheunert, I. and Korte, F. (1989). Reaction of the soil pollutant 4-chloroaniline with humic acid monomer catechol, *Chemosphere*, **18**, 1599.

227. Graveel, J. G., Sommers, L. E. and Nelson, D. W. (1985). Sites of benzidine, α-naphthylamine and *p*-toluidine retention in soils, *Environ. Toxicol. Chem.*, **4**, 607.

228. Suflita, J. M. and Bollag, J.-M. (1981). Polymerization of phenolic compounds by a soil–enzyme complex, *Soil Sci. Soc. Am. J.*, **45**, 297.

229. Bollag, J.-M., Minard, R. D., and Liu, S.-Y. (1983). Cross-linkage between anilines and phenolic humus constituents, *Environ. Sci. Technol.*, **17**, 72.

230. Bollag, J.-M. (1983). Cross-coupling of humus constituents and xenobiotics substances. In *Aquatic and Terrestrial Humic Materials*, ed. Christman, R. F. and Gjessing, E. T., Ann Arbor Science, Ann Arbor, MI, p. 127.

231. Berry, D. F. and Boyd, S. A. (1984). Oxidative coupling of phenols and anilines by peroxidase: Structure–activity relationships, *Soil Sci. Soc. Am. J.*, **48**, 565.

232. Berry, D. F. and Boyd, S. A. (1985). Reaction rates between phenolic humus constituents and anilines during cross-coupling, *Soil Biol. Biochem.*, **17**, 631.

233. Liu, S. Y. and Bollag, J.-M. (1985). Enzymatic binding of the pollutant 2,6-xylenol to a humus constituent, *Wat. Air Soil Pollut.*, **25**, 97.

234. Leenheer, J. A. and Aldrichs, J. L. (1971). A kinetic and equilibrium study of the adsorption of carbaryl and parathion upon soil organic matter surfaces, *Soil Sci. Soc. Am. Proc.*, **35**, 700.

235. Pierce, R. H., Olney, C. E., and Felbeck, G. T. (1971). Pesticide adsorption in soils and sediments, *Environ. Lett.*, **1**, 157.

236. Briggs, G. G. (1969). Molecular structure of herbicides and their sorption by soils, *Nature*, **223**, 1288.

237. Matsuda, K. and Schnitzer, M. (1971). Reaction between fulvic acid, a soil humic material, and dialkylphthalates, *Bull. Environ. Contam. Toxicol.*, **6**, 200.

238. Chiou, C. T., Porter, P. E., and Schmedding, D. W. (1983). Partition equilibria of nonionic organic compounds between soil organic matter and water, *Environ. Sci. Technol.*, **17**, 227.

239. Haider, K. (1992). Problems related to the humification processes in soils of temperate climates. In: *Soil Biochemistry*, Vol. 7, ed. Stotzky, G. and Bollag, J.-M., Marcel Dekker, New York, p. 55.

240. Ogram, A. V., Jessup, R. E., Lou, L. T. and Rao P. S. C. (1985). Effects of sorption on biological degradation rates of (2,4-dichlorophenoxy) acetic acid, *Appl. Environ. Microbiol.*, **49**, 582.

241. Miller, M., and Alexander, M. (1991). Kinetics of bacterial degradation of benzylamine in a montmorillonite suspension, *Environ. Sci. Technol.*, **25**, 240.

242. Speitel, G. E., Lu, C. J., Turakhia, M. and Sho, X.-J. (1988). Biodegradation of trace concentrations of substituted phenols in granular activated carbon columns, *Environ. Sci. Technol.*, **22**, 68.

243. Schäffer, A. (1993). Pesticide effects on enzyme activities in the soil ecosystem. In *Soil Biochemistry*, Vol. 8, ed. Bollag, J.-M. and Stotzky, G., Marcel Dekker, New York, p. 273.

244. Benoit, P., Barriuso, E. and Soulas, G. (1999). Degradation of 2,4-D, 2,4-dichlorophenol, and 4-chlorophenol in soil after sorption on humified and nonhumified organic matter, *J. Environ. Qual.*, **28**, 1127.

245. Smith, S. C., Ainsworth, C. C., Traina, S. J. and Hicks, R. J. (1992). Effect of sorption on the biodegradation of quinoline, *Soil Sci. Soc. Am. J.*, **56**, 737.

246. Guerin, W. F. and Boyd, S. A. (1992). Differential bioavailability of soil-sorbed naphthalene to two bacterial species, *Appl. Environ. Microbiol.*, **58**, 1142.

247. Shimp, R. J., and Young, R. L. (1988) Availability of organic chemicals for biodegradation in settled bottom sediments, *Ecotoxicol. Environ. Safety*, **15**, 31.

248. Weber, J. B. and Coble, H. D. (1968). Microbial decomposition of diquat adsorbed on montmorillonite and kaolinite clays, *J. Agric. Food Chem.*, **16**, 475.

249. Hatzinger, P. B. and Alexander, M. (1995). Effect of aging of chemicals in soil on their biodegradability and extractability, *Environ. Sci. Technol.*, **29**, 537.

250. Crocker, F. H., Guerin, W. F. and S. A. Boyd. (1995). Bioavailability of naphthalene sorbed to cationic surfactant-modified smectite clay, *Environ. Sci. Technol.*, **29**, 2953.

251. Stotzky, G. (1986). Influence of soil mineral colloids on metabolic processes, growth, adhesion, and ecology of microbes and viruses. In *Interactions of Soil Minerals with Natural Organics and Microbes*, ed. Huang, P. M. and Schnitzer, M., Soil Science Society of America, Madison, WI, p. 305.

252. Daughton, C. G., Cook, A. M. and Alexander, M. (1979). Phosphate and soil binding: Factors limiting bacterial degradation of ionic phosphorus-containing pesticide metabolites, *Appl. Environ. Microbiol.*, **37**, 605.

253. Cook, A. M., Daughton, C. G. and Alexander, M. (1978). Phosphonate utilization by bacteria, *J. Bacteriol.*, **133**, 85.

254. Ortega-Calvo, J.-J. and Saiz-Jimenez, C. (1998). Effect of humic fractions and clay on biodegradation of phenanthrene by a *Pseudomonas fluorescens* strain isolated from soil, *Appl. Environ. Microbiol.*, **64**, 3123.

255. Kästner, M. and Mahro, B. (1996). Microbial degradation of polycyclic aromatic hydrocarbons in soils affected by the organic matrix of compost, *Appl. Microbial. Biotechnol.*, **44**, 668.

256. Erhardt, H. M. and Rehm, H. J. (1985). Phenol degradation by microorganisms adsorbed on activated carbon, *Appl. Microbiol. Biotechnol.*, **21**, 32.

257. Amador, J. A. and Alexander, M. (1988). Effect of humic acids on the mineralization of low concentrations of organic compounds, *Soil Biol. Biochem.*, **20**, 185.

258. Dec, J., and Bollag, J.-M. (1988). Microbial release and degradation of catechol and chlorophenols bound to synthetic humic acid, *Soil Sci. Soc. Am. J.*, **52**, 1366.

259. Dec, J., Shuttleworth, K. L. and Bollag, J.-M. (1990). Microbial release of 2,4-dichlorophenol bound to humic acid or incorporated during humification, *J. Environ. Qual.*, **19**, 546.

260. Fletcher, M. (1996). *Bacterial Adhesion: Molecular and Ecological Diversity*, ed. Fletcher M., Wiley-Liss, New York.
261. van Schie, P. and M. Fletcher. (1999). Adhesion of biodegradative anaerobic bacteria to solid surfaces, *Appl. Environ. Microbiol.*, **65**, 5082.
262. Di Grazia, P. M., Blackburn, J. W., Bienkowski, P. R., Hilton B., Reed, G. D., King, J. M. H. and Sayler, G. S. (1990). Development of a system analysis approach for resolving the structure of biodegrading systems, *Appl. Biochem. Biotechnol.*, **24/25**, 237.
263. Kefford, B., Kjelleberg, S. and Marshall, K. C. (1982). Bacterial scavenging: Utilization of fatty acids localized at a solid-liquid interface, *Arch. Microbiol.*, **133**, 257.
264. Gordon, A. S., Gerchakov, S. M. and Millero, F. J. (1983). Effects of inorganic particles on metabolism by a periphytic marine bacterium, *Appl. Environ. Microbiol.*, **45**, 411.
265. Harms, H. and Zehnder, A. J. B. (1995). Bioavailability of sorbed 3-chlorodibenzofuran, *Appl. Environ. Microbiol.*, **61**, 27.
266. Bell, C. R. and Albright, L. J. (1982). Attached and free-floating bacteria in a diverse selection of water bodies, *Appl. Environ. Microbiol.*, **43**, 1227.
267. Thomas, J. M. and Alexander, M. (1987). Colonization and mineralization of palmitic acid by *Pseudomonas pseudoflava, Microbiol. Ecol.*, **14**, 75.
268. Haider, K. (1995). Sorption phenomena between inorganic and organic compounds in soils: Impacts on transformation processes. In *Environmental Impact of Soil Component Interactions—Natural and Anthropogenic Organics*, ed. Huang, P. M., Berthelin, J. B., Bollag, J.-M., McGill, W. B. and Page, A. L., CRC Press, Boca Raton, FL, p. 21.
269. Filip, Z., Haider, K. and Martin, J. P. (1972). Influence of clay minerals on growth and metabolic activity of *Epicoccum nigrum* and *Stachybotrys chartarum, Soil Biol. Biochem.*, **4**, 147.
270. Haider, K., Filip, Z. and Martin, J. P. (1970). Einfluss von Montmorillonit auf die Bildung von Biomasse und Stoffwechselzwischenprodukten durch einige Mikroorganismen, *Arch. Mikrobiol.*, **73**, 201.
271. Anderson, T.-H. and Gray, T. R. G. (1990). Soil microbial carbon uptake characteristics in relation to soil management, *FEMS Microbiol. Ecol.*, **74**, 11.
272. Burns, R. G. (1986). Interactions of enzymes with soil mineral and organic colloids. In *Interactions of Soil Minerals with Natural Organics and Microbes*, ed. Huang, P. M. and Schnitzer, M. Special Publication No. 17, Soil Science Society of America, Madison, WI, p. 429.
273. Boyd, S. A., and Mortland, M. M. (1990). Enzyme interactions with clays and clay-organic matter complexes. In *Soil Biochemistry*, Vol. 6. ed. Bollag, J.-M. and Stotzky, G., Marcel Dekker, New York, p. 1.
274. Gianfreda, L., and Bollag, J.-M. (1996). Influence of natural and anthropogenic factors on enzyme activity in soil. In *Soil Biochemistry*, Vol. 9, ed Stotzky, G. and Bollag, J.-M., Marcel Dekker, New York, p. 123.
275. Harter, R. D. and Stotzky, G. (1971). Formation of clay–protein complex, *Soil Sci. Soc. Am. Proc.*, **35**, 383.
276. Harter, R. D. and Stotzky, G. (1973). X-ray diffraction, electron microscopy, electrophoretic mobility, and pH of some smectite-protein complexes, *Soil. Sci. Soc. Am. Proc.*, **37**, 116.
277. Harter, R. D. (1975) Effect of exchange cations and solution ionic strength on formation and stability of smectite-protein complexes, *Soil Sci.*, **120**, 174.
278. Quiquampoix, H. (1987). A stepwise approach to the understanding of extracellular enzyme activity in soil. I. Effect of electrostatic interactions on the conforma-

tion of a β-D-glucosidase adsorbed on different mineral surfaces, *Biochemie*, **69**, 753.

279. Hamzehi, E., and Pflug W. (1981). Sorption and binding mechanisms of polysaccharide cleaving soil enzymes by clay minerals, *Z. Pflanzenernaehr. Bodenkd.*, **144**, 505.

280. Morgan, H. W. and Corke, C. T. (1976). Adsorption, desorption, and activity of glucose oxidase on selected clay species, *Can. J. Microbiol.*, **22**, 684.

281. Claus, H., and Filip, Z. (1988). Behaviour of phenoloxidases in the presence of clays and other soil-related adsorbents, *Appl. Microbiol. Biotechnol.*, **28**, 506.

282. Martinek, K., and Mozhaev, V. V. (1985). Immobilization of enzymes: An approach to fundamental studies in biochemistry, *Adv. Enzymol.*, **57**, 179.

283. Garwood, G. A., Mortland, M. M. and Pinnavaia, T. J. (1983). Immobilization of glucose oxidase on montmorillonite clay: Hydrophobic and ionic modes of binding, *J. Molec. Catal.*, **22**, 153.

284. Boyd, S. A., and Mortland, M. M. (1985). Manipulating the activity of immobilized enzymes with different organo-smectite complexes, *Experientia*, **41**, 1564.

285. Ruggiero, P., Sarkar, J. M. and Bollag, J.-M. (1989). Detoxification of 2,4-dichlorophenol by a laccase immobilized on soil or clay, *Soil Sci.*, **147**, 361.

286. Sarkar, J. M., Leonowicz, A. and Bollag, J.-M. (1989). Immobilization of enzymes on clays and soils, *Soil Biol. Biochem.*, **21**, 223.

287. Ruggiero, P., Dec, J. and Bollag J.-M. (1996). Soil as a catalytic system. In *Soil Biochemistry*, Vol. 9, ed. Stotzky, G. and Bollag, J.-M., Marcel Dekker, New York, p. 79.

288. Goldstein, L. (1976). Kinetic behavior of immobilized enzyme systems. In *Methods in Enzymology*, Vol. XLIV, ed. Mosbach, K., Academic Press, Orlando, FL, p. 397.

289. Nannipieri, P., Ceccanti, B. and Bianchi, D. (1988). Characterization of humus–phosphatase complexes extracted from soil, *Soil. Biol. Biochem.*, **20**, 683.

290. Serban, A. and Nissenbaum, A. (1986). Humic acid association with peroxidase and catalase, *Soil Biol. Biochem.*, **18**, 41.

291. Ruggiero, P. and Radogna, V. M. (1987). Tyrosinase activity in a humic–enzyme complex, *Sci. Tot. Environ.*, **62**, 365.

292. Gosewinkel, U. and Broadbent, F. E. (1986). Decomplexation of phosphatase from extracted soil humic substances with electron donating reagents, *Soil Sci.*, **141**, 261.

293. Mackboul, H. E. and Ottow, I. C. G. (1979). Einfluss von zwei- und Dreischichttonmineralen auf die Dehydrogenase-, saure Phosphatase- und Urease-Aktivität in Modellversuchen, *Z. Pflanzenernaehr. Bodenkd.*, **142**, 500.

294. Ruggiero, P. and Radogna, V. M. (1988). Humic acid–tyrosinase interactions as a model of soil humic–enzyme complexes, *Soil Biol. Biochem.*, **20**, 353.

295. Sarkar, J. M. and Bollag, J.-M. (1987). Inhibitory effect of humic and fulvic acids on oxidoreductases as measured by the coupling of 2,4-dichlorophenol to humic substances, *Sci. Tot. Environ.*, **62**, 367.

(B) *Rhizosphere Chemistry and Biology*

9 Chemical and Biological Processes in the Rhizosphere: Metal Pollutants

P. M. HUANG AND J. J. GERMIDA

University of Saskatchewan, Canada

Interactions between Soil Particles and Microorganisms
Edited by P. M. Huang, J.-M. Bollag and N. Senesi. © 2002 John Wiley & Sons, Ltd

1. INTRODUCTION

The term 'rhizosphere' was first used by Hiltner [1] but has since been modified and redefined. It is the narrow zone of soil influenced by the root and exudates. The extent of the rhizosphere may vary with soil type, plant species, age, and many other factors [2], but it is usually considered to extend from the root surface out into the soil for a few millimeters. The rhizosphere can be divided into the ectorhizosphere, which is the soil region, and the endorhizosphere, which includes the rhizosphere and the epidermis and cortical cells of the root that are invaded by microorganisms (Figure 1). The ectorhizosphere can extend a substantial distance from the root with the development of mycorrhizal fungal associations [4].

More intense microbial activity and larger microbial populations occur in this zone than in bulk soil, in response to the release by roots of large amounts of organic compounds. Up to 18 % of the C assimilated through photosynthesis can be released from roots. Microbial populations in the rhizosphere can be 10–100 times larger than the populations in bulk soil [5]. Therefore, the rhizosphere is bathed in root exudates and microbial metabolites. Thus, the chemistry and biology at the soil–root interface differs significantly from those in bulk soil.

The soil rhizosphere is the bottleneck of metal contamination of the terrestrial food chain. The dynamics, transformations, bioavailability, and toxicity of

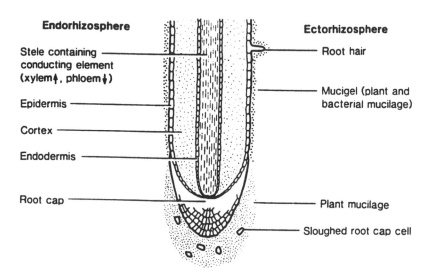

Figure 1. Root region. From [3]. Reproduced by permission of Blackwell Science Inc.

metal pollutants are influenced substantially by chemistry and biology of the rhizosphere.

2. ROOT-INDUCED CHANGES IN SOIL PROPERTIES

2.1 PHYSICOCHEMICAL AND BIOCHEMICAL PROPERTIES OF SOILS

2.1.1 Rhizosphere pH

The pH of the rhizosphere generally differs from that of the bulk soil. The changes in pH are often higher than one unit and may be more than two units [6,7]. The direction (increase or decrease) of changes in pH in the rhizosphere is determined by plant factors. The degree of pH changes and their extension from the root surface toward the bulk soil depend on both plant and soil factors.

The important soil factors that influence changes in rhizosphere pH are the initial soil pH and the pH buffering capacity. In most soils, buffer capacity is lowest at about pH 5. The root-induced pH changes in the rhizosphere are, thus, maximal at bulk soil pH between 5 and 6 [6,8].

The plant factors that cause changes in rhizosphere pH are excretion or reabsorption of H^+ or HCO_3^-, evolution of CO_2 by root respiration, and release of low molar mass (LMM) root exudates (e.g., organic acids). Indirectly, production of organic acids and CO_2 by microorganisms from organic carbon released by roots also contributes to the changes in rhizosphere pH. CO_2 is presumably of minor importance for changes in rhizosphere pH in aerated soils, because it rapidly diffuses away from the root through air-filled pores. Rhizosphere pH changes are predominantly the result of differences in net excretion of H^+ and $HCO_3^-(OH^-)$ due to an imbalance between cation and anion uptake [9]. The form of N supply (NH_4^+;NO_3^-;N_2 fixation) has the most prominent influence on rhizosphere pH [10]. Nitrate supply is usually correlated with a higher rate of net release of HCO_3 (or H^+ consumption) than net excretion of H^+; with NH_4^+ supply, the reverse is the case. Striking differences in rhizosphere pH also exist between plant species growing in the same soil and supplied with NO_3^-N [11].

Changes in solution pH at the soil–root interface may have substantial impact on the dynamics, transformation, and bioavailability of metal pollutants. The observation that Cd availability in soil decreases with increasing soil pH [12] suggests that rhizosphere acidification will result in an increase in Cd uptake. NH_4^+ supply has been shown to increase Cd uptake in wheat grain [13], which is attributable to rhizosphere acidification in response to NH_4^+ supply. Further, it has been demonstrated that the more acidifying N fertilizers, such as ammonium sulfate, increase Cd uptake over those (for example, calcium nitrate) having an alkaline effect in soil [14,15]. Therefore, the impact of changes of rhizosphere pH on metal transformation and food chain contamination merits close attention.

2.1.2 Rhizosphere Redox Potential

The information on gradients in the redox potential around the roots in aerated soils is limited. Oxygen consumption by roots and rhizosphere biomass most likely decrease the redox potential in the rhizosphere. However, in one of the few studies on redox potential in the rhizosphere zone of aerated soils, gradients in redox potential around alfalfa (*Medicago sativa* L.) roots were not observed [16]. Nevertheless, in aerated soils, anaerobic microsites do occur and vary in location and time. Such microsites are most likely more abundant in the rhizosphere than in the bulk soil [17] because of O_2 consumption by roots and microbial respiration.

In contrast, in waterlogged soils, the redox potential of the bulk soil is low leading to high and phytotoxic concentrations of Mn^{2+} and Fe^{2+} and organic solutes in the soil solution. In plants adapted to waterlogging and to submerged soils, high redox potentials in the rhizosphere are maintained by O_2 transport from shoot through aerenchyma to the roots and release of O_2 into the rhizosphere. This process is termed 'internal ventilation' [18]. The formation of aerenchyma is a typical root response to low O_2 levels in the rhizosphere, most likely induced by an increase in the ethylene concentrations [19,20]. The capacity for such adaptation differs among plant species and even among cultivars[18,21].

Root-induced oxidation of heavy metal, such as Fe^{2+}, is associated with production of H^+ and the resulting decrease in rhizosphere pH (Figure 2). The capacity of plant roots to release O_2 into the rhizosphere by wetland species is shown by substantial precipitation of Fe (III) oxides at the external surface of rhizodermal cells. In wetland rice, striking differences in genotype exist in this respect, and at maturity, about 500 kg of Fe OOH per hectare is present as root coating each season [23].

Compared with the oxidation rate of Fe^{2+} in the rhizosphere of wetland species, the oxidation rate of Mn^{2+} is relatively low [24,25]. An important factor for lowering the Mn^{2+} concentration in the soil solution of the rhizosphere is adsorption of Mn^{2+} to the freshly precipitated Fe (III) oxides [25].

2.1.3 Root Exudates

Plant roots release a wide variety and considerable amounts of organic compounds into the rhizosphere. On average, between 30 and 60% of the net photosynthetic C is allocated to the roots in annual species; of this C, an appreciable portion (4–70%) is released as organic compounds into the rhizosphere [26,27]. Over the whole vegetation period in soil-grown crop species,

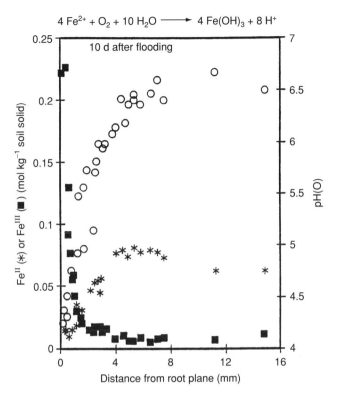

Figure 2. Root-induced Fe oxidation and pH changes in the rhizosphere of lowland rice. From [22]. Reproduced by permission of the Executive Editor, New Phytologist

more than twice as much organic C is released into the rhizosphere than retained in the root system at harvest [28]. Compared with root exudation under sterile condition, losses of C from roots are much greater under conditions where microorganisms are present [29].

The major components of rhizodeposition are LMM organic compounds, mucilage, and sloughed-off cells and tissues:

(1) LMM organic compounds: The compounds of this group are organic acids, amino acids, sugars, phenolics, and phytosiderophores [30–39]. The occurrence, nature and concentration of LMM organic acids (LMMOAs) common in the soil solution are summarized in Table 1. Both the amounts and the proportion of the various organic compounds of root exudates vary substantially with plant species and cultivars. Further, the same plant cultivar grown in different soils varies in the kind and amount LMMOAs present in the rhizosphere (Table 2). These LMMOAs can interact with

metals to influence their transformations, mobility, and toxicity, as discussed in section 3.

(2) Mucilage: Mucilage consists mainly of polysaccharides, such as polygalacturonic acids [53]. It is secreted by the root cap and rhizodermal cells [54].

Table 1. Concentration and occurrence of organic acids common in the soil solution and in nature. Modified from [40]

Organic acid*	Concentration in soil solutions ($\times 10^5$ mol L^{-1})	Occurrence in nature	References
Acetic	265–570	Microbial metabolities. Accumulates if microbial respiration is anaerobic. Commonly found in root exudates of many grasses and herbaceous plants and in green manures. Volatile but can be adsorbed by soil clays.	[41,42]
Amino	8–60	Building blocks of all plant protein. Omnipresent. Concentration in soils that are under cultivation is higher than in fallow fields.	[43]
Benzoic	7.5 or less	Accumulates if microbial action is intense. The hydroxy derivative (p-hydroxybenzoic acid) is more common and inhibits plant growth and seed germination. p- hydroxybenzoic acid has been identified in Podzol B horizons.	[44–47]
Citric	1.4	Important intermediate in the tricarboxylic acid cycle of all higher plants and organisms. Identified in root exudates of mustard plants. Present in high concentrations in many fruits and leaves. Produced by soil fungi.	[41,48–50]
Formic	250–435	Produced by bacteria in the rhizosphere. Has been isolated from root exudate of corn. Volatile.	[41]
Malic, tartaric, malonic	100–400	Excreted by roots of many cereals and solanaceous crops.	[41,43]
Oxalic	6.2	Produced during lysis of microbial cells. May make up 50% of the dry weight of the leaves of plants. Commonly present in root exudates of cereals. Nonvolatile.	[41,49]
Tannic acid related compounds (gallic acid, tannins and other phenolic acids)	5–30	Common in tea leaves and barks of trees. Related compounds isolated in root exudates of mustard. Common in waters near tanneries. If the phenolic acids in senescent leaves are polymerized and condensed, tannic acid is produced. Tannins are believed to be building blocks of humus because of the similarity and stability of the functional groups.	[47,48,50–52]

* In addition to the above list, other LMM aliphatic and aromatic acids have been identified in Podzol B horizons [47]. Fifty percent is believed to be of phenolic origin. The native vegetation in podzolic areas are coniferous and deciduous forests. In nonforested regions, native vegetation is usually heath in temperate climate or savanna grasslands in the tropics.

Table 2. Amount of low molar mass organic acids ($\mu g\,kg^{-1}$ soil)* in rhizosphere soil of durum wheat cv. Kyle grown in three different soils as determined by gas chromatography ($n = 3$). Reproduced from [38] by courtesy of Marcel Dekker Inc.

Acid	Soil		
	Yorkton	Sutherland	Waitville
Malonic	99a	56a	68a
Succinic	22a	35476c	10826b
Fumaric	12a	150b	71ab
Malic	45a	898c	370b
Tartaric	ND†	665b	214a
trans-Aconitic	ND	13a	3a
Citric	ND	195b	81a
Acetic	865a	29245c	12240b
Propionic	ND	499a	ND
Butyric	ND	7604b	2127a
Total	1043a	74801c	26000b

* On the oven-dry basis of soil.
† ND = not detected.
Means within the same row having the same letter are not significantly different ($p \leq 0.05$).

Mucilage has a series of biological functions [55], which include protection of apical root zones of plants from desiccation and improvement of root–soil contact, especially in dry soils. Plant roots growing through a layer of soil with a water potential below the wilting point can uptake a significant amount of Zn provided that other parts of the root system have access to water elsewhere, for example, in the subsoil [56]. Under such conditions, mucilage may facilitate the transport of Zn from the soil particles within the mucilage to the plasma membrane of root cells [57]. In acid mineral soils, mucilage may protect root apical zones from the harmful effects of monomeric and polynuclear hydroxy Al species on cell division and expansion [58,59]. Further, mucilage may protect root apical meristems from heavy metal toxicity.

(3) Sloughed-off cells and their lysates: These materials are primarily a C substrate for rhizosphere microorganisms. They can become important in metal transformation and detoxification indirectly via the activity of rhizosphere microorganisms [34]. The microbial population density, especially that of bacteria, is higher in the rhizosphere than that of the bulk soil because of the high supply of organic C by roots [26]. At least some of the end-products of microbial activity (e.g., organic acids, phenolics, siderophores) may have effects on metal mobilization similar to those of the (LMM) root exudates.

Root exudation is increased by various forms of stress, such as anaerobiosis, drought, mechanical impedance, and nutrient deficiency [9,26]. A schematic presentation of root exudation as affected by nutrient deficiency and mechanical impedance is shown in Figure 3 [60]. Root exudation is much higher from roots growing in solid substrates, such as quartz sand or glass beads, or soils than in nutrient solution. Root length is usually strongly negatively related to the bulk density of the substrate. Therefore, allocation of photosynthates to roots is enhanced by mechanical impedance; the consumption of photosynthates for respiration and exudation per unit root length may increase by a factor of 2 in soils with high bulk density as compared with those of low bulk density [61]. Mechanical impedance increases root exudation of sugars and vitamins by a factor of 3 [62] and of phenolics by a factor of 10 [63].

The increase in root exudation as a result of mechanical impedance has an important effect on the ability of plants to tolerate high Al concentrations. In nutrient solution, a concentration of only 74 μmol L^{-1} Al greatly inhibited root elongation; the same concentration in the percolating nutrient solution in sand had no inhibitory effect [64]. This effect was attributed to higher root exudation and the corresponding decrease in the concentration of the toxic

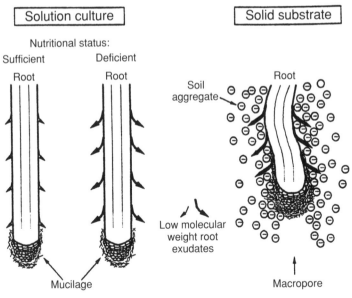

Figure 3. Schematic presentation of root exudation as affected by mineral nutrient deficiency and mechanical impedance. From [60]. Reproduced by permission of Academic Press Ltd

Al species in the rhizosphere. Besides tolerance to Al toxicity, the increase in root exudation brought about by mechanical impedance has important implications for the dynamics and transformations of heavy metals [9].

The amount of LMM root exudates often increases and the composition of these exudates is changed when plants are nutrient deficient [9]. For example, enhanced root exudation of organic acids is often observed under P deficiency in dicots, and particularly in legumes. In alfalfa (*Medicago sativa* L.) even latent P deficiency, when the total dry weight has not yet been depressed, increases root exudation of citric acid by about twofold [65]. Under K deficiency in maize the amounts of exudates increase and the proportion of organic acids to sugars increases [66]. The enhanced release of phenolics, such as caeffeic acid, can be observed in many dicots in response to Fe deficiency [67,68]. The impact of nutrient deficiency on the increase of root exudates and the alteration of the composition of organic molecules in the rhizosphere merits close attention in the transformation, mobility and bioavailability of metals and their subsequent food chain contamination.

2.1.4 Soil Solution Composition and Ionic Strength in the Rhizosphere

The soil solution is the medium through which metals are taken up, either actively or passively, by plant roots and microbial cells. The composition of the soil solution has been treated in several reviews [69–73]. The soil solution is the center of chemical and biological activity in soil. In the absence of sufficient water, soil organisms become dormant or die, mineral transformations become imperceptibly slow, and soil chemical weathering and formation processes become greatly impeded. Addition of moisture to previously dry systems reinitiates these reactions.

The soil solution is the resultant of many dynamic equilibria that occur in soils [72]. Some of the reactions attain equilibrium rapidly, while others do not. The soil solution is the key for understanding and interpreting which chemical reactions can occur and which ones do occur. Monitoring the composition of the soil solution with time provides valuable information on how rapidly equilibrium is approached. The dynamic equilibria that occur in soils are depicted in Figure 4. The soil solution is the focal point of the dynamic equilibria. Plants take up nutrients and pollutants from the soil solution and decrease their activities in the immediate vicinity of absorbing roots. Plants differ in their ability to take up nutrients and pollutants and to produce exudates in the rhizosphere. Ions are retained by surface adsorption and exchange sites in soils. The precipitation–dissolution reactions of minerals can ultimately control the activity of certain ions in the soil solution, which in turn affects the activity of these ions on exchange sites. Ion activity gradients are the driving forces that move ions into and out of the exchange sites. The nature of exchange sites and the exchange capacity of soils affect the chemistry of the soil solution,

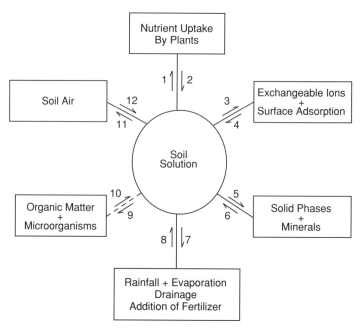

Figure 4. Dynamic equilibria in soil that affect the composition of the soil solution. From [74]. Reproduced by permission of The Blackburn Press

especially when ion activities are not controlled by precipitation and dissolution reactions of the solid phases.

Evapotranspiration concentrates the soil solution, whereas rainfall dilutes it (Figure 4). Water draining through soil carries with it dissolved salts, transporting them deeper into the soil profile and finally to groundwater. Water-soluble fertilizers (e.g., monoammonium phosphate) applied to soils readily dissolve, and the reaction products form in the environment of soils.

Microorganisms and organic matter also have an important role in influencing the composition of the soil solution in soil (Figure 4). Microorganisms accelerate the decomposition of organic matter and the synthesis of new biochemical reaction products. The dashed line (Figure 4) indicates that the equilibrium for the reactions involving organic matter is generally not attained. Enzymes [75] and abiotic catalysts [76] are important factors that affect the rates of the formation and degradation of organic components. Organic matter, especially LMMOAs, greatly enhance the solubility and mobility of metals in soil environments.

Gas phase molecules continuously exchange between soil air and soil solution (Figure 4). Consumption of O_2 and release of CO_2 in respiration processes occur in soils. When the water content of soils increases, gaseous diffusion becomes restricted due to the decrease in air spaces. Therefore, oxidized soils become reduced, which results in the modification of the composition of the soil solution. Consequently, the transformation, mobility, and bioavailability of redox-sensitive metals will be affected.

The most common inorganic constituents in the soil solution are summarized in Table 3. Predominant cations are usually Ca^{2+}, Mg^{2+}, and K^+, with a large number of minor cations such as Fe, Cu, and Zn. The most prevalent anions are HCO_3^-, Cl^-, and SO_4^{2-}. Dissolved Si is predominantly in the form of a neutral molecule, $Si(OH)_4^0$. Soils that become flooded are strongly influenced by reducing conditions and microbial activities and Fe^{2+}, HS^-, and HCO_3^- may take on greater importance. The sum of soluble cations and anions is usually less than 10^{-2} mol L^{-1}. Among the cations in alkaline, neutral, and near neutral soils, Ca^{2+} comprises at least half of this total. For the anions, the division among HCO_3^-, Cl^-, and SO_4^{2-} will depend on the pH and the composition of the solids in soil.

Soluble organics originate from organisms, including exudates from plant roots and soil microorganisms. The concentrations of these organics in the soil solution are more dynamic and more variable than the inorganic constituents. The composition of root exudates is discussed in section 2.1.3. The occurrence and concentration of LMMOAs common in the soil solution are summarized in Table 1. The exudation of soluble organic compounds by plants is perhaps one of the more important processes that affect the dynamics of metals in the rhizosphere. However, little is known about speciation of metal contaminants in the rhizosphere.

Stotzky's group has conducted extensive research work and published a series of review articles on interactions between heavy metals, microorganisms,and surface-active particles in the environment, especially in soil [77–81]. These interactions should influence the chemistry of the soil solution of the

Table 3. Major inorganic components found in soil solutions. Concentration ranges are estimates and could change depending upon specific environments. The components listed are not necessarily the dominant solution species. Reprinted with permission from [73]. Copyright CRC Press, Boca Raton, Florida

Category	Major components (10^{-4}–10^{-2} mol L^{-1})	Minor components (10^{-6}–10^{-4} mol L^{-1})	Others*
Cations	Ca^{2+}, Mg^{2+}, Na^+, K^+	Fe^{2+}, Mn^{2+}, Zn^{2+}, Cu^{2+}, NH_4^+, Al^{3+}	Cr^{3+}, Ni^{2+}, Cd^{2+}, Pb^{2+}
Anions	HCO_3^-, Cl^-, SO_4^{2-}	$H_2PO_4^-$, F^-, HS^-	CrO_4^{2-}, $HMoO_4^-$
Neutral	$Si(OH)_4^0$	$B(OH)_3^0$	

* Components normally found in concentrations $< 10^{-6}$ mol L^{-1} unless in a contaminated environment.

rhizosphere. The main abiotic and environmental factors, which influence metal speciation and toxicity, include temperature [82], clay minerals [83,84], humic substances [85], pH [83,86], and chloride and salinity [82,87]. Although these findings were not obtained from the rhizosphere system, they are relevant to understanding the solution chemistry of metal transformation in the rhizosphere.

The chemistry of the soil solution and its effect on metal retention by soil have implications for the rhizosphere environment. Increasing salt concentrations in the solution affects metal retention by soils through effect on the surface potential of soil particles, effect of the cation in competition for sorption sites, effect of ionic strength on the metal activity coefficient, and the effect of ionic strength on dissociation constants [88]. However, there has been no comprehensive investigation of the extent to which the ionic strength of the soil solution of the rhizosphere differs from that of bulk soil. Plants transpire approximately 250 ml water for each gram of dry matter produced and contain approximately 300 cmol ions kg^{-1} dry matter. If the concentrations of ions in the bulk soil solution is > 1.2 cmol L^{-1}, the ionic concentration in the rhizosphere solution will be lower than that in the bulk soil [7]. However, exudation of organics and the accumulation of certain ions (e.g., Ca) and organic material through mass flow at the root surface in excess of plant requirement could increase the ionic strength of the soil solution in the rhizosphere. Further, because plants and microorganisms rapidly remove ions from solution and exude substances into soils, it is unlikely that any chemical reaction can attain equilibrium. Therefore, the chemistry of the soil solution in the rhizosphere is a challenging area for future research.

2.2 DYNAMICS OF MICROBIAL POPULATIONS

One of the major components of soil is the living biota [89]. This biota consists of organisms of all sizes and shapes. For example, the megafauna, which includes such organisms as earthworms and insect larvae, is capable of moving large soil particles, of creating tunnels and cracks in the soil fabric, and breaking down the structural components of plant material. Hence, these organisms contribute significantly to transformations of organic matter and soil structure. The mesofauna, which includes such organisms as springtails and nematodes, is primarily responsible for feeding on smaller organisms. The activity of these organisms increases nutrient turnover and elemental cycling in soil. The smallest organisms are the microbiota. This includes a wide array of microorganisms, such as protozoa, bacteria, actinomycetes, and fungi. These microorganisms can be considered the driving force, or the catalytic engine, behind the key elemental and organic matter transformations in soil. These microorganisms, through various mechanisms and metabolic processes, affect metal availability and toxicity.

Microorganisms under the influence of the plant roots may be categorized into three groups [90]. The first group, i.e., rhizosphere microorganisms, consists of selected soil organisms in very close proximity, within 1–2 mm, of the root surface. These microorganisms are under the direct and indirect influence of root metabolic processes. Differences between numbers and types of rhizosphere microorganisms and those living in bulk soil are well documented [91–93]. The second group, i.e., rhizoplane microorganisms, are those organisms living on the root surface. In many instances, these organisms are physically attached to the plant root. The third group consists of organisms termed endophytes, which live inside the root. Plant growth influences all these organisms, either directly or indirectly, and hence, any factors that influence plant growth affect the activity of these organisms. For the purpose of this discussion, microorganisms in the rhizosphere and rhizoplane will be referred to as 'noninfecting rhizosphere microorganisms', and endophytes will be considered as 'infecting rhizosphere microorganisms'.

2.2.1 Noninfecting Rhizosphere Microorganisms

The relative composition and number of microbial populations in the rhizosphere constantly changes in response to growth of and exudation by plant roots. Hence, environmental factors that influence root exudation may influence microbial population numbers and activity. These interrelations in the rhizosphere are illustrated in Figure 5. This fluctuation in numbers in microbial populations due to root activity is referred to as the 'rhizosphere effect'. This rhizosphere effect diminishes with increasing distance from the root surface, and many studies have tried to quantify this phenomenon [91,94–101].

A large number of chemical and biological interactions affecting metal bioavailability and toxicity occur in the rhizosphere. Since Hiltner [1] first defined the rhizosphere and investigated the influence of roots on bacterial activity, many controversies have been raised regarding the true nature of the microbial populations on roots. Does the rhizosphere harbor a special microbiota or does the root system merely stimulate the general native population? Recent studies [97–100,102–104] suggest that plant roots select for, and stimulate the growth of, specific microbial populations but that this process may be mitigated by soil factors. The consequence of this is that plant–microbe systems may be utilized to mobilize and transport metals in soils specifically, especially soils contaminated with toxic heavy metals [105].

A great deal of effort has been devoted to enumeration and characterization of microorganisms in the rhizosphere. This is justified by the idea that understanding the nature and population dynamics of rhizosphere microorganisms would provide insights as to their functionality and allow the prediction of their ability to transform organic materials, toxic and essential metals, and, ultimately, predict how they might influence plant growth. Unfortunately, many of

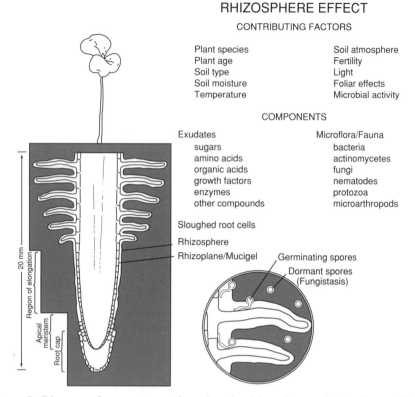

Figure 5. Diagram of a young root featuring the rhizosphere and rhizoplane. Major organic materials released by the root and groups of the microbiota affected are indicated along with factors governing the extent of root influence. *Inset* shows fungal spores germinating in the rhizosphere but not outside this nutrient zone. From [2]. Reproduced by permission of Springer-Verlag

these studies are flawed, because they tended to access only those microorganisms that were easily cultured and detected on artificial laboratory media. It is estimated that over 90 % of the soil microbiota cannot be cultured, and, thus, it is not possible to study their influence on organic matter transformations or cycling of elements [89,106]. Nevertheless, estimates of microbial population numbers and dynamics do provide some useful information. Standardized techniques for estimating and determining population dynamics of rhizosphere microorganisms are summarized elsewhere [107–109]. These estimates provide relative numbers that may be compared with other soil samples if the estimates have been based on the same determinations or the same methodologies.

An alternative to culture methods involves the use of direct microscopic examination of rhizosphere soils [89,110]. This provides larger estimates of the number of microorganisms in soil samples, but it does not enable living cells to be distinguished from dead ones. The use of specific fluorescent dyes, or fluorochromes, makes it possible to assign viable or nonviable status to some of these directly counted microorganisms [89]. More recently, molecular techniques have been used to extract and analyze microbial DNA from the rhizosphere [106]. This provides insight on the genetic information found in the population and in some cases is linked to information on the presence of specific catabolic genes involved in nutrient or elemental transformations.

Irrespective of the techniques used to estimate the microbial population in the rhizosphere, it is clear that the population number and dynamics in the rhizosphere are significantly greater than in bulk soil not under the influence of the root. Comparative estimates for microbial populations in rhizosphere (R) and non-rhizosphere soils (S) can be used to calculate the R/S ratio of soil [95,101]. The R/S ratio will vary from one crop to another and from soil to soil. The rhizosphere effect exerted by a growing plant is typically more pronounced for bacterial populations (including actinomycetes) than for fungi. Examples of these estimates are shown in Tables 4 and 5. In the case of culturable bacteria, the rhizosphere effect of actively growing plants usually corresponds to R/S ratios of somewhere between 2 and 20, but ratios of over 100 have been recorded [2]. The wide variation of values is because of the different rhizosphere effects of different plant species, and it may be associated with different stage of plant growth. Thus, any R/S comparisons made at any one point in time will only be valid when the same plant growing in different soils is compared at the same point in its physiological growth stage.

Bacteria are by far the most abundant microorganisms in field or rhizosphere soils, whether determined by plate counts or direct microscopy or even using

Table 4. A comparison of the colony counts of bacteria in the rhizosphere and rhizoplane of different crop plants and in root-free soil. From [101]. Copyright The Society for Applied Bacteriology

	Colony count ($10^6\,g^{-1}$ of soil based on oven dry weight)		R/S*	Colony count ($10^6\,g^{-1}$ of rhizoplane based on root oven-dry weight)
	Rhizosphere	Root-free		
Red clover	3,255	134	24	3,844
Oats	1,090	184	6	3,588
Flax	1,015	184	5	2,450
Wheat	710	120	6	4,119
Maize	614	184	3	4,500
Barley	505	140	3	3,216

* The ratio of the colony counts of bacteria in the rhizosphere to those in the root-free soil.

Table 5. Comparison of the numbers of various groups of organisms g^{-1} (oven-dry basis) soil in the rhizosphere of spring wheat and in control soil. From [95]. Reproduced by permission of Soil Science Society of America

	Rhizosphere soil	Control soil	R/S ratio*
Bacteria	$1,200 \times 10^6$	53×10^6	23
Actinomycetes	46×10^6	7×10^6	7
Fungi	12×10^6	1×10^5	12
Algae	5×10^6	27×10^3	0.2

* The ratio of the numbers of organisms in the rhizosphere to those in the control soil.

molecular techniques [2,90]. Typically, population counts indicate levels of approximately 10^6–10^9 cells per g of soil. These numbers are generally an order of magnitude greater in the rhizosphere than in bulk soils. Unfortunately, studies on population dynamics and activity of actinomycetes in rhizosphere soil have lagged behind those on other groups of bacteria and fungi. Interest in these organisms is primarily due to their metabolic capabilities. Many actinomycetes produce an inventory of organic molecules such as antibiotics. Hence, they may influence the activity of bacteria and fungi. More recently, it has become apparent that actinomycetes are involved in transformations of some recalcitrant materials. In addition, some actinomycetes are responsible for the production of secondary metabolites, some of which may have the ability to bind and mobilize trace elements [111].

Populations of soil fungi are typically much lower than those estimated for bacteria and actinomycetes. Rhizosphere soils may contain fungal populations on the order of 10^3–10^5 viable propagules per g of soil. The R/S ratios for fungi in soil may range in the order of 3 to over 200, but they are frequently found to be somewhere between 10 and 20 for crop plants [2]. It is important to note, however, that the fungal biomass in soil is much greater than that of bacteria and actinomycetes. This fungal biomass may accumulate and store large amounts of metals or, alternatively, produce substantial amounts of organic acids that affect metal availability. In addition, many fungi are endophytes, living inside the root and extending extraradical hyphae into the soil that affect transformation and mobilization of elements.

2.2.2 Infecting Rhizosphere Microorganisms/Endophytes

Many bacteria and fungi form very specific associations or symbiosis with plant roots. The *Rhizobium*–legume symbiosis is probably the best-known example of a plant–microbe association [89]. This association provides a direct benefit to the plant, as the bacterium is capable of fixing atmospheric nitrogen and providing the resultant NH_4^+ to the host, and the host provides a protected environment and a food source (i.e., photosynthate) to the rhizobial endophyte.

Another example of symbiosis is between plants and fungi. This fungal association is referred to as mycorrhizae [112,113]. The mycorrhizal association may be one of the most important and one of the least understood biological associations, as most plants grown in soil are colonized by mycorrhizal fungi. Studies on the population dynamics of mycorrhizae in soil are limited and typically reflect changes in number of fungal resting spores or mycelium inside plant roots. The role of mycorrhizae on uptake of nutrients, such as phosphorus, is well documented, but little is known about the uptake of metals in soil [112,114]. The most common mycorrhizae found in agricultural soils are arbuscular mycorrhizae , whereas ectomycorrhizae are more common in forest soils. Both types of mycorrhizae are responsible for the uptake and increased availability of phosphorus, ammonium, nitrate, potassium, calcium, sulfate, copper, zinc and iron. In addition to mobilizing nutrients directly and transporting them to the root, mycorrhizae are capable of modifying the rhizosphere environment by direct and indirect means (i.e., increasing aggregate stability thereby altering pH and Eh; releasing organic acids that solubilizing elements; binding elements to hyphae, etc.), which can influence elemental availability [112].

It has become apparent that a large number of bacterial endophytes colonize the interior of roots of many plants including grasses and tree species [97,103,115,116]. The function and activity of these bacteria are not understood. It appears that these organisms reside inside the roots receive protection from the soil and a readily available source of food in the form of photosynthate. Little is known about the population dynamics of these bacteria or their ability to transform nutrient and pollutant elements. One can only imagine that they have an important role in both mobilization and transport of different kinds of metal contaminants.

For all endophytes, population dynamics are directly affected by the nature of the plant and the soil. For example, the *Rhizobium*–legume symbiosis is a host-specific phenomenon, and the population of rhizobia and the occurrence of the symbiosis in a soil are only evident when the specific host is present [89]. Nevertheless, populations of rhizobia in soil do change in response to the organic acids and amino acids typically found in root exudates, even in the rhizosphere of nonlegumes [117]. Mycorrhizal fungi are capable of colonizing the roots of many different susceptible host plants, and their population levels will vary depending on the mycotrophic nature of the host and edaphic soil factors. For example, plants with fibrous root systems tend to support lower populations and less active mycorrhizal populations than plants with coarse root systems. Mycorrhizal fungi appear to respond to some root exudates, such as flavinoids [112]. There is no evidence of growth of arbuscular mycorrhizal fungi in soil in the absence of host plant roots.

3. PHYSICOCHEMICAL REACTIONS OF METAL TRANSFORMATION IN THE RHIZOSPHERE

3.1 REDOX REACTIONS AND METAL TRANSFORMATION

Redox reactions are important in controlling the chemical speciation of a number of contaminant metals, notably As, Se, Cr, Pu, Co, Pb, Ni, and Cu [118]. Redox reactions also are important in controlling the transformation and reactivity of Mn and Fe oxides in soils, which have enormous capacities to adsorb metal pollutants and are the major sinks of these pollutants. Further, reduction of sulfate to sulfide in anaerobic environments also affects the transformation, solubility, and availability of metal pollutants through the formation of highly insoluble metal sulfides [119].

Metals such as As, Se, Cr, Pu, Co, Pb, Ni, and Cu exist in more than one oxidation state. Therefore, redox reactions are critical in controlling their transformation, mobility, and toxicity. Arsenic exists in two main oxidation states, namely, As(III) (arsenite) and As(V) (arsenate). Its bioavailability and toxicity depend on its chemical state, with As(III) being more toxic than As(V) [120]. Although As(V) is a thermodynamically stable species in oxygenated water at pH values common in natural environments [121], the kinetics of oxidation of As(III) with O_2 is very slow at near neutral pH values [122]. Manganese oxides are capable of oxidizing As(III) to As(V) in soil [123,124]. The ability of Mn oxides to oxidize As(III) to As(V) varies with their crystallinity and specific surface [124]. The mechanism of conversion of As(III) to As(V) catalyzed by Mn oxides has been treated [123–125]. The conversion of As(III) to As(V) by Mn oxides has important implications for the transport, fate, and toxicity of As in soil environments. Arsenious acid is weak and only dissociates at high pH ($pK = 9.23$ at $25\,°C$), and arsenite is less firmly bound by soil particles than arsenate. Reducing conditions may release As into solution through reduction of Fe and Mn oxides and reduction of arsenate to arsenite. Various organic forms of As, e.g., mono- and dimethyl arsenic acid, also occur in soils [126].

Selenium can exist in a wide range of oxidation states, namely, Se($-$II), Se(0), Se(II), Se(III), Se(IV), and Se(VI); it can also exist as organic Se compounds [127]. Their concentration and biogeochemical transformations determine the activity of Se in the environment. Although Se is essential to human and animal life at low concentration, Se compounds at high concentrations are carcinogenic and teratogenic [127]. The higher valent forms of Se are more soluble and their reduction in soil to the less reactive Se(0) form has generally been considered to be facilitated primarily by soil organic acids [128] and microorganisms [129,130]. However, many suboxic soil environments contain green rust (GR), which is a mixed Fe(II)–Fe(III) oxide, and it has been shown to catalyze redox reactions [131–133]. X-ray absorption near-edge structure spectroscopic evidence shows that Se reduces from an oxidation state of VI to 0 in the presence of GR [134]. These redox reactions represent an abiotic pathway for Se cycling in natural environments. Similar GR-mediated abiotic redox reactions are probably involved in the mobility of several other trace metals in the environment.

Chromium and Pu are similar in chemical behavior in the environment [135,136]. Both elements can exist in multiple oxidation states and as cationic and anionic species in aqueous systems. Chromium occurs in the II, III, and VI oxidation states in the environment. Chromium (II) is unstable. Chromium(III) has broad stability, exists as the cation Cr^{3+} and its hydrolysis products, or as the anion CrO_2^-. Chromium(VI) exists under strongly oxidizing conditions, occurs as dichromate $Cr_2O_7^{2-}$ or chromate $HCrO_4^-$ and CrO_4^{2-} anions. Plutonium exists in the III to VI oxidation states as Pu^{3+}, Pu^{4+}, PuO_2^+, and PuO_2^{2+} in strong acid conditions. Chromium(III) and Pu(III)/(IV) cations are sorbed to soil constituents and, therefore, immobile in most soil environments. In contrast, Cr(VI) and Pu(VI) are quite mobile in soils, because they are not sorbed by temperate soils to any extent. These elements in the hexavalent form are readily bioavailable and extremely toxic [137] and, thus, of concern in food chain contamination and ecosystem health. Manganese(III/IV) oxides can oxidize Cr(III) and Pu(III/IV) [137,138]. Therefore, Mn oxides can enhance the mobility and toxicity of Cr and Pu in soil ecosystem.

Manganese oxides and oxyhydroxides can also catalyze the oxidation of other trace metals, such as Co, Pb, Ni, and Cu [76,139,140]. The catalytic transformation of these metal ions may be directly influenced by redox processes coupled to disproportionation of Mn mixed-valence oxide, to catalyzed oxidation by aqueous O_2, or to other redox reactions involving changes from one Mn oxide species to another. When the oxidized form of the element has a lower solubility than the reduced form, this effect can be of major significance. More information on the kinetics and mechanisms of redox reactions of these trace metals on the surfaces of Mn oxides is needed.

Besides Mn oxides, heterogeneous oxidation/reduction reactions involving electron transfer between transition metals and Fe-containing minerals have been investigated. Fe(III) oxyhydroxides can oxidize V(II) and V(IV) [141]. Reduction of Cr(VI) by biotite has been reported by Eary and Rai [142] and Ilton and Veblen [143]. Direct evidence of Cr(VI) reduction on magnetite surfaces has been shown by using X-ray absorption fine structure spectroscopy [144]. Recent experimental evidence has demonstrated that structural Fe(II) in magnetite and ilmenite heterogeneously reduces aqueous Cu(II), V(IV), and Cr(VI) at the oxide surfaces [145]. Compared with the catalytic oxidation of transition metals, far fewer investigations have addressed the reduction of transition metal contaminants on the surfaces of natural Fe(II)-bearing minerals [146].

Massechelyn and Patrick [147] have summarized the critical redox potentials for the transformation of some metal contaminants in soils. There has been little study of how changes in soil redox potential in the rhizosphere could affect the chemistry of metal contamination. However, the creation of an oxidized zone adjacent to the plant root in wetland soils has been identified as one process affecting the chemistry of As, Zn, and Cu in soils. In wetland soils, it has been well established that steep gradients in redox potentials develop around plant roots as a result of O_2 release from the roots. This process is reflected in precipitation of FeOOH (iron plaque) on the roots [148,149]. Compared with the surrounding soil, these Fe-rich plaques on the roots of the salt-marsh plant, *Aster tripolium*, are enriched in Zn and Cu [148]. Zinc also accumulates in the rhizosphere of rice (*Oryza sativa* L.), which is the result of the formation of a zone of oxidation of Fe^{2+} to Fe^{3+} adjacent to the roots [149]. Zinc concentration in red roots (with iron plaque) is higher than in white roots; a positive effect of the Fe concentration on the root surface, up to a certain level, on Zn uptake into the xylem fluid has been demonstrated [148]. Above this level of Fe coating, Zn uptake by the plant is reduced, due possibly to complete coating of the root surface by FeOOH and blocking of absorption sites. Therefore, Fe coatings on roots up to a certain level could enhance uptake of heavy metal contaminants by plants.

Arsenic is mobilized in reduced conditions as a result of reduction of Fe and Mn oxides and reduction of As(V) to As(III). However, in the rhizosphere in wetlands, As is immobilized owing to the oxidation to As(V) and adsorption to FeOOH [150]. Therefore, As has been found to accumulate in the rhizosphere of many plant species [150,151]. Arsenic uptake from solution is increased with increasing Fe concentration on the roots, but most of the As is likely to be retained on the root surface.

3.2 SOLUTION COMPLEXATION REACTIONS AND METAL TRANSFORMATION

A series of complexation reactions in the soil solution affect metal transformation in the rhizosphere. Metals very often act as the central components that attract neutral or negatively charged ligands. In the hydrolysis of a metal ion, such as Al^{3+} to form $Al(H_2O)_5(OH)^+$, the Al^{3+} acts as the central unit, and water and hydroxyl ion act as the ligands. This is a special case of complex formation. There are two other categories of complexes that can be formed depending on the strength of bonding between the central unit and the ligand. If the interaction between the cation and the ligand is strong enough for the ligand to displace the hydration sphere, the resulting complexation is termed the formation of an inner sphere complex. If the interaction between the cation and the ligand is not strong enough to displace the inner hydration sphere, the association is termed the formation of an outer sphere complex or an ion pair.

In an outer sphere complex or an ion pair, the metal ion or the ligand or both retain the coordinated water when forming the complex compound. Therefore, the metal ion and the ligand are separated by one or more water molecules. In contrast, the interacting ligand is immediately adjacent to the metal ion in an inner sphere complex. If a ligand that combines with the metal ion contains two or more electron donor groups so that one or more rings are formed, the complex is referred to as a multidentate complex and a metal chelate. The reaction is termed chelation.

The hard and soft acid–base rules [152] are useful concepts in predicting the association between cations and ligands. According to the Pearson terminology, Lewis acids and bases can be classified on a scale of 'hard' to 'soft'. Soft bases (ligands) are large molecules that are easily polarized; they readily donate electrons to form covalent bonds. They selectively bond with soft acids, typically metal ions of relatively large radius and low charge. Hard bases tend to be small molecules that are not easily polarized and form less covalent and more ionic bonds, preferentially with hard acids—typically metal ions of small radius and high charge. Therefore, as a rule, soft acids bond preferentially with soft bases and hard acids bond preferentially with hard bases. The classification of Lewis acids and bases according to the Pearson Hardness Concept is listed in Table 6.

Complexation reactions of metals with ligands in the soil solution are significant in determining the chemical behavior and toxicity of metals in the rhizosphere. In view of the occurrence of organic and inorganic ligands in the rhizosphere [7,9,37,38] and the stability constants of the complexes of metals with these ligands [153], a large fraction of the soluble metal ions in the soil solution may actually be complexed with a series of organic and inorganic ligands commonly present in the rhizosphere.

The study of metal speciation in the soil solution has been encouraged by the free metal hypothesis in environmental toxicology [154]. This hypothesis states

Table 6. Hard and soft acids and bases. Reproduced with permission from [152]. Copyright (1963) American Chemical Society

	Hard	Intermediate	Soft
Bases	F^-, CO_3^{2-}, OH^-, CH_3COO^-, PO_4^{3-}, SO_4^{2-}, NH_3, $R-NH_2$, H_2O, $R-OH$, NO_3^-	SO_3^{2-}, NO_2^-, $C_6H_5NH_2$	S^{2-}, CN^-, Cl^-, I^-, $R-SH$
Acids	H^+, Li^+, Na^+, K^+, Mg^{2+}, Ca^{2+}, Sr^{2+}, Al^{3+}, La^{3+}, Si^{4+}, Zr^{4+}, Th^{4+}, Cr^{3+}, Mn^{3+}, Fe^{3+}	Mn^{2+}, Fe^{2+}, Ni^{2+}, Cu^{2+}, Zn^{2+}, Pb^{2+}, Bi^{3+}	Ag^+, Cu^+, Cd^{2+}, Hg^{2+}, Sn^{2+}

that the toxicity or bioavailability of a metal is related to the activity of the free aquo ion. Although this hypothesis is gaining popularity in studies of soil–plant relations [155] some evidence is now emerging that the free metal ion hypothesis may not be valid in all situation [156]. Differences in plant uptake of metals at the same free metal activity were noted when different chelators were used in solution culture studies [157–159]. Bell *et al.* [158] reported that critical Fe^{3+} activities in solution for growth of barley (*Hordeum vulgare* L.) depend on the kind of the chelate in the system. Further, given the same chelate, total Fe concentration in solution affects plant uptake of Fe. Either kinetic limitations to dissociation of the chelate or uptake of the intact chelate could explain these observations. Römheld and Marschner [160] found that maize (*Zea may* L.) could absorb intact FeEDDHA (Fe-ethylenediamine-dio-hydroxyphenyl acetate) at breaks in the endodermis where lateral roots budded. Taylor and Foy [157] were able to induce Cu^{2+} toxicity in wheat (*Triticum aestivum* L.) by increasing the concentration of CuEDTA (Cu-ethylenediamine tetraacetate) while maintaining the Cu^{2+} activity of the solution constant. This also suggests uptake of the intact chelate or better buffering of free Cu^{2+} activities at the site of uptake, which leads to increased Cu concentrations in the plant. The latter hypothesis assumes that there is a large diffusive limitation to Cu uptake in the unstirred layer adjacent to the root or in the root apoplast. The possible reactions of complexed metals at the soil–root interface are diagramatically presented in Figure 6.

The complexation of Cd, Cu, and Pb by 8-hydroxyquinoline (8-HQ), diethyldithio-carbamate (DDC), EDTA, and sulfoxine (SO_x) in relation to the uptake of these metals by diatoms (*Thalassiosira weissflogii*) was reported by Phinney and Bruland [161]. Although the metals actually transferred across the cell membrane and the metals adsorbed to the cell surface were not differentiated, their data indicated that, in comparison with free metal or negatively charged metal-organic complexes (EDTA, SO_x), organic ligands (DDC, 8-HQ) forming uncharged complexes with metals increase metal uptake dramatically. These data suggested

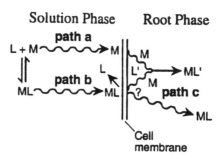

Figure 6. Conceptual model for potential uptake by plants of metal ions (M) complexed by organic ligands (L). Modified from [159]

that uncharged metal–organic complexes can cross cell mem branes. Therefore, the role of metal–organic complexes in metal uptake merits attentions [156]. Further research is warranted to investigate the mechanisms of the uptake of metals as influenced by the size and charge of metal–organic complexes.

Exudates of various kinds isolated from axenically grown plants have been shown to complex metals [162–165]. As evidenced by ^{14}C-labeling, root-derived compounds have been shown to be able to complex Co, Mn, and Zn during plant growth [166]. Hamon et al. [167] also reported that Cd and Zn were complexed in the soil solutions of the rhizosphere after radish growth (Table 7). In this experiment, Ca concentrations in solution were low as a result of plant growth depleting Ca concentrations in solution. This allowed greater complexation of both Zn and Cd than would be expected in the rhizosphere where Ca concentrations are usually higher than in the bulk soil [7]. Using a new Donnan equilibrium method to measure activities of free Cd^{2+} in solution, Helmke et al. [168] have clearly demonstrated that the activity of Cd^{2+} in sludge-amended soils was strongly dependent on Ca and Mg concentrations in the soil solution.Their data indicated that the extent of complexation of Cd in soil is influenced by the activity of Ca and Mg in the soil solution.

Seasonal changes of the concentrations of such metals as Cu, Mn, Zn, and Co in the rhizosphere are related to the presence of complexing agents of biological origin [169,170]. Krishnamurti et al. [171] reported variations in pH and the cadmium availability index (CAI) of the bulk and rhizosphere soils collected after 2 weeks and 7 weeks of crop growth (Table 8). At the 2-week growth stage the pH of the rhizosphere soil was lower than that of the corresponding bulk

Table 7. Percentages of total Cd or Zn in soil solution present as the free ion in relation to days of growth of radish plants. Note the soil solution at Day 0 is assumed to be equivalent to bulk soil solution and that at 30 days rhizosphere soil solution as a result of root exploration throughout the pot. From [167]. Reproduced by permission of Blackwell Science Inc.

Days of radish growth	Cd^{2+} (%)	Zn^{2+} (%)
0	95	100
14	15	17
16	11	3
18	15	8
20	25	21
22	31	11
24	50	0
26	36	11
28	28	0
30	28	13

Table 8. The influence of application of Idaho phosphate fertilizer on pH and cadmium availability index (CAI)* of the bulk and rhizosphere soils collected at 2-week and 7-week crop growth stages. Modified from [171]

Soil and cultivar		Bulk soil		Rhizosphere soil			LSD‡	
			Control†		Idaho§			
			2-week	7-week	2-week	7-week	0.01	0.05
Luseland soil								
Kyle	pH	7.95	7.75	7.90	7.38	7.90	0.11	0.08
	CAI	87	97	88	152	88	85	63
Arcola	pH	7.90	7.80	7.85	7.38	7.85	0.11	0.08
	CAI	87	102	87	208	89	85	63
Jedbergh soil								
Kyle	pH	8.15	7.95	8.10	7.38	8.10	0.14	0.10
	CAI	9	12	9	80	10	27	19
Arcola	pH	8.10	7.90	8.05	7.70	8.05	0.14	0.10
	CAI	9	16	9	84	8	27	19

* CAI = Cadmium availability index (in $\mu g\,kg^{-1}$) determined by $1\,mol\,L^{-1}$ NH_4Cl extraction method of Krishnamurti et al. [172].
† Without application of Idaho phosphate fertilizer.
§ With application of Idaho phosphate fertilizer.
‡ LSD = Least significant difference at $p = 0.01$ and $p = 0.05$.

soil and the CAI values, which were determined by the method of Krishnamurti et al. [172], were higher in the rhizosphere soil, indicating that more Cd was complexed with the LMMOAs at the soil–root interface. Compared with the bulk soils, the CAI values were 2–9 times higher in the rhizosphere of field plots fertilized with Idaho monoammonium phosphate fertilizer at the 2-week growth stage, which was attributed to the combined effects of the Cd introduced into the rhizosphere from the fertilizer and of complexation reactions of phosphate and LMMOAs with soil Cd. At the 7-week plant growth stage, such differences were not observed. Appreciable amounts of LMMOAs were detected in the root exudates in the rhizosphere soils collected at the 2-week plant growth stage, and no LMMOAs were detected in the rhizosphere soils collected at the 7-week plant growth stage (P. M. Huang, unpublished data, 1996). The high value of CAI observed in the rhizosphere soils at the 2-week growth stage was attributed to the result of complexation of the particulate-bound Cd with solution LMMOAs at the soil–root interface. The enhanced root growth of the plants, particularly in the soils treated with Idaho phosphate fertilizer, might have resulted in high amounts of metal–organic complex-bound Cd species in the soil solution by complexation. The prolific plant and microbial activity with the application of phosphate fertilizer is expected to result in increased amounts of LMMOAs in solution at the soil–root interface. Therefore, a larger fraction of the metal contaminants will be in a complexed and usually soluble form in the rhizosphere than in the bulk soil.

Inorganic ligands also have a role in forming metal complexes in the soil solution. For example, the toxic metal, Cd(II), has a tendency to complex with as many as four Cl^- ions [153]. Metal complexes with inorganic ligands such as Cl^- may also be taken up by plants [173,174]. In these studies, both solutions unbuffered and buffered with respect to Cd^{2+} activity were used. Cadmium uptake by Swiss chard [*Beta vulgaris* (L.) Koch] was found to be related not only to Cd^{2+} activity in solution, but also to activities of a series of Cd–chloride complexes formed in the presence of increasing concentrations of Cl in solution. These results can be explained based on two possible mechanisms: (1) a series of Cd–chloride complexes are taken up directly by the root and/or (2) the presence of Cd–chloride complexes allows greater availability (and, hence, greater buffering) of Cd^{2+} in the unstirred liquid layer adjacent to the root or in the root apoplast to sites of ion uptake. Although the hypothetical efficiency of uptake of Cd–chloride complexes is less than for the free metal ion, Cd^{2+}, the results help to explain the large increase in Cd uptake by field crops observed under conditions of chloride salinity [175,176]. Cadmium–chloride complexes constituted the dominant species of inorganic Cd in soil solutions from irrigated horticultural soils in South Australia; Cd–chloride and–sulfate complexes accounted for 75 and 20%, respectively, of the inorganic Cd in soil solution [177]. In view of the composition of saturation extracts of a wide range of saline soils in the USA reported by Jurinak and Suarez [178], the complexation of Cd^{2+} with Cl^- and SO_4^{2-} is probably significant in many saline soils (Table 9). It

Table 9. Median concentrations of Cl^- and SO_4^{2-} anions in well waters, river waters, and saturation extracts of a range of salt-affected soils in the USA and impact on speciation of inorganic Cd in soil solution

Ligand	Number of samples analyzed	Median concentration $(mmol\,L^{-1})$*	Percentage of inorganic Cd complexed by ligand at median concentration†
Cl^-			
Saturated extract	139	34.8	65
Well water	115	2.5	17
River water	58	1.5	11
SO_4^{-2}			
Saturated extract	134	29.4	50
Well water	23	3.6	23
River water	58	4.1	25

* Jurinak and Suarez [178].
† McLaughlin *et al.* [179].

has been reported that complexation of Cd^{2+} by SO_4^{2-} in solution culture, unbuffered with respect to Cd^{2+} activity, has no effect on Cd uptake [7]. Similar to a series of Cd–chloride complexes, this may be as a result of uptake of the uncharged $CdSO_4^0$ complex or the presence of $CdSO_4^0$ complexes, which enhances greater availability and, hence, greater buffering of Cd^{2+} in the unstirred liquid layer adjacent to the root or in the root apoplast.

Besides saline soils, chloride and sulfate can be introduced to soil environments through applications of fertilizers, deicing salts, sewage effluents, and animal wastes. The impact of such anthropogenic activities on the formation of metal–chloride and–sulfate complexes in the soil solution and the subsequent mobility and food chain contamination merits attention.

3.3 ADSORPTION AND DESORPTION REACTIONS OF METALS

Dynamics and equilibria of metal ions in soil environments are significantly influenced by adsorption–desorption reactions [7,180–183]. Soils are remarkable for their abilities to remove metal ions from the soil solution by adsorption reactions. Those elements that take the form of cations are retained in soils by cation exchange reactions on mineral colloids and organic matter. More selective and less reversible adsorption reactions, such as surface complexation with organic functional groups and bonding with the reactive sites on variable-charge minerals, can also retain and even fix metal cations. Metals and metalloids also can exist in anionic form in solution. They can be retained in soils, primarily by chemisorption at reactive sites of variable charge mineral surfaces and edges of phyllosilicates. These types of cation and anion adsorption are collectively referred to as specific adsorption, which is distinctly different from non-specific adsorption. The specific adsorption is inherently less reversible than the non-specific adsorption.

Crystalline and, especially, noncrystalline oxides, hydroxides and oxyhydroxides of Fe, Al, and Mn, noncrystalline aluminosilicates (allophanes and imogolite), and edges of phyllosilicates are surface reactive sites of soils for the chemisorption of metals and metalloids. All of these mineral surfaces have a similar type of adsorptive sites to metals and metalloids in the soil solution. Ligands, such as OH^- and H_2O, are bound to a metal ion (usually Fe^{3+}, Al^{3+}, or $Mn^{3+,4+}$) on the terminal surface sites of these minerals. A metal (M) may be bound to these mineral surfaces as shown in equation (1):

$$X\text{-}OH^{0.5-} + M(H_2O)_6^{n+} = X\text{-}O\text{-}M(H_2O)_5^{(n-1.5)+} + H_3O^+ \qquad (1)$$

where X is the metal ion at the surface sites. This type of reaction is distinctly different from cation exchange reaction: (1) a high degree of specificity of particular minerals for particular metals, (2) a rate of desorption that is orders of magnitude slower than that of the adsorption rate, and (3) change in the measured surface charge toward a more positive value for a metal ion valency of more than 1 [183].

Metals, which exist as anions, can be chemisorbed on the mineral surfaces as shown in equations (2) and (3):

$$X - OH^{0.5-} + A^{n-} \rightleftharpoons X - A^{(n-0.5)-} + OH^- \tag{2}$$

$$2X - OH^{0.5-} + A^{n-} \rightleftharpoons \begin{array}{c} X \\ \diagdown \\ \diagup \\ X \end{array} A^{(n-1)-} + 2OH^- \tag{3}$$

where $XOH^{0.5-}$ is a reactive surface hydroxyl group and A^{n-} is an anion of charge $-n$. Reactions (2) and (3) are referred to as ligand exchange reactions in which the anion displaces OH^- or H_2O from the terminal surface sites of these minerals. The reactions are promoted by low pH, which causes surface OH^- groups to accept protons. This facilitates the ligand exchange reaction because H_2O is an easier ligand to displace from terminal surface sites than OH^-. This reaction has a high degree of specificity toward particular metals and tends to be nonreversible or, at least, for desorption to be much slower than adsorption.

Except for some noncrystalline minerals that have very high specific surface charge density with very reactive sites, soil organic matter generally has the greatest capacity and strength of binding metals, compared with other soil components [183]. Soil organic matter contains many different types of functional groups that act as Lewis bases in binding metals. These include: (1) hard bases (ligands preferred by hard cations)—carboxylic, phenolic, ester, alcoholic, ether, phosphate, sulfate, (2) intermediate bases (ligands preferred by borderline cations)—primary, secondary and tertiary amines and amides, and (3) soft bases (ligands preferred by soft cations)—sulfhydryl, sulfide, disulfide, and thioether [184]. Carboxylic and phenolic groups are the most abundant functional groups of soil organic matter.

The metal binding reaction by soil organic matter can be described as an ion exchange reaction between H^+ and the metal at acidic functional groups (L):

$$M^{n+} + XLH_Y = XLM^{(n-Y)^+} + Y^{H^+} \tag{4}$$

where X is the organic matrix, L is the acidic functional group, and M^{n+} is a metal cation with a charge of $n+$. The high degree of selectivity of soil organic matter for certain metals indicates that some metals form inner-sphere complexes with the functional groups. The formation of these complexes involves strong ionic or covalent bonds.

For soft metals (e.g., $Cu^{2+}, Cd^{2+}, Hg^{2+}$), the donor atom affinity follows the order: $O < N < S$, whereas the reverse order is observed for hard metals. In general, the competitive reactions for a given ligand essentially involve hard cations and intermediate metals for O sites and intermediate and soft metals for N and S sites, with competition between hard and soft metals being weak [185].

The typical sequence of affinity of soil organic matter for divalent metals (at pH 5) generally parallels their electronegativity values [43,186,187]. However, the relative affinities are often dependent on the method used to measure metal bonding and on pH [43]. The type, source, and concentration of organic matter can affect metal binding affinity. In view of the diversity in functional groups of soil organic matter, a wide range of bonding strengths for metals is expected [188]. This will depend on the nature of the organic matter and metals. Selectivity coefficients for metal binding also vary with the amount of the metal bound. The strength of metal binding by soil organic matter decreases with an increase in the amount of the metal bound. This is, in part, related to the nature of binding sites.

The free radicals present in humic substances have an important role in metal–organic interactions [189–191]. The free radical sites contribute directly to metal binding by soil organic matter. Aromatic free radicals are predominantly involved in reactions with metal ions at high cation (Fe^{3+} or Cu^{2+}) to humic acid (HA) ratios, whereas at low metal concentrations, the predominant reaction is with free radicals of a prevalently aliphatic nature [191]. Electron spin resonance (ESR) spectroscopy has been widely used to study complexes formed by HA and fulvic acids (FA) of different origins and natures with paramagnetic transition metal ions (Fe, Cu, Mn, V, and Mo) that are of great chemical and biological importance to agricultural sustainability and environmental protection. Although ESR spectroscopy is applicable only to paramagnetic metal ions that give a detectable ESR signal, modeling chemical bonding in binding sites using suitable paramagnetic metal ion probes and/or modeling ligand compounds will add to the knowledge of the nature of complexation reactions of nonparamagnetic metal ions of environmental and/or agricultural importance with organic matter in nature [184].

Krishnamurti and Huang [192] reported the relationship between ESR data of HA of soils and their Cd binding ability and availability. No free radicals were observed in the spectrum of the HA extracted by $0.1\,mol\,L^{-1}$ Na pyrophosphate from the Luseland soil, whereas the spectrum of the HA of the Jedburgh soil exhibited free radicals with g values of 2.0025. The HA extracted by $0.5\,mol\,L^{-1}$ NaOH from the Jedburgh soil had a free radical content almost four times higher than that extracted from the Luseland soil. The higher concentration of free radicals in the Jedburgh soil, in comparison with that of the Luseland soil, indicates their role in metal complexation resulting in the less availability of the heavy metal associated with HA, as discussed below. They further reported that the HA of the Jedburgh soil, obtained after extraction with $0.5\,mol\,L^{-1}$ NaOH and particularly with $0.1\,mol\,L^{-1}$ Na pyrophosphate, showed high proportion of carboxylic C (172 ppm) and aliphatic structures bearing OH groups (72 and 83 ppm) resonances in comparison with that of the Luseland soil. The metals complexed with these groups were apparently more stable. The evidence obtained from the ESR and ^{13}C cross-polarization magic angle spinning (CPMAS) NMR spectroscopies [192] were in agreement with the CAI and the Cd concentration of grain of durum wheat grown on the Jedburgh and Luseland soils. The CAI, determined as $1 \, mol \, L^{-1}$ NH$_4$Cl-extractable Cd [172], of the Luseland soil was almost 17 times higher than that of the Jedburgh soil, and the Cd content of the grain of durum wheat, cultivar Kyle, grown in the Luseland soil was nine times that of the Jedburgh soils [172]. Therefore, the nature of soil organic matter has an important bearing on the binding strength of such metals as Cd. The spectroscopic evidence demonstrated differences in the nature of the organic matter of the soils which differed greatly in Cd availability. Further work on a series of soils with widely different metal bioavailability characteristics should provide greater insight on the nature of organic matter and its role in binding of heavy metals in soils.

Organic components also influence the formation and transformation of Al and Fe oxides, hydroxides, and oxyhydroxides [40,193]. Organic matter and other anionic ligands cause structural distortion of these sequioxides and, thus, increase their specific surface and reactive sites [194,195]. Therefore, organic components have an important role in promoting the formation of short-range ordered sesquioxides. Hydrous metal oxides are very reactive constituents for the adsorption of metals [195–198]. The formation of short-range ordered sesquioxides is expected to be especially significant in the rhizosphere where LMMOAs are present in relatively high concentrations as a result of exudation and microbial metabolism.

The adsorption–desorption of metals in the rhizosphere is influenced by a series of factors. These include solution pH, LMMOAs, inorganic ligands, solution ionic strength, cation concentrations, and other compositions of the solution as discussed in section 2.1. Few studies have investigated the adsorption–desorption reactions in the rhizosphere. Most studies examined these reactions in bulk soils and under modifying conditions to simulate rhizosphere conditions.

Excretion products of roots include a variety of LMMOAs, many of which (e.g., citric, oxalic, tartaric, and acetic) are capable of forming complexes with metal ions [48,199]. The relative abundance of LMMOAs in the rhizosphere is both species and cultivar dependent [37,39]. There is increasing evidence to show that LMMOAs secreted by plant roots may contribute to the formation of soluble complexes and chelates [7,166,200] and modify the transformation and dynamics of heavy metals in the rhizosphere [7,201,202]. The biomolecules in metabolites from rhizosphere microorganisms [203] may also have a similar effect on metal transformation and dynamics. Krishnamurti et al. [204] reported that the increase in Cd release in the presence of LMMOAs can be explained by the surface complexation of the particulate-bound Cd in soil with LMMOAs, which is reflected in the increase in the release of Cd from the soils with the increase in the stability constant of Cd–LMMOA complexes (Figure 7).

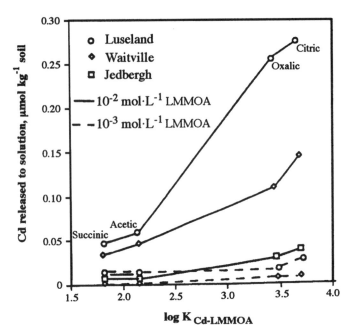

Figure 7. Relationship between Cd released from soils by selected low molar mass organic acids (LMMOAs) (acetic, citric, oxalic, and succinic acids) during the reaction period of 0.25 h and the logarithm of the stability constant (log K) of Cd–LMMOA complexes. From [204]. Reproduced by permission of the American Society of Agriculture, Crop Science Society of America, Soil Science Society of America

Further, the rate coefficients of Cd release from the soils, calculated from the parabolic diffusion law, are substantially influenced by LMMOAs (Table 10). The rate coefficients of Cd release within each ligand vary from soil to soil. The complexibility of soil Cd should vary with the nature of the particulate-bound Cd of the soil. Therefore, the rate of Cd release by each ligand should vary with the nature of the particulate-bound Cd of the soils.

The concentration of Cd species in the soil solution of the soil–root interface governs the amount of labile soil Cd. The importance of the metal–organic complex-bound particulate Cd species in determining the bioavailability of Cd species in soils has been shown by Krishnamurti *et al.* [205]. The rate coefficients of Cd release from the soils by the LMMOAs (Table 10), which is a measure of the rate of the release of soil Cd to soil solution through complexation of soil Cd with LMMOAs, follow the same order as that of the CAI values of the soils [172]. Furthermore, the amounts of the Cd released from the soils by renewal of LMMOAs [204], which is an indication of Cd sustaining power of the soils, also follow the same trend as the CAI values of the soils. The data indicate the importance of the kinetics of Cd release from

Table 10. Rate coefficients of Cd release from soils by low molar mass organic acids (LMMOAs) at 10^{-2} mol L^{-1} during a reaction period of 0.25–1 h. From [204]. Reproduced by permission of the American Society of Agriculture, Crop Science Society of America, Soil Science Society of America

Soil	Rate coefficient* (μmol kg^{-1} h$^{-0.5}$)				
	Acetic acid	Citric acid	Fumaric acid	Oxalic acid	Succinic acid
Luseland	0.112 ± 0.010	0.200 ± 0.015	0.199 ± 0.012	0.079 ± 0.006	0.090 ± 0.005
Waitville	0.046 ± 0.004	0.049 ± 0.003	0.050 ± 0.005	0.036 ± 0.004	0.019 ± 0.003
Jedbergh	0.036 ± 0.005	0.196 ± 0.009	0.041 ± 0.003	0.026 ± 0.004	0.009 ± 0.003

* Calculated on the basis of the parabolic diffusion equation.

soils by LMMOAs found in the rhizosphere soils in understanding Cd availability.

In the rhizosphere, where plants and microorganisms rapidly remove ions from solution and exude substances into soil, not all physicochemical reactions proceed to equilibrium. Progress has been made in the description of the kinetics of adsorption–desorption reactions [206]. However, it is important to note the time scales on which these studies were performed; some reactions may occur in nanoseconds while others may take days, weeks, or months. In terms of plant uptake of metals, reactions that have time scales measured in hours or days are relevant in terms of reaction kinetics controlling availability of metal uptake. Little is known about the dynamics of metals in the rhizosphere. Therefore, kinetics and mechanisms of the adsorption–desorption of heavy metals in the rhizosphere soils as influenced by LMMOAs and other ionic factors prevailing at the root–soil interface merit increased attention. Such information is fundamental to understanding the pathways of the contamination of heavy metals to the terrestrial food chain and to developing innovative management strategies to protect ecosystem health.

3.4 PRECIPITATION AND DISSOLUTION REACTIONS OF METALS

Chemical precipitation may influence the concentration of metals in soil solutions. For many of the more abundant elements such as Al, Fe, and Mn, precipitation of mineral forms is common and may control their solubility. For most of the trace metals, direct precipitation from solution through homogeneous nucleation appears to be less likely than adsorption–desorption by virtue of the low concentrations of these metals in soil solutions in well aerated dryland soils. When soils become heavily polluted, metal solubility may reach a level to cause precipitation. In addition, precipitation may occur in the immediate vicinity of the phosphate fertilizer zone where the concentration of heavy metals present as impurities may be sufficiently high. In reduced environments where the sulfide concentration is sufficiently high, precipitation of trace metals as sulfides may have a significant role in metal transformation[199].

When the concentration of a metal ion in a solution is increased, precipitation of a new solid phase will not occur until the solubility product of that phase has been exceeded. Therefore, some degree of supersaturation is required because crystal nuclei can only be formed after an energy barrier has been overcome. Extreme supersaturation leads to rapid formation of crystal nuclei and produces many small crystallites or even noncrystalline solids. Minimal supersaturation, in contrast, results in exceedingly slow nuclei formation; crystal growth occurs at these few nuclei only, and a highly crystalline product with large crystals, if any, is formed.

The solubility products of metal carbonates, oxides, and sulfides, which are most likely to precipitate in soil environments are shown in Table 11. The solubility of Hg may be controlled by sulfide precipitation in reducing soils, whereas the relatively insoluble Hg oxide can form in nonacidic aerobic soils. The solubility of Cd and Pb may be limited by carbonate and the redox potential of the soils. In isolated instances, silicates, phosphates, and sulfates may limit the solubility of some trace metals. For instance, phosphate concentrations in well fertilized soils may be high enough to induce the formation of Pb phosphate and Zn phosphate precipitates over precipitates of oxides and carbonates [183]. Apatite, a calcium phosphate, can provide phosphate to precipitate Pb for the formation of pyromorphite. Therefore, apatite is a rather effective sink for trace metals, especially Pb, even at low pH [207,208]. Application of soluble phosphate to Pb-contaminated soils is a useful way to reduce Pb solubility and bioavailablity in severely contaminated soils [209]. However, separate Pb phosphate phases do not appear to account for a significant

Table 11. Solubility products of metal carbonates, oxides, and sulfides. Modified from [183]

Carbonates: $K_{so}* = (M^{2+})(CO_3^{2-})$

	Pb	Cd	Fe	Mn	Zn	Ca
$-\log K_{so}$	13.1	11.7	10.7	10.4	10.2	8.42

Oxides and hydroxides: $K_{so} = (M^{n+})(OH^-)^n$

	Fe^{3+}	Al^{3+}	Hg^{2+}	Cu^{2+}	Zn^{2+}	Pb^{2+}	Fe^{2+}	Cd^{2+}	Mn^{2+}	Mg^{2+}
$-\log K_{so}$	39	31.2	25.4	20.3	16.9	15.3	15.2	14.4	12.8	11.2

Sulfides: $K_{so} = (M^{2+})(S^{2-})$

	Hg	Cu	Pb	Cd	Zn	Fe	Mn
$-\log K_{so}$	52.1	36.1	27.5	27.0	24.7	18.1	13.5

* Solubility product constant.

fraction of the total Pb in heavily polluted soils. Lead is more likely to disperse over colloidal mineral and organic surfaces in soils [210].

Metal concentrations in the soil solution are seldom high enough to exceed the solubility product of any trace metal precipitate in aerobic soils [211–213]. Evidence for the control of Cd and Pb solubility by precipitation–dissolution derives from experiments with soils equilibrated in the laboratory with high concentrations of metals in solution, e.g., equilibrium concentrations of $10^{-7}-10^{-3}$ mol L^{-1} Cd and Pb [214] and $10^{-6}-10^{-2}$ mol L^{-1} Cd [215,216]. These concentration ranges are higher than those in most soils, even heavily contaminated soils [71]. In studying a range of heavily contaminated soils (total Cd concentration from 1 to 1638 mg kg^{-1} and total Pb concentration from 330 to 38178 mg kg^{-1}) in England, Jopony and Young [217] reported that metal concentrations in the soil solution could not be adequately described by the solubility product principle. The emerging data obtained from recent improvements in analytical sensitivity and the ability to measure free metal ion activities in true soil solutions [218,219] render support to the view that trace metal concentration in soil solutions appears to be governed by adsorption–desorption reactions rather than by precipitation through homogeneous nucleation and dissolution reactions in aerobic soils [168].

However, in soil environments, heterogeneous nucleation is more likely than homogeneous nucleation because of the presence of mineral, organic, and microbial surfaces that can catalyze the nucleation set of crystallization. The energy barrier to nucleation is reduced or removed by these surfaces. This is especially true in cases where there are crystallographic similarities between the surface and the precipitating phase. This catalytic process reduces the extent of supersaturation necessary for precipitation to occur. For example, $CaCO_3$ in soils seems to promote the heterogeneous nucleation of $CdCO_3$ [211]. However, precipitation reactions are often much slower than adsorption–desorption reactions in soil environments. Besides physicochemical reactions, metals have easy access to bacterial surfaces through diffusion. Metal sorption and precipitation on bacterial surfaces are the interfacial effect. Surface metal concentrations frequently exceed the stoichiometry expected per reactive chemical sites within the cell walls [220, also see Chapter 6 of this book]. The sorption of metals can be so great that precipitates can be formed and distinct metallic minerals are eventually formed, as discussed in section 4.1 of this chapter.

In the rhizosphere, activities of metal contaminants may be decreased through the uptake by plants and microorganisms. Further, concentrations of complexing organic ligands in rhizosphere are higher than in bulk soils. Metal contaminants are, thus, substantially complexed with organic ligands. Compared with bulk soils, the concentration of trace metals in the soil solution of the rhizosphere in aerobic dryland soils appears to be even less controlled by precipitation through homogeneous nucleation. However, in the rhizosphere, precipitation of metals through heterogeneous nucleation on microbial surfaces and metal mobilization by biomolecules as a result of intense biological activity warrant in-depth research.

4. MICROBIAL ACTIVITY AND METAL TRANSFORMATION IN THE RHIZOSPHERE

Microorganisms are the catalytic engines behind the cycling of elements in the biosphere, most of which are essential for the growth of living organisms (Table 12). Major bioelements such as nitrogen, phosphorus, and sulfur, are required in high concentrations and form important constituents of cell components or store energy through chemical bonds. Some major bioelements, such as sulfur or iron, may be used by bacteria to generate energy, e.g., oxidation of Fe^{2+} to Fe^{3+}. Minor bioelements, such as Mn, Se, and Zn, are required at low concentrations (less than $10^{-4} \, mol \, L^{-1}$) and act as important co-factors in cellular metabolism. Some of these elements are toxic at higher concentrations. All bioelements are subject to various microbial transformations that can alter the bioavailability of the elements, ecological flux, or toxicity (Table 13).

Some of these metals are essential for life, and others are potentially toxic depending on their oxidation state and form. The metabolic activity of soil microorganisms may increase or decrease the toxicity and bioavailability of these metals (Table 14). The key metabolic processes involved in the transformation of both macro- and microelements that affect bioavailability are summarized as follows. Mineralization and immobilization reactions are processes that involve the transformation of an element from an organic form to an inorganic form and from an inorganic to an organic form, respectively. Redox reactions involve the oxidation or reduction of an element. Solubilization reactions refer to processes where soil microorganisms increase the solubility of a soil mineral, such as insoluble phosphate minerals, through the production of organic acids or ligands. Methylation reactions are processes where elements such as mercury (Hg) are transformed to methyl Hg through the addition of a methyl group (-CH_3) by microorganisms. These reactions may increase the availability or the toxicity of some metals in the environment. Microbial transformations are mediated and regulated by a number of biotic and abiotic factors in the soil. Consequently, Eh, pH, availability of organic nutrients, biological interactions (e.g., predation, symbiosis) will influence the rate of these transformations or determine whether these transformations will occur.

Microbial transformations of metals regulate the bioavailability, the toxicity, and environmental impact of these elements in the biosphere. All these transformations are closely linked to microbial metabolism and provide the organism with an essential nutrient, energy, or detoxification process.

Table 12. Major bioelements (required at $> 10^{-4}\,\text{mol}\,\text{L}^{-1}$), their sources, and some of their functions in microorganisms. Reproduced from [221] by permission of Springer-Verlag New York, Inc.

Element	Source	Role in organism
C	organic compounds, CO_2	main constituents of cellular material
O	O_2, H_2O, organic compounds, CO_2	
H	H_2, H_2O, organic compounds	
N	NH_4^+, NO_3^-, N_2, organic compounds	
S	SO_4^{2-}, HS^-, S^0, $S_2O_3^{2-}$	constituent of cysteine, methionine, thiamine pyrophosphate, coenzyme A, biotin, and lipoic acid
P	HPO_4^{2-}	constituent of nucleic acids, phospholipids, and nucleotide
K	K^+	cofactor of many enzymes (e.g., kinases): present in cell walls, membranes, and phosphate esters
Ca	Ca^{2+}	cofactor of enzymes; present in exoenzymes (amylases, proteases): Ca-dipicolinate is an important component of endospores
Fe	Fe^{2+}, Fe^{3+}	present in cytochromes, ferredoxins; and other iron-sulfur proteins; cofactor of enzymes (some dehydratases)
Zn	Zn^{2+}	present in alcohol dehydrogenase, alkaline phosphatase, adolase, RNA and DNA polymerase
Mn	Mn^{2+}	present in bacterial superoxide dismutase; cofactor of some enzymes (phosphoenolpyruvate carboxykinase, citrate synthase)
Mo	MoO_4^{2-}	present in nitrate reductase, nitrogenase, and formate dehydrogenase
Se	SeO_3^{2-}	present in glycine reductase and formate dehydrogenase
Co	Co^{2+}	present in coenzyme B_{12}-containing enzymes (glutamate mutase, methylmalonyl-CoA mutase)
Cu	Cu^{2+}	present in cytochrome oxidase and oxygenases
W	WO_4^{2-}	present in some formate dehydrogenases
Ni	Ni^{2+}	present in urease, required for autotrophic growth of hydrogen-oxidizing bacteria

Rhizosphere microorganisms may accumulate substantial amounts of metals during normal metabolic activities and cell growth. These accumulated metals are essentially immobilized and unavailable until the microorganisms die. Upon

Table 13. Microbially mediated transformations of several essential and nonessential elements in soil. Reproduced from [222] by permission of Prentice-Hall

Element	Microbial transformation		
	Mineralization–immobilization	Oxidation–reduction	Methylation
Phosphorus, potassium, calcium, magnesium, copper, zinc	Yes	No	No
Iron, manganese	Yes	Yes	No
Arsenic, mercury, selenium	Yes*	Yes	Yes

* Mercury is transformed between inorganic and organic forms; arsenic and selenium can act as phosphorus and sulfur in biological systems.

Table 14. Elements subject to microbial oxidation–reduction reactions in soils and sediments and examples of bacterial genera involved with each reaction. Reproduced from [222] by permission of Prentice-Hall

Elements and their common oxidation states	Reaction, significance, and redox couple		Some bacterial genera reported to be involved
Cr^{6+}, Cr^{3+}	Oxidation-NR* Reduction-AR, D	$Cr^{6+} + 3e^- \rightarrow Cr^{3+}$	*Aeromonas, Bacillus, Chlorella, Pseudomonas*
Fe^{3+}, Fe^{2+}	Oxidation-E Reduction-AR	$2Fe^{2+} \rightarrow 2Fe^{3+} + 2e^-$ $2Fe^{3+} + 2e \rightarrow 2Fe^{2+}$	*Thiobacillus* *Geobacter, Desulfovibrio, Pseudomonas, Thiobacillus*
Hg^{2+}, Hg^0	Oxidation-NE Reduction-D	$Hg^0 \rightarrow Hg^{2+} + 2e^-$ $Hg^{2+} + 2e^- \rightarrow Hg^0$	*Bacillus, Pseudomonas* *Chlorellas, Pseudomonas, Streptomyces*
Mn^{4+}, Mn^{2+}	Oxidation-ED Reduction-AR	$Mn^{2+} \rightarrow Mn^{4+} + 2e^-$ $Mn^{4+} + 2e^- \rightarrow Mn^{2+}$	*Arthrobacter, Pseudomonas* *Bacillus, Geobacter, Pseudomonas*
$Se^{6+}, Se^{4+}, Se^0, Se^{2-}$	Oxidation-E Reduction-AR	$Se^{2-} \rightarrow Se^0 + 2e^-$ $SeO_4^{2-} + 8e^- \rightarrow Se^{2-}$	*Bacillus, Thiobacillus* *Clostridium, Desulfovibrio, Micrococcus*

* NR, not reported to be biologically mediated; AR, element used as a terminal-electron acceptor in anaerobic respiration; D, detoxification mechanisms; E, energy source, NE, nonenzymatic reaction, microorganism alters the physicochemical environment.

their death, decomposition of microbial biomass results in the mineralization or release of the metal back into the environment. These processes are much more intense in the rhizosphere because of the influence of the growing plants and roots on the environment. Root exudation and dying root cells may stimulate microbial activity in the rhizosphere to four- to fivefold greater than that observed in bulk soil.

4.1 METAL SORPTION AND UPTAKE BY MICROORGANISMS AND BIOMINERALIZATION

As cells grow and increase their biomass, they assimilate, i.e., biologically immobilize, elements by using these elements for the construction of cellular components such as phospholipids or proteins. During growth, metabolic by-products, e.g., organic acids and H^+, exuded by microorganisms alter the surrounding environment and often solubilize elements [111]. In some cases, microorganisms produce metabolites, e.g., siderophores, enzymes, that specifically function to acquire or increase the bioavailability of certain elements. During degradation of dead biomass, microorganisms with a surplus of nutrients (i.e., elements) will convert biologically immobilized elements back into an inorganic form, i.e., mineralization.

All microorganisms contain biopolymers, such as proteins, nucleic acids, and polysaccharides, which provide sites where metal ions can bind (Figure 8). These binding sites include negatively charged groups, such as carboxylate, thiolate, or phosphate, and groups such as amines, which coordinate to the metal center through lone pairs of electrons. Because of the inherent ability of these biopolymers to bind metals, large concentrations of metals are frequently associated, not only with living microbial biomass, but also with dead cells [224].

Many metals bind with various degrees of tenacity to the largely anionic outer surface layers of microbial cells. In a series of detailed studies, isolated cell walls from the common soil bacterium, *Bacillus subtilis* were shown to bind substantial amounts of numerous metals [225]. Binding of metals by microbial cells alters the cell wall composition and induces morphological, ultrastructural, and surface charge changes [226–230]. Some metals were bound by cell walls to a greater extent than by clay minerals (Table 15), suggesting that bacterial cell walls and membranes may act as foci for accumulation of metals or minerals in soil. Binding apparently results largely from the many anionic sites on the wall, especially (i) the phosphodiester groups of teichoic acids, (ii) free carboxyl groups of peptidoglycan, (iii) the sugar hydroxyl group of both major classes of wall polymers, and (iv) the amide groups of peptide chains. Peptidoglycan, however, is especially significant in binding metal ions and forming nucleation sites for secondary deposition.

All metals required by microorganisms are transported into the cell via membrane transport systems with some degree of specificity, to assure the cell of an adequate supply of the metal. A bacterial transport system for inorganic cations is illustrated in Figure 9. Metal ions may diffuse through channels or the lipid phase of cell membranes, but most metal ions are actively (i.e., energy requiring) imported or exported from cells against concentration gradients of considerable magnitude. The uptake of metals usually involves two stages: an apparently instantaneous metabolism-independent binding of the metal to the microbial surface and, in some cases, a metabolism-dependent translocation of

Residue	Substituent	Metals bound
Histidine	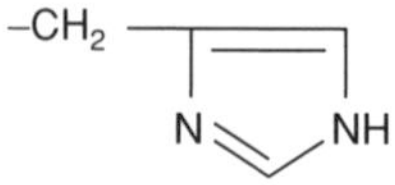 $-CH_2$ (imidazole ring with N, NH)	Zn, Fe, (axial), Cu
Cysteine	$-CH_2SH$	Zn, Fe, Cu
Methionine	$-CH_2CH_2SCH_3$	Fe, Cu
Glutamic acid	$-CH_2CH_2COOH$	Ca, Zn, Fe
Aspartic acid	$-CH_2COOH$	Ca, Zn, Fe
γ-carboxyglutamic acid	$-CH_2CH$(COOH)(COOH)	Ca
β-hydroxyaspartic acid	$-CH$(COOH)(COOH)	Ca
Tyrosine	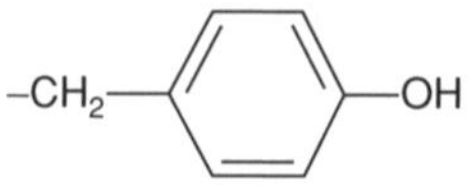 $-CH_2-$(benzene ring)$-OH$	Zn, Fe, Mn, Ca
Serine	$-CH_2OH$	Ca
Asparagine	$-CH_2CONH_2$	Ca
Glutamine	$-CH_2CH_2CONH_2$	Fe
Threonine	$-CH(OH)CH_3$	Ca

Figure 8. Metal-binding groups in proteins. Reproduced from [223] copyright International Thompson Publishing Services

solute from the surface to the interior. Once inside the cell, the metal may be accumulated in particular cell compartments.

Microbial biomineralization (the formation of minerals by microorganisms) is another important activity of microorganisms that is now being defined [220].

Table 15. Metals bound by native *B. subtilis* walls, *E. coli* envelopes, kaolinite, and smectite*. From [225]. Reproduced by permission of American Society for Microbiology

Metal	Amount of metal bound (μ mol g^{-1}) (oven-day weight)			
	Walls	Envelopes	Kaolinite	Smectite
Ag	423 $\pm$ 15	176 $\pm$ 3	0.46 $\pm$ 0.02	43 $\pm$ 0.3
Cu	530 $\pm$ 13	172 $\pm$ 9	5 $\pm$ 0.03	197 $\pm$ 4
Ni	654 $\pm$ 25	190 $\pm$ 3	4 $\pm$ 0.2	173 $\pm$ 10
Cd	683 $\pm$ 19	221 $\pm$ 6	6 $\pm$ 0.2	1 $\pm$ 0.02
Pb	543 $\pm$ 11	254 $\pm$ 5	3 $\pm$ 0.2	118 $\pm$ 6
Zn	973 $\pm$ 13	529 $\pm$ 32	37 $\pm$ 1	65 $\pm$ 2
Cr	435 $\pm$ 37	102 $\pm$ 2	8 $\pm$ 0.5	39 $\pm$ 5

* The data represent the average of three to five determinations for each sample from duplicate experiments and the standard error.

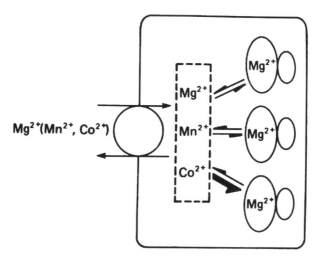

Figure 9. Magnesium transport in *E. coli*. The carrier also transports Mn^{2+} and Co^{2+} which, from an intracellular pool (dashed), can displace Mg^{2+} bound to ribosomes (oval structures). Reproduced from [223] copyright International Thompson Publishing Services

Its scope is much larger than initially thought, as it involves metal transformation and the development of fine-grain minerals of tremendous range and kind (cf. Chapter 6 of this book). Biomineralization has global consequences in dynamics, transformations, and toxicity of metal pollutants. Small quantities of metals are normally associated with many bacterial surfaces and their integral components.

Recent investigations of biomineralization indicate that specific molecular interactions at inorganic–organic interfaces can result in the controlled nucleation and growth of inorganic crystals [231]. A central tenet of biomineralization is that the nucleation, growth, morphology, and aggregation (assembly) of the inorganic crystals are regulated by organized assemblies of organic macromolecules, 'the organic matrix'. Control over the crystallochemical properties of the biominerals is achieved by specific processes involving molecular recognition at inorganic–organic interfaces. Electrostatic binding or association, geometric matching (epitaxis), and stereochemical correspondence are important in these recognition processes. The subtle differences in the kinetics of these recognition processes on different crystal faces lead to specific changes in crystal morphology.

Compared with any other life form, bacteria may have a greater capacity to sorb and precipitate metals from solution leading to mineral development as they have the highest surface area to volume ratio [220]. Although in most environments, soluble metal ions are present at low concentrations, bacterial cells show a remarkable ability to concentrate metal ions from solutions [232]. The dynamics and chemical heterogeneity of these biological interfaces are complicated, but progress in modeling by interfacial theorists is being made[233; personal communication with H.P. van Leeuwen, 2000]. The research on the pH-dependent binding of protons [234] and metal ions to bacterial cell walls and the effect on metal bioavailability is currently emerging as discussed by Huang and Bollag [235].

4.2 METAL MOBILIZATION BY MICROBIAL EXCRETIONS

Microorganisms have a range of metal transport systems that are often highly specific for certain metals and capable of accumulating metals against large concentration gradients (cf. Chapter 5 of this book). However, certain microorganisms synthesise compounds that bind specific metals with high affinity [111]. For example, some microorganisms make iron-binding siderophores, which are organic molecules usually with a phenolate or hydroxamate ligand [236]. Other microorganisms make compounds called metallothioneins, which are small cystine rich proteins that strongly bind copper, zinc, and cadmium. Ligands of this type or related organic molecules are of great interest, as they may facilitate transport and uptake of metals in the rhizosphere.

The use of microbial metabolites to facilitate uptake of metals is best illustrated through the discussion of sideropheres. Iron is essential for all living organisms; however, the solubility product of $Fe(OH)_3$ or $Fe(O)(OH)$ is near 10^{-38} and the maximum concentration of free iron at pH 7 is about 10^{-17} mol L^{-1} [223]. In soil, iron is generally in the oxidized form, Fe(III), which is easily precipitated out from the soil solution, and, thus, is largely unavailable for assimilation by microorganisms. The microbial solution to this problem is the production of specific ligands now generally called side-rophores (Figure 10).

Diverse bacteria and fungi produce siderophores as illustrated in Table 16. Siderophores are recognized by specific receptors on the cell surface, trans-ported back into the cell, and the Fe is released as Fe^{2+} into the cytoplasm by highly elaborate transport systems [237]. Some bacteria have multiple transport systems that use a number of different siderophores produced by other micro-organisms [237]. In other cases, microorganisms are restricted to and are capable of using only one siderophore. Some siderophores are able to complex with other metals such as Cr^{3+}, but interactions between siderophores and metals other than iron are rare[90]. Large amounts of siderophores are excreted by microorganisms into the soil and their impact on the mobility of Fe and other metals merits attention [237]. It should be noted that some plants produce phytosiderophores. These compounds, which are diamino-carboxylic acids, may be excreted into the rhizosphere and might have an important role in the uptake of iron and various other metals that they are capable of binding.

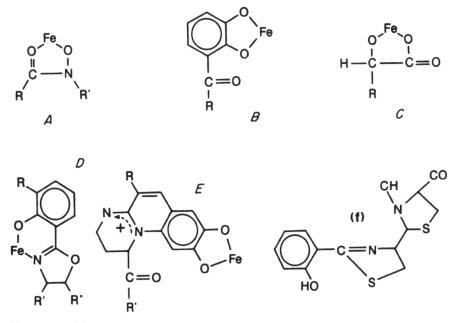

Figure 10. Bidentate ligand systems of siderophores: (A) hydroxamate (ferrichromes, rhodotorulic acid, other hydroxamates); (B) catechol (enterobactin, other catechols); (C) α-hydroxyacid (citrate-containing siderophores, pseudobactin); (D) 2-(2-hydroxyphe-nyl)-oxazoline (mycobactins, agrobactin, parabactin, vibriobactin); (E) fluorescent qui-nolinyl chromophore (pseudobactins); (F) 2-(2-o-hydroxyphenyl-2-thiazolin-4-yl)-3-methylthiazolidine-4-carboxylic acid. From [111]. Reproduced by permission of John Wiley & Sons, Ltd

Table 16. Partial list of sidereophores in microbial species. From [236]. Reproduced with permission from Annual Reviews (www.AnnualReviews.org)

Organism	Siderophore(s)
Bacteria	
Enteric spp.	Enterobactin, aerobactin
Paracoccus denitrificans	Parabactin
Vibrio cholerae	Vibriobactin
Bacillus megaterium	Schizokinen
Anabaena spp.	Schizokinen
Azotobacter vinelandii	α ε-*bis*-2, 3-dihydroxybenzolyllysine
Pseudomonas spp.	Pyochelin, pyoverdine, pseudobactin, ferribactin
Actinomyces spp.	Ferrioxamines
Mycobacteria	Mycobactins
Fungi	
Widely distributed in Ascomycetes and Basidiomycetes	Ferrichromes, coprogen
Rhodotorula spp.	Rhodotorulic acid
Ectomycorrhizal spp.	Hydroxamate type

4.3 MYCORRHIZAL INFECTION AND METAL TRANSFORMATION

Most plants in natural habitats form associations with mycorrhizae. Arbuscular mycorrhizal fungi are obligate symbionts, and infection of plant roots exerts a metabolic load on the host plant. The plant gains a number of benefits from the fungus, including the uptake of metals [238].

During infection and colonization of host plant roots, mycorrhizal fungi produce mycelium inside root cortical cells. They may form storage structures, referred to as vesicles, and they also form other structures termed arbuscules, which serve as the site of nutrient exchange between the host plant and the mycorrhizal fungus. The fungi also form extraradical hyphae that penetrate out of the root and explore the soil in search of various nutrients, including metals. Thus, mycorrhizal associations enhance nutrient solubilization and nutrient uptake in the rhizosphere and expand the volume of soil the root can explore. These benefits include increased rates of nutrient absorption and selective ion uptake.

When a nutrient is deficient in the soil solution, the critical root parameter controlling its uptake is surface area. Hyphae of mycorrhizal fungi have the potential to increase greatly the absorbing surface area of the root. It is also important to consider the distribution and function of the extraradical hyphae. If mycorrhizae are to be effective in nutrient uptake, the hyphae must be distributed beyond the nutrient depletion zone that develops around the root.

Uptake of micronutrients, such as zinc and copper, is also provided by mycorrhizae because these elements are also diffusion limited in soils. In addition, mycorrhizal fungi release organic acids (e.g., oxalic acid) into soil, which can dissolve metal oxides or complex metals which might facilitate uptake of metals. The function of all mycorrhizal systems depends on the ability of the fungal symbiont to absorb nutrients available in inorganic and/or organic forms in soil and to translocate them to the symbiotic roots through the extensive vegetative mycelium. Unfortunately, there are considerable gaps in our understanding of how mycorrhizal fungi facilitate the uptake of metals.

5   IMPACT OF INTERACTIONS OF PHYSICOCHEMICAL, BIOCHEMICAL, AND BIOLOGICAL PROCESSES IN THE RHIZOSPHERE ON METAL CONTAMINATION AND ECOSYSTEM HEALTH

In the rhizosphere, the kinds and concentrations of substrates are different from those in the bulk soil because of root exudation. This leads to colonization by different populations of bacteria, fungi, protozoa, and nematodes. Plant–microbe interactions result in intense biological processes in the rhizosphere. These interactions, in turn, affect physicochemical reactions in the rhizosphere. Physicochemical properties that can be different in the rhizosphere include acidity, concentration of complexing biomolecules, redox potential, ionic strength, moisture, and nutrient status. The total rhizosphere environment is governed by an interacting trinity of the soil, the plant, and the organism associated with the roots (Figure 11). Therefore, the reactions and processes in the rhizosphere can only be interpreted satisfactorily with interdisciplinary approaches. It is difficult to separate the impacts on the chemistry of the contaminant of microbial activity from those of plant root activity in the rhizosphere. Microorganisms can act in a way similar to plant roots. They can accumulate metal contaminants through adsorption and uptake. Also they can mobilize metal contaminants through the action of microbial excretions. Bacteria are especially abundant in the rhizosphere (Figure 12), and they have a large capacity to sorb metals because of their high surface area to volume ratio [240]. Most of the information on the uptake of metal contaminants by microorganisms has been obtained in vitro. The effect of the free-living rhizosphere biota on metal uptake by plants remains to be established and is even less understood for the effect of the rhizosphere biota. Further, the effect of the nature and properties of soil particles on plant–microorganism interactions and the impact on metal uptake remain to be established.

Much of the research on physicochemical reactions of metal contaminants in soil has used well-defined model systems that simulate bulk soil characteristics.

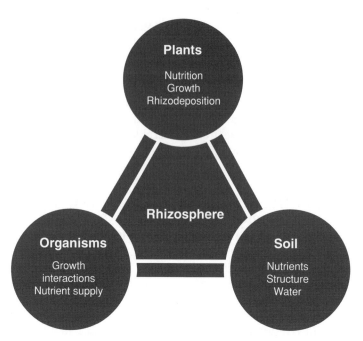

Figure 11. The rhizosphere trinity. The total rhizosphere environment is determined by an interacting trinity of the soil, the plant, and the organism associated with the roots. From [4]. Reproduced by permission of John Wiley & Sons, Ltd

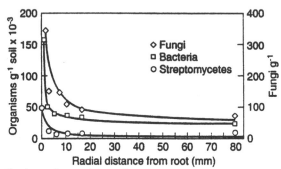

Figure 12. Distribution of organisms with distance from the roots of 18-day old lupin (*Lupinus angusti-folius* L.) seedlings. Modified from [239]

The impacts of physicochemical, biochemical, and biological processes in the rhizosphere on metal uptake, food chain contamination, and ecosystem health merit increasing attention.

6 SUMMARY AND CONCLUSIONS

In the rhizosphere, larger microbial populations and more intense microbial activity occur than in the bulk soil, in response to the release by roots of large amounts of biomolecules. Intense biological activity alters significantly the physicochemical properties of soils.

Physicochemical reactions affecting the chemistry of metal contaminants in the rhizosphere include oxidation–reduction, solution complexation, adsorption–desorption, and precipitation–dissolution reactions. Precipitation reactions may proceed through homogeneous nucleation from soil solutions and, especially through heterogeneous nucleation on surfaces of minerals, organic matter and microorganisms. As a result of root-induced changes in pH, redox potential, concentration of complexing biomolecules, soil solution ionic strength and composition in the rhizosphere, the chemistry of metal contaminants should differ significantly from the bulk soil. More research is needed in this extremely important and challenging area of environmental soil chemistry.

Compared with the bulk soil, microbial population dynamics and activity are significantly enhanced in the rhizosphere. Microbial populations in the rhizosphere can be 10–100 times larger than the populations that occur in the bulk soil. Microorganisms in the rhizosphere can mobilize metal contaminants because of excretion of extracellular complexing biomolecules. Microorganisms also have a large capacity to sorb metals because of their high surface area to volume ratio and their ability to biomineralize. Mycorrhizal infection of plants may either increase or decrease uptake of metals by plants.

Further, the rhizosphere environment is governed by an interacting soil–plant–microorganism trinity. The transformation of metals is, thus, controlled by the interaction of the soil, the plant, and the microorganism associated with the roots. The dynamics and mechanisms of transformation of metals in the rhizosphere and the impact on their mobility, phytoavailability, toxicity, food chain contamination, and ecosystem health should be of increasing concern.

REFERENCES

1. Hiltner, L. (1904). Uber neuere Erfahrungen und Probleme auf dem Gebiet der Bodenbakteriologie und unter besonderer Berucksichtigung und Brache, *Arb Dtsch. Landwirt. Ges.*, **98**, 59.
2. Curl, E. A. and Truelove, B. (1986). *The Rhizosphere*, Springer, Berlin.
3. Lynch, J. M. (1983). *Soil Biotechnology. Microbiological Factors in Crop Productivity*, Blackwell Scientific, Oxford.
4. Lynch, J. M. (1990). Introduction: Some consequences of microbial rhizosphere competence for plant and soil. In *The Rhizosphere*, ed. Lynch, J. M., John Wiley & Sons, Chichester, p.1.

5. Sposito, G. and Reginato, R. J. (1992). *Opportunities in Basic Soil Science Research,* Soil Science Society of America, Madison, WI.
6. Marschner, H. and Römheld, V. (1996). Root-induced changes in the availability of micronutrients in the rhizosphere. In *Plant Roots. The Hidden Half,* 2nd edn, ed. Waisel, Y. Eshel, A. and Kafkafi, U., Marcel Dekker, New York, p. 557.
7. McLaughlin, M. J., Smolders, E. and Merckx, R. (1998). Soil-root interface: Physicochemical Processes. In *Soil Chemistry and Ecosystem Health,* ed. Huang, P. M., Adriano, D. C., Logan, T. J. and Checkai, R. T., SSSA Special Publication Number 52, Soil Science Society of America, Madison, WI, p. 233.
8. Nye, P. H. (1986). Acid–base changes in the rhizosphere, *Adv. Plant Nutr.,* **2**, 129.
9. Marschner, H. 1998. Soil–root interface: Biological and biochemical processes. In *Soil Chemistry and Ecosystem Health,* ed. Huang, P. M., Adriano, D. C., Logan, T. J. and Checkai, R. T., SSSA Special Publication Number 52, Soil Science Society of America, Madison, WI, p. 191.
10. Thomson, C. J., Marschner, H. and Römheld, V. (1993). Effect of nitrogen fertilizer form on pH of the bulk soil and rhizosphere, and on the growth, phosphorus, and micronutrient uptake of bean, *J. Plant Nutr.,* **16**, 493.
11. Marschner, H. and Römheld, V. (1983), *In vivo* measurement of root-induced pH changes at the soil-root interface: Effect of plant species and nitrogen source, *Z. Pflanzenphysiol.,* **111**, 241.
12. Chaney, R. L. and Hornick, S. B. (1978). Accumulation and effects of cadmium on crops. In *Proc. the Int. Cadmium Conf., 1st,* San Francisco, 1977, Metals Bulletin, London, p. 125.
13. Williams, C. H. and David, D. J. (1976). The accumulation in soil of cadmium residues from phosphate fertilizers and their effects on the cadmium content of plants, *Soil Sci.,* **121**, 861.
14. Eriksson, J. E. (1990). Effects of nitrogen-containing fertilziers on solubility and plant uptake of cadmium, *Wat. Air Soil Pollut.,* **49**, 355.
15. Willaert, G. and Verloo, M. (1992). Effects of various nitrogen fertilizers on the chemical and biological activity of major and trace elements in a cadmium contaminated soil, *Pedologie,* **43**, 83.
16. Blanchar, R. W. and Lipton, D. S. (1986). The pe and pH in alfalfa seedling rhizospheres, *Agron. J.,* **78**, 216.
17. Fischer, W., Felssa, H. and Schaller, G. (1989). pH values and redox potentials in microsites of the rhizosphere, *Z. Pflanzenernähr Bodenk.,* **152**, 191.
18. Armstrong, W. (1979). Aeration in higher plants, *Adv. Bot. Res.,* **7**, 225.
19. Drew, M. C., Giffards, S. and Jackson, M. B. (1979). Ethylene-promoted adventitious rooting and development of cortical air spaces (aerenchyma) in roots may be adaptive responses to flooding in *Zea Mays,* L, *Planta,* **147**, 83.
20. Drew, M. C. (1990). Sensing soil oxygen, *Plant Cell Environ.* **13**, 681.
21. Yu, P. T., Letey, J. and Stolzy, L. H. (1969). Survival of plants under prolonged flooded conditions, *Agron. J.,* **61**, 844.
22. Begg, C. B. M., Kirk, G. J. D., Mackenzie, A. F. and Neue, H.-U. (1994). Root-induced iron oxidation and pH changes in the lowland rice rhizosphere, *New Phytol.,* **128**, 469.
23. Chen, C. C., Dixon, J. B. and Turner, F. T. (1980). Iron coatings on rice roots: mineralogy and quantity influencing factors, *Soil Sci. Soc. Am. J.,* **44**, 635.
24. Bacha, R. E. and Hossner, L. R. (1977). Characteristics of coatings formed on rice roots as affected by iron and manganese additions, *Soil Sci. Soc. Am. J.,* **41**, 931.

25. Marschner, H. (1988). Mechanisms of manganese acquisition by roots from soils. In *Manganese in Soils and Plants*, ed. Graham, R., Hannam, R. J. and Uren, N. C., Kluwer, Norwell, MA, p. 191.

26. Lynch, J. M. and Whipps, J. M. (1990). Substrate flow in the rhizosphere, *Plant Soil*, **129**, 1.

27. Liljeroth, E., Kuikman, P. and Van Veen, J. A. (1994). Carbon translocation to the rhizosphere of maize and wheat and influence on the turnover of native soil organic matter at different soil nitrogen levels, *Plant Soil*, **161**, 233.

28. Sauerbeck, D., Nonnen, S. and Allard, J. L. (1981). Assimilateverbruch und -umsaltz im Wurzelraum in Abhängigkeit von Pflanzenart und-anzucht, *Landwirtsch. Forsch. Sonderh.*, **37**, 207.

29. Barber, D. A. and Martin, J. K. (1976). The release of organic substances by cereal roots into soils, *New Phytol*, **76**, 69.

30. Rovira, A. D. (1965). Plant root exudates and their influence upon soil microorganisms. In *Ecology of Soil-Borne Plant Pathogens*, ed. Baker, K. F. and Snyder, W. C., Univ. of California Press, Berkeley, p. 107.

31. Rovira, A. D. and McDougal, B. M. (1967). Microbial and biochemical aspects of the rhizosphere. In *Soil Biochemistry*, Vol. 1, ed. McLaren, A. D. and Peterson, G. H., Marcel Dekker, New York, p. 417.

32. Hale, M. G., Foy, C. L. and Shay, F. J. (1971). Factors affecting root exudation, *Adv. Agron.*, **23**, 89.

33. Smith, W. H., (1976). Character and significance of forest tree root exudates, *Ecology*, **57**, 324.

34. Warembourg, F. R. and Billes, G. (1979). Estimation of carbon transfers in the plant rhizosphere. In *The Soil–Root Interface*, ed. Harley, J. L. and Scott-Russell, R., Academic Press, London, p. 183.

35. Uren, N. C. and Reisenauer, H. M. (1988). The role of root exudates in nutrient acquisiotn, *Adv. Plant Nutr.*, **3**, 79.

36. Szmigielska, A. M., Van Rees, K. C. J., Cieslinski, G., Huang, P. M. and Knott, D. R. (1995). Determination of low- molecular-weight dicarboxylic acids in root exudates by gas chromatography, *J. Agric. Food Chem.*, **43**, 956.

37. Szmigielska, A. M., Cieslinski, G., Van Rees, K. C. J. and Huang, P. M. (1996). Low molecular weight dicarboxylic acids in rhizosphere soil of durum wheat, *J. Agric. Food Chem.*, **44**, 1036.

38. Szmigielska, A. M., Van Rees, K. C. J., Cieslinski, G., Huang, P. M (1997). Comparison of liquid and gas chromatography for analysis of low molecular weight organic acids in rhizosphere soil, *Commun. Soil Sci. Plant Anal.*, **28**, 99.

39. Cieslinski, G., Van Rees, K. C. J., Szmigielska, A. M., Krishnamurti, G. S. R. and Huang, P. M. (1998). Low-molecular-weight organic acids in rhizosphere soils of durum wheat and their effects on cadmium bioaccumulation, *Plant Soil*, **203**, 109.

40. Huang, P. M. and Violante, A. (1986). Influence of organic acids on crystallization and surface properties of precipitation products of aluminum. In *Interactions of Soil Minerals with Natural Organics and Microbes*, ed. Huang, P. M. and Schnitzer, M., SSSA Special Publication Number 17, Soil Science Society of America, Madison, WI, p. 159.

41. Stevenson, F. J. (1967). Organic acids in soil. In *Soil Biochemistry*, Vol. 1, ed. McLaren, A.D. and Peterson, G. H., Marcel Dekker, New York, p. 119.

42. Rao, D. N. and Mikkelsen, D. S. (1977). Effect of rice straw additions on production of organic acids in a flooded soil, *Plant Soil*, **47**, 303.

43. Stevenson, F. J. and Ardakani, M. S. (1972). Organic matter reactions involving micronutrients in soils. In *Micronutrients in Agriculture*, ed. Mortvedt J. J, *et al.*, Soil Science Society of America, Madison, WI, p. 79.
44. Shorey, E. C. (1913). Some organic soil constituents, *USDA Bur. Soils Bull.*, **88**, 5, US Government Printing Office, Washington, D.C.
45. Whitehead, D. C. (1964). Identification of *p*-hydroxybenzoic, vanillic, *p*-coumaric, and ferulic acids in soils, *Nature (London)*, **202**, 417.
46. Wang, T. S. C., Yang, T. K. and Chuang, T. T. (1967). Soil phenolic acids as plant growth inhibitors, *Soil Sci.*, **103**, 239.
47. Davies, R. J. (1971). Relation of polyphenols to decomposiiton of organic matter and to pedogenic processes, *Soil Sci.*, **111**, 80.
48. Stevenson, F. J. (1994), *Humus Chemistry: Genesis, Composition, Reactions*, 2nd edn, John Wiley & Sons, New York.
49. Bruckert, S. (1970). Influence des composés organiques solubles sur la pèdogenése en millieu acide. II. Expériences de laboratoire, *Ann. Agron.*, **21**, 725.
50. Förstner, U. (1981). Metal transfer between solid and aqueous phases. In *Metal Pollution in the Aquatic Environment*, ed. Förstner, U. and Wittman, G. T. W., Spinger, New York, p. 197.
51. Coulson C. B., Davies, R. I. and Lewis, D. A. (1960). Polyphenols in plant, humus and soil. I. Polyphenols of leaves, litter and superficial humus from mull and mor site, *J. Soil. Sci.*, **11**, 20.
52. Ladd, J. N. and Butler, J. H. A. (1975). Humus-enzyme systems and synthetic, organic polymers-enzyme analogs. In *Soil Biochemistry*, ed. Paul, E. A. and McLaren, A. D., Marcel Dekker, New York, p. 143.
53. Rovira, A. D., Foster, R. C. and Martin, J. K. (1979). Origin, nature, and nomenclature of the organic materials in the rhizosphere. In *The Soil–Root Interface*, ed. Harley, J. L. and Russell, R. S., Academic Press, London, p. 1.
54. Vermeer, J. and McCully, M. E. (1982). The rhizosphere in *Zea mays*: new insight into its structure and development, *Planta*, **156**, 45.
55. Ray, T. C., Callow, J. A. and Kennedy, J. F. (1988). Composition of root mucilage polysaccharides from *Lepidium sativum*, *J. Exp. Bot.*, **39**, 1249.
56. Nambiar, E. K. S. (1976a). Uptake of 65 Zn from dry soil by plants, *Plant Soil*, **44**, 267.
57. Nambiar, E. K. S. (1976b). The uptake of zinc 65 by oats in relation to soil water content and root growth, *Aust. J. Soil Res.*, **14**, 67.
58. Horst, W. J., Wagner, A. and Marschner, H. (1982). Mucilage protects root meristems for aluminum injury, *Z. Pflanzenphysiol*, **105**, 435.
59. Horst, W. J., Asher, C. J. and Cakmak, I. (1991). Short-term responses of soybean roots to aluminium. In *Soil–Plant Interactions at Low pH*, eds. Wright, R. J, *et al.*, Kluwer, Dordrecht, The Netherlands, p. 733.
60. Marschner, H. (1995), *Mineral Nutrition of Higher Plants*, 2nd edn, Academic Press, London.
61. Sauerbeck, D. and Helal, H. M. (1986). Plant root development and photosynthate consumption depending on soil compaction. In *Trans. 13th Congr. Int. Soc. Soil Sci.*, Hamburg, Vol. III, p. 948.
62. Schönwitz, R. and Ziegler, H. (1982). Exudation of water-soluble vitamins and of some carbohydrates by intact roots of maize seedlings (*Zea mays* L.) into mineral nutrient solution, *Z. Pflanzenphysiol.*, **107**, 7.
63. D'Arcy, L. A. (1982). Etude des exsudats racinaire de soja et de lentille: l. cinétique d'excudation des composés phénoliques, des amino acides at des sucres, au cours de premiers jours de la vie des plantules, *Plant Soil*, **68**, 399.

64. Horst, W. J., Klotz, F. and Szulkiewicz, P. (1990). Mechanical impedance increases aluminum tolerance of soybean (*Glycine max*) roots, *Plant Soil*, **124**, 227.

65. Lipton, D. S., Blanchar, R. W. and Blevins, D. G. (1987). Citrate, malate, and succinate concentration in exudates from P-sufficient and P-stress *Medicago sativa* L. seedlings, *Plant Physiol*, **85**, 315.

66. Kraffczk, I., Trolldenier, G. and Beringer, H. (1984). Soluble root exudates of maize: Influence of potassium supply and rhizosphere microorganisms, *Soil Biol. Biochem.*, **16**, 315.

67. Olsen, R. A., Bennet, J. H., Blume, D. and Brown, J. C. (1981). Chemical aspects of the Fe stress response mechanism in tomatoes, *J. Plant Nutr.*, **3**, 905.

68. Jolley, D. and Brown, J. C. (1987). Soybean response to iron-deficiency stress as related to iron supply in the growth medium, *J. Plant Nutr.*, **10**, 637.

69. Adams, F. (1974). Soil solution. In *The Plant Root and Its Environment I*, ed. Carson, E.W., University of Virginia Press, Charlottesville, VA, p. 441.

70. Sposito, G. (1989), *The Chemistry of Soils*, Oxford University Press, New York.

71. Ritchie, G. S. P. and Sposito, G. (1995). Speciation in soils. In *Chemical Speciation in the Environment*, ed. Ure, A. M. and Davidson, C. M., Blackie Academic and Professional, London, p. 201.

72. Lindsay, W. L. and Catlett, K. M. (1998). Chemistry of the soil solution. In *Future Prospects for Soil Chemistry*, ed. Huang, P. M., Sparks, D. L. and Boyd, S. A., SSSA Special Publication Number 55, Soil Science Society of America, Madison, WI, p. 123.

73. Schwab, A. P. (2000). The soil solution. In *The Handbook of Soil Science*, editor-in-chief Sumner, M. E., CRC Press, Boca Raton, FL, p. B85.

74. Lindsay, W. L. (1979), *Chemical Equilibria in Soils*, Wiley-Interscience, New York.

75. Bollag, J.-M., Dec., J. and Huang, P. M. (1998). Formation mechanisms of complex organic structures in soil habitats, *Adv. Agron.*, **63**, 237.

76. Huang, P. M. (2000). Abiotic catalysis. In *The Handbook of Soil Science*, editor-in-chief Sumner, M. E., CRC Press, Boca Raton, FL, p. B302.

77. Babich, H. and Stotzky, G. (1978). Effect of cadmium on microbes *in vitro* and *in vivo*: influence of clay minerals. In *Microbial Ecology*, ed. Loutit, M. W. and Miles, J.A.R., Springer, Berlin, p. 412.

78. Babich, H. and Stotzky, G. (1980). Environmental factors that influence the toxicity of heavy metals and gaseous pollutants to microorganisms, *Crit. Rev. Microbiol.* **9**, 99.

79. Babich, H. and Stotzky, G. (1983). Influence of chemical speciation on the toxicity of heavy metals to the microbiota. In *Aquatic Toxicology*, ed. Nriagu, J. O., John Wiley & Sons, New York, p. 1.

80. Stotzky, G. and Babich, H. (1980). Physicochemical factors that affect the toxicity of heavy metals to microbes in aquatic habitats. In *Proceedings of the ASM Confernce on Aquatic Microbial Ecology*, ed. Colwell, R.R. and Foster, J., Univ. Maryland, College Park, MD, p. 181.

81. Stotzky, G. and Babich, H. (1986). Physicochemical environmental factors affect the response of microorganisms to heavy metals: implications for the application of microbiology to mineral exploration. In *Mineral Exploration: Biological Systems and Organic Matter*, ed. Carlisle, D., Berry, W. L., Kaplan, I. R. and Watterson, J. R., Prentice-Hall, Englewood Cliffs, NJ, p. 238.

82. Babich, H. and Stotzky, G. (1983). Temperature, pH, salinity, hardness, and particulates mediate nickel toxicity to eubacteria, an actinomycete, and yeasts in lake, simulated estuarine, and sea water, *Aquatic Toxicol.* **3**, 195.

83. Babich, H. and Stotzky, G. (1977). Effect of cadmium on fungi and on interactions between fungi and bacteria in soil: influence of clay minerals and pH, *Appl. Environ. Microbiol.* **33**,1059.

84. Bewley, R. J. F. and Stotzky, G. (1983). Effects of cadmium and zinc on microbial activity in soil; influence of clay minerals Part II: Metals added simultaneously, *Sci. Total Environ.* **31**, 57.
85. Debosz, K., Babich, H. and Stotzky, G. (1985). Toxicity of lead to soil respiration: mediation by clay minerals, humic acids, and compost, *Bull. Environ. Contamin. Toxicol.* **35**, 517.
86. Babich, H. and Stotzky, G. (1981). Manganese toxicity to fungi: influence of pH, *Bull. Environ. Contamin. Toxicol.* **27**, 474.
87. Babich, H. and Stotzky, G. (1978). Toxicity of zinc to fungi, bacteria, and coliphages: influence of chloride ions, *Appl. Environ. Microbiol.* **36**, 906.
88. Barrow, N. J. (1987), *Reactions with Variable Charge Soils*, Martinus Nijhoff, Dordrecht, The Netherlands.
89. Paul, E. A. and Clark F. E. (1996), *Soil Microbiology and Biochemistry*, 2nd edn, Academic Press, San Diego, CA.
90. Lynch, J. M. (ed.) (1990), *The Rhizosphere*, John Wiley & Sons, New York.
91. Katznelson, H., Lochhead, A. G. and Timonin, M. I. (1948). Soil microorganisms and the rhizosphere, *Bot. Rev.* **14**, 543.
92. Lochhead, A. G. and Rouatt, J. W. (1955). The rhizosphere effect on the nutritional groups of soil bacteria, *Soil Sci. Soc. Am. Proc.* **19**, 48.
93. Rovira, A. D. and McDougall, B. M. (1967). Microbiological and biochemical aspects of the rhizosphere. In *Soil Biochemistry*, ed. McLaren A. D. and Peterson G. H., Marcel Dekker, New York, p. 417.
94. Lochhead, A. G. and Chase, F. E. (1943). Qualitative studies of soil microorganisms. V. Nutritional requirements of the predominant bacterial flora, *Soil Sci.* **55**, 185.
95. Rouatt, J. W., Katznelson, H. and Payne, T. M. B. (1960). Statistical evaluation of the rhizosphere effect, *Soil Sci. Soc. Am. Proc.* **24**, 271.
96. Ivarson, K. C. and Katznelson, H. (1960). Studies on the rhizosphere microflora of yellow birch seedlings, *Plant Soil* **12**, 30.
97. Germida, J. J., Siciliano, S. D., de Freitas, J. R. and Seib, A. M. (1998). Diversity of rhizosphere and endophytic bacteria associated with field grown canola (*Brassica napus* L.), *FEMS Microbiol. Ecol.* **26**, 43.
98. Grayston, S. J., Wang, S., Campbell., C. D. and Edwards, A. C. (1998). Selective influence of plant species on microbial diversity in the rhizosphere, *Soil Biol. Biochem.* **30**, 369.
99. Lemanceau, P., Corberand, T., Gardan, L., Latour, X., Laguerre, G., Boeufgras, J.-M, and Alabouvette, C. (1997). Effect of two plant species Flax (*Linum usitatissimum* L.) and tomato (*Lycopersicon esculentum* Mill.) on the diversity of soilborne populations of fluorescent pseudomonads, *Appl. Environ. Microbiol.* **61**, 1004.
100. Marilley, L., Vogt, G., Blanc, M. and Aragno, M. (1998). Bacterial diversity in the bulk soil and rhizosphere fractions of *Lolium perenne* and *Trifolium repens* as revealed by PCR restriction analysis of 16S rDNA, *Plant Soil* **198**, 219.
101. Rouatt, J. W. and Katznelson, H. (1961). A study of bacteria on the root surface and the rhizosphere of soil and crop plants, *J. Appl. Bacteriol.* **24**, 164.
102. Siciliano, S. D., Theoret, C. M., de Freitas, J. R., Hucl, P. J. and Germida, J. J. (1998). Differences in the microbial communities associated with roots of different cultivars of canola and wheat, *Can. J. Microbiol.* **44**, 1.
103. Siciliano, S. D. and Germida, J. J. (1999). Taxonomic diversity of bacteria associated with the roots of field grown transgenic *Brassica napus* cv. Quest, compared to the non-transgenic *B. napus* c. Excel and *B. rapa* cv. Parkland, *FEMS Microbiol. Ecol.* **29**, 263.

104. Westover, K. M., Kennedy, A. C. and Kelleys, S. E. (1997). Patterns of rhizosphere microbial community structure associated with co-occurring plant species, *J. Ecol.* **85**, 863.
105. Chaney, R. L., Malik, M., Li, Y. M., Brown, S. L., Brewer, E. P., Angle, S. J. and Baker, A. J. M. (1997). Phytoremediation of soil metals, *Curr. Opinion Biotechnol.* **8**, 279.
106. Torsvik, V., GokSoyr, J. and Daae, F. L. (1990). High diversity in DNA of soil bacteria, *Appl. Environ. Microbiol.* **53**, 782.
107. Angle, J. S., McGrath, S. P. and Chaney, R. L. (1991). New culture medium containing ionic concentrations of nutrients similar to concentrations found in the soil solution, *Appl. Environ. Microbiol.* **57**, 3674.
108. Germida, J. J. (1993). Cultural methods for soil microorganisms. In *Soil Sampling and Methods of Analysis*, ed. Carter, M. R., a special publication of the Canadian Society of Soil Science, CRC Press/Lewis, Boca Raton, FL, p. 263.
109. Martin, J. K., 1975. Comparison of agar for counts of viable soil bacteria, *Soil Biol. Biochem.* **7**, 401.
110. Skinner, F. A., Jones, P. C. T. and Mollison, J. E. (1952). A comparison of direct and a plating-counting technique, *J. Gen. Microbiol.* **6**, 261.
111. Lynch, J. M. (1990). Microbial metabolites. In *The Rhizosphere*, ed. Lynch, J. M., John Wiley & Sons, New York, p. 177.
112. Smith, S. E. and Read, D. J. (1997), *Mycorrhizal Symbiosis*, 2nd edn, Academic Press, San Diego, CA.
113. Sylvia, D. M., Fuhrmann, J. J., Hartel, P. G. and Zuberer, D. A. (1998), *Principles and Applications of Soil Microbiology*, Prentice Hall, Upper Saddle River, NJ.
114. Joner, E. J. and Jakobsen, I. (1995). Growth and extracellular phosphatase activity of arbuscular mycorrhizal hyphae as influenced by soil organic matter, *Soil Biol. Biochem.* **27**, 1153.
115. McInroy, A. and Kloepper, J. W. (1995). Survey of indigenous bacterial endophytes from cotton and sweet corn, *Plant Soil* **173**, *337*.
116. Hallmann, J., Quadt-Hallmann, A., Mahaffee, W. F. and Kloepper, J.W. (1997). Bacterial endophytes in agricultural crops, *Can. J. Microbiol.* **43**, 895.
117. Germida, J. J. (1988). Growth of indigenous *Rhizobium leguminosarum* and *Rhizobium melioti* in soils amended with organic nutrients, *Appl. Environ. Microbiol.*, **54**, 257.
118. Sparks, D. L. (1995), *Environmental Soil Chemistry*, Academic Press, San Diego, CA.
119. Alloway, B. J. 1995. Soil processes and the behavior of metals. In *Heavy Metals in Soils*, ed. Alloway, B. J., Blackie Academic and Professional, London, p. 11.
120. Huang, P. M. and Fujii, R. (1996). Selenium and arsenic. In *Methods of Soil Analysis: Part 3. Chemical Methods,* ed. Sparks, D. L, *et al.*, Soil Science Society of America, Madison, WI, p. 793.
121. Penrose, W. R. (1974). Arsenic in the marine and aquatic sediments: analysis, occurrence and significance, *CRC Crit. Rev. Environ. Control* **4**, 465.
122. Kolthoff, I. M. (1921). Iodometric studies. VII. Reactions between arsenic trioxide and iodine, *Anal. Chem.*, **60**, 393.
123. Oscarson, D. W., Huang, P. M., Defose, C. and Herbillon, A. (1981). Oxidative power of Mn (IV) and Fe (III) oxides with respect to As (III) in terrestrial and aquatic environments, *Nature* (London), **291**, 50.
124. Oscarson, D. W., Huang, P. M., Liaw, W. K. and Hammer, U. T. (1983). Kinetics of oxidation of arsenic by various manganese dioxides, *Soil Sci. Soc. Am. J.*, **47**, 644.
125. Huang, P. M. (1991). Kinetics of redox reactions on surfaces of Mn oxides and its impact on environmental quality. In *Rates of Chemical Processes in Soils*, ed.

Sparks, D. L. and Suares, D. L., SSSA Special Publication Number 17, Soil Science Society of America, Madison, WI, p. 191.

126. O'Neill, P. (1995). Arsenic. In *Heavy Metals in Soils*, ed. Alloway, B.J., Blackie Academic and Professional, London, p. 105.

127. Ohlendorf, H. M., Hoffman, D. J., Saiki, M. K. and Aldrich, T.W (1986). Embryonic mortality and abnormalities of aquatic birds: Apparent impacts of selenium from irrigation drainwater, *Sci. Total Environ.*, **52**, 49.

128. Shaker, A. M (1996). Kinetics of the reduction of Se (IV) to Se-Sol, *J. Colloid Interface Sci.*, **180**, 225.

129. Oremland, R. S., Steinberg, N. A., Maest, A. S., Miller, L.G. and Hollibaugh, J.T. (1990). Measurement of in situ rates of selenate removal by dissimilatory bacterial reduction in sediments, *Environ. Sci. Technol.*, **24**, 1157.

130. Garbisu, C., Ishii, T., Leighton, T. and Buchanan, B. B. (1996). Bacterial reduction of selenite to elemental selenium, *Chem. Geol.*, **132**, 199.

131. Hansen, H. C. B., Koch, C. B., Nancke-Krogh, H., Borggaard, O. K. and Sorenson, J. (1996). Abiotic nitrate reduction to ammonium: Key role of green rust, *Environ. Sci. Technol.*, **30**, 2053.

132. Ottley, C. J., Davison, W., Edmunds, W. (1997). Chemical catalysis of nitrate reduction by iron (II), Geochim. Cosmochim. Acta, **61**, 1819.

133. Trolard, F., Génin, J.-M. R., Abdelmoula, M., Bourrie, G., Humbert, B. and Herbillon, A. (1997). Identification of a green rust mineral in a reductomorphic soil by Mössbauer and Raman spectroscopies, *Geochim. Cosmochim.* Acta, **61**, 1107.

134. Myneni, S. C. B., Tokunaga, T. K. and Brown, G. E. Jr. (1997). Abiotic selenium redox transformation in the presence of Fe (II, III) oxides, *Science*, **278**, 1106.

135. Rai, D. and Serne, R. J. 1977. Plutonium activities in soil solutions and the stability and formation of selected plutonium minerals, *J. Environ. Qual.*, **6**, 89.

136. Bartlett, R. J. and James, B. 1979. Behavior of chromium in soils. III. Oxidation, *J. Environ. Qual.*, **8**, 31.

137. Amacher, M. L. and Baker, D. E. (1982), *Redox Reactions involving Chromium, Plutonium, and Manganese in Soils*, DOE/DP/04515–1, Pennsylvania State University, University Park, PA.

138. Cleveland, J. M. (1970), *The Chemistry of Plutonium*, Gordon and Breach, New York.

139. Hem, J. D. (1978). Redox processes at surfaces of manganese oxide and their effects on aqueous metal ions, *Chem. Geol.*, **21**, 199.

140. Murray, J. W. and J. G. Dillard (1979). The oxidation of cobalt (III) adsorbed on manganese dioxide, *Geochim, Cosmochim. Acta*, **43**, 781.

141. Wehrli, B. and Stumm, W. (1989). Vanadyl in natural waters: adsorption and hydrolysis promote oxygenation, *Geochim. Cosmochim. Acta,* **53**, 69.

142. Eary, L. E. and Rai, D. (1989). Kinetics of chromate reduction by ferrous ions derived from hematite and biotite at 25 °C, *Am. J. Sci.*, **289**, 180.

143. Ilton, E. S. and Veblen, D. R. (1994). Chromium sorption by phlogopite and biotite in acidic solutions at 25 °C: Insights from X-ray photoelectron spectroscopy and electron microscopy, *Geochim. Cosmochim. Acta*, **58**, 2777.

144. Peterson, M. L., Brown, G. E. and Parks, G. A. (1996). Direct XAFS evidence for heterogeneous redox reactions at the aqueous chromium/magnetite interface, *Colloid Surf. A*, **107**, 77.

145. White, A. F. and Peterson, M. L. (1996). Reduction of aqueous transition metal species on the surfaces of Fe(II)- containing oxides, *Geochim. Cosmochim. Acta*, **60**, 3799.

146. White, A. F. (1990). Heterogeneous electrochemical reactions associated with oxidation of ferrous oxides and silicate surfaces, *Rev. Mineral.* **23**, 467.

147. Masschelyn, P. H. and Patrick, W. H. Jr. (1994). Selenium, arsenic, and chromium redox chemistry in wetland soils and sediments. In *Biogeochemistry of Trace Elements,* ed. Adriano, D. C., Science and Technology Letters, Northwood, p. 615.

148. Otte, M. L., Rozem, J., Koster, L., Haarsma, M. S. and Broekman, R. A. (1989). Iron plaque on roots of *Aster tripolium* L.: Interaction with Zn uptake, *New Phytol,* **111**, 309.

149. Kirk, G. J. D. and Bajita, J. B. (1995). Root-induced iron oxidation, pH changes, and zinc solubilization in the rhizosphere of lowland rice, *New Phytol.,* **131**, 129.

150. Otte, M. L., Dekkers, M. J., Rozema, J. and Broekman, R.A. (1991). Uptake of arsenic by *Aster tripolium* in relation to rhizosphere oxidation, *Can. J. Bot.,* **69**, 2670.

151. Otte, M. L., Kearns, C. C. and Doyle, M. O. (1995). Accumulation of arsenic and zinc in the rhizosphere of wetland plants, *Bull. Environ. Contamin. Toxicol,* **55**, 154.

152. Pearson, R.G. (1963). Hard and soft acids and bases, *J. Am. Chem. Soc.,* **85**, 3533.

153. NIST (1997), *Standard Reference Database 46. Critically Selected Stability Constants of Metal Complexes,* Version 5.0, National Institute of Standards and Technology, US Department of Commerce, Gaithersburg, MD.

154. Lund, W. (1990). Speciation analysis—why and how? *Fresenius J. Anal. Chem.,* **337**, 557.

155. Parker, D. R., Chaney, R. L. and Norvell, W. A. (1995). Chemical equilibrium models: Applications to plant nutrition. In *Soil Chemical Equilibrium and Reaction Models,* ed. R. H. Loeppert *et al.,* SSSA Special Publication Number 42, American Society of Agronomy and Soil Science Society of America, Madison, WI, p. 163.

156. Tessier, A. and D. R. Turner (1995), *Metal Speciation and Bioavailability in Aquatic Systems,* IUPAC Series on Analytical and Physical Chemistry of Environmental Systems, Vol. 3, John Wiley & Sons, Chichester.

157. Taylor, G. J. and Foy, C. D. (1985). Differential uptake and toxicity of ionic and chelated copper in *Triticum aestivum, Can. J. Bot.,* **63**, 1271.

158. Bell, P. F., Chaney, R. L. and Angle, J. S. (1991). Free metal activity and total metal concentrations as indices of micronutrient availability to barley (*Hordeum vulgare* L. 'Klages') *Plant Soil,* **130**, 51.

159. Laurie, S. H., Tancock, N., McGrath, S. P. and Sanders, J. R. (1991). Influence of complexation on the uptake by plants of iron, manganese, copper and zinc: Effect of DTPA in a multi-metal and computer simulation study, *J. Exp. Bot.,* **42**, 509.

160. Römheld, V. and Marschner, H. (1981). Effect of Fe stress on utilization of Fe chelates by efficient and inefficient plant species, *J. Plant Nutr.,* **3**, 1.

161. Phinney, J. T. and Bruland, K. W. (1994). Uptake of lipophilic organic Cu, Cd, and Pb complexes in the coastal diatom *Thallassiosira weissflogii. Environ. Sci. Technol.,* **28**, 1781.

162. Morel, J. L., Mench, M. and Guckert, A. (1986). Measurement of Pb^{2+}, Cu^{2+}, and Cd^{2+} binding with mucilage exudates from maize (*Zea mays* L.) roots, *Biol. Fertil. Soils,* **2**, 29.

163. Mench, M., Morel, J. L. and Guckert, A. (1987). Metal binding properties of high molecular weight soluble exudates from maize (*Zea mays* L.) roots, *Biol. Fertil. Soils,* **3**, 165.

164. Mench, M., Morel, J. L., Guckert, A. and Guillet, B. (1988). Metal binding with root exudates of low molecular weight, *J. Soil Sci.,* **39**, 521.

165. Gries, D., Brunor, S., Crowley, D. E. and Parker, D. R. (1995). Phytosiderophore release in relation to micronutrient metal deficiency in barley, *Plant Soil*, **172**, 299.
166. Merckx, R. van Ginkel, J.H. Sinnaeve, J. and Cremers, A. (1986). Plant-induced changes in the rhizosphere of maize and wheat: II. Complexation of cobalt, zinc and manganese in the rhizosphere of maize and wheat, *Plant Soil*, **96**, 95.
167. Hamon, R. E., Lorenz, S. E., Holm, P. E. Christensen, T. H. and McGrath, S. P. (1995). Changes in trace metal species and other components of the rhizosphere during growth of radish, *Plant Cell Environ.*, **18**, 749.
168. Helmke, P. A., Salam, A. B. and Li, Y. 1998. Measurement and behavior of indigenous levels of the free, hydrated cations of Cu, Zn, and Cd in the soil-water system. In *Proc. 3rd Int. Conf. on the Biogeochemistry of Trace Elements in the Environment, 3rd* edn Prost, R., INRA, Versailles, France, CD-ROM 008.PDF.
169. Nielsen, N. E. (1976). The effect of plants on the copper concentration in the soil solution, *Plant Soil*, **45**, 679.
170. Linehan, D. J., Sinclair, A. H. and Mitchell, M. C. (1989). Seasonal changes in Cu, Mn, Zn, and Co concentrations in soil in the root zone of barley (*Hordeum vulgare* L.), *J. Soil Sci.*, **40**, 103.
171. Krishnamurti, G. S. R., Huang, P. M. and Van Rees, K. C. J. (1996). Studies on soil rhizosphere: speciation and availability of Cd, *Chem. Spec. Bioavail.*, **8**, 23.
172. Krishnamurti, G. S. R., Huang, P. M., Van Rees, K. C. J., Kozak, L. M. and Rostad, H. P. W. (1995). A new soil test method for the determination of plant-available cadmium in soils, *Commun. Soil Sci. Plant Anal.*, **26**, 2857.
173. Smolders, E. and McLaughlin, M. J. (1996a). Effect of Cd on Cd uptake by Swiss chard in unbuffered and chelator buffered nutrient solutions, *Plant Soil*, **179**, 57.
174. Smolders, E. and McLaughlin, M. J. (1996b). Influence of chloride on Cd availability to Swiss chard: A resin buffered solution culture system, *Soil Sci. Soc. Am. J.*, **60**, 1443.
175. Li, Y.-M., Chaney, R. L. and Schneiter, A. A. (1994). Effect of soil chloride level on cadmium concentration in sunflower kernels, *Plant Soil*, **167**, 275.
176. McLaughlin, M. J., Tiller, K. G., Beech, T. A. and Smart, M. K. (1994). Soil salinity causes elevated cadmium concentrations in field grown potato tubers, *J. Environ. Qual.*, **23**, 1013.
177. McLaughlin, M. J. and Tiller, K. G. (1994). Chloro-complexation of cadmium in soil solutions of saline/sodic soils increases phyto-availability of cadmium. In *Trans. World Congress Soil Sci., 15th*, Acapulco, Mexico, 1994, ISSS, Wageningen, The Netherlands, p. 195.
178. Jurinak, J. J. and Suarez, D. L. (1990). The chemistry of salt-affected soils and waters. In *Agricultural Salinity Assessment and Management*, ed. Tanji, K. K., American Society of Civil Engineering, New York, p. 42.
179. McLaughlin, M. J., Tiller, K. G., Naidu, R. and Stevens, D. P. (1996). The behavior and environmental impact of contaminants in fertilziers, *Aust. J. Soil Res.* **34**, 1.
180. Schindler, P. W. and Sposito, G. (1991). Surface complexation at (hydr)oxide surfaces. In *Interactions at the Soil Colloid–Solution Interface*, ed. Bolt, G. H., Kluwer, Dordrecht, The Netherlands, p. 115.
181. Hayes, K. F. and Traina, S. J. (1998). Metal ion speciation and its significance in ecosystem health. In *Soil Chemistry and Ecosystem Health*, ed. Huang, P. M., Adriano, D. C., Logan, T. J. and Checkai, R. T., SSSA Special Publication Number 52, Soil Science Society of America, Madison, WI, p. 45.

182. Fellows, R. J., Ainsworth, C. C., Driver, C. J. and Cataldo, D. A. (1998). Dynamics and transformations of radionuclides in soils and ecosystem health. In *Soil Chemistry and Ecosystem Health*, ed. Huang, P. M. Adriano, D. C., Logan, T. J. and Checkai, R. T., SSSA Special Publication Number 52, Soil Science Society of America, Madison, WI, p. 85.

183. McBride, M. B. (2000). Chemisorption and precipitation reactions. In the *Handbook of Soil Science*, editor-in-chief, Sumner, M. E., CRC Press, Boca Raton, FL, p. B265.

184. Senesi, N. (1992). Metal–humic substance complexes in the environment. Molecular and mechanistic aspects by multiple spectroscopic approach. In *Biogeochemistry of Trace Elements*, ed. Adriano, D. C., Lewis, Boca Raton, FL, p. 429.

185. Buffle, J. (1988), *Complexation Reactions in Aquatic Systems: An Analytical Approach*, Ellis Horwood, Chichester.

186. Schnitzer, M. and Skinner, S. I. M. (1966). Organo-metallic interactions in soils. 5. Stability constants of Cu^{2+}, Fe^{2+}, and Zn^{2+}–fulvic acid complexes, *Soil Sci.*, **102**, 361.

187. Schnitzer, M. and Skinner, S. I. M. (1967). Organo-metallic interactions in soils. 7. Stability constants of Pb^{2+}, Ni^{2+}, Mn^{2+}, Co^{2+}, Ca^{2+}, and Mg^{2+}–fulvic acid complexes, *Soil Sci.*, **103**, 247.

188. Waller, P. A. and Pickering, W. F. (1993). The effect of pH on the lability of lead and cadmium sorbed on humic acid particles, *Chem. Spec. Bioavail.*, **5**, 11.

189. Schnitzer, M. (1970). Characteristics of organic matter extracted from podzol B horizons, *Can. J. Soil Sci.*, **50**, 199.

190. Senesi, N. (1996). Electron spin (or paramagnetic) resonance spectroscopy. In *Methods of Soil Analysis. Part 3. Chemical Methods*, eds. Sparks *et al.*, SSSA Book Series No. 5., Soil Science Society of America, Madison, WI, p. 323.

191. Cheshire, M. V. and Senesi, N. (1999). Electron spin resonance spectroscopy of organic and mineral soil particles. In *Structure and Surface Reactions of Soil Particles*, eds. Huang, P. M., Senesi, N. and Buffle, J., IUPAC Series on Analytical and Physical Chemistry of Environmental Systems, Vol. 4, John Wiley & Sons, Chichester, p. 325.

192. Krishnamurti, G. S. R. and Huang, P. M. (2001). The nature of organic matter of soils with contrasting cadmium phytoavailability. In *Proc. Int. Meeting of International Humic Substances Society, 9th*, Adelaide, Australia.

193. Schwertmann, U., Kodama, H. and Fisher, W. R. (1986). Mutual interactions between organics and iron oxides. In *Interactions of Soil Minerals with Natural Organics and Microbes*, ed. Huang, P. M. and Schnitzer, M., SSSA Special Publication Number 17, Soil Science Society of America, Madison, WI, p. 223.

194. Kwong, K. F., Ng Kee and Huang, P. M., 1979. Surface reactivity of aluminum hydroxides precipitated in the presence of low molecular weight organic acids, *Soil Sci. Soc. Am. J.*, **43**, 1107.

195. Huang, P. M. (1988). Ionic factors affecting aluminum transformations and the impact on soil and environmental sciences, *Adv. Soil Sci.*, **8**, 1.

196. Tiller, K. G., Gerth, J. and Brummer, G. 1984. The relative affinities of Cd, Ni, and Zn for different clay fractions and goethite, *Geoderma*, **34**, 17.

197. Helmke, P. A. and Naidu, R. (1996). Fate of contaminants in the soil environment: Metal contaminants. In *Contaminants and the Soil Environment in the Australasia–Pacific Region*, ed. Naidu R, *et al.*, Kluwer, Dordrecht, The Netherlands, p. 69.

198. Sposito, G. (1996), *The Environmental Chemistry of Aluminum*, CRC/Lewis, Boca Raton, FL.

199. Robert, M. and Berthelin, J. (1986). Role of biological and biochemical factors in soil mineral weathering. In *Interactions of Soil Minerals with Natural Organics and*

Microbes, ed. Huang, P. M. and Schnitzer, M., SSSA Special Publication Number 17, Soil Science Society of America, Madison, WI, p. 453.
200. Mench, M. and Martin, E. (1991). Mobilization of cadmium and other metals from two soils by root exudates of *Zea mays* L., *Nicotiana tabacum* L. and *Nicotinana rustica* L, *Plant Soil*, **132**, 187.
201. Marschner, H., Treeby, M. and Römheld, V. (1989). Role of root-induced changes in the rhizosphere for iron acquisiton in higher plants, *Z. pflanzenernähr. Bodenk.*, **152**, 197.
202. Zhang, F., Römheld, V. and Marschner, H. (1989). Effect of zinc deficiency in wheat on the release of zinc and iron mobilizing root exudates, *Z. pflanzenernähr. Bodenk.* **152**, 205.
203. Huang, P. M. and Schnitzer, M. (ed.) (1986), *Interactions of Soil Minerals with Natural Organics and Microbes*, SSSA Special Publication Number 17, Soil Science Society of America, Madison, WI.
204. Krishnamurti, G. S. R., Cieslinski, G., Huang, P. M. and Van Rees, K. C. J. (1997). Kinetics of cadmium release from soils as influenced by organic acids: Implication in cadmium availability, *J. Environ. Qual.*, **26**, 271.
205. Krishnamurti, G. S. R., Huang, P. M., Van Rees, K. C. J., Kozak, L. M. and Rostad, H. P. W. (1995). Speciation of particulate-bound cadmium of soils and its bioavailability, *The Analyst*, **120**, 659.
206. Sparks, D. L. (1999). Kinetics of Sorption/Release Reactions on Natural Particles. In *Structure and Surface Reactions of Soil Particles*, Vol. 4., ed. Huang, P. M., Senesi, N. and Buffle, J., John Wiley & Sons, Chichester, p. 413.
207. Xu, T., Schwartz, F. W. and Traina, S. J., (1994). Sorption of Zn^{2+} and Cd^{2+} on hydroxyapatite surfaces, *Environ. Sci. Technol.*, **28**, 1472.
208. Chen, X., Wright, J. V., Conca, J. L. and Peurrung, L. M. (1997). Effects of pH on heavy metal sorption on mineral apatite, *Environ. Sci. Technol.*, **31**, 624.
209. Berti, W. R. and Cunningham, S. D. (1997). In-place inactivation of Pb in Pb-contaminated soils, *Environ. Sci. Technol.*, **31**, 1359.
210. Cotter-Howells, J. (1996). Lead phosphate formation in soils, *Environ. Pollut.*, **93**, 9.
211. McBride, M. B. (1989). Reactions controlling heavy metal solubility in soils, *Adv. Soil Sci.*, **10**, 1.
212. Tiller, K. G. (1996). Soil contamination issues: past, present and future, a personal perspective. In *Contaminants and the Soil Environment in the Australasia–Pacific Region*, ed. Naidu, R, *et al.*, Kluwer, Dordrecht, The Netherlands, p. 1.
213. Christensen, T. H. and Huang, P. M. (1999). Solid phase cadmium and the reactions of aqueous cadmium with soil surfaces. In *Cadmium in Soils and Plants*, ed. McLaughlin, M. J. and Singh, B. R., Kluwer, Dordrecht, The Netherlands, p. 65.
214. Santillan-Medrano, J. and Jurinak, J. J. (1975). The chemistry of lead and cadmium in soil: solid phase formation, *Soil Sci. Soc. Am. Proc.*, **39**, 851.
215. Cavallaro, N. and McBride, M. B. (1978). Copper and cadmium adsorption characteristics of selected acid and calcareous soils, *Soil Sci. Soc. Am. J.*, **42**, 550.
216. Papadopoulos, P. and Rowell, R. L. (1988). The reactions of cadmium with calcium carbonate surfaces, *J. Soil Sci.*, **39**, 23.
217. Jopony, M. and Young, S. D. (1994). The soil: solution equilibria of lead and cadmium in polluted soils, *Eur. J. Soil Sci.*, **45**, 59.
218. Fitch, A. and Helmke, P. A. (1989). Donnan equilibrium/graphite furnace atomic absorption estimates of soil extract complexation capacity, *Anal. Chem.*, **61**, 1295.

219. Berggren, D. (1990). Speciation of cadmium (II) using Donan dialysis and differential-pulse anodic stripping voltammetry in a flow-injection system, *Int. J. Environ. Anal. Chem.*, **41**, 133.
220. Beveridge, T. J. (1989). Role of cellular design in bacterial metal accumulation and mineralization, *Annu. Rev. Microbiol.*, **43**, 147.
221. Gottschalk, G. (1979), *Bacterial Metabolism*, Springer, New York.
222. Mullen, D. D. (1998). Transformations of other elements. In *Principles and Applications of Soil Microbiology*, ed. Sylvia, D. M., Fuhrman, J. J., Hartel, P. G. and Zuberer, D. A., Prentice Hall, Upper Saddle River, NJ, p. 369.
223. Hughes, M. N. and Poole, R. K. (1989), *Metals and Micro-organisms*, Chapman and Hall, New York.
224. Berthelin, J., Munier-Lamy, C. and Leyval, C. (1995). Effect of microorganisms on mobility of heavy metals in soils. In *Environmental Impact of Soil Component Interactions. Vol. II—Metals, Other Inorganics and Microbial Activities*, ed. Huang, P. M., Berthelin, J., Bollag, J.-M., McGill, W. B. and Page, A. L., CRC Press/Lewis, Boca Raton, FL, p. 3.
225. Walker, S. G., Flemming, C. A., Ferris, F. G., Beveridge, T. J. and Bailey, G. W. (1989). Physicochemical interaction of *Escherichia coli* cell envelopes and *Bacillus subtilis* cell walls with two clays and ability of the composite to immobilize heavy metals from solution, *Appl. Environ. Microbiol.* **55**, 2976.
226. Venkateswerlu, G. and Stotzky, G. (1986). Copper and cobalt alter the cell wall composition of *Cunninghamella blakesleeana. Can. J. Microbiol.* **32**, 654.
227. Venkateswerlu, G. and Stotzky, G. (1989). Binding of metals by cell walls of *Cunninghamella blakesleeana* grown in the presence of copper or cobalt, *Appl. Microbiol. Biotechnol.* **31**, 619.
228. Venkateswerlu, G., Yoder, M. J. and Stotzky, G. (1989). Morphological, ultrastructural, and chemical changes induced in *Cunninghamella blakesleeana* by copper and cobalt, *Appl. Microbiol. Biotechnol.* **31**, 204.
229. Collins, Y. E. and Stotzky, G. (1992). Heavy metals alter the electrokinetic properties of bacteria, yeasts, and clay minerals, *Appl. Environ. Microbiol.* **58**, 1592.
230. Collins, Y. E. and Stotzky, G. (1996). Changes in the surface charge of bacteria caused by heavy metals does not affect survival, *Can. J. Microbiol.* **42**, 621.
231. Mann, S., Archibald, D. A., Didymus, J. M., Douglas, T., Heywood, R. R., Meldrum, F. C. and Reeves, N. J. (1993). Crystallization at inorganic–organic interfaces: Biominerals and biomimetic synthesis, *Science*, **261**, 1286.
232. Remacle, J., Houba, C. and Hambuckers-Berlin, F. (1986). Cadmium accumulation by two gram-negative bacteria: *Alcaligenes eutrophus* CH34 and *Pseudomona aeruginosa* AK1800 preliminary results. In *Perspectives in Microbial Ecology*, ed. Megusar, F. and Gantar, M., Slov. Soc. Microbiol., Ljubljana, Yugoslavia, p. 668.
233. Huang, P. M. (1994). Role of organics and microbes in mineral transformations. In *Trans. 15th World Congr. of Soil Sci.*, Vol. **8a**, 68.
234. Plette, A. C. C., van Riemsdijk, W. H., Benedetti, M. F. and van der Wal, A. (1995). pH dependent charging behavior of isolated cell walls of a gram-positive soil bacterium, *J. Colloid Interface Sci.*, **173**, 354.
235. Huang, P. M. and Bollag, J.-M. (1999). Minerals-organics-microorganisms interactions in the soil. In *Structure and Surface Reactions of Soil Particles*, ed. Huang, P. M., Senesi, N. and Buffle, J., IUPAC Series on Analytical and Physical Chemistry of Environmental Systems, Vol. 4, John Wiley & Sons, Chichester.
236. Neilands, J. B. (1981). Microbial iron compounds, *Annu. Rev. Biochem.* **50**, 715.

237. Powell, P. E., Szaniszlo, P. J. and Reid, C. P. P. (1983). Confirmation of occurrence of hydroxamate siderophores in soil by a novel *Escherichia coli* bioassay, *Appl. Environ. Microbiol.* **46**, 1080.
238. Reid, C. P. P. (1990). Mycorrhizas. In *The Rhizosphere*, ed. Lynch, J. M. John Wiley & Sons, New York, p. 281.
239. Rovira, A. D. and Davey, C. B. (1974). Biology of the rhizosphere. In *The Plant Root and Its Environment*, ed. Cursova, E. W., Univ. of Virginia Press, Charlottesville, VA, p. 153.
240. Beveridge, T. J. (1988). The bacterial surface: General consideration towards design and function, *Can. J. Microbiol.*, **34**, 363.

10 Biotic and Abiotic Interactions in the Rhizosphere: Organic Pollutants

T. A. ANDERSON
Texas Tech University, USA

D. P. SHUPACK
Bradburne, Briller & Johnson, LLC, USA

H. AWATA
CH2M Hill, USA

1 INTRODUCTION

The multiple interactions between microorganisms and soil which take place at the root–soil interface, or rhizosphere, can have a dramatic impact on the terrestrial environment. Interest in the interactions in the rhizosphere began in the early 1900s when it was recognized, although not yet entirely understood, that the relationship between microorganisms and plants in soil was critical to crop production. For the next several decades, research on root zone interactions of microorganisms and plants was guided by agronomics. Elucidation of the symbiotic relationship between nitrogen-fixing rhizobia and leguminous plants was one of the most significant discoveries made during this period.

Interactions between Soil Particles and Microorganisms
Edited by P. M. Huang, J.-M. Bollag and N. Senesi. © 2002 John Wiley & Sons, Ltd

Long-term studies on this relationship have contributed to making it arguably one of the most differentiated symbiotic relationships known.

As the legume–rhizobium relationship was further characterized at chemical and molecular levels, interest in other relationships between plants and microorganisms in the root zone began to emerge. The substrates available for microbial growth in soil mainly originate from primary production by higher plants. The interaction between plants and microbial communities in the rhizosphere appears to have evolved to the mutual benefit of both groups [1]. This has led to speculation about other possible relationships, such as the coevolution of plant–microbe defenses [2] (Figure 1). Whereas previous research was focused on the agricultural benefits of root zone interactions between plants and microorganisms, this curiosity allowed research to expand to a broader focus including the interactions between xenobiotic chemicals and rhizosphere microorganisms.

Early research guided by this new curiosity showed that plant root exudates responded to the presence of xenobiotic chemicals [3,4] and that increased rates of xenobiotic metabolism were observed in the root zone compared with non-vegetated soil (originally reviewed by Anderson *et al.*, [5]; Shimp *et al.*, [6]). It appeared that the interaction of the complex microbial communities often found in the rhizosphere may have facilitated the enhanced biodegradation over single strain metabolism [7–11]. The structural diversity of contaminants was such that the microbial interactions fostered in the rhizosphere appeared to contribute to enhanced and/or complete biodegradation of xenobiotics [12]. In addition, survival of soil-inoculated microorganisms (developed for the specific

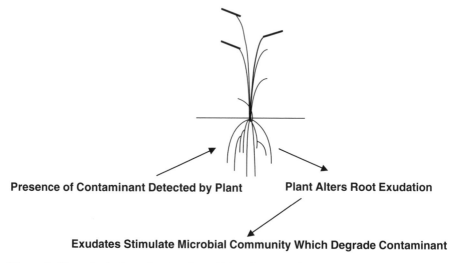

Presence of Contaminant Detected by Plant   **Plant Alters Root Exudation**

Exudates Stimulate Microbial Community Which Degrade Contaminant

Figure 1. Hypothetical mechanism by which microorganisms in the rhizosphere may protect host plants from chemical injury. Adapted with permission from [2]. Copyright (1994) American Chemical Society

purpose of degrading xenobiotic contaminants) could be facilitated by the presence of plant roots [13–17]. Inoculum survival is often a limiting factor in the use of laboratory cultured organisms in bioremediation.

It has been previously speculated that the humification activity of microorganisms may be particularly pronounced in the rhizosphere and could be important for immobilizing and degrading organic pollutants in soil [2,18–21]. Laccases produced by plants, bacteria, and fungi in the rhizosphere are of interest with respect to their role in contaminant transformation. Additionally, humic substances in soils may facilitate the transformation of organic substrates (for example, [22]) and function as electron acceptors during low-oxygen conditions [23].

It is within the context of xenobiotic metabolism and sorption that the biotic and abiotic interactions in the rhizosphere are currently being explored (for example, Chapter 8 in this volume). Current research is driven by the widespread contamination of soil with organic pollutants, and the desire to apply the theories developed during earlier rhizosphere research to practical applications in the field. The benefits of these explorations are not only in increasing our understanding of the interactions between microorganisms and plants in the rhizosphere, but also in establishing and applying methods for waste remediation and cleanup of contaminated soils.

2 BIOTIC AND ABIOTIC CHARACTERISTICS OF THE RHIZOSPHERE

The unique physical environment of the rhizosphere is created by differences in carbon dioxide and oxygen concentrations, osmotic and redox potentials, pH, and moisture content between rhizosphere soil and the surrounding nonvegetated (edaphosphere) soil (for a more detailed review, see Chapter 9 of this volume). These differences allow the rhizosphere to support a vast microbial community that is often greater in biomass, microbial activity, and diversity than in nonvegetated soil [1].

Using both light and electron microscopy, various researchers [24–26] originally characterized the abundance of microorganisms in the rhizosphere. This 'rhizosphere effect' is often quantified as the ratio of microorganisms in rhizosphere soil to those in non-rhizosphere soil, the R/S ratio [27]. R/S ratios from 5 to 20 are common, but occasionally are as high as 100 and above [28,29]. In addition, nitrogen-fixing bacteria often associated with leguminous plants enrich root zone soil leading to a further increase in the rhizosphere effect for legumes [30]. R/S ratios usually decline as plants senesce.

The plant root zone hosts not only an increase in the numbers of microorganisms, but also a rich variety of microbial species. The actual composition of the rhizosphere microbial community is dependent on host plant species, plant age, and soil type [31] as well as other selection pressures including the presence of xenobiotic compounds [32–35]. Rhizosphere communities typically consist of (1) bacteria which are very abundant and reproduce rapidly under favorable conditions, and (2) fungi which form mycelia on plant roots and extend into the surrounding soil to seek nutrients. Aerobic bacteria, particularly Pseudomonads, are generally abundant in the root zone and play an important role in organic contaminant degradation. The composition of the microbial community in the rhizosphere undergoes successional changes as the plant matures, corresponding to changes in materials released by the roots [1].

Root cells excrete a gelatinous mixture to aid the growing root in penetrating the soil matrix. This root exudate is composed of several classes of compounds (Table 1) including carbohydrates, enzymes, and amino acids, but whose actual composition varies among plant species [31,36–38]. Several factors can control the extent of root exudation (Table 2) including plant growth stage, soil structure, moisture, light intensity, temperature and the presence or absence of microorganisms [36,39–41]. Root exudates ultimately provide carbon and energy for microorganisms in the root zone. While these important root exudates are primarily released from the root cap, additional substrate is supplied to the rhizosphere through the sloughing off of decaying root hairs along the length of the root [42].

Table 1. Common exudates from plants grown under axenic conditions

Compound Class	Examples
Amino acids	all naturally occurring amino acids
Carbohydrates	fructose, glucose, sucrose
Enzymes	amylase, phosphatase
Growth factors	choline, thiamine
Nucleic acids	adenine, guanine
Organic acids	acetic, citric, propionic
Other compounds	auxins, glutamine

Compiled from refs [36–41].

Table 2. Some of the environmental factors controlling root exudation in plants

Factor	Effect
Plant growth stage	Exudation is usually increased in younger plants. Exudate composition changes with plant age.
Soil structure	Particle size distribution (% sand, silt, and clay) influences the extent of root exudation.
Light intensity	Increases in light intensity increase the extent of root exudation.
Presence of microorganisms	Microorganisms can act as sinks for exudates with the absence of bacteria leading to less exudate production.

Compiled from refs [36–41].

Research on plant root exudates has primarily been conducted in the laboratory using seedling plants grown in axenic culture. However, these studies have had limited relevance as the presence or absence of microorganisms can affect root exudation. Small-scale field investigations using $^{14}CO_2$, which is fixed by the plant and released as radiolabeled root exudate, has allowed easier detection and more sensitive measurements. In addition, microbial population responses have been used as indirect endpoints to measure factors effecting exudation [1]. The research on interactions in the rhizosphere should be expanded to explore the role of quantity and composition of root exudate on microbial composition and in contaminant degradation in the rhizosphere. Recent studies have shown that the exudates themselves have enzymatic activity against certain xenobiotics [43]. The ability to control and alter variables that influence exudation (and in turn microbial activity) such as plant species and soil physical parameters may permit further field application of this knowledge to cleanup of soils contaminated with organic pollutants.

3 BIOTIC INTERACTIONS WITH ORGANIC POLLUTANTS

Degradation of organic pollutants in the rhizosphere is largely a biotic phenomenon. The enhanced rates of pollutant metabolism observed in the root zone are often due to the favorable conditions for microbial growth and activity [44]. There are several general mechanisms (Figure 2) responsible for the enhanced microbial degradation observed in the root zone [45]. In most instances, the enhanced degradation appears to be the result of a combination of mechanisms rather than a single mechanism. The ability of plants to select for different microbial communities can translate into differences in the rates and extent of microbial degradation and the classes of compounds that can be degraded. Often this selection is critical to plant survival, as the microorganisms may help protect the plant by degrading the toxic chemical [32,46–49].

3.1 MICROBIAL DIVERSITY

A diverse population of microorganisms is supported in the rhizosphere by the variety of potential substrates exuded by plant roots. This increased diversity may enhance microbial degradation of organic contaminants in soil because of the presence of a key group of organisms involved in carrying out a key step in the contaminant transformation. Biodegradation by rhizosphere microorganisms primarily occurs because these organisms benefit metabolically by using the organic pollutants as a carbon source.

Although the microbial community is diverse, it is generally colonized by a predominantly gram-negative bacterial community [30]. Interestingly, gram-negative bacteria appear to have some important metabolic capabilities for

SYNERGISM. The plant root zone fosters interactions among microorganisms which may ultimately lead to complete transformation of xenobiotics.

BIOMASS. The presence of more microorganisms in the rhizosphere compared with nonvegetated soil may lead to increased rates of chemical transformation.

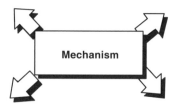

DIVERSITY. The diverse population of microorganisms supported in the rhizosphere may enhance microbial degradation because of the presence of a key group of organisms involved in the metabolism of the contaminant.

EXUDATES. Root exudates may serve as structural analogs to contaminants as well as enhance cometabolism of contaminants.

Figure 2. Summary of potential mechanisms for the enhanced microbial degradation of organic contaminants in the root zone or rhizosphere. Adapted from [86]. By permission of American Association for the Advancement of Science (Environmental Science & Engineering Program)

degrading xenobiotic chemicals not found in gram-positive microorganisms. Glutathione-S-transferase, an enzyme responsible for conjugation of xenobiotics in mammals, plants, and microorganisms, is particularly abundant among gram-negative bacteria including the genera *Pseudomonas* and *Enterobacter* [50]. Certain rhizosphere bacteria also have high activities of other detoxification enzymes such as aryl acylamidase [51]. Laccases, an oxidative enzyme, are also produced by plants, bacteria, and fungi in the rhizosphere and are of interest with respect to their role in contaminant transformation. Degradation of nitroaromatics, commonly used in herbicides, may be accomplished by rhizosphere communities that produce nitroreductase enzymes [52].

3.2 MICROBIAL BIOMASS

The increased microbial biomass produced by the rhizosphere effect is an important biotic interaction in the root zone that likely contributes to enhanced microbial degradation for organic pollutants which are relatively easy to degrade. These biodegradation pathways may not require an extensive array of enzymes, but rather just an organism in need of a carbon source, as is the case with deicing fluids (ethylene and propylene glycol [53]) and 2,4-D [32] in rhizosphere soils. Propylene glycol, for example, is transformed by alcohol dehydrogenase to produce lactic acid and pyruvate, which continue on through the citric acid cycle [54]. Since alcohol dehydrogenase is a fairly ubiquitous enzyme, contaminants such as glycols or other alcohols may be readily degraded in the presence of any microbial community provided the population is sufficient in size to impact the degradation rate. Additionally, the xenobiotic of interest may prove toxic to the microbial community. A large biomass may allow for increased degradation despite the toxicity of the chemical. However, several studies have indicated that microbial activity or another parameter (Table 3) are much better predictors of enhanced degradation in the root zone than simply biomass [10,55].

3.3 SYNERGISM

The complex structure of many organic pollutants encountered by microorganisms can require interaction of multiple microbial species to achieve complete biochemical transformation [12]. As indicated earlier, the variety of root exudates in the rhizosphere environment fosters these types of interactions. The degradation pathway for some organic constituents may involve a number of enzymes, not all of which are produced by any single strain of microorganism [10,11]. Collectively, a diverse microbial community may provide the spectrum of degradative enzymes necessary to achieve complete contaminant transformation [12,56–58].

Degradation of tetrachloroethylene (PCE), a common solvent pollutant, is a multi-step process that begins with the reductive dechlorination of PCE to trichloroethylene (TCE). While the initial step in the biotransformation of PCE to TCE will occur only under anaerobic conditions, the transformation of

Table 3. Microbial parameters used as potential predictors for the degradation of trichloroethylene (TCE) in rhizosphere and nonvegetated soils from a solvent-contaminated site

Sample	TCE Degradation*	Biomass[†] (μg g^{-1} soil)	Activity[§] (pmol g^{-1}soil^{-1}h)	Toxicity[‡]	Methanotrophs** (cells g^{-1} soil)
L. cuneata	Good	1449	1407	0	2.45×10^7
Solidago sp.	Poor	5825	1727	–	
P. notatum	Poor	3624	599	–	
P. taeda	Good	1252	905	0	2.0×10^7
Nonvegetated	Fair	680	445	0	3.25×10^5

* Ability to mineralize ^{14}C-TCE (from [84]).
[†] Determined using phospholipid fatty acid analysis (from [10]).
[§] Determined using ^{14}C-acetate incorporation into microbial lipids (from [10]).
[‡] Microbial respiration ('O' = no change; '–' = depression) with the addition of 500 ppm TCE (from [85]).
** Determined using fluorescent antibodies (from [59]).
Soils were Typic Udorthents, low in %OC, and slightly acidic.

TCE can be carried out both aerobically and anaerobically [7]. Under anaerobic conditions, successive removal of chlorine results in the production of vinyl chloride and ethane. Under aerobic conditions, TCE is converted to an epoxide (probably by methanotrophic organisms [59]) which spontaneously breaks down to a series of more water soluble metabolites (Figure 3). These are utilized by a variety of heterotrophic organisms commonly found in soil. Pockets of aerobic and anaerobic activity in the rhizosphere will be occupied by different interacting microbial communities which may facilitate complete degradation of compounds such as PCE and TCE over time.

Related to increased microbial synergism in the rhizosphere is the idea that mycorrhizal fungi associated with plant roots have recently been studied for their ability to degrade contaminants [60]. Much like bacteria, mycorrhizal fungi can use organic pollutants as a carbon source. However, due to their increased contact with soil through increased surface area, mycorrhizae may increase the overall metabolic capacity at the root–soil interface. Ectomycorrhizae modify the plant–soil interface through a hyphal network extending beyond the host's rhizosphere. In addition, mycorrhizal fungi may improve

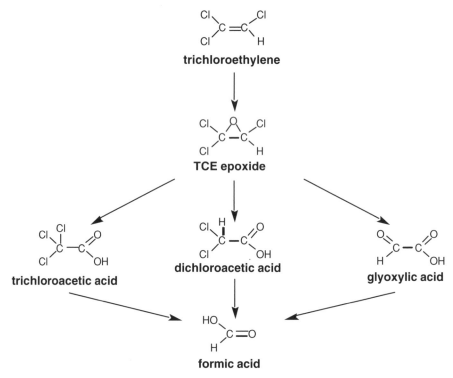

Figure 3. Aerobic metabolism of trichloroethylene by microorganisms. Adapted from [87]. By permission of American Society for Microbiology

soil structure and indirectly influence contaminant transformation through better soil aeration.

3.4 STRUCTURAL ANALOGS/COMETABOLISM

The presence of compounds in the soil (root exudates) that act as structural analogs to an organic pollutant of interest may enhance microbial degradation of that pollutant. Microbial communities that are capable of growing on the root exudate substrate may produce the enzymes necessary to degrade the structurally similar xenobiotic. Such was the case with polychlorinated biphenyl (PCB) degradation, where the presence of biphenyl in the soil was used to enhance the activity of the PCB-degrading bacteria [61]. This concept was expanded to include compounds produced by the plants themselves such as those present in the root exudates [62]. Studies revealed that certain phenolic compounds in root exudate could support growth of PCB-degrading bacteria enhancing the degradation of various PCB congeners. Sandmann and Loos [32] suggested that the increase in 2,4–D-degrading bacteria in sugarcane rhizosphere soil was supported by phenolic analogs in the exudate which selected for the microbial community responsible for degrading 2,4–D. In addition, red mulberry, *Morus rubra*, is capable of producing large amounts of root phenolics [63] and continues to be the focus of further studies on the influence of growth stage on the release of root exudates [64].

The presence of exudates diverse in nutritional quality makes the rhizosphere a good place for cometabolism of chemical contaminants [65,66]. This mechanism is closely related to exudates serving as structural analogs. Cometabolism is the process of chemical biotransformation that does not support growth (i.e., the microorganism does not derive energy, carbon, or nutrients) and therefore is not directly beneficial. The microorganism can transform the substrate into intermediate degradation products but does not multiply as a direct result. Early studies on organophosphate insecticides were useful in identifying this mechanism [67]. The observed increase in degradation of insecticides such as diazinon and parathion in the rhizosphere of bush bean (*Phaseolus vulgaris*) is primarily due to enhanced cometabolic activity rather than community composition or size [44]. The importance of root exudates, either as structural analogs or cometabolites, was further illustrated in studies by Rasolomanana and Balandreau [68] where a *Bacillus* sp. isolated from rice rhizosphere was able to grow on oil residues only in the presence of root exudate.

4 ABIOTIC INTERACTIONS WITH ORGANIC POLLUTANTS

While the physicochemical processes contributing to sorption and availability of organic contaminants in rhizosphere and nonrhizosphere soils are similar, the magnitude of such processes may vary due to the continual changes that occur at the root–soil interface. As indicated earlier, differences in carbon dioxide and oxygen concentrations, osmotic and redox potentials, pH, and moisture content exist between rhizosphere and bulk soil. Ultimately, these differences may contribute to variation in organic chemical fate in rhizosphere and nonrhizosphere soils. Mathematical models such as the Pesticide Root Zone Model [69] have been developed to compare and contrast chemical movement in vegetated and nonvegetated systems; however, the resolution of such models is not likely to be fine enough to evaluate the complexities of molecular processes. The abiotic processes described below are presented conceptually for soils. However, the magnitude of each process in rhizosphere soil compared with nonrhizosphere soil is not yet well understood.

Sorption is one of the key factors controlling the bioavailability and fate of organic contaminants in the rhizosphere and other areas of soil. In recent years, sequestration of chemicals with increasing residence time in soil, referred to as 'aging', has been documented by several laboratories [70–75]. The term 'aging' does not refer to reactions that change the chemical structure of organic contaminants such as polymerization or covalent binding to humic substances through biological activities, but rather the movement of the chemical into soil particles making it less available for chemical or biotic reactions [72]. Although the amount of chemical in soil remains fairly constant, the fraction of that chemical available to soil organisms is reported to decrease significantly with time [72,76].

Many mathematical models have been developed to explain the kinetics of sorption and desorption of hydrophobic contaminants in soil and sediment. There is general agreement in sorption kinetic data for soil as well as aquifer solids and sediment. The initial phase of sorption takes place rapidly, and is followed by slow uptake to an equilibrium [77]. This two-phase sorption process is often explained with a three-dimensional sorption model. The initial phase, adsorption, takes place on the exterior surface of soil particles, which is easily accessed by contaminants; accordingly, adsorption of chemicals to soil particles occurs rapidly. This process is followed by a second phase, absorption, in which a continuous slow sorption takes place at the inner part of the soil particle; therefore, absorption to soil is relatively slow. The longer compounds are in contact with soil, the more likely they are to diffuse to remote sites in soil particles. Many chemicals show a desorption pattern similar to their sorption pattern; rapid initial desorption followed by a slow release from soil particles [78]. In the same manner, contaminant release from the external surface of soil particles accounts for the rapid initial desorption. Chemicals at more remote sites in soil micropores take longer to diffuse (slow desorption).

The slow phase desorption is an extremely important element in chemical availability to both soil invertebrate species and microorganisms. The increasing remoteness of the contaminant in soil micropores with time (desorption resistant fraction) can explain this decreasing bioavailability of chemicals in soil. Soil particles have extremely small micropores (effective diameter $< 2.0\,nm$). Therefore, even the smallest bacterium, animal, or root hair will not come in contact with compounds sorbed in the pore spaces for extended periods [71,75].

To describe this slow sorption–desorption phase, several hypotheses have been proposed. One involves partitioning of hydrophobic organic chemicals into soil organic matter (intraorganic matter diffusion). In this scenario, rapid chemical sorption takes place on the external soil surface followed by a slow chemical partitioning into organic matter existing within the interior region of the soil solid [77,79]. Another model involves slow chemical diffusion by entrapment of chemicals in micropores of soil solids (intraparticle diffusion). This hypothesis suggests that diffusion of chemicals is retarded by microscale partitioning of the chemicals between pore water and organic matter on the interior surface and by long and tortuous diffusion pathways [77,80–82]. In addition, Isaacson and Frink [83] suggest hydrogen-bond interactions between sediment organic matter and phenolic compounds account for nonreversible sorption.

At present, regulatory decisions for risk assessment and site remediation of organic pollutants in soils are made based on chemical concentrations determined by vigorous chemical extraction with organic solvents. These extraction methods are designed to remove the maximum amount of chemical from the soil matrix because it is assumed that the amount of chemical that is available to organisms is simply related to the concentration of that chemical in soil [71,75]. However, accumulating evidence suggests that the chemical available to soil organisms, bioavailability, can be widely affected by the nature of soil structure and physical and chemical properties of the contaminant. Conventional extraction methods that remove as much of the chemical as possible do not account for this phenomenon. Therefore, data based on these extraction methods may overestimate risk and mandate unnecessary remediation work at contaminated sites. The widespread adoption of remediation technologies based on vegetation (phytoremediation) will, to a large part, depend on the recognition of aging and bioavailability by the regulatory community.

5 CONCLUSIONS

The characteristics of the rhizosphere allow for a unique environment which encourages biotic and abiotic interactions in greater magnitude and diversity than in bulk nonvegetated soil. These interactions are responsible for the decrease in toxicity of polluted soils observed for plants in soil through increased degradation and decreased availability of organic compounds.

Biotic reactions generally result in a breakdown of potentially hazardous compounds as microorganisms metabolize these organics as an energy source or an electron acceptor or donor. Byproducts of this process are generally less toxic than the parent compound. These reactions are more prevalent in the rhizosphere and in densely vegetated areas make a significant contribution to an overall decrease in contaminant concentration in soil.

The abiotic reactions occurring at the root–soil interface may serve to sequester organic pollutants decreasing their bioavailability. As the importance of bioavailability has increased in regulatory decision making, a greater understanding of the factors affecting bioavailability of contaminants is warranted. To begin to fill in the gaps between the existing regulatory framework and real world phenomena, several investigators have begun to investigate intensively the process of chemical aging in soil and bioavailability of contaminants to soil invertebrates and microorganisms. Those studies intend to provide more realistic information which takes bioavailability and toxicity into consideration for risk assessments and site remediation.

As the details of biotic and abiotic interactions in the rhizosphere unfold, research has shifted to the practical application of these interactions. Mechanical methods of soil remediation can often be supplemented or replaced with phytoremediation techniques. Remediation objectives can be altered to coincide with the actual exposure by incorporating the beneficial effects of plants and microorganisms on sorption. Research on these interactions continues to impact the field of environmental remediation and regulation through development of new techniques and an expanding base of knowledge of the relationship between microorganisms and soil particles.

REFERENCES

1. Curl, E. A. and Truelove, B. (1986). *The Rhizosphere*, Springer, Berlin, 288 pp.
2. Walton, B. T., Hoylman, A. M., Perez, M. M., Anderson, T. A., Johnson, T. R., Guthrie, E. A. and Christman R. F. (1994). *Rhizosphere Microbial Communities as a Plant Defense Against Toxic Substances in Soils*. In *Bioremediation Through Rhizosphere Technology*, ed. Anderson, T. A. and Coats, J. R., American Chemical Society, Washington, DC, pp. 82–92.
3. Hale, M. G., Foy, C. L. and Shay F. J. (1971). Factors affecting root exudation, *Adv. Agron.*, **23**, 89.
4. Hale, M. G. and Moore, L. D. (1979). Factors affecting root exudation II: 1970–1978. *Adv. Agron.*, **31**, 93–124.
5. Anderson, T. A., Guthrie, E. A. and Walton, B. T. (1993). Bioremediation in the rhizosphere, *Environ. Sci. Technol*, **27**, 2630–2636.
6. Shimp, J. F., Tracy, J. C., Davis, L. C., Lee, E., Huang, W., Erickson, L. E. and Schnoor, J. L. (1993). Beneficial effects of plants in the remediation of soil and groundwater contaminated with organic materials, *Crit. Rev. Environ. Sci. Technol.*, **23**, 41.
7. Walton, B. T. and Anderson, T. A. (1990). Microbial degradation of trichloroethylene in the rhizosphere: potential application to biological remediation of waste sites, *Appl. Environ. Microbiol.*, **56**, 1012.

8. Aprill, W. and Sims, R. C. (1990). Evaluation of the use of prairie grasses for stimulating polycyclic aromatic hydrocarbon treatment in soil, *Chemosphere*, **20**, 253.
9. Ferro, A. M., Sims, R. C. and Bugbee, B. (1994). Hycrest crested wheatgrass accelerates the degradation of pentachlorophenol in soil, *J. Environ. Qual.*, **23**, 272.
10. Anderson, T. A., White, D. C. and Walton, B. T. (1995). Degradation of hazardous organic compounds by rhizosphere microbial communities. In *Biotransformations: Microbial Degradation of Health-Risk Compounds*, ed. Singh, V. P., Vol. 32, Elsevier, Amsterdam, pp. 205–225.
11. Crowley, D. E., Alvey, S. and Gilbert, E. S. (1997). Rhizosphere ecology of xenobiotic-degrading microorganisms. In Phytoremediation of Soil and Water Contaminants, ed. Kruger, E. L., Anderson, T. A. and Coats, J. R. American Chemical Society, Washington, DC, pp. 20–36.
12. Lappin, H. M., Greaves, M. P. and Slater, J. H. (1985). Degradation of the herbicide mecoprop [2-(2-methyl-4-chlorophenoxy) propionic acid] by a synergistic microbial community, *Appl. Environ. Microbiol.*, **49**, 429.
13. Brazil, G. M., Kenefick, L. Callanan, M. Haro, A. DeLorenzo, V. Dowling, D. N. and O'Gara, F. (1995). Construction of a rhizosphere pseudomonad with potential to degrade polychlorinated biphenyls and detection of bph gene expression in the rhizosphere, *Appl. Environ. Microbiol.*, **61**, 1946.
14. Crowley, D. E., Brennerova, M. V., Irwin, C. Brenner, V. and Focht, D. D. (1996). Rhizosphere effects on biodegradation of 2,5-dichlorobenzoate by a bioluminescent strain of root-colonizing *Pseudomonas fluorescens*, *FEMS Microbiol. Ecol.*, **20**, 79.
15. Alvey, S. and Crowley, D. E. (1996). Survival and activity of an atrazine-mineralizing bacterial consortium in rhizosphere soil, *Environ. Sci. Technol.*, **30**, 1596.
16. Siciliano, S. D. and Germida, J. J. (1997). Bacterial inoculants of forage grasses that enhance degradation of 2-chlorobenzoic acid in soil, *Environ. Toxicol. Chem.* **16**, 1098.
17. Yee, D. C., Maynard, J. A. and Wood, T. K. (1998). Rhizoremediation of trichloroethylene by a recombinant, root-colonizing *Pseudomonas fluorescens* strain expressing toluene ortho-monooxygenase constitutively, *Appl. Environ. Microbiol.*, **64**, 112.
18. Senesi, N. (1993). Nature of Interactions between organic chemicals and dissolved humic substances and the influence of environmental factors. In *Organic Substances in Soil and Water: Natural Constituents and Their Influences on Contaminant Behavior*, ed. Beck, A. J., Jones, K. C., Hayes, M. H. B. and Mingelgrin, U. The Royal Society of Chemistry, Cambridge, pp. 73–101.
19. Senesi, N. (1993). Organic pollutant migration in soils as affected by soil organic matter: molecular and mechanistic aspects. In *Migration and Fate of Pollutants in Soils and Subsoils*, ed. Petruzzelli, D. and Helfferich, F. G. Springer, Berlin. pp. 47–74.
20. Bollag, J-M., Mertz, T. and Otjen, L. (1994). Role of microorganisms in soil bioremediation. In *Bioremediation Through Rhizosphere Technology*, ed. Anderson, T. A. and Coats, J. R. American Chemical Society, Washington, DC, pp. 2–10.
21. Senesi, N. and Miano, T. M. (1995). The role of abiotic interactions with humic substances on the environmental impact of organic pollutants. In *Environmental Impact of Soil Component Interactions: Natural and Anthropogenic Organics*, ed. Huang, P. M., Berthelin, J. Bollag, J-M., McGill, W. B. and Page, A. L. Lewis, Boca Raton, FL, pp. 311–335.
22. Dec, J. and Bollag, J-M. (1994). Dehalogenation of chlorinated phenols during binding to humus. In *Bioremediation Through Rhizosphere Technology*, ed. Anderson, T. A. and Coats, J. R. American Chemical Society, Washington, DC, pp. 102–111.

23. Barkovskii, A. L., Boullant, M-L. and Balandreau, J. (1994). Polyphenolic compounds respired by bacteria. In *Bioremediation Through Rhizosphere Technology*, ed. Anderson, T. A. and Coats, J. R. American Chemical Society, Washington, DC, pp. 28–42.
24. Linford, M. B. (1942). Methods of observing soil flora and fauna associated with roots, *Soil Sci.*, **53**, 93.
25. Campbell, R. and Rovira, A. D. (1973). The study of the rhizosphere by scanning electron microscopy, *Soil Biol. Biochem.*, **5**, 747.
26. Rovira, A. D., Newman, E. I., Bowen, H. J. and Campbell, R. (1974). Quantitative assessment of the rhizosphere microflora by direct microscopy, *Soil Biol. Biochem.*, **6**, 211.
27. Katznelson, H. (1946). The rhizosphere effect of mangels on certain groups of micro-organisms, *Soil Sci.* **62**, 343.
28. Katznelson, H. (1965). Nature and importance of the rhizosphere. In *Ecology of Soil-Borne Plant Pathogens*, ed. Baker, K. F. and Snyder, W. C. University of California Press, Berkeley, CA, pp. 187–207.
29. Gray, T. R. G. and Parkinson, D. (1968). *The Ecology of Soil Bacteria*, University of Toronto Press, Toronto.
30. Atlas, R. M. and Bartha, R. (1993). *Microbial Ecology: Fundamentals and Applications*, 3rd edn, Benjamin/Cummings, Menlo Park, CA.
31. Campbell, R. (1985). *Plant Microbiology*, Edward Arnold, Baltimore, MA, 191 pp.
32. Sandmann, E. R. I. C. and Loos, M. A. (1984). Enumeration of 2,4-D-degrading microorganisms in soils and crop plant rhizospheres using indicator media: high populations associated with sugarcane (*Saccharum officinarum*), *Chemosphere*, **13**, 1073.
33. Abdel-Nasser, M., Makawi, A. A. and Abdel-Moneim, A. A. (1979). Occurrence of certain microorganisms in rhizosphere soils of common bean and cotton as affected by the application of Temik or orthocide pesticides, *Egypt. J. Microbiol*, **14**, 37.
34. Abueva, A. A. and Bagaev, V. B. (1975). Transformation of nitrogen fertilizer in soddy-podzolic soil under the effect of 2,4-D, *Izv. Timiiryazevskoi Selskokhozyaistvennoi Akad.*, **2**, 127.
35. Gavrilova, E. A., Kruglov, Y. V. and Garankina, N. G. (1983). Effects of plants and microflora of the rhizosphere on the degradation of Basudin in the soil, *Trudy Vsesoyuznyi Nauchno-Issledovatelskii Institut Selskokhozyaistevennoi Mikrobiologii.*, **52**, 67.
36. Rovira, A. D. (1959). Root excretions in relation to the rhizosphere effect. IV. Influence of plant species, age of plants, light, temperature and calcium nutrition on exudation, *Plant Soil*, **11**, 53.
37. Vancura, V. (1964). Root exudates of plants. 1. Analysis of root exudates of barley and wheat in their initial phases of growth, *Plant Soil*, **21**, 231–248.
38. Bowen, G. D. (1969). Nutrient status effects on loss of amides and amino acids from pine roots, *Plant Soil*, **30**, 139.
39. Rovira, A. D. and Davey, C. B. (1974). Biology of the Rhizosphere. In *The Plant Root and Its Environment*, ed. Carson, E. W. University of Virginia Press, Charlottesville, VA, pp. 153–204.
40. Katznelson, H., Rouatt, J. W. and Payne, T. M. B. (1955). The liberation of amino acids and reducing compounds by plant roots, *Plant Soil*, **7**, 35–48.
41. Boulter, D., Jeremy, J. J. and Wilding, M. (1966). Amino acids liberated into the culture medium by pea seedling roots, *Plant Soil*, **24**, 121.

42. Davis, L. C., Erickson, L. E., Lee, E. Shimp, J. F. and Tracy, J. C. (1993). Modeling the effects of plants on the bioremediation of contaminated soil and ground water, *Environ. Prog.*, **12**, 67.
43. Siciliano, S. D., Goldie, H. and Germida, J. J. (1998). Enzymatic activity in root exudates of dahurian wild rye (*Elymus dauricus*) that degrades 2-chlorobenzoic acid, *J. Agric. Food Chem.*, **46**, 5.
44. Hsu, T. S. and Bartha, R. (1979). Accelerated mineralization of two organophosphate insecticides in the rhizosphere, *Appl. Environ. Microbiol.*, **37**, 36.
45. Siciliano, S. D. and Germida, J. J. (1998). Mechanisms of phytoremediation: Biochemical and ecological interactions between plants and bacteria, *Environ. Rev.*, **6**, 65.
46. Herring, R. and Berring, C. L. (1988). Effects of phthalate esters on plant seedlings and reversal by a soil microorganism, *Bull. Environ. Contamin. Toxicol.*, **40**, 626.
47. Krueger, J. P., Butz, R. G. and Cork, D. J. (1991). Use of dicamba-degrading microorganisms to protect dicamba susceptible plant species, *J. Agric. Food Chem.*, **39**, 1000.
48. Pfender, W. F. (1996). Bioremediation bacteria to protect plants in pentachlorophenol-contaminated soil, *J. Environ. Qual.*, **25**, 1256.
49. Siciliano, S. D. and Germida, J. J. (1998). Degradation of chlorinated benzoic acid mixtures by plant-bacteria associations, *Environ. Toxicol. Chem.*, **17**, 728.
50. Zablotowicz, R. M., Hoagland, R. E. and Locke, M. A. (1994). Glutathione S-transferase activity in rhizosphere bacteria and the potential for herbicide detoxification. In *Bioremediation Through Rhizosphere Technology*, ed. Anderson, T. A. and Coats, J. R. American Chemical Society, Washington, DC, pp. 184–198.
51. Hoagland, R. E. and Zablotowicz, R. M. (1995). Rhizobacteria with exceptionally high aryl acylamidase activity, *Pesticide Biochem. Physiol.*, **52**, 190.
52. Zablotowicz, R. M., Locke, M. A. and Hoagland, R. E. (1997). Aromatic nitroreduction of acifluorfen in soils, rhizospheres, and pure cultures of rhizobacteria. In *Phytoremediation of Soil and Water Contaminants*, ed. Kruger, E. L., Anderson, T. A. and Coats, J. R. American Chemical Society, Washington, DC, pp. 38–53.
53. Rice, P. J., Anderson, T. A. and Coats, J. R. (1997). Evaluation of the use of vegetation for reducing the environmental impact of deicing agents. In *Phytoremediation of Soil and Water Contaminants*, ed. Kruger, E. L., Anderson, T. A. and Coats, J. R. American Chemical Society, Washington, DC, pp. 162–176.
54. Raja, L. M. V., Elamvaluthy, G., Palaniappan, R. and Krishnana, R. M. (1991). Novel biotreatment procedures for glycol waters, *Appl. Biochem. Biotechnol.*, **28/29**, 827.
55. Knaebel, D. B. and Vestal, J. R. (1992). Effects of intact rhizosphere microbial communities on the mineralization of surfactants in surface soils, *Can. J. Microbiol*, **38**, 643.
56. Bordeleau, L. M. and Bartha, R. (1968). Ecology of a pesticide transformation: synergism of two soil fungi, *Soil Biol. Biochem.*, **3**, 281.
57. Gunner, H. B. and Zuckerman, B. M. (1968). Degradation of diazinon by synergistic microbial action, *Nature (London)*. **217**, 1183.
58. Senior, E., Bull, A. T. and Slater, J. H. (1976). Enzyme evolution in a microbial community growing on the herbicide dalapon, *Nature (London)*, **263**, 476.
59. Brigmon, R. L., Anderson, T. A. and Fliermans, C. B. (1999). Methanotrophic bacteria in the rhizosphere of trichloroethylene-degrading plants, *Int. J. Phytoremediation.*, **1**, 241.
60. Donnelly, P. K. and Fletcher, J. S. (1995). PCB metabolism by ectomycorrhizal fungi, *Bull. Environ. Contamin. Toxicol.*, **54**, 507.

61. Bedard, D. L., Wagner, R. E., Brennen, M. J., Haberl, M. L. and Brown, J. F. Jr. (1987). Extensive degradation of aroclors and environmentally transformed polychlorinated biphenyls by Alcaligenes eutrophus H850, *Appl. Environ. Microbiol.*, **53**, 1094.

62. Donnelly, P. K., Hegde, R. S. and Fletcher, J. S. (1994). Growth of PCB-degrading bacteria on compounds from photosynthetic plants, *Chemosphere*, **28**, 981.

63. Fletcher, J. S. and Hegde, R. S. (1995). Release of phenols by perrenial plant roots and their potential importance in bioremediation, *Chemosphere*, **31**, 3009.

64. Hegde, R. S. and Fletcher, J. S. (1996). Influence of plant growth stage and season on the release of root phenolics by mulberry as related to development of phytoremediation technology, *Chemosphere*, **32**, 2471.

65. Liu, C. M., McLean, P. A., Sookdeo, C. C. and Cannon, F. C. (1991). Degradation of the herbicide glyphosate by members of the family Rhizobiaceae, *Appl. Environ. Microbiol*, **57**, 1799.

66. Haby, P. A. and Crowley, D. E. (1996). Biodegradation of 3-chlorobenzoate as affected by rhizodeposition and selected carbon substrates, *J. Environ. Qual.*, **25**, 304.

67. Reddy, B. R. and Sethunathan, N. (1983). Mineralization of parathion in the rice rhizosphere, *Appl. Environ. Microbiol.*, **45**, 826.

68. Rasolomanana, J. L. and Balandreau, J. (1987). Role de la rhizosphere dans la biodegradation de composes recalcitrants: Cas d'une riziere polluee par des residus petroliers, *Rev. Ecol. Biol. Sol.*, **24**, 443.

69. Carsel, R. F., Smith, C. N., Mulkey, L. A., Dean, J. D. and Jowise, P. (1984). *Users Manual for the Pesticide Root Zone Model (PRZM)*, U.S. Environmental Protection Agency, Athens, GA, EPA/600/3–84–109.

70. Pignatello, J. J., Ferrandino, F. J. and Huang, L. Q. (1993). Elution of aged and freshly added herbicides from a soil, *Environ. Sci. Technol.*, **27**, 1563.

71. Alexander, M. (1995). How toxic are toxic chemicals in soil? *Environ. Sci. Technol.*, **29**, 2713.

72. Hatzinger, P. B. and Alexander, M. (1995). Effect of aging of chemicals in soil on their biodegradability and extractability, *Environ. Sci. Technol.*, **29**, 537.

73. Nam, K., Chung, N. and Alexander, M. (1998). Relationship between organic matter content of soil and the sequestration of phenanthrene, *Environ. Sci. Technol.* **32**, 3785.

74. Tang, J., Carroquino, M. J., Robertson, B. K. and Alexander, M. (1998). Combined effect of sequestration and bioremediation in reducing the bioavailability of polycyclic aromatic hydrocarbons in soil, *Environ. Sci. Technol*, **32**, 3586.

75. Alexander, M. (2000). Aging, bioavailability, and overestimation of risk from environmental pollutants, *Environ. Sci. Technol.*, **34**, 4259.

76. Kelsey, J. W. and Alexander, M. (1997). Declining bioavailability and inappropriate estimation of risk of persistent compounds, *Environ. Toxicol. Chem.*, **16**, 582.

77. Wu, S-C. and Gschwend, P. M. (1986). Sorption kinetics of hydrophobic organic compounds to natural sediments and soil, *Environ. Sci. Technol.*, **20**, 717.

78. Pavlostathis, S. G. and Mathaven, G. N. (1992). Desorption kinetics of selected volatile organic compounds from field contaminated soils, *Environ. Sci. Technol.*, **26**, 532.

79. Chiou, C. T. (1985). Partition coefficient of organic compounds in lipid-water systems and correlations with fish bioaccumulation factors, *Environ. Sci. Technol.*, **19**, 57.

80. Steinberg, S. M., Pignatello, J. J. and Sawhney, B. L. (1987). Persistence of 1,2-dibromoethane in soils: entrapment in intraparticle micropores, *Environ. Sci. Technol.*, **21**, 1201.

81. Brusseau, M. L., Jessup, R. E. and Rao, P. S. C. (1991). Nonequilibrium sorption of organic chemicals: Elucidation of rate-limiting processes, *Environ. Sci. Technol.*, **25**, 134.
82. Chiou, C. T., Poter, P. E. and Schmedding, D. W. (1983). Partition equilibria of nonionic organic compounds between soil organic matter and water, *Environ. Sci. Technol.*, **17**, 227.
83. Isaacson, P. J. and Frink, C. R. (1984). Nonreversible sorption of phenolic compounds by sediment fractions: The role of sediment organic matter, *Environ. Sci. Technol*, **18**, 43.
84. Anderson, T. A. and Walton, B. T. (1995). Comparative fate of 14C-trichloroethylene in the root zone of plants from a former solvent disposal site, *Environ. Toxicol. Chem*, **14**, 2041.
85. Anderson, T. A. (1991). *Comparative plant uptake and microbial degradation of trichloroethylene in the rhizospheres of five plant species: Implications for bioremediation of contaminated surface soils*, Ph.D. Thesis, The Univeristy of Tennessee, Knoxville, TN, 187 pp.
86. Anderson, T. A. (1997). *Development of a phytoremediation handbook: considerations for enhancing microbial degradation in the rhizosphere*, Environmental Science and Engineering Fellows Program, 1997 Reports, American Association for the Advancement of Science, Washington, DC, pp. 1–13.
87. Little, C. D., Palumbo, A. V., Herbes, S. E., Lindstrom, M. E., Tyndall, R. L. and Gilmer, P. J. (1988). Trichloroethylene biodegradation by a methane-oxidizing bacterium, *Appl. Environ. Microbiol.*, **54**, 951.

(C) *Anaerobic and Transport Processes, and the Subsurface Terrestrial Environment*

11 Interactions between Microorganisms and Minerals under Anaerobic Conditions

HENRY L. EHRLICH
Rensselaer Polytechnic Institute, USA

Interactions between Soil Particles and Microorganisms
Edited by P. M. Huang, J.-M. Bollag and N. Senesi. © 2002 John Wiley & Sons, Ltd

1 INTRODUCTION

As shown in other chapters in this book, soil minerals are not inert entities insofar as microbiological activity is concerned. They can profoundly affect the activity of soil microbes, especially bacteria, fungi, and, in zones where light is accessible to them, cyanobacteria and algae. Conversely, soil microorganisms can alter soil minerals. The microbes involved in these interactions may be aerobic (see Chapter 5), facultative or anaerobic. The nature of their interactions with minerals is chemical and physical (Table 1). Soil ecosystem health can be significantly affected by them in terms of production or elimination of

Table 1. Possible interactions of microorganisms with soil minerals under anaerobic conditions

1. Mineral degradation
2. Mineral diagenesis
3. Mobilization of trace metal nutrients from soil minerals
4. Organic carbon mineralization through anaerobic respiration with Fe(III)-, Mn(IV)-, and SO_4^{2-}-containing minerals as terminal electron acceptors
5. Mobilization of toxic mineral constituents
6. Detoxification of arsenate, arsenite, selenate, selenite and heavy metal pollutants

inorganic and organic toxicants. Although the following discussion will stress anaerobic interactions, only some of them are uniquely anaerobic.

2 MINERAL EFFECTS ON MICROBES

2.1 MINERAL SURFACES AS A MICROBIAL HABITAT

Surfaces of soil minerals offer habitats for aerobic or anaerobic microbes. Bacteria may occupy mineral surfaces as scattered individual cells, as micro-colonies, or as biofilms (Figure 1) (see also Chapter 6). Depending on physical and/or chemical properties of a mineral surface, a microbe may attach to it with the help of hydrophilic exopolymer produced by it or which another organism laid down before it [1–3], or via hydrophobic cell surface structures if the mineral surface is hydrophobic [4–6]. Hydrophilic attachment may involve electrostatic interaction between negative charges at the cell surface or exopolymer surrounding a cell and positively charged regions at a mineral surface. Once attached, individual microbes may form microcolonies, which may be embedded in an exopolymer matrix produced by the colony. In the case of hydrophilic attachment, an assemblage of cells and/or microcolonies embedded in exopolymer may ultimately become a continuous biofilm [7,8]. The extent to

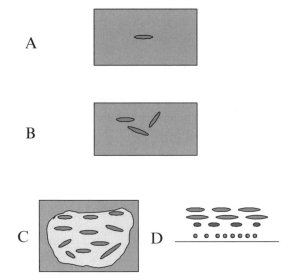

Figure 1. Diagrammatic representation of different states of bacterial attachment to solid surfaces. (A) single cell attachment; (B) multiple cell attachment without embedment in slime; (C) multiple cell attachment in the form of a biofilm; (D) vertical section through a multilayered biofilm showing positioning of three different microbial types in the biofilm. See text for further discussion

which bacteria attach to a mineral surface may be determined by the composition of the mineral [9]. A biofilm, once formed, may promote significant alteration (weathering) of the mineral surface to which it is attached. This alteration is the result of interaction with metabolic products formed by the bacteria in the biofilm from nutrients they consume.

Microbes may be attracted to mineral surfaces because nutrients essential for their growth occur in higher concentration on these surfaces than in the surrounding solution. The accumulation of organic nutrients at the mineral surface is most likely the result of adsorption. Bacteria in a biofilm may obtain adsorbed nutrients as a result of solubilization by convective fluid in the biofilm matrix [7]. Microbes may also be attracted to a mineral surface because one or more components of the mineral can serve as electron donor or acceptor in bacterial energy metabolism, or one or more mineral components may satisfy trace mineral requirements.

Biofilm on a mineral surface may be the result of growth of a single species of bacteria or a mixed bacterial population (Figure 1). Residence in a biofilm offers protection and nutritional advantages to bacteria that they would not encounter if dispersed in a liquid environment [7]. A mixed biofilm facilitates vital synergistic interactions among various bacterial members in it. Such a biofilm may arise from invasion by different bacterial species at different times. The different species are apt to occupy specific locations in the biofilm where growth conditions are optimal for them. Thus, in mixed biofilms having sufficient thickness, an oxygen gradient may be generated by aerobic bacteria at the surface in direct contact with oxygen-carrying solution. These bacteria may consume oxygen so rapidly that little can diffuse below the surface. As a result, the region below the biofilm surface becomes oxygen depleted, and organisms in this location must be microaerophilic or anaerobic to thrive [10]. The different organisms in a mixed biofilm may be metabolically interdependent in that metabolic products of one biofilm member become a major substrate for one or more of the others. The close proximity of the members of a metabolically interdependent mixed population in a biofilm ensures a relatively concentrated and sustained supply of these nutrients, which would not be the case if the same organisms were widely dispersed in a liquid. The organisms that are located on the bottom of a biofilm next to the mineral surface may mobilize essential mineral nutrients from the mineral surface for consumption by all biofilm members and by unattached microbes dispersed in the surrounding solution.

To date, studies of microbial attachment to mineral surfaces have dealt mainly with aerobes. Our knowledge of surface attachment to minerals by anaerobes is rather limited [4,11,12] and often only inferred, as from observations of reductive attack of insoluble oxides of iron or manganese. In the case of mixed biofilms with aerobic and anaerobic members, the aerobes probably initiate formation of the biofilm, and only later, after the biofilm has achieved sufficient thickness, do anaerobes such as sulfate-reducing bacteria (SRB) invade it. In cases in which bacterial cells attached to a mineral surface by exopolymer enzymatically oxidize or reduce a mineral component without its prior solubilization, metallic ions that can undergo rapid, reversible oxidation and reduction, such as Fe^{2+}/Fe^{3+} or Mn^{2+}/Mn^{3+}, need to be bound in the exopolymer to conduct electrons between the cell and the mineral component [e.g., 13–15]. In the case of redox reactions by cells attached to the mineral surface hydrophobically [4], redox enzyme at the cell surface may be able to interact directly with the reactive mineral component [16].

2.2 MINERALS AS ELECTRON DONORS OR TERMINAL ELECTRON ACCEPTORS IN MICROBIAL METABOLISM

As also discussed in Chapter 6, some bacteria may use one or more constituents of a soil mineral as a source of energy or as terminal electron acceptors in their energy-generating metabolism. To date, the best known microbial processes in which a mineral serves as energy source, require aerobic conditions (for detailed discussion see Chapter 5). However, anaerobic microbial oxidation of a sulfide mineral, covellite (CuS), has been described by Donati *et al.* [17]. These workers demonstrated in a laboratory experiment that *T. ferrooxidans* was able to grow and solubilize copper in an atmosphere of oxygen-free nitrogen from CuS in the presence of Fe(III). They explained their findings by postulating that CuS underwent abiological oxidation by the Fe(III) present in the medium,

$$CuS + 2Fe^{3+} \rightarrow Cu^{2+} + S^0 + 2Fe^{2+} \tag{1}$$

Reaction (1) was first experimentally demonstrated by Sullivan [18]. Donati *et al.* [17] assumed that the S^0 in reaction (1) accumulated on the surface of the CuS particles, and that *T. ferrooxidans* removed this sulfur by oxidizing it with Fe^{3+} in the medium, a reaction first observed by Brock and Gustafson [19],

$$S^0 + 6Fe^{3+} + 4H_2O \rightarrow 6Fe^{2+} + SO_4^{2-} + 8H^+ \tag{2}$$

The S^0 in reaction (2) is clearly the energy source utilized by *T. ferrooxidans*. It is conceivable that some metal sulfides are also susceptible to anaerobic microbial oxidation with nitrate as terminal electron acceptor because dissolved sulfide has been shown to be oxidized with nitrate by *T. denitrificans* (see summary in [20], p. 523). Furthermore, it has been shown that Fe^{2+} can be oxidized anaerobically with nitrate by some nitrate-reducing bacteria [21–23].

Microbial reduction of minerals (Table 2) is usually thought of as an anaerobic process because the reducible mineral constituents serve as terminal electron acceptors in place of oxygen. Examples in which a mineral serves as a terminal electron acceptor only under anaerobic conditions are indeed common. Thus, iron(III) oxides and manganese(IV) oxides are reduced anaerobically but not aerobically by organisms such as *Geobacter metallireducens*

Table 2. Examples of microbially influenced formation, transformation, and degradation reactions of minerals in soil under anaerobic conditions

Type reactions	References
Redox reactions with acetate as reductant	
$8FeOOH+CH_3COO^-+15H^+ \rightarrow 8Fe^{2+}+2HCO_3^-+12H_2O$ (ferrihydrite reduction)	[91,193]
$4MnO_2+CH_3COO^-+7H^+ \rightarrow 4Mn^{2+}+2HCO_3^- + 4H_2O$ (Mn(IV) oxide reduction)	[14,15,49,69]
$CaSO_4+CH_3COO^- \rightarrow HS^-+2HCO_3^-+Ca^{2+}$ (gypsum reduction)	[194]
$2SeO_3^{2-}+CH_3COO^-+3H^+ \rightarrow 2Se^0+2HCO_3^-+2H_2O$ (reduction of selenite)	[195]
$2SeO_4^{2-}+3CH_3CH(OH)COO^-+H^+ \rightarrow 2Se^0+3CH_3COO^-$ $+3HCO_3^-+2H_2O$ (reduction of selenate)	[73,74]
Diagenetic transformation	
$24FeOOH+CH_3COO^- \rightarrow 8Fe_3O_4+2HCO_3^- + H^++12H_2O$ (transformation of ferrihydrite magnetite)	[48]
$8FeOOH+CH_3COO^- + 13H^+ \rightarrow 2FeCO_3 + 6Fe^{2+}+12H_2O$ (transformation of ferrihydrite to siderite)	[48]
$4MnO_2+CH_3COO^-+5H^+ \rightarrow 2MnCO_3 + 2Mn^{2+}+4H_2O$ (transformation of vernadite or pyrolusite to rhodochrosite)	[48]
Weathering of minerals via acidolysis	
$CaCO_3 + 2H^+ \rightarrow Ca^{2+} + H_2O + CO_2$ (calcite dissolution)	[20,Chapter 6]
$KAlSi_3O_8 + H^+ + H_2O \rightarrow K^+ + Al^{3+} + 3HSiO_3^-$ (orthoclase dissolution)[*]	[62]
$Fe_3(PO_4)_2+3H_2S \rightarrow 3FeS+2HPO_4^{2-} + 4H^+$ (vivianite dissolution by biogenic hydrogen sulfide)	[106]
Weathering by complexation with ligands	
$CaSiO_3+2\alpha\text{-ketogluconate} \rightarrow Ca(\alpha\text{-ketogluconate})_2+SiO_3^{2-}$ (wollastonite dissolution)	[34]

[*] Idealized.

and *Shewanella putrefaciens* (see summary in [20], pp. 347–348; 413). The former is a strict anaerobe, but the latter is a facultative organism that can also use oxygen as terminal electron acceptor in place of Fe(III) and Mn(IV). However, it reduces Fe(III) and Mn(IV) only anaerobically. A few instances of aerobic reduction of minerals containing reducible components are known. Two examples are the reduction of MnO_2 with glucose by *Bacillus* GJ33, and with acetate by marine pseudomonad BIII88 [14,15,24]. Some others were reported by Di-Ruggiero and Gounot [25].

2.3 MINERALS AS A SOURCE OF TRACE NUTRIENTS

All cellular life, including aerobic and anaerobic bacteria, requires traces of some elements such as Ca, Mg, Fe, Mn, Cu, Mo, Ni, Zn, and Se in their nutrition. Minerals containing any of these are their primary source. The trace elements become available to cells through mobilization by biological and/or chemical weathering of these minerals under aerobic and anaerobic conditions. If the mobilized elements reach sufficiently elevated concentrations, they may become toxic [26]. Clays, when present, can exert some control on the concentration of cations in soil by binding them exchangeably.

2.4 MINERALS AS DETOXIFYING AGENTS

Soil minerals such as clays may bind inorganic and/or organic toxicants in soil solution [27], thereby detoxifying the solution and promoting microbial growth in it. This may happen aerobically or anaerobically, depending on the stability of the reactant under a given set of redox conditions. Clays carry a net negative charge and thus have a significant affinity for cations, including toxic metal ions such as Al^{3+}, Cu^{2+}, Ni^{2+}, Co^{2+}, Pb^{2+}, Zn^{2+}, Ag^+, Cd^{2+}, and Hg^{2+}. Another important mechanism for relieving soil toxicity due to Hg^{2+} is its bacterial reduction to volatile Hg^0. Toxic anions, such as arsenate, arsenite, selenate, and selenite, may be immobilized through precipitation by base metal ions such as those of iron, copper, cobalt, and nickel under appropriate redox conditions, but they are more likely to be removed anaerobically through microbial reduction to their elemental state, e.g., Se^0, or to a volatile, methylated state, e.g., $(CH_3)_2AsH$, $(CH_3)_3As$, $(CH_3)_2Se$ (see summary in [20], pp. 283–286 and 618–621; discussion below; and Chapter 5).

3 EFFECTS OF MICROBES ON MINERALS

3.1 MINERAL WEATHERING

Some bacteria, including anaerobic species (Table 3), and fungi in soil can attack mineral surfaces and thereby mobilize constituents of the mineral (Table 2). These constituents may include trace elements required in microbial nutrition. They may also include toxic mineral constituents, which may inhibit or kill members of the soil microflora. The modes of attack of minerals include (1) direct enzymatic oxidation or reduction of a reduced or oxidized mineral component (see summary in [20], chapters 14, 15, 17, 18), (2) indirect attack with a metabolically produced redox agent or inorganic or organic acid, [e.g., 28–31], (3), indirect attack by metabolically produced alkali, usually in the form of ammonia [32,33], (4) indirect attack with a metabolically produced ligand that forms a highly soluble product with a mineral component [34; see also

Table 3. Some anaerobic microorganisms contributing to mineral and rock weathering

Name of organism	Response to O_2	Mode of action	Referenc
Bacillus polymyxa	facultative	MnO_2 reduction	[61]
Clostridium acetobutylicum	anaerobic	acetic, butyric acids from hexoses such as glucose for silicate and carbonate solubilization*	[62]
Clostridium butyricum		reduction of Fe(III) oxides	[89]
Geobacter metallireducens	anaerobic	ferrihydride and Mn(IV) oxide reduction	[48,85]
Mixed culture	anaerobic	acid production from carbon/ energy source resulting in silicate weathering	[41]
Shewanella putrefaciens	facultative	Fe(III) oxide, Fe_3O_4, and Mn(IV) oxide reduction	[90]
Sulfate-reducing bacteria	anaerobic	dissolution of $FePO_4$ $Fe_3(PO4)_2$ with H_2S	[106,107]
		Dissolution of $CaSO_4$ and $CaSO4 \cdot 2H_2O$	[124]
Thiobacillus ferrooxidans	facultative	anaerobic oxidation CuS anaerobic oxidation of S^0 with Fe(III)	[17] [19]

* By inference.

Chapter 5], and (5) indirect attack by biopolymer [31,35–38]. The mode of microbial attack of a mineral may involve a combination of some of these mechanisms. In some cases of attack, the microorganisms producing weathering agents may be dispersed in the soil solution, in others they may grow in biofilms on the surface of the susceptible minerals. Although to date, the microbes most frequently studied in mineral weathering have been aerobes, some anaerobes also have this capacity. Various bacteria have been shown to reduce Fe(III) and Mn(IV) in minerals [e.g., 39–41]. Some of these were strict anaerobes, others were facultative. Fermenting bacteria, which are strict anaerobes, produce organic acids that are important weathering agents of aluminosilicates [e.g., 40, 41]. Many fermenters and anaerobic respirers also produce CO_2, which dissolves carbonate minerals (see summary in [20], p. 199–201).

3.2 MINERAL DIAGENESIS

Some microbes can promote the transformation of one mineral into another by a process called diagenesis (Table 2). A special example which involves both aerobic and anaerobic phases is the transformation of chalcopyrite into ferrous sulfides and soluble copper by the combined activity of iron-oxidizing thiobacilli and anaerobic sulfate-reducing bacteria (SRB). Although not directly observed in soil, it has been noted in ore leaching [42]. In this case, acidophilic, metal-sulfide-oxidizing thiobacilli and SRB may form a natural association in a leach heap or dump of sulfidic ore. In the oxidizing zone of the leach heap or dump, one of the bacterial partners (e.g., *T. ferrooxidans*) mobilizes the metal constituents of the ore by oxidizing the metal sulfide by direct attack of the mineral surface. Or it may oxidize ferrous iron dissolved in the interstitial solution in contact with the ore particles to ferric iron, which then oxidizes the sulfidic ore chemically (see Chapter 5). In the reducing zone of the leach dump, the other partner, the SRB, reduces the sulfate formed in the oxidation of the metal sulfide to H_2S [42–44]. At low sulfate concentration or in its absence, Fe(III) produced by the thiobacilli in the oxidizing zone may also serve as an electron acceptor for the SRB in the reducing zone, forming Fe^{2+}. At stoichiometric excess, Fe^{2+}, whether generated by SRB or resulting from chemical reduction of Fe^{3+} by some of the H_2S generated by SRB, will react with remaining H_2S to form iron sulfide precipitate [42]. The Fe^{2+} assists in controlling the concentration of dissolved sulfide, which might otherwise inhibit the SRB. The thiobacilli can grow aerobically or anaerobically, depending on which energy source and electron acceptor is available, whereas the SRB normally grow only anaerobically. In leach dumps operated for mobilization of metals from sulfidic ore, aerobic conditions are essential in the oxidizing zone, as discussed in Chapter 5. Unlike the thiobacilli, the SRB are generally not acidophilic, so the downward migrating interstitial solution carrying Fe(III) must undergo a rise in pH as a result of as yet poorly defined reactions [43].

4 SPECIFIC FUNCTIONS OF MINERALS IN ANAEROBIC BACTERIAL METABOLISM

4.1 ANAEROBIC RESPIRATION VERSUS FERMENTATION

When accessible, mineral components that are reducible, such as Fe(III) and Mn(IV), can serve as terminal electron acceptors to some bacteria anaerobically (Table 3), and in a few instances even aerobically. Although the bacteria are mostly heterotrophic, they also include a few autotrophic bacteria. Of the heterotrophic bacteria that are capable of anaerobic respiration, some are restricted to this mode of metabolism, while others may also be capable of fermentation. In the latter instance, viable cells that are in direct contact with a mineral containing reducible components will preferentially respire, while the remaining cells suspended in the aqueous (bulk) phase that bathes the mineral will ferment if a fermentable substrate is present and dissolved external electron acceptors are absent. Many anaerobically respiring, heterotrophic bacteria can completely degrade (mineralize) their organic electron donors, whereas fermenting bacteria cannot. As a result, respiring bacteria can conserve more energy per mole of an electron donor consumed than fermenting bacteria (for a more detailed discussion, see, for instance, [45] and [20, Chapter 6]).

Reducible mineral components may have to be mobilized first before they can be reduced by bacteria. The mobilization may itself be microbially mediated through production of ligands, as in the case of iron(III) oxide reduction [46, but see also 47].

4.1.1 Bacteria Capable of Anaerobic Respiration: Organic Electron Donor

A range of bacteria have been isolated and shown to be capable of anaerobic respiration using an organic electron donor and an inorganic electron acceptor, the latter either dissolved or in mineral form. Most of these bacteria are Fe (III)-, Mn(IV)- and sulfate-reducing bacteria [39,48–51], but some others are selenate- and selenite-, arsenate-, chromate-, and vanadate-reducers [52; see also summary in 20].

4.1.2 Bacteria Capable of Anaerobic Respiration: Inorganic Electron Donor

Only very few bacteria have been described to date that can use an inorganic electron donor to reduce an inorganic electron acceptor anaerobically. Hydrogen is the most common electron donor and energy source in these cases and may be used autotrophically or mixotrophically [53–56]. In soil environments, the source of this hydrogen is mostly bacterial fermentation. Fe^{2+} can also serve as an electron donor anaerobically when coupled to the reduction of nitrate [21–23,57], which suggests that the mineral siderite ($FeCO_3$) may be bacterially oxidizable by nitrate. Elemental sulfur may serve as an electron donor in the reduction of Fe^{3+} by *T. ferrooxidans* [19,58]. A recent report by Hulth *et al.* [59] suggests that ammonium may reduce Mn(IV) oxides (e.g., MnO_2) anaerobically, the ammonium being oxidized to nitrate. Specific bacterial involvement in this process remains to be demonstrated, however. This would be an example of anaerobic nitrification.

4.1.3 Bacteria Capable of Anaerobic Respiration and Fermentation

This type of organism was already discussed in the introduction to this section. One probable example of such an organism is *Bacillus polymyxa* [60], and another is *B. polymyxa* D1 [61].

5 SOME CONSEQUENCES OF INTERACTION BETWEEN MICROBES AND MINERALS

5.5 INCREASE IN PERMEABILITY OF ROCK STRATA

Limestone is a type of a rock whose porosity can be increased by microbial activity. Both aerobic and anaerobic microbes can enhance limestone porosity by generating lixiviants that can dissolve $CaCO_3$. Anaerobically, such agents include any metabolically formed acid, whether H_2CO_3 resulting from the reaction of CO_2 with H_2O, or organic acids such as acetic, propionic, lactic, pyruvic, butyric, and others produced mainly in fermentations. The dissolution mechanism involves the dissociation of the acids and the subsequent reaction of $CaCO_3$ with the protons,

$$CaCO_3 + H^+ \rightarrow Ca^{2+} + HCO_3^- \tag{3}$$

Below pH 5, the following reaction becomes important,

$$HCO_3^- + H^+ \rightarrow CO_2 + H_2O \tag{4}$$

The porosity of sandstone can also be increased by aerobic or anaerobic microbial activity. If the substance that cements the sand grains in sandstone to one another is calcite ($CaCO_3$), the same mechanism as for limestone would operate. If some form of ferric oxide cements the sand grains to one another, solubilization of the oxide by natural ligands such as humic acids, possibly followed by bacterial reduction of the Fe(III) to Fe(II) could be the operating mechanism. Ligands have been shown to stimulate microbial reduction of oxides of iron and manganese [25,46]. If the sand grains are held together in a silicate matrix, acidolysis and/or complexation by microbial products generated anaerobically would be the mechanism by which gradual destruction of the silicate matrix can occur [31,62,63].

Increase in porosity in permeable rock strata will enhance groundwater migration and the dispersion of any solutes, beneficial or harmful, in the groundwater.

5.2 BLOCKAGE OF CHANNELS DUE TO MICROBIAL MINERAL PRECIPITATION OR BIOMASS ACCUMULATION

Channels in porous sedimentary strata may become blocked as a result of microbially promoted precipitation of one or more minerals from groundwater solution saturated with corresponding solutes. Examples of this under anaerobic conditions are (1) the precipitation of $CaCO_3$ by SRB during reduction of $CaSO_4$, and (2) the precipitation of iron sulfides from microbially generated

Fe^{2+} and S^{2-}. $CaCO_3$ is also precipitated by some aerobic microbes [see 20, pp. 182–195].

Channels in porous strata may also become blocked by biomass accumulation in pore throats. Such blockage will impede groundwater circulation. In microbially enhanced oil recovery, such blockage by microbial biomass may be artificially induced to enhance sweep efficiency in removing petroleum from its natural reservoir [64,65]. Owing to conditions at the site of action, the microflora responsible will be strictly or facultatively anaerobic.

5.3 DIAGENETIC CHANGES

The discovery by Fortin and Beveridge [42] of secondary iron sulfide precipitation from portions of the iron released by bioleaching from chalcopyrite, pyrite and other metal sulfides in a polymetallic sufidic ore was mentioned in section 3.2. Other examples of microbially enhanced diagenetic change include the transformation of Fe_2O_3 and MnO_2 into $FeCO_3$ and $MnCO_3$, respectively [66–69] (Table 2). In these instances, appropriate bacteria anaerobically reduce Fe_2O_3 and MnO_2 to Fe^{2+} and Mn^{2+}, respectively, which then precipitate by reaction with dissolved carbonate under neutral to slightly alkaline conditions. The source of the carbonate is very likely microbial mineralization of organic matter coupled to the reduction of the iron and manganese oxides. Another example of microbially promoted diagentic change is the transformation of $FeOOH$ to Fe_3O_4 [69] (Table 2).

5.4 MICROBIAL MOBILIZATION OF TRACE MINERAL NUTRIENTS

In promoting mineral weathering, microbes mobilize mineral elements, at least some of which are essential in nutrition not only of microbes but of all forms of life (see section 2.3 and Chapter 5). The microbes promoting this weathering may be aerobes or anaerobes.

5.5 MICROBIAL MOBILIZATION OF INORGANIC TOXICANTS AND SUBSEQUENT DETOXIFICATION

Microbial weathering may mobilize toxic mineral constituents in cationic or anionic form. Examples are the release of Pb^{2+}, AsO_4^{2-}, SeO_4^{2-}, and SeO_3^{2-} from corresponding minerals. Although the microbial mobilization of these toxic ions is most frequently the result of aerobic activity (see Chapter 5), especially the oxidation of sulfide minerals containing these toxic species, their subsequent microbial detoxification through immobilization is often the result of anaerobic activity. Thus, Pb^{2+} can be anaerobically immmobilized by precipitation as PbS with sulfide generated from sulfate by SRB [see 20, Chapter 18). Arsenate can be anaerobically immobilized by reduction to

As$_2$S$_3$ by *Desulfotomaculum auripigmentum* [70]. Selenate and selenite can be anaerobically reduced to elemental red selenium by *Thauera selenatis* [71,72], by strain SES-3 [73], and by *Bacillus selenitireducens* [74] (Table 2).

6 SOME SPECIFIC EXAMPLES OF INTERACTION OF MINERALS WITH ANAEROBIC MICROBES

6.1 FERRIC OXIDES

As mentioned in a different context in Chapter 6, a number of different anaerobic bacteria have been shown to be able to use ferric iron as a terminal electron acceptor. These include strict anaerobes such as *Geobacter metallireducens* [75], *G. acetoxidans* [76], *G. sulfurreducens* [55], *Geospirillum barnesi* [76], *Geovibrio ferrireducens* [56], *Desulfovibrio desulfuricans* [77], *Desulfuromonas acetoxidans* [78], *Pelobacter carbinolicus* [76], *Desulfuromonas* spp. [76], *Bacillus infernus* [79], and *Clostridium* spp. [80]. They also include facultative anaerobes such as *Shewanella* (formerly *Pseudomonas* or *Alteromonas*) *putrefaciens* [49,53], *S. alga* [54], and *Ferrimonas balearica* [81]. These facultative organisms reduce Fe(III) only anaerobically. Some acidophiles of the genus *Acidiphilium* can reduce Fe(III) aerobically and anaerobically [82–84].

Whereas all of the above named organisms have been shown to reduce Fe(III) when it is dissolved as a complex in solution, for instance as a citrate complex [85] or as a nitriloacetic acid complex [46,86,87], only some of these organisms have been tested in the laboratory for their ability to attack ferric iron contained in solid phases, whether amorphous FeOOH or as crystalline hematite, goethite, magnetite, or other iron oxide mineral [e.g, 88–91]. Ottow [88], Munch and Ottow [92], and Lovley and Phillips [85] concluded from their experiments that crystalline iron oxides were less readily attacked than amorphous iron oxides. However, Ehrlich *et al.* [40] and Ehrlich and Wickert [41] found that bacteria associated with bauxites containing significant amounts of hematite, goethite, and/or aluminian goethite reduced the Fe(III) in this ore readily under anaerobic conditions.

Ottow [88] indicated that bacterial reduction of ferric iron in iron oxide solids requires physical contact between the bacteria and the oxide. This was clearly demonstrated in experiments by Munch and Ottow [89]. Magnetite (Fe$_3$O$_4$), which has been shown to be formed diagenetically from amorphous Fe(III) oxide *by G. metallireducens* GS-15 [69], is reduced to Fe(II) by *S. putrefaciens* at weakly acid pH [90]. Recognition of the requirement for physical contact leaves, however, unexplained how a bacterial cell that reduces Fe(III) in an oxide conveys electrons from the electron donor to the Fe(III). A clue probably resides in the discovery by Myers and Myers [16] that in anaerobically grown *S. putrefaciens* MR-1, which is capable of reducing Fe(III) anaerobically but not aerobically, cytochromes appear in significant quantity in the outer membrane of the cell envelope while in aerobically grown cells they appear in the plasma membrane. Cytochromes of the c type can be expected to serve as conveyors of electrons to Fe(III), either directly or via a distinct Fe(III) reductase.

Very recent work by Gaspard et al. [93] with G. sulfurreducens suggests that c-type cytochrome is part of the Fe(III) reductase in that organism. They found about one-third of this enzyme in the periplasmic fraction from G. sulfurreducens and two-thirds in the membrane fraction, 80% of which was associated with the outer cell membrane of the cell envelope. According to Seeliger et al. [94], G. sulfurreducens does not have to be in contact with Fe(III) oxide since it excretes c-type cytochrome with Fe(III) reductase activity into the bulk phase. Lloyd et al. [95] have, however, been unable to confirm this claim. In any case, the finding of Seeliger et al. [94] does not explain how any Fe(III) reducers that cannot attach at all to Fe(III) oxide particles reduce the iron. Any Fe(III) reductase the organisms excrete into the bulk phase would be rapidly diluted and dispersed. It is difficult to understand how this dispersed extracellular cytochrome could be re-reduced once it was oxidized. A more likely explanation of this instance of bacterial reduction of mineral-contained Fe(III) without cellular attachment to the mineral would be prior solubilization of the Fe(III) by ligands that may or may not be of biological origin, and subsequent bacterial reduction of the dissolved Fe(III) complex [e.g., 46,87]. It is interesting that Holmgrén et al. [47] found that neither acetohydroxamate nor nitrilotriacetate stimulated cell growth of G. metallireducens in the presence of goethite or amorphous iron hydroxide. The authors suggested that in nature, other bacteria might degrade the ligands of chelated iron to make the ferric iron in the chelates more accessible to G. metallireducens.

Except in the diagenetic formation of magnetite and siderite (Table 2), bacteria usually reduce the iron in Fe(III) oxides to soluble Fe^{2+} [96]. However, when S. putrefaciens reduces the Fe(III) in smectites, it does not solubilize it [97–99]. The reduction of Fe(III) in smectites does alter the surface properties of the clays [100].

A special mechanism of Fe(III) reduction in bacterial attack of iron(III) oxides in bauxite was postulated by Ehrlich and Wickert [41]. They noted that, when the Fe(III) of pisolites of pisolitic bauxite was reduced, the reduction proceeded from the surface inward. Since the pisolites did not appear to be porous, H.L. Ehrlich (unpublished, 1998) proposes, based on an earlier suggestion by Ehrlich and Wickert [41], that each ferrous ion generated by the bacteria on the pisolite surface transmits an electron to the interior of the pisolites to reduce the Fe(III) of iron oxide below the surface,

$$(2Fe^{2+})_{surface} + Fe_2O_3 + 6H^+ \rightarrow (2Fe^{3+})_{surface} + 2Fe^{2+}$$
$$+ 3H_2O \ (\Delta G^\circ = +1.99 \, kcal) \quad (5)$$

Iron-reducing bacteria at the surface quickly re-reduce any $(Fe^{3+})_{surface}$ that results from the transmission of an electron to the interior by $(Fe^{2+})_{surface}$,

$$(2Fe^{3+})_{surface} + H_2 \rightarrow 2Fe^{2+} + 2H^+ \ (\Delta G° = -35.56 \, kcal) \qquad (6a)$$

$$(2Fe^{3+})_{surface} + \frac{1}{4}CH_3COO^- + H_2O \rightarrow 2Fe^{2+} + \frac{1}{2}HCO_3^-$$
$$+ \frac{9}{4}H^+ \ (\Delta G° = -26.9 \, kcal) \quad (6b)$$

It is the rapid bacterial reduction of the Fe(III) at the surface of the pisolites that drives the reduction of the Fe(III) in the interior of the pisolites. As the free energy changes show, reaction (5) is thermodynamically unfavorable by itself, but when coupled to reactions (6a) or (6b), it becomes favorable. Although these reactions seem to resemble those involved in the reduction of Fe(III) in smectite, a major difference is that most of the Fe(II) produced by the Fe(III)-reducing bacteria when acting on the pisolites appears ultimately as Fe^{2+} in solution.

Fe(III) oxides become important terminal electron acceptors in anaerobic zones in soil or sediment when the levels of nitrate and sulfate are too low to permit them to serve as terminal electron acceptors in mineralization of organic matter. This was first shown by Lovley and Phillips [69,85]. Organic compounds that may be anaerobically mineralized with Fe(III) as terminal electron acceptor include some that are aromatic and some that are aliphatic [48,86,87,96,101]. However, the relative concentration of Fe(III) oxides, nitrates or sulfate does not always determine which will be the dominant terminal electron acceptor in a natural system. The nature of the dominant organism(s), the concentration and composition of organic matter (electron donor), the reactivity of Fe(III), the H_2 availability, and the pH are other factors that may determine it [102].

6.2 FERRIC PHOSPHATE (STRENGITE) AND FERROUS PHOSPHATE (VIVIANITE)

Strengite ($FePO_4 \cdot 2H_2O$) may form in soil and sediment pore waters in which the concentrations of phosphate and iron(III) reach saturation levels [103]. Similarly vivianite ($Fe_3(PO_4)_2 \cdot 8H_2O$) may form if phosphate and Fe(II) concentrations in pore water reach saturation levels at circumneutral pH. The elevated iron concentrations may be the result of microbial weathering of iron-containing minerals, and the elevated phosphate concentrations may be the result of microbial degradation of organic phosphorus compounds in debris. Both processes occur aerobically or anaerobically, but precipitation of the iron as a phosphate mineral is more likely to occur aerobically because the prevailing pH may be too low anaerobically. Since phosphate is an essential nutrient for all forms of life, its immobilization in strengite or vivianite may be detrimental by limiting its bioavailability. Nature has provided means of reversing this process, which depend largely on microbial activity.

The solubility of iron phosphates is pH dependent. Microbially produced acids, such as H_2SO_4 from bacterial oxidation of reduced forms of sulfur, and HNO_3 from the bacterial oxidation of ammonia [104;105, p. 230], will solubilize the phosphate minerals. Whereas HNO_3 is generally formed only aerobically, H_2SO_4 can be formed either aerobically or anaerobically, but its aerobic formation is more common. Anaerobic, microbial solubilization of iron phosphates by organic acids, such as acetic, propionic, butyric, and lactic acids that are formed as a result of fermentation of sugars or more complex carbohydrates [105, p. 262] is common. Microbially formed ligands with an affinity for Fe(III) or Fe(II) may also function as lixiviants, either by themselves or in conjunction with acids (see also Chapter 5).

SRB have been reported to liberate phosphate from strengite and vivianite by generating H_2S anaerobically when they reduce sulfate. The H_2S liberates phosphate from the iron phosphates by reacting abiologically with the iron and precipitating it as iron sulfide. In the case of strengite, the H_2S reduces the iron(III) to iron(II), before the iron(II) is precipitated with unreacted sulfide [106,107],

$$2FePO_4 + HS^- + H^+ \rightarrow 2Fe^{2+} + 2HPO_4^{2-} + S^0 \qquad (7a)$$

$$2Fe^{2+} + 2HS^- \rightarrow 2FeS + 2H^+ \qquad (7b)$$

In the case of vivianite, no prior reduction of iron is required before its precipitation as a sulfide,

$$Fe_3(PO_4)_2 + 3HS^- \rightarrow 3FeS + 2HPO_4^{2-} + H^+ \qquad (8)$$

Fe(III)-reducing bacteria may contribute to solubilization of both iron and phosphate in strengite under anaerobic conditions [108],

$$FePO_4 + (H) \rightarrow Fe^{2+} + HPO_4^{2-} \qquad (9)$$

In this case, vivianite may form diagenetically when the dissolved ferrous iron and phosphate reach critical concentrations [108].

6.3 MANGANESE OXIDES

As mentioned in a different context in Chapter 6, oxides of Mn(IV) and Mn(III) that occur in soils can be effective terminal electron acceptors in anaerobic respiration, the Mn being reduced to the $+2$ oxidation state [20, Chapter 15; 109] Manganese oxides have been identified in Australian soils and include birnessite, lithiophorite, and hollandite. Todorokite and pyrolusite were of rare occurrence [110]. However, in laboratory studies on microbial manganese reduction, pyrolusite (MnO_2) has been most widely employed because of its well-defined composition. In general, the higher oxides of manganese are extremely insoluble in water. To bring about their solution, their manganese has to be reduced to Mn^{2+}. Such reduction may be a microbiological process in which the manganese oxide serves as a terminal electron acceptor in respiration, or it may be an abiologic process in which the reductant may or may not have a microbiological origin. In the latter instance, manganese(IV) oxides may be abiologically (chemically) reduced to Mn^{2+} by H_2S formed by SRB or by Fe^{2+} formed from ferric iron by iron reducers [111–113,113a,114],

$$MnO_2 + H_2S + 2H^+ \rightarrow Mn^{2+} + 2H_2O + S° \qquad (10)$$

$$MnO_2 + 2Fe^{2+} + 4H^+ \rightarrow Mn^{2+} + 2Fe^{3+} + 2H_2O \qquad (11)$$

Equations (10) and (11) show clearly that these reactions are favored by acidic conditions in soils since both reactions consume protons.

When MnO_2 or other manganese oxide serves as a terminal electron acceptor in microbial respiration, surface contact between the bacterial cells and the oxide is essential because of the extreme water insolubility of the oxides. Only extracellular electron shuttles can eliminate a need for physical contact. Ehrlich [14,15] has proposed a detailed model for bacterial MnO_2 reduction by organisms that can carry out the reduction aerobically and anaerobically, e.g., marine pseudomonad BIII 88. The model is based on laboratory observations. It assumes intimate contact between an MnO_2 surface and manganese-reducing cells. This process can operate with acetate as electron donor. In this model, $\{Mn^{2+}\}$, which is manganese bound in the cell envelope, acts as electron carrier to MnO_2,

$$0.5MnO_2 + 0.5\{Mn^{2+}\} + H_2O \rightarrow \{Mn^{3+}\} + 2OH^- \qquad (12)$$

The resultant bound $\{Mn^{3+}\}$ is then re-reduced to $\{Mn^{2+}\}$ by the cell with an electron donor such as acetate formed by the cell,

$$0.125CH_3COOH + \{Mn^{3+}\} + 0.25H_2O \rightarrow 0.25CO_2 + \{Mn^{2+}\} + H^+ \qquad (13)$$

Most of the bound manganese $\{Mn^{2+}\}$ in reaction (13) is released into the bulk phase as Mn^{2+}.

In gram-negative bacteria that are either facultative or strict anaerobes but reduce MnO_2 or other higher oxide of manganese only anaerobically, c-type cytochrome in the outer membrane appears to function either directly or indirectly as an electron carrier to MnO_2, as in Fe(III) reduction (see section 6.1) [16,93]. In gram-positive bacteria that reduce MnO_2 only anaerobically, $\{Mn^{2+}\}$ bound in their cell wall and, if present, their exopolysaccharide may be the electron carrier, but that has yet to be determined.

Local Mn oxide accumulations in soil are not as common as those of iron oxide, but where they do occur, they may be the chief terminal electron acceptor in organic matter mineralization in the absence of alternative acceptors. Nitrate in sufficient concentration can inhibit bacterial Mn(IV) oxide reduction [e.g., 115] because it then competes as an alternative terminal electron acceptor. Thermodynamics for nitrate reduction are more favorable than for Mn(IV) oxide reduction [45]. Many types of organic compounds can be mineralized with Mn(IV) oxide as electron acceptor, including toluene [116] and other aromatic compounds [117].

To what extent Mn(III) oxides are susceptible to chemical reduction by H_2S or Fe^{2+} seems not to have been investigated. Only very limited work has been done on bacterial reduction of such oxides [109,118,119]. Water-soluble pyrophosphate complexes of Mn(III), and probably others not as yet tested, are susceptible to bacterial reduction and to chemical reduction by Fe^{2+} and H_2S [109].

6.4 MANGANESE SILICATE (RHODONITE)

Rhodonite is a manganese silicate mineral in which the oxidation state of manganese is $+2$. As discussed in the section on weathering (3.1), the manganese can be mobilized by microbes that generate weathering agents. Under anaerobic conditions this would include acids and ligands. The mobilized manganese can satisfy a trace-metal requirement of soil microbes for manganese. If enough is mobilized in a soil of circumneutral pH, it could be oxidized microbiologically under aerobic conditions. It could serve as a reductant of nitrate in an anaerobic, abiologic reaction [120],

$$5Mn^{2+} + 2NO_3^- + 4H_2O \rightarrow 5MnO_2 + N_2 + 8H^+$$
$$(\Delta G° = -3.4 \, \text{kcal mol}^{-1}) \tag{14a}$$

$$5Mn^{2+} + 2NO_3^- + 8OH^- \rightarrow 5MnO_2 + N_2 + 4H_2O$$
$$(\Delta G° = -156.3 \, \text{kcal mol}^{-1}) \tag{14b}$$

Reactions (14a) and (14b) indicate that they are favored by alkaline pH, which under aerobic conditions is unfavorable for biooxidation of Mn^{2+}. Luther et al. [120] indicate that this abiologic reduction of nitrate can occur aerobically and anaerobically. However, in air, autoxidation of Mn^{2+} to MnO_2 may overtake its oxidation by nitrate at pH values above 8.0.

6.5 GYPSUM AND ANHYDRITE

Gypsum ($CaSO_4 \cdot 2H_2O$) and anhydrite ($CaSO_4$) are sources of sulfate in groundwater in contact with these minerals. Anaerobically, sulfate is one of the important terminal electron acceptors in microbial mineralization of organic matter, although in soil, Fe^{2+}, MnO_2, and NO_3^- are in most instances more important because sulfate concentrations are usually very low. More free energy per mole of carbon is obtainable when Fe^{2+}, NO_3^-, or MnO_2 are the electron acceptors than when sulfate is [45].

Sulfate serves as terminal electron acceptor for a taxonomically diverse group of bacteria [121]. It has so far not been clearly established whether any SRB can attack sulfate in the crystal lattice of gypsum or anhydrite directly. However, since these two minerals are sufficiently water soluble (e.g., K_{sol} of calcium sulfate is 1.96×10^{-4} at 25°C; solubility of gypsum is 0.241 mg per 100 ml [122]), direct microbial attack may not be necessary. Microbially produced ligands of calcium should enhance dissociation of the two minerals.

Whereas the concentration of sulfate in the majority of soils is low, its concentration in salt marshes is significant, its primary source in this environment being seawater. In some desert soils, gypsum and anhydrite are products of rock weathering and may be a primary source of sulfate. Gypsum and anhydrite are commonly found in evaporites, which result from the evaporation of seawater. Gypsum and anhydrite are common features of salt domes, which derived from deeply buried evaporites [123], and occur in other types of buried evaporite deposits [124]. SRB associated with these deposits play a central role in converting sulfate, mobilized by plutonic (anoxic) water in contact with anhydrite and gypsum, to H_2S. The H_2S is then oxidized to $S°$ by bacteria such as *Thiobacillus thioparus* at the interface where oxygenated surface water meets the H_2S-charged plutonic water (for a more detailed summary see [124]).

6.6 SELENIFEROUS MINERALS

Selenium-containing minerals occur in rocks associated with uranium roll-front deposits [125,126]. These deposits, containing minerals of iron, selenium, and uranium, are weathering products from the action of oxidizing meteoritic solution on sandstone. Selenium also occurs in high-temperature sulfide ores in which it may appear as selenide minerals or incorporated in diagenetic pyrite [127]. Throughout the Western United States, selenium is associated with fine-grained sediments of marine origin in Upper Cretaceous formations [126,127]. As a consequence, soils derived from these Cretaceous rocks in the Western United States are seleniferous.

Of the oxyanions of selenium, those of Se(IV) ($HSeO_3^-$; SeO_3^{2-}) tend to be adsorbed by iron oxides under oxidizing conditions in the pH range 2–8 [127]. Those of Se(VI) ($HSeO_4^-$; SeO_4^{2-}), by contrast, are not strongly adsorbed by ferric oxide in this pH range. They also do not tend to form insoluble compounds [127]. Thus, in the presence of ferric oxides, Se(VI) is much more mobile than Se(IV). Very little appears to be known about microbial weathering of selenium-containing minerals, i.e., the role of microbes in mobilizing Se in minerals. Those microbes that have been described are all aerobic. The only recorded case of microbial oxidation of a selenide mineral involves a laboratory experiment in which CuSe was oxidized with O_2 to Cu^{2+} and $Se°$ by the acidophile *T. ferrooxidans* [128],

$$CuSe + 2H^+ + 0.5O_2 \rightarrow Cu^{2+} + Se° + H_2O \tag{15}$$

Elemental Se has been reported to be oxidized aerobically by *Micrococcus selenicus* from mud [129] and by a group of autotrophs [130], but these cultures have been lost. A soil strain of *Bacillus megaterium* was more recently reported to oxidize $Se°$ [131]. Se has also been reported to be oxidized anaerobically during photosynthesis by a purple bacterium [132].

Much more is now known about the reduction of oxyanions of selenium in groundwater (see also Chapter 6). Under anaerobic conditions, the red form of $Se°$ is formed from Se(IV) and Se(VI). Among the most recent reports on selenium reduction may be cited the findings of Rech and Macy [72] and Macy *et al.* [71] with *Thauera selenatis*, and those with *Bacillus arsenicoselenatis* and *B. selenitireducens* by Blum *et al.* [74]. Earlier examples were summarized by Ehrlich [20]. Despite its toxicity at elevated concentration, selenium is an essential trace element for various microorganisms, plants and animals (see [20] for references).

6.7 TUNGSTATES

In ore, tungsten occurs predominantly either as calcium tungstate, of which scheelite is the most common form, or as iron and/or manganese tungstate, collectively known as the wolframite group [133]. Other less common mineral forms of tungstate are also known [134]. In soil, the average concentration of tungsten has been listed as $1.5 \, mg \, kg^{-1}$, with a range of $0.5 \, mg \, kg^{-1}$) [135]. Its average concentration in freshwater has been reported to be $3 \times 10^{-5} \, mg \, kg^{-1}$ [135]. Although tungsten can exist in a number of different oxidation states $(+2, +3, +4, +5, \text{ and } +6)$, the $+6$ oxidation state is the most stable.

Tungsten is nutritionally essential for at least some bacteria. *Clostridium thermoaceticum* forms an NADP-dependent formate dehydrogenase, which is a tungsten–selenium–iron protein [136]. *Clostridium acidiurici* and *Cl. cylindrosporum* also require tungsten nutritionally. Based on [185]W labeling, tungsten was incorporated into at least three different proteins. One of these proteins exhibited formate dehydrogenase activity, and at least one may be a tungsten-storage protein [137]. A third example of a W-assimilating bacterium is *Methanocorpusculum parvum* [138]. All the organisms are strict anaerobes, and all but *M. parvum* are eubacteria; *M. parvum* is an archeon.

Since in laboratory culture, tungsten has been taken up as tungstate (WO_4^{2-}) from a solution containing its sodium salt (10^{-3} to $10 \, \mu mol \, L^{-1}$), it may be inferred that in soil, tungstate is mobilized from tungstate minerals by acidolysis. Active mobilizing agents are likely to include microbially formed acids. The only systematic study of tungsten mobilization from wolframite to date is one by Chashina and Lyalikova [139] (see Chapter 5). Athough this study was run under aerobic conditions, there is reason to believe that such mobilization would also occur anaerobically, possibly more effectively because of greater production of fermentation acids, the mobilizing agents. A systematic investigation of anaerobic tungsten mobilization is urgently needed.

6.8 MOLYBDATES AND VANADATES

The primary mineral of molybdenum is molybdenite (MoS_2). Secondary molybdenum minerals are powellite ($CaMoO_4$) and wulfenite ($PbMoO_4$). They may form in oxidizing zones of molybdenum deposits [140]. Molybdenum may partially replace tungsten in scheelite [141]. In soil, olivine and clay minerals may contain traces of molybdenum [142]. Secondary, immobilized molybdenum in soil can be expected to take the form of $CaMoO_4$. In reducing environments in which H_2S is present, MoS_2 can also be expected. The average concentration of Mo in soil has been given as $1.2\,mg\,kg^{-1}$ although actual concentrations can vary widely. In freshwater, the average Mo concentration has been given as $5 \times 10^{-4}\,mg\,kg^{-1}$ [135].

Vanadium ore minerals include complex sulfides, patronite (VS_4), roscolite (a vanadium mica), carnotite (hydrous potassium uranium vanadate), and lead vanadates [143,144]. In soil, secondary, immobilized forms of vanadium can be expected to be vanadium oxy-salts (vanadates). The vanadate ion, VO_4^{3-}, tends to be adsorbed by Fe–Al–Mn hydroxides in laterites, bauxites, Fe–Mn nodules, and iron-containing clay minerals [144]. The average concentration of V in soil has been given as $90\,mg\,kg^{-1}$ and in freshwater as $5 \times 10^{-4}\,mg\,kg^{-1}$ [135].

Molybdenum is an essential structural and functional component of many nitrogenase enzymes. Vanadium or iron may replace molybdenum in some alternative nitrogenases [145–148]. Vanadium has so far only been found in alternative nitrogenases of some *Azotobacter* species and in the cyanobacterium *Nostoc muscorum* (see [149]), which are aerobes. Only very limited evidence exists so far for vanadium-based alternative nitrogenase in strict anaerobes such as *Clostridium* [150].

Besides being a component of most nitrogenases, molybdenum is also a component of nitrate reductase [151], sulfide reductase, and arsenite oxidase [152]. Most methanogens require molybdenum and/or tungsten for growth [153]. Dissolved molybdate at appropriate concentration can inhibit bacterial sulfate reduction [154].

Immobilized molybdate and vanadate in soil are probably solubilized by the action of acids, especially organic acids formed microbiologically. Such mobilization should be possible anaerobically as well as aerobically. So far, however, this seems not to have been systematically studied.

Mo and V can exist in various oxidation states, Mo in the $0, +2, +3, +4, +5$ and $+6$ states, and V in the $0, +2, +3, +4$ and $+5$ states. Although aerobic oxidation and reduction of some forms of Mo by a few different organisms has been observed (see brief review in [20]; see also Chapter 5), only one report of anaerobic reduction of molybdate $(+6)$ to molybdenum blue $(+5)$ with glucose as electron donor by *Enterobacter cloacae* strain 48 has appeared to date [155]. Anaerobic vanadate reduction to vanadyl $(+3)$ with hydrogen by *Veillonella (Micrococcus) lactolyticus, Desulfovibrio desulfuricans* and *Clostridium pasteurianum* was reported by Woolfolk and Whiteley [156]. Yurkova and Lyalikova [157] subsequently discovered *Pseudomonas vanadiumreductans* and *P. issachenkovii*, which can grow anaerobically as chemolithotrophs with hydrogen as energy source (electron donor) and vanadate as terminal electron acceptor. These organisms can also grow anaerobically as mixotrophs with glucose, maltose, ribose, galactose, lactose, arabinose, lactate, proline, histidine, threonine, and serine replacing hydrogen as electron donor and vanadate as electron acceptor. The vanadium was reduced to a $+4$ intermediate (blue) and then to a $+3$ oxidation state in the form of a black precipitate (V_2O_3).

6.9 URANIUM (GRANITE, URANINITE)

Common rock types in which uranium occurs naturally include granites, pegmatites. black shales, limestones, and gneisses among others [135,158]. Its average concentration is highest in granites $(4.4\,\text{mg}\,\text{kg}^{-1})$ [135]. One of the most common uranium minerals in ore is uraninite (UO_2) [159].

As summarized by Ehrlich [20, pp. 601–602], heterotrophic bacteria such as *Pseudomonas fluorescens, P. putida*, and *Achromobacter* have been found to be able to mobilize uranium in granitic rock, ore and sand by weathering these materials with products of their energy metabolism, especially organic acids and ligands. Microbes were also shown to be able to reprecipitate uranium by destroying the organic moiety of organic uranium complexes [160,161]. Although these experiments were not done under strictly anaerobic conditions (they were done in flasks and in perfusion columns), similar results would be expected, had they been done strictly anaerobically. The organisms responsible would, however, have been different.

Mobile, uncomplexed uranium exists mostly in the form of UO_2^{2+} (uranyl). A number of organisms have been found able to immobilize the uranium in uranyl by reducing it to UO_2 (uranium dioxide) involving a change in oxidation state from $+6$ to $+4$ [156,162–164]. The electron donor in this reduction can be as simple as H_2 or acetate. Active organisms include the anaerobes *G. metallireducens* GS-15, *S. putrefaciens, Desulfovibrio* species [101] and *Clostridium* sp. [164a,b]. The *Desulfovibrio* species, however, do not grow when UO_2^{2+} is the sole terminal electron acceptor. Since natural phosphorite may contain as much as 90 mg U kg^{-1} [135], it may introduce significant amounts of U into soil when it is used as a phosphorus fertilizer. Soil bacteria of the type listed above can immobilize this polluting uranium.

6.10 ALUMINUM HYDROXIDES AND OXYHYDROXIDES (GIBBSITE, BOEHMITE, DIASPORE)

Aluminum hydroxides and oxyhydroxides are especially abundant in lateritic soils in humid subtropical and tropical regions [165–167] and in bauxites, whose development is related to that of laterites [167]. The aluminum hydroxide gibbsite (γ-Al(OH)$_3$ or γ-Al$_2$O$_3 \cdot 3$H$_2$O), and the oxyhydroxides boehmite (γ-AlO(OH) or γ-Al$_2$O$_3 \cdot$ H$_2$O), and diaspore (α-AlO(OH) or α-Al$_2$O$_3 \cdot$ H$_2$O) are the products of weathering of aluminosilicate-containing rocks [168,169]. This weathering usually has a biological component [170]. For the aluminum oxides to form, the pH of the environment has to be above 4–5. Below this pH range, aluminum oxides are soluble [171,172]. Recent work has shown that bauxites harbor a mixed bacterial flora that under anaerobic conditions can reduce the ferric oxide of the ore with sucrose as carbon/energy source in a pH range from 6.5 to 3.5. The flora generated acidity which began to mobilize the aluminum in the bauxite when the pH dropped below 4.5 [40,41]. Dissolved aluminum concentrations as high as 285 mg L^{-1} at pH 3.5 were noted in these experiments. These aluminum concentrations were tolerated by the active bacteria that were attacking the bauxite.

Dissolved aluminum is toxic to plants. Jute is sensitive to about 300 mg L^{-1} whereas rice, cocoa, citrus, black eye peas, corn, and sugar cane may tolerate 300–500 mg L^{-1} [173]. The toxicity of dissolved aluminum is, however, dependent on the extent to which it is complexed by naturally occurring ligands in soil solution [174], complexed Al^{3+} being significantly less toxic than uncomplexed Al^{3+}. Aluminum toxicity in bacteria and other microbes has been discussed by Robert [175].

6.11 FELDSPARS

Feldspars are igneous minerals that include the alkali feldspars albite (NaAlSi$_3$O$_8$), orthoclase (KAlSi$_3$O$_8$) and plagioclase (CaAl$_2$Si$_2$O$_8$) as end members [28]. They are readily weathered through protonation [30,31,62,176], although complexation by some of organic acids [31,63,177] and acidic polysaccharides of microbial origin can also play a role [35,178]. The source of the acids may be biotic or abiotic, but more frequently the former. The acids may be organic or inorganic. Many kinds of organic acids with a potential for weathering feldspars are produced anaerobically through bacterial fermentation. The weathering results in solubilization of the cations Na$^+$, K$^+$, Ca^{2+} of feldspars and some of the framework Si and Al. It also results in diagenetic formation of clays such as kaolinite.

6.12 CLAYS

Some clays are secondary minerals that arise diagenetically as products of weathering of primary rock minerals such as feldspars [179]. Other clays arise diagenetically from preexistent clays [180] or authigenically from reaction of dissolved silica, alumina, and various cations [see e.g., 181]. Based on their structure, clays have been assigned to as many as six categories [182]: (1) kaolinite group, (2) mica group, (3) montmorillonite group, (4) chlorite group, (5) vermiculite group, and (6) mixed-layer clay minerals. The difference in structure among the groups accounts for the difference in some of their properties, for example, the extent of expandability (swelling) and ion-exchange capacity. Clay particles vary in shape and size but are generally very small, $< 2 \mu m$ [183].

Clays can have beneficial as well as detrimental effects on the growth of microorganisms in soil [184]. They may enhance or inhibit microbial activity by pH buffering or by binding nutrient molecules. They may detoxify the environment by adsorbing toxic metals. In general, these effects should be manifested anaerobically as well as aerobically, although many of the experimental observations were made only aerobically.

From experiments on the dissolution of clays by dilute acids [185–189], it can be inferred that microbes can promote breakdown of clays by their production of organic acids that can serve as agents for protonation and chelation. Such attack of clays should be favored in anaerobic soil environments containing fermentable organic matter. The presence of humic substances, which can complex clays [190,191], may modify this microbial action.

6.13 CALCITE AND OTHER CARBONATE MINERALS

Calcite $(CaCO_3)$ occurs commonly in neutral to alkaline soils. It may be the product of microbial activity in which respiratory CO_2 as CO_3^{2-} reacts with Ca^{2+} in soil solution under conditions of stoichiometric excess. It can serve as a buffering agent that neutralizes acids, such as those formed in an anaerobic microbial metabolism such as fermentation. In sulfate reduction, $CaCO_3$ has been shown to be formed anaerobically when the source of the sulfate is gypsum $(CaSO_4 \cdot 2H_2O)$ or anhydrite $(CaSO_4)$. Rhodochrosite $(MnCO_3)$ and siderite $(FeCO_3)$ have been shown to be formed anaerobically by G. metallireducens GS-15, the former in the reduction of MnO_2 [69] and the latter in the reduction of Fe^{3+} [85,192]. SRB can also form siderite anaerobically from Fe^{3+} in sediments [77]. Ehrlich and Wickert [41] observed siderite formation when a mixed bacterial flora reduced the iron in the hematite of a bauxite from Brazil anaerobically in a percolation column.

7 CONCLUSION

The foregoing discussion shows that extensive interaction can occur anaerobically between microbes and minerals. Specific minerals may determine which
microbes will thrive in a given environment. They may make soil environments
more or less habitable for microbes and other organisms. Minerals may participate in microbial metabolism. Some of the microbe–mineral interactions are
enzymatic, others involve interaction between products of microbial metabolism and minerals. Some enzymatic reactions, such as the respiratory reduction
of Fe(III), seem to occur only anaerobically, whereas others, such as the
respiratory reduction of Mn(IV), can occur aerobically as well as anaerobically.
Anaerobic respiration that depends on the use of inorganic electron acceptors,
has come to be recognized to play a vital role in the mineralization of organic
carbon in the absence of air. Soil minerals are frequently a source of these
inorganic electron acceptors. Some anaerobic interactions between minerals
and microbes deserve more detailed study than they have received heretofore.
One example is the mobilization of Mo and V, which are vital constituents of
nitrogenases, from minerals that contain them.

Other topics also need further investigation on a broader scale. Thus, we need
to determine more extensively which anaerobes in soil are able to attach to
mineral surfaces, form microcolonies, or invade preexistent biofilms, and to
what extent? Are their mechanisms for attachment similar to those of aerobes in
all instances and to what extent and by what means do they contribute to
weathering of mineral surfaces?

We need more information on the morphological and physiological basis that
enable enzyme-catalyzed redox reactions to occur at the interface between the
surface of cells of different facultative and anaerobic bacteria capable of such
reaction and the surface of minerals containing redox-reactive constituents. In
what respects do active gram-positive species with this capacity differ from gram-
negative species? Furthermore, how does the action on mineral surfaces of
bacteria capable of only anaerobic respiration differ from that of bacteria capable
of both anaerobic respiration and fermentation? Consideration of the works of
Myers and Myers [16], Gaspard et al. [93], and Ehrlich [14,15], Ehrlich et al. [40],
and Ehrlich and Wickert [41] may be an appropriate point of departure.

We need to determine if metal sulfides and siderite ($FeCO_3$) can be oxidized
anaerobically by bacteria using nitrate as a terminal electron acceptor and
which bacteria exhibit such a capacity. In addition, we need to determine if
the anaerobic reduction of manganese(IV) oxide by ammonia reported by
Hulth et al. [59] can be catalyzed by bacteria and which bacteria are able to
perform this reaction. We need more detailed information on the anaerobic
enzymatic reduction of Mo(VI).

It would be useful to know if biogenic H_2S and Fe^{2+} can reduce Mn(III)
oxide minerals.

Clearly, many questions about the interactions between microorganisms and minerals under anaerobic conditions remain unanswered and offer challenging opportunities for research.

REFERENCES

1. Fletcher, M. and Floodgate, G. D. (1973). An electron-microscopic demonstration of an acidic polysaccharide involved in the adhesion of a marine bacterium to solid surfaces, *J. Gen. Microbiol.*, **74**, 325.
2. Marshall, K. C. (1980). Microorganisms and interfaces, *BioScience*, **30**, 246.
3. Marshall, K. C. and Cruickshank, R. H. (1973). Cell surface hydrophobicity and the orientation of certain bacteria at interfaces, *Arch. Mikrobiol.*, **91**, 29.
4. Caccavo, F. Jr., Schamberger, P. C., Keiding, K. and Nielsen, P. H. (1997). Role of hydrophobicity in adhesion of the dissimilatory Fe(III)-reducing bacterium *Shewanella alga* BrY. *Appl. Environ. Microbiol.*, **63**, 3837.
5. Caccavo, F. Jr. (1999). Protein-mediated adhesion of the dissimilatory Fe(III)-reducing bacterium *Shewanella alga* BrY to hydrous ferrous oxide, *Appl. Environ. Microbiol.*, **65**, 5017.
6. Van Schie, P. M. and Fletcher, M. (1999). Adhesion of biodegradative anaerobic bacteria to solid surfaces, *Appl. Environ. Microbiol.*, **65**, 5082.
7. Costerton, J. W., Lewandowski, Z., Caldwell, D. E., Korber, D. R. and Lappin-Scott, H. M. (1995). Microbial biofilms, *Annu. Rev. Microbiol.*, **49**, 711.
8. Lawrence, J. R., Korber, D. R., Wolfaard, G. M. and Caldwell, D. E. (1995). Behavioral strategies of surface colonizing bacteria, *Adv. Microb. Ecol.*, **14**, 1.
9. Mills, A. L. and Maubrey, R. (1981). Effect of mineral composition on bacterial attachment to submerged rock surfaces, *Microb. Ecol.*, **7**, 315.
10. Wimpenny, J. W. T. and Kinniment, S. L. (1995). Biochemical reactions and the establishment of gradients within biofilms. In *Microbial Biofilms*, ed. Lappin-Scott H. and Costerton J. W., Cambridge University Press, Cambridge, p. 99–117.
11. Bass, C., Sanders, P. and Lappin-Scott, H. (1998). Study of biofilms of sulfidogens from North Sea oil production facilities using continuous-flow apparatus, *Geomicrobiol. J.*, **15**, 101.
12. Pederson, K. (1993). The deep subterranean biosphere, *Earth-Sci. Rev.*, **34**, 243.
13. Ingledew, W. J. (1982). *Thiobacillus ferrooxidans*. The bioenergetics of an acidophilic chemolithotroph, *Biochim Biophys. Acta*, **683**, 89.
14. Ehrlich, H. L. (1993). Electron transfer from acetate to the surface of MnO_2 particles by a marine bacterium, *J. Ind. Microbiol.*, **12**, 121.
15. Ehrlich, H. L. (1993). A possible mechanism for the transfer of reducing power to insoluble mineral oxide in bacterial respiration. In *Biohydrometallurgical Technologies*, Vol. II, ed. Torma AE, Apel ML, and Brierley CL, The Minerals, Metals and Materials Society, Warrendale, PA, pp. 415–422.
16. Myers, C. R. and Myers, J. M. (1992). Localization of cytochromes to the outer membrane of anaerobically grown *Shewanella putrefaciens*. MR-1. *J. Bacteriol.* **174**, 3429.
17. Donati, E., Pogliani, C. and Boiardi, J. L. (1997). Anaerobic leaching of covellite by *Thiobacillus ferrooxidans. Appl. Microbiol. Biotechnol.*, **47**, 636.
18. Sullivan, J. D. (1930). *Chemistry of leaching covellite*, Technical Paper 487, U. S. Department of Commerce, Bureau of Mines, Washington, DC , 18 pp.

19. Brock, T. D. and Gustafson, H. (1976). Ferric iron reduction by sulfur- and iron-oxidizing bacteria, *Appl. Environ. Microbiol.*, **32**, 567.
20. Ehrlich, H. L. (1996). *Geomicrobiology*, 3rd edn, Marcel Dekker, New York.
21. Benz, M., Brune, A. and Schink, B. (1998). Anaerobic and aerobic oxidation of ferrous iron at neutral pH by chemoheterotrophic nitrate-reducing bacteria, *Arch. Microbiol.*, **169**, 159.
22. Hafenbrandl, D., Keller, M., Dirmeier, R., Rachel, R., Roßnagel, P., Burggraf, S., Huber, H. and Stetter, H. O. (1996). *Ferroglobus placidus* gen nov., sp. nov., a novel hyperthermophilic archaeum that oxidizes Fe^{2+} at neutral pH under anoxic conditions, *Arch. Microbiol.*, **166**, 308.
23. Straub, K. L., Benz, M., Schink, B. and Widdel, F. (1996). Anaerobic, nitrate-dependent microbial oxidation of ferrous iron, *Appl. Environ. Microbiol.*, **62**, 1458.
24. Ehrlich, H. L. (1988). Bioleaching of manganese by marine bacteria. In *8th International Biotechnology Symposium, Proceedings*, Vol. II, ed. Durand G, Bobichon L and Florent J., Société Française de Microbiologie, Paris, pp 1094–1105.
25. Di-Ruggiero, J. and Gounot, A. M. (1990). Microbial manganese reduction mediated by bacterial strains isolated from aquifer sediments, *Microb. Ecol.*, **20**, 53.
26. Nies, D. H. (1999). Microbial heavy metal resistance, *Appl. Microbiol. Biotechnol.*, **51**, 730.
27. Koppelman, M. H. and Dillard, J. G. (1977). A study of the adsorption of Ni(II) and Cu(II) by clay minerals, *Clay Clay Miner.*, **25**, 457.
28. Banfield, J. F. and Hamers, R. J. (1997). Processes at minerals and surfaces with relevance to microorganisms and prebiotic synthesis. In *Geomicrobiology: Interactions Between Microbes and Minerals*, ed. Banfiled JF and Nealson KH, *Rev. Mineral.*, **35**, 81.
29. Silverman, M. P. and Ehrlich, H. L. (1964). Microbial formation and degradation of minerals, *Adv. Appl. Microbiol.*, **6**, 153.
30. Ullman, W. J., Kirchman, D. L., Welch, S. A. and Vandevivere, P. (1996). Laboratory evidence for microbially mediated silicate mineral dissolution in nature, *Chem. Geol.*, **132**, 11.
31. Welch, S. A. and Ullman, W. J. (1999). The effect of microbial glucose metabolism on bytownite feldspar dissolution rates between 5° and 35° C, *Geochim. Cosmochim. Acta*, **63**, 3247.
32. Aristovskaya, T. V. and Kutuzova, R. S. (1968). Microbiological factors in the mobilization of silicon from poorly soluble natural compounds, *Pochvovedenie*, no. 12, p. 59.
33. Karavaiko, G. N., Belkanova, N. P., Eroshchev-Shak, V. A. and Avakyan, Z. A. (1984). Role of microorganisms and some physicochemical factors of the medium in quartz destruction, *Mikrobiologiya*, **53**, 976 (Engl. transl., p. 795).
34. Duff, R. B., Webley, D. M. and Scott, R. O. (1963). Solubilization of minerals and related materials by 2–ketogluconic acid-producing bacteria, *Soil Sci.*, **95**, 105.
35. Avakyan, Z. A., Pivovarova, T. A. and Karavaiko, G. I. (1986). Properties of a new species, *Bacillus mucilaginosus, Mikrobiologiya*, **55**, 477 (Engl. transl., p. 369).
36. Holzapfel, L. (1951). Siliziumverbindungen in biologischen Systemen. Organische Kieselsaureverbindungen. XX. Mitteilung, *Z. Elektrochem.*, **55**, 577.
37. Malinovskaya, M., Kosenko, L. V., Votselko, S. K. and Podgorskii, V. S. (1990). Role of *Bacillus mucilaginosus* polysaccharide in degradation of silicate minerals, *Mikrobiologiya*, **59**, 70 (Engl. transl., p. 49).
38. Welch, S. A. and Vandevivere, P. (1994). Effect of microbial and other naturally occurring polymers on mineral dissolution, *Geomicrobiol. J.* **12**, 227.
39. Lovley, D. R. (1993). Dissimilatory metal reduction, *Annu. Rev. Microbiol.*, **47**, 263.

40. Ehrlich, H. L., Wickert, L. M., Noteboom, D. and Doucet, J. (1995). Weathering of pisolitic bauxite by heterotrophic bacteria. In *Biohydrometallurgical Processing*, Vol. I, ed. Vargas T, Jerez, C. A., Wiertz, J. V. and Toledo, H. University of Chile, Santiago, pp. 395–403.

41. Ehrlich, H. L. and Wickert, L. M. (1997). Bacterial action on bauxites in columns fed with full-strength and dilute sucrose-mineral salts medium, In *Biotechnology and the Mining Environment. Proceedings, 13th Annu. Meet.*, ed. Lortie, L., Bédard, P. and Gould, W. D., BIOMINET, Natural Resources Canada. SP 97–1, Ottawa, Canada, pp. 74–89.

42. Fortin, D. and Beveridge, T. J. (1997). Microbial sulfate reduction within sulfidic mine tailings: formation of diagenetic Fe sulfides, *Geomicrobiol. J.*, **14**, 1.

43. Fortin, D., Davis, B., Southam, G. and Beveridge, T. J. (1995). Biogeochemical phenomena induced by bacteria within sulfidic mine tailings, *J. Ind. Microbiol.*, **14**, 178.

44. Southam, G., Ferris, F. G. and Beveridge, T. J. (1995). Mineralized bacterial biofilms in sulfide tailings and in acid mine drainage systems. In *Microbial Biofilms*, eds. Lappin-Scott H. and Costerton J. W., Cambridge University Press, Cambridge, pp 148–170.

45. Ehrlich, H. L. (1993). Bacterial mineralization of organic carbon under anaerobic conditions, In *Soil Biochemistry*, Vol. 8, eds. Bollag, J.-M. and Stotzky, G., Marcel Dekker, New York, pp 219–247.

46. Lovley, D. R. and Woodward, J. C. (1996). Mechanisms for chelator stimulation of microbial Fe(III)-oxide reduction, *Chem. Geol.*, **132**, 19.

47. Holmgrén, B. A., Sison, J. D., Nelson, D. C. and Casey, W. H. (1999). Hydroxamate siderophores, cell growth and Fe(III) cycling in two anaerobic iron oxide media containing *Geobacter metallireducens, Geochim. Cosmochim. Acta*, **63**, 227.

48. Lovley, D. R. (1991). Dissimilatory Fe(III) and Mn(IV) reduction, *Microbiol. Rev.*, **55**, 259.

49. Myers, C. R. and Nealson, K. H. (1988). Bacterial manganese reduction and growth with manganese oxide as sole electron acceptor, *Science*, **240**, 1319.

50. Nealson, K. H. and Little, B. (1997). Breathing manganese and iron: solid-state respiration, *Adv. Appl. Microbiol.*, **45**, 213.

51. Odom, J. M. and Singleton, R. Jr., (ed.) (1993). *The Sulfate-Reducing Bacteria: Contemporary perspectives*, Springer, New York.

52. Knight, V. and Blakemore, R. (1998). Reduction of diverse electron acceptors by *Aeromonas hydrophila. Arch. Microbiol.*, **169**, 239.

53. Lovley, D. R., Phillips, E. J. P. and Lonergan, D. J. (1989). Hydrogen and formate oxidation coupled to dissimilatory reduction of iron and manganese by *Alteromonas putrefaciens. Appl. Environ. Microbiol.*, **55**, 700.

54. Caccavo, F. Jr., Blakemore, R. P. and Lovley, D. R. (1992). A hydrogen-oxidizing, Fe(III)-reducing microorganism from the Great Bay estuary, New Hampshire, *Appl. Environ. Microbiol.*, **58**, 3211.

55. Caccavo, F. Jr., Lonergan, D. J., Lovley, D. R., Davis, M., Stolz, J. F. and McInerney, M. J. (1994). *Geobacter sulfurreducens* sp. nov., a hydrogen- and acetate-oxidizing dissimilatory metal-reducing microorganism, *Appl. Environ. Microbiol.*, **60**, 3752.

56. Caccavo, F. Jr., Coates, J. D., Rossello-Mora, R. A., Ludwig, W., Schleifer, K. H., Lovley, D. R. and McInerney, M. J. (1996). *Geovibrio ferrireducens*, a phylogenetically distinct dissimilatory Fe(III) reducing bacterium, *Arch. Microbiol.*, **165**, 370.

57. Straub, K. L. and Buchholz-Cleven, B. E. E. (1998). Enumeration and detection of anaerobic ferrous iron-oxidizing, nitrate-reducing bacteria from diverse European sediments, *Appl. Environ. Microbiol.*, **64**, 4846.

58. Corbett, C. M. and Ingledew, W. J. (1987). Is Fe^{3+}/Fe^{2+} cycling an intermediate in sulfur oxidation by *Thiobacillus ferrooxidans? FEMS Microbiol.* Lett. **41**, 1.
59. Hulth, S., Aller, R. C. and Gilbert, F. (1999). Coupled anoxic nitrification/manganese reduction in marine sediments, *Geochim. Cosmochim. Acta*, **63**, 49.
60. Roberts, J. L. (1947). Reduction of ferric hydroxide by strains of *Bacillus polymyxa. Soil Sci.*, **63**, 135.
61. Rusin, P. A., Quintana, L., Sinclair, N. A., Arnold, R. G. and Oden, K. L. (1991). Physiology and kinetics of manganese-reducing *Bacillus polymyxa* strain D1 isolated from manganiferous silver ore, *Geomicrobiol. J*, **9**, 13.
62. Welch, S. A. and Ullman, W. J. (1993). The effect of organic acids on plagioclase dissolution rates and stoichiometry, *Geochim. Cosmochim. Acta*, **57**, 2725.
63. Welch, S. A. and Ullman, W. J. (1996). Feldspar dissolution in acidic and organic solutions: Compositional and pH dependence of dissolution rate, *Geochim. Cosmochim. Acta*, **60**, 2939.
64. McInerney, M. J. and Westlake, W. S. (1990). Microbially enhanced oil recovery. In *Microbial Mineral Recovery*, ed. Ehrlich HL and Brierley CL, McGraw-Hill, New York, pp. 409–445.
65. Davey, M. E., Gevertz, D., Wood, W. A., Clark, J. B. and Jenneman, G. E. (1998). Microbial selective plugging of sandstone through stimulation of indigenous bacteria in a hypersaline oil reservoir, *Geomicrobiol. J*, **15**, 335.
66. Sokolova-Dubinina, G. A. and Deryugina, Z. P. (1967). On the role of microorganisms in the formation of rhodochrosite in Punnus-Yarvi Lake, *Mikrobiologiya*, **36**, 535 (Engl. Transl., p. 445).
67. Pye, K. (1984). SEM analysis of siderite cements in intertidal marsh sediments, Norfolk, England, *Mar. Geol.*, **56**, 1.
68. Pye, K. I., Dickson, J. A. D., Schiavon, N., Coleman, M. L. and Cox, M. (1990). Formation of siderite-Mg-calcite-iron sulfide concretions in intertidal marsh and sandflat sediments, North Norfolk, England, *Sedimentology*, **37**, 325.
69. Lovley, D. R. and Phillips, E. J. P. (1988). A novel mode of microbial energy metabolism: Organic carbon oxidation coupled to dissimilatory reduction of iron and manganese, *Appl. Environ. Microbiol.*, **54**, 1472.
70. Newman, D. K., Beveridge, T. J. and Morel, F. M. M. (1997). Precipitation of arsenic trisulfide by *Desulfotomaculum auripigmentum, Appl. Environ. Microbiol.*, **63**, 2022.
71. Macy, J. M., Rech, S., Auling, G., Dorsch, M., Stackebrandt, E. and Sly, L. (1993). *Thauera selenatis* gen. nov sp. nov., a member of the subclass of *Proteobacteria* with a novel type of anaerobic respiration, *Int. J. Syst. Bacteriol.*, **43**, 135.
72. Rech, S. and Macy, J. M. (1992). The terminal reductases for selenate and nitrate respiration in *Thauera selenatis* are two distinct enzymes, *J. Bacteriol.*, **174**, 7316.
73. Oremland, R. S., Switzer Blum, J., Culbertson, C. W., Visscher, P. T., Miller, L. G., Dowdle, P. and Strohmeier, F. E. (1994). Isolation, growth, and metabolism of an obligately anaerobic, selenate-respiring bacterium, SES-3, *Appl. Environ. Microbiol.*, **60**, 3011.
74. Blum, J. S., Bindi, A. B., Buzzell, J., Stolz, J. F. and Oremland, R. S. (1998). *Bacillus arsenicoselenatis* sp. nov., and *Bacilllus selenitireducens* sp. nov., two haloalkaliphiles from Mono Lake, California that respire oxyanions of selenium and arsenic, *Arch. Microbiol.*, **121**, 19.
75. Lovley, D. R., Giovannoni, S. J., White, D. C., Champine, J. E., Phillips, E. J., Gorby, Y. A. and Goodwin, S. (1993). *Geobacter metallireducens* gen. nov. spec. nov., a microorganism capable of coupling the complete oxidation of organic compounds to the reduction of iron and other metals, *Arch. Microbiol.*, **159**, 336.

76. Lonergan, D. J., Jenter, H., Coates, J., Phillips, E. J. P., Schmidt, T. and Lovley, D. R. (1996). Phylogenetic analysis of dissimilatory Fe(III)-reducing bacteria, *J. Bacteriol.*, **178**, 2402.
77. Coleman, M. L., Hedrick, D. B., Lovley, D. R., White, D. C., and Pye, K. (1993). Reduction of Fe(III) in sediments by sulfate reducing bacteria, *Nature (London)*, **361**, 436.
78. Roden, E. E. and Lovley, D. R. (1993). Dissimilatory Fe(III) reduction by the marine microorganism *Desulfuromonas acetoxidans, Appl. Environ. Microbiol.*, **59**, 734.
79. Boone, D. R., Liu, Y., Zhao, Z.-J., Balkwill, D. L., Drake, G. R., Stevens, T. O. and Aldrich, H. C. (1995). *Bacillus infernus* sp. nov., an Fe(III) and Mn(IV)-reducing anaerobe from the deep terrestrial subsurface, *Int. J. Syst. Bacteriol.*, **45**, 441.
80. Ottow, J. C. G. (1971). Iron reduction and gley formation by nitrogen-fixing Clostridia, *Oecologia (Berlin)*, **6**, 164.
81. Rosello-Mora, R. A., Ludwig, W., Kämpfer, P., Amann, R. and Schleifer, K.-H. (1995). *Ferrimonas balearica* gen nov., spec. nov., a new marine facultative Fe(III)-reducing bacterium, *Syst. Appl. Microbiol.*, **18**, 196.
82. Johnson, D. B., McGinness, S. (1991). Ferric iron reduction by acidophilic heterotrophic bacteria, *Appl. Environ. Microbiol.*, **57**, 207.
83. Johnson, D. B., Ghauri, M. A. and McGinness, S. (1993). Biogeochemical cycling of iron and sulfur in leaching environments, *FEMS Microbiol. Rev.*, **11**, 63.
84. Küsel, K., Dorsch, T., Acker, G. and Stackebrandt, E. (1999). Microbial reduction of Fe(III) in acidic sediments: Isolation of *Acidiphilium cryptum* JF-5 capable of coupling the reduction of Fe(III) to the oxidation of glucose, *Appl. Environ. Microbiol.*, **65**, 3633.
85. Lovley, D. R. and Phillips, E. J. P. (1986). Availability of ferric iron for microbial reduction in bottom sediments of the freshwater tidal Potomac River, *Appl. Environ. Microbiol.*, **52**, 751.
86. Lovley, D. R., Woodward, J. C. and Chapelle, F. H. (1994). Stimulated anoxic biodegradation of aromatic hydrocarbons using Fe(III) ligands, *Nature (London)*, **370**, 128.
87. Lovley, D. R. (1997). Microbial Fe(III) reduction in subsurface environments, *FEMS Microbiol. Rev.*, **20**, 305.
88. Ottow, J. C. G. (1969). Der Einfluß von Nitrat, Chlorat, Sulfat, Eisenoxydform und Wachstumsbedingungen auf das Ausmaß der bakteriellen Eisenreduktion, *Z. Pflanzenern. Bodenk.*, **124**, 238.
89. Munch, J. C. and Ottow, J. C. G. (1982). Einfluß von Zellkontakt und Eisen(III)-Oxidform auf die Bakterielle Eisenreduktion, *Z. Planzenern. Bodenk.*, **145**, 66.
90. Kostka, J. E. and Nealson, K. H. (1995). Dissolution and reduction of magnetite by bacteria, *Environ. Sci. Technol.*, **29**, 2535.
91. Roden, E. E. and Zachara, J. M. (1996). Microbial reduction of crystalline oxides: Influence of oxide surface area and potential for cell growth, *Environ. Sci. Technol.*, **30**, 1618.
92. Munch, J. C. and Ottow, J. C. G. (1980). Preferential reduction of amorphous to crystalline iron oxides by bacterial activity, *Soil Sci.*, **129**, 15.
93. Gaspard, S., Vazquez, F. and Holliger, C. (1998). Localization and solubilization of the iron(III) reductase of *Geobacter sulfurreducens, Appl. Environ. Microbiol.*, **64**, 3188.
94. Seeliger, S., Cord-Ruwisch, R. and Schink, B. (1998). A periplasmic and extracellular c-type cytochrome of *Geobacter sulfurreducens* acts as a ferric iron reductase and as an electron carrier to other electron acceptors or to partner bacteria, *J. Bacteriol.*, **180**, 3686.

95. Lloyd, J. R., Blunt-Harris, E. L., Lovley, D. R. (1999). The periplasmic 9.6-kilo-dalton c-type cytochrome of *Geobacter sulfurreducens* is not an electron shuttle to Fe(III), *J. Bacteriol.*, **181**, 7647.

96. Lovley, D. R., Roden, E. E., Phillips, E. J. P. and Woodward, J. C. (1993). Enzymatic iron and uranium reduction by sulfate-reducing bacteria, *Mar. Geol.*, **113**, 41.

97. Stucki, J. W., Komadel, P. and Wilkinson, H. T. (1987). Microbial reduction of structural iron(III) in smectites, *Soil Sci. Soc. Am. J.*, **51**, 1663.

98. Kostka, J. E., Stucki, J. W., Nealson, K. H. and Wu, J. (1996). Reduction of structural Fe(III) in smectite by a pure culture of *Shewanella putrfaciens* strain MR-1, *Clays Clay Miner.*, **44**, 522–529.

99. Ernstsen, V., Gates, W. P. and Stucki, J. W. (1998). Microbial reduction of structural iron in clays—a renewable source of reduction capacity, *J. Environ. Qual.*, **27**, 761.

100. Kostka, J. E., Wu, J., Nealson, K. H., Stucki, J. W. (1999). The impact of structural Fe(III) reduction by bacteria on the surface chemistry of smectite clay minerals, *Geochim. Cosmochim. Acta*, **63**, 3705.

101. Lovley, D. R. (1995). Bioremediation of organic and metal contaminants with dissimilatory metal reduction, *J. Ind. Microbiol.*, **14**, 85.

102. Jakobsen, R. and Postma, D. (1999). Redox zoning, rates of sulfate reduction and interactions with Fe-reduction and methanogenesis in a shallow sandy aquifer, Romo, Denmark, *Geochim. Cosmochim. Acta*, **63**, 137.

103. Konhauser, K. O., Fyfe, W. S., Schultze-Lam, S., Ferris, F. G. and Beveridge, T. J. (1994). Iron phosphate precipitation by epilithic microbial biofilms in Arctic Canada, *Can. J. Earth Sci.*, **31**, 1320.

104. Lipman, J. G. and McLean, H. (1916). The oxidation of sulfur soils as a means of increasing the availability of mineral phosphates, *Soil Sci.*, **1**, 533.

105. Dommergues, Y. and Mangenot, F. (1970). *Ecologie Microbiènne du Sol.*, Masson, Paris.

106. Sperber, J. I. (1958). The release of phosphates from soil minerals by hydrogen sulfide, *Nature (London)*, **181**, 934.

107. Swaby, R. J. and Sperber, J. (1958). Phosphate-dissolving microorganisms in the rhizosphere of legumes. In *Proc., Univ. Nottingham Easter School Agric. Sci, 5th*, Academic Press, London, pp. 289–294.

108. Willett, I. R. (1985). The reductive dissolution of phosphate ferrihydrite and strengite, *Aust. J. Soil Res.*, **23**, 237.

109. Kostka, J. E., Luther III, G. W. and Nealson, K. H. (1995). Chemical and biological reduction of Mn(III)-pyrophosphate complexes: Potential importance of dissolved Mn(III) as an environmental oxidant, *Geochim. Cosmochim. Acta*, **59**, 885.

110. Taylor, R. M., McKenzie, R. M. and Norrish, K. (1964). The mineralogy and chemistry of manganese in some Australian soils, *Aust. J. Soil Res.*, **2**, 235.

111. Burdige, D. J. and Nealson, K. H. (1986). Chemical and microbiological studies of sulfide-mediated manganese reduction. *Geomicrobiol. J.* **4**, 361.

112. Mulder, E. G. (1972). Le cycle biologique tellurique et aquatique du fer et du manganèse, *Rev. Biol. Sol*, **9**, 321.

113. Lovely, D. R. and Phillips, E. J. P. (1988). Manganese inhibition of microbial iron reduction in anaerobic sediments, *Geomicrobiol. J.* **6**, 145.

113a. Myers, C. R. and Nealson, K. H. (1988). Microbial reduction of manganese oxides: Interactions with iron and sulfur, *Geochim. Cosmochim. Acta*, **52**, 2727.

114. Nealson, K. H. and Saffarini, D. (1994). Iron and manganese in anaerobic respiration: environmental significance, physiology, regulation, *Annu. Rev. Microbiol.*, **48**, 311.

115. Vandenabeele, J., De Beer, D., Germonpré, R., Van De Sande, R. and Verstraete, W. (1995). Influence of nitrate on manganese removing microbial consortia from sand filters, *Wat. Res.*, **29**, 579.
116. Langenhoff, A. A. M., Brouwers-Ceiler, D. L., Engelberting, J. H. L., Quist, J. J., Wolkenfeld, J. G. P. N., Zehnder, A. J. B. and Schraa, G. (1997). Microbial reduction of manganese coupled to toluene oxidation, *FEMS Microbiol. Ecol.*, **22**, 119.
117. Myers, C. R., Alatalo, L. J. and Myers, J. M. (1994). Microbial potential for the anaerobic degradation of simple aromatic compounds in sediments from the Milwaukee harbor, Green Bay and Lake Erie, *Environ. Toxicol. Chem.*, **13**, 461.
118. Gottfreund, J. and Schweisfurth, R. (1983). Mikrobiologische Oxidation und Reduktion von Manganspecies, *Fresenius Z. Anal. Chem.*, **316**, 634.
119. Gottfreund, J., Schmitt, G. and Schweisfurth, R. (1985). Wertigkeitswechsel von Manganspecies durch Bakterien in Nährlösungen und in Lockergestein, *Landwirtsch. Forsch.*, **38**, 80.
120. Luther, G. W., Sundby, B., Brent, L. L., Brendel, P. J. and Silverberg, N. (1997). Interactions of manganese with the nitrogen cycle: Alternative pathways to dinitrogen, *Geochim. Cosmochim. Acta* **61**, 4043.
121. Devereux, R. and Stahl, D. A. (1993). Phylogeny of sulfate-reducing bacteria and a perspective for analyzing their natural communities. In *The Sulfate-Reducing Bacteria: Contemporary Perspectives*, ed. Odom JR, and Singleton R Jr, Springer, New York, pp. 131–160.
122. Weast, R. C. (ed.) (1982). *CRC Handbook of Chemistry and Physics*, 63rd edn, CRC Press, Boca Raton, FL.
123. Martinez, J. D. (1991). Salt domes, *Am. Sci.*, **79**, 420.
124. Ivanov, M. V. (1968). *Microbiological Processes at the Formation of Sulfur Deposits*, Israel Program for Scientific Translations, U. S. Department of Agriculture and National Science Foundation, Washington, DC.
125. Levinson, A. A. (1980). *Introduction to Exploration Geochemistry, 2nd edn, Applied Publishing, Wilmette, IL.*
126. Naftz, D. L. and Barkley, C. S. V. (1991). *Selenium and associated trace elements in soil, rock, water and streamed sediment of the proposed Sandstone Reservoir, south-central Wyoming*, U. S. Geological Survey, Water-Resources Investigations Report 91–4000.
127. Howard, J. H. III (1977). Geochemistry of selenium: formation of ferroselite and selenium behavior in the vicinity of oxidizing sulfide and uranium deposits, *Geochim. Cosmochim. Acta*, **41**, 1665.
128. Torma, A. E. and Habashi, F. (1972). Oxidation of copper (II) selenide by *Thiobacillus ferrooxidans, Can. J. Microbiol.*, **18**, 1780.
129. Breed, R. S., Murray, E. G. D. and Smith, N. R. (1957). *Bergey's Manual of Determinative Bacteriology*, 7th edn., Williams & Wilkins, Baltimore, MD.
130. Lipman, J. G. and Waksman, S. A. (1923). The oxidation of selenium by a new group of autotrophic microorganisms, *Science*, **57**, 60.
131. Sarathchandra, S. U. and Watkinson, J. H. (1981). Oxidation of elemental selenium to selenite by *Bacillus megaterium, Science*, **211**, 600.
132. Sapozhnikov, D. I. (1937). The substitution of selenium for sulfur in the photoreduction of carbonic acid by purple sulfur bacteria, *Mikrobiologiya*, **6**, 643.
133. Smith, R. K., Curtis, C. D. (1972). Tungsten: Element and geochemistry. In *The Encyclopedia of Geochemistry and Environmental Sciences*, ed. Fairbridge RW, Encyclopedia of Earth Sciences Series, Vol. IVA, Van Nostrand Reinhold, New York, pp. 1212–1214.

134. Stevens, R. F., Jr., (1965). Tungsten. In *Mineral Facts and Problems*, 1965 edn, Bureau of Mines, U. S. Department of the Interior, Washington, DC.
135. Bowen, H. J. M. (1979). *Environmental Chemistry of the Elements*, Academic Press, London.
136. Yamamoto, I., Sakai, T., Liu, S.-M. and Ljungdahl, L. G. (1983). Purification and properties of NADP-dependent formate dehydrogenase from *Clostidium thermoaceticum*, a tungsten–selenium–iron protein, *J. Biol. Chem.*, **258**, 1826.
137. Wagner, R. and Andreesen, J. R. (1987). Accumulation and incorporation of [185]W-tungsten into proteins of *Clostridium acidiurici* and *Clostridim cylindrosporum*, *Arch. Microbiol.*, **147**, 295.
138. Zellner, G., Alten, C., Stackebrandt, E., Conway de Macario, E. and Winter, J. (1987). Isolation and characterization of *Methanocorpusculum parvum*, gen. nov., spec. nov., a new tungsten requiring, coccoid methanogen, *Arch. Microbiol.*, **147**, 13.
139. Chashchina, N. M. and Lyalikova, N. N. (1989). Role of bacteria in transformation of tungsten minerals, *Mikrobiologiya*, **58**, 122 (Engl. transl., p. 104).
140. Holliday, R. W. (1965). Molybdenum. In *Mineral Facts and Problens*, United States Department of the Interior, Bureau of Mines, Washington, DC, pp 595–606.
141. Staples, L. W. (1972). Molybdates and tungstates. In *The Encyclopedia of Geochemistry and Environmental Sciences. Encyclopedia of Earth Sciences Series*, Vol. IVA, ed. Fairbridge RW, Van Nostrand Reinhold, New York, p 752.
142. Mitchell, R. L. (1955). Trace elements. In *Chemistry of the Soil*, ed. Bear FE, Reinhold, New York, pp. 253–285.
143. DeHuff, G. L. (1965). Vanadium. In *Mineral Facts and Problems*, United States Department of the Interior, Bureau of Mines, Washington, DC, pp. 1039–1046.
144. Burkart-Baumann, I. (1972). Vanadium: element and geochemistry. In *The Encyclopedia of Geochemistry and Environmental Sciences. Encyclopedia of Earth Sciences Series*, Vol. IVA, ed. Fairbridge, R. W. Van Nostrand Reinhold, New York, pp.1234–1237.
145. Chiswell, J. R., Premakumar, R. and Bishop, P. E. (1988). Purification of a second alternative nitrogenase from a *nif*HDK deletion strain of *Azotobacter vinelandii.*, *J. Bacteriol.*, **170**, 27.
146. Joerger, R. D. and Bishop, P. E. (1988). Bacterial alternative nitrogen fixation systems, *CRC Crit. Rev. Microbiol.*, **16**, 1.
147. Pau, R. N. (1991). The alternative nitrogenases. In *Biology and Biochemistry of Nitrogen Fixation*, ed. Dilworth, M. J. and Glenn, A. R. Elsevier, Amsterdam, pp. 37–57.
148. Sprent, J. I. and Sprent, P. (1990). *Nitrogen Fixing Organisms*, Chapman and Hall, London.
149. Rowell, P. and Kerby, N. W. (1991). Cyanobacteria and their symbionts. In *Biology and Biochemistry of Nitgrogen Fixation*, ed. Dilworth, M. J. and Glenn, A. R., Elsevier, Amsterdam, pp. 373–407.
150. Robson, R. L. (1991). The genetics and regulation of alternative nitrogenases, In *Biology and Biochemistry of Nitgrogen Fixation*, ed. Dilworth, M. J. and Glenn, A. R., Elsevier, Amsterdam, pp. 142–161
151. Brock, T. D. and Madiagan, M. T. (1991). *Biology of Microorganisms*, 6th edn, Prentice Hall, Englewood Cliffs, NJ.
152. Anderson, G. L., Williams, J. and Hille, R. (1992). The purification and characterization of arsenite oxidase of *Alcaligenes faecalis*, a molybdenum-containing hydroxylase, *J. Biol. Chem.*, **267**, 23674.
153. Ferry, F. G. (1993). *Methanogens. Ecology, Physiology, Biochemistry and Genetics*, Chapman and Hall, New York.

154. Oremland, R. S. and Capone, D. G. (1988). Use of 'specific' inhibitors in biogeochemistry and microbial ecology, *Adv. Microb. Ecol.*, **10**, 285.
155. Ghani, B., Takai, M., Hisham, N. Z., Kishimoto, N., Ismail, A. K. M., Tano, T. and Sugio, T. (1993). Isolation and characterization of a Mo^{6+}-reducing bacterium, *Appl. Environ. Microbiol.*, **59**, 1176.
156. Woolfolk, C. A. and Whiteley, H. R. (1962). Reduction of inorganic compounds with molecular hydrogen by *Micrococcus lactilyticus, J. Bacteriol.*, **84**, 647.
157. Yourkova, N. A. and Lyalikova, N. N. (1990). New vanadate-reducing facultative chemolithotrophic bacteria, *Mikrobiologiya*, **59**, 968 (Engl. transl., p. 672).
158. Haglund, D. S. (1972). Uranium: element and geochemistry. In *The Encyclopedia of Geochemistry and Environmental Sciences. Encyclopedia of Earth Sciences Series, Vol. IVA*, ed. Fairbridge, R. W., Van Nostrand Reinhold, New York, pp. 1215–1222.
159. Wright, R. J. (1972). Uranium: economic deposits. In *The Encyclopedia of Geochemistry and Environmental Sciences. Encyclopedia of Earth Sciences Series*, Vol. IVA, ed. Fairbridge, R. W., Van Nostrand Reinhold, New York, p. 1222–1224.
160. Magne, R., Berthelin, J. and Dommergues, Y. (1973). Solubilisation de l'uranium dans les roches par des bactéries n'appartenant pas au genre *Thiobacillus, C. R. Acad. Sci. (Paris)*, **276**, 2625.
161. Magne, R., Berthelin, J. R. and Dommergues, Y. (1974). Solubilisation et insolubilisation de l'uranium des granites par des bactéries heterotrophes. In *Formation of Uranium Ore Deposits*, International Atomic Energy Agency, Vienna, pp. 73–88.
162. Gorby, Y. A. and Lovley, D. R. (1991). Enzymatic uranium precipitation, *Environ. Sci. Technol.*, **26**, 205.
163. Lovley, D. R., Phillips, E. J. P., Gorby, Y. A. and Landa, E. R. (1991). Microbial reduction of uranium, *Nature (London)*, **350**, 413.
164. Lovley, D. R., Widman, P., Woodward, J. and Phillips, E. J. P. (1993). Reduction of uranium by cytochrome c₃ of *Desulfovibrio vulgaris, Appl. Environ. Microbiol.*, **59**, 3572.
164a. Kauffman, J. W., Lauglin, W. C. and Baldwin, R. A. (1986). Microbial treatment of uranium mine waters, *Eviron. Sci. Technol.*, **20**, 243.
164b. Francis, A. J., Dodge, C. J., Lu, F., Halada, G. P. and Clayton, C. R. (1994). XPS and XANES studies of uranium reduction by *Clostridium* sp., *Environ. Sci. Technol.*, **28**, 636.
165. Lawton, K. (1955). Chemical composition of soils. In *Chemistry of the Soil*, ed. Bear FE, Reinhold, New York, pp. 53–84.
166. Persons, B. S. (1970). *Laterite; Genesis, Location, Use*, Plenum Press, New York.
167. Schnellmann, W. (1994). Geochemical differentiation in laterite and bauxite formation, *Catena*, **21**, 131.
168. Allen, J. B. (1972). Aluminum ore deposits. In *The Encyclopedia of Geochemistry and Environmental Sciences. Encyclopedia of Earth Sciences Series*, Vol. IVA, ed. Fairbridge, R. W. Van Nostrand Reinhold, New York, pp. 23–27.
169. Gornitz, V. (1972). Aluminum: element and geochemistry. In *The Encyclopedia of Geochemistry and Environmental Sciences. Encyclopedia of Earth Sciences Series*, Vol. IVA, ed. Fairbridge, R. W., Van Nostrand Reinhold, New York, pp 21–23.
170. Vance, G. F., Stevenson, F. J., Sikora, F. J. (1996). Environmental chemistry of aluminum-organic complexes. In *The Environmental Chemistry of Aluminum*, ed. Sposito G, CRC Press, Boca Raton, FL, pp. 171–173.
171. Kardos, L. T. (1955). Soil fixation of plant nutrients. In *Chemistry of the Soil*, ed. Bear, F. E., Reinhold, New York, p. 183.

172. Lindsay, W. L. and Walthall, P. M. (1996). The solubility of aluminum in soil. In *The Environmental Chemistry of Aluminum*, ed. Sposito G, CRC Lewis, Boca Raton, FL, pp. 337, 344.

173. Harris, S. A. (1972). Aluminum toxicity. In *The Encyclopedia of Geochemistry and Environmental Sciences. Encyclopedia of Earth Sciences Series*, Vol. IVA, Fairbridge, R. W., Van Nostrand Reinhold, New York, pp. 27–29.

174. Huang, P. M. and Violante, A. (1986). Influence of organic acids on crystallization and surface properties of precipitation products of aluminum. In *Interactions of Soil Minerals with Natural Organics and Microbes*, ed. Huang, P. M. and Schnitzer, M., SSSA Special Publication No. 17, Soil Science Society of America, Madison, WI, pp. 159–221.

175. Robert, M. (1995). Aluminum toxicity, a major stress for microbes in the environment. In *Environmental Impact of Soil Component Interactions*; Vol II. *Metals, Other Inorganics, and Microbial Activities*, ed. Huang, P. M., Berthelin, J., Bollag, J.-M., McGill, W. B. and Page, A. L., Lewis, Boca Raton, F. L., pp. 227–242.

176. Stillings, L. L., Drever, J. I., Brantley, S. L., Sun, Y. and Oxburgh, R. (1996). Rates of feldspar dissolution at pH 3–7 with 0–8 mM oxalic acid. *Chem.Geol.*, **132**, 79–89.

177. Vandevivere, P., Welch, S. A., Ullman, W. J. and Kirchman, D. L. (1994). Enhanced dissolution of silicate minerals by bacteria at near-neutral pH, *Microb. Ecol.*, **27**, 241.

178. Barker, W. W. and Banfield, J. F. (1996). Biological versus inorganically mediated weathering reactions: relationships between minerals and extracellular microbial polymers in lithobionitic communities, *Chem. Geol.*, **132**, 55.

179. Righi, D. and Meunier, A. (1995). Origin of clays by rock weathering and soil formation. In *Origin and Mineralogy of Clays*, ed. Velde B, Springer, Berlin, pp. 52, 78.

180. Velde, B. (1955). Compaction and diagenesis. In *Origin and Mineralogy of Clays*, ed. Velde B, Springer, Berlin, pp. 234–235.

181. Michalopoulos, P. and Aller, R. C. (1995). Rapid clay mineral formation in Amazon Delta sediments: Reverse weathering and oceanic cycles, *Science*, **270**, 614.

182. Millot, G. (1970). *Geology of Clays*, translators, Farrand WR and Paquet H, Springer, New York.

183. Velde, B. (1955). Composition and mineralogy of clay minerals. In *Origin and Mineralogy of Clays*, ed. Velde B, Springer, Berlin, pp. 11, 14.

184. Stotzky, G. (1986). Influence of soil mineral colloids on metabolic processes, growth, adhesion, and ecology of microbes and viruses. In *Interactions of Soil Minerals with Natural Organics and Microbes*, ed. Huang, P. M. and Schnitzer, M., SSSA Special Publication No. 17, Soil Science Society of America, Madison, WI, pp. 305–428.

185. Huang, W. H. and Keller, W. D. (1971). Dissolution of minerals in dilute organic acids at room temperature, *Am. Mineral.*, **56**, 1082.

186. Huang, W. H. and Keller, W. D. (1972). Organic acids as agents of chemical weathering of silicate minerals, *Nature (London)*, **239**, 149.

187. Huertas, J. F., Chou, L. and Wollast, R. (1998). Mechanism of kaolinite dissolution at room temperature and pressure: Part 1. Surface speciation, *Geochim. Cosmochim. Acta*, **62**, 417.

188. Polzer, W. L. and Hem, J. D. (1965). The dissolution of kaolinite, *J. Geophys. Res.*, **70**, 6233.

189. Zysset, M. and Schindler, P. W. (1996). The proton promoted dissolution kinetics of K-montmorillonite, *Geochim. Cosmochim., Acta*, **60**, 921.

190. Fraser, G. K. (1955). Soil organic matter. In *Chemistry of the Soil*, ed. Bear FE, Reinhold, New York, p. 174.
191. Stevenson, F. J. (1994). *Humus Chemistry, Genesis, Composition, Reactions*, 2nd edn, John Wiley & Sons, New York, pp. 441–445.
192. Mortimer, R. J. G. and Coleman, M. L. (1997). Microbial influence on the oxygen isotopic composition of diagenetic siderite, *Geochim. Cosmochim. Acta*, **61**, 1705.
193. Roden, E. E., Urrutia, M. M. (1999). Ferrous iron removal promotes microbial reduction of crystalline iron(III) oxides. *Environ Sci. Technol.*, **33**, 1847.
194. Widdel, F., Pfennig, N. (1977). A new anaerobic, sporing, acetate-oxidizing, sulfate-reducing bacterium, *Desulfotomaculum* (emend.) *acetoxidans, Arch. Microbiol.*, **112**, 119.
195. DeMoll-Decker, H., Macy, J. M. (1993). The periplasmic nitrite reductase of *Thauera selenatis* may catalyze the reduction of selenite to elemental selenium, *Arch. Microbiol.*, **160**, 241.

12 The Role of Mineral Colloids in the Facilitated Transport of Contaminants in Saturated Porous Media

MENACHEM ELIMELECH
Yale University, USA

JOSEPH N. RYAN
University of Colorado, USA

Interactions between Soil Particles and Microorganisms
Edited by P. M. Huang, J.-M. Bollag and N. Senesi. © 2002 John Wiley & Sons, Ltd

1 COLLOID-FACILITATED TRANSPORT OF CONTAMINANTS IN LABORATORY AND FIELD EXPERIMENTS

1.1 OVERVIEW OF COLLOID-FACILITATED TRANSPORT OF CONTAMINANTS

In 1989, McCarthy and Zachara [1] gave credence to the growing suspicion that 'mobile colloids in the subsurface environment may alter the transport of contaminants.' Their review brought widespread recognition to the concept of 'colloid-facilitated transport,' the enhancement of contaminant transport resulting from the strong association of contaminants with mobile colloids (Figure 1). Although evidence of the role of colloids in facilitating the subsurface transport of contaminants appeared as early as 1968 [2–4], most predictions of contaminant transport were still based on two-phase equilibrium adsorption models (a dissolved phase and a sorbed, immobile phase). In some cases [5,6], these models were severely under-predicting the transport of certain strongly sorbing contaminants. McCarthy and Zachara [1] clearly emphasized the need to consider three phases in contaminant transport models (dissolved, sorbed to immobile subsurface materials, and sorbed to mobile colloids). Most importantly, their review spurred and guided a strong body of research over the last decade on colloid occurrence and properties, contaminant association with colloids, colloid transport, and colloid-based mitigation and remediation of contamination.

The goal of this review is to summarize our progress in addressing these colloid-facilitated transport issues over the last decade. The review is organized according to the three basic criteria for colloid-facilitated transport [1,7]:

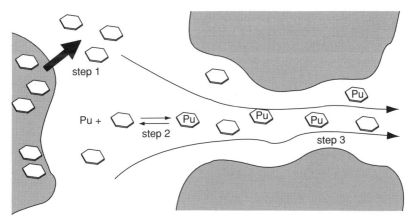

Figure 1. A schematic summary of the processes controlling colloid-facilitated transport: (1) colloid generation, (2) contaminant association with colloids, and (3) colloid transport

(1) Colloids must be present in sufficient amount to compete with the immobile subsurface materials for contaminant sorption.
(2) Contaminants must associate extensively and virtually irreversibly with colloids.
(3) Colloids must transport the contaminants into contaminant-free regions of the subsurface.

In the remainder of this first section, we examine laboratory and field evidence of colloid-facilitated transport presented in the last decade. The second section of this review covers the types of colloids and their mechanisms of generation in subsurface environments. The third section outlines the basic mechanisms and kinetics of contaminant association with colloids. The fourth section describes the role of hydrodynamics and surface chemistry in the transport of colloids. Over the past decade, colloid-facilitated transport of contaminants has been reviewed by Buffle and van Leeuwen [8,9], Swanton [10], Ouyang et al. [11], Ryan and Elimelech [7], and Kretzschmar et al. [12].

In this review, we focus on the genesis and transport of inorganic mineral colloids and their natural organic matter coatings. Dissolved natural organic matter is also important in facilitating the transport of metals [13–16] and nonpolar organic compounds [13,17–20], but we do not consider dissolved organic matter in this review because the genesis and transport of dissolved organic matter are governed by different mechanisms. Colloid-facilitated transport has also been attributed to bacteria [20–23] and their extracellular polymers [24], but we do not consider contaminant transport facilitated by 'biocolloids' because it is unlikely that such materials would be sufficiently abundant in natural subsurface environments to significantly affect contaminant transport. We also focus on colloid generation and transport processes that occur in saturated porous media. In the unsaturated zone, the same basic criteria for colloid-facilitated transport apply, but the processes of colloid generation and transport are complicated by variable degrees of soil wetness, periodic flushing of soils by infiltrating rainfall, preferential flow path development, and other issues [25–28].

1.2 COLLOID-FACILITATED TRANSPORT OF CONTAMINANTS IN LABORATORY EXPERIMENTS

Over the last decade, colloid-facilitated transport experiments have been conducted in the laboratory with a wide variety of contaminants and colloids (Table 1). Colloid-facilitated transport has been demonstrated for alkaline and alkali earth cations (Cs and Sr), transition metals (Ni, Co, Cu, and Pb), oxyanions (arsenate and iodate), non-polar organic compounds (phenanthrene and pyrene) and a polar organic compound (the pesticide prochloraz). The colloids responsible for facilitated transport of these contaminants include polystyrene latex microspheres, kaolinite, goethite, hematite, and colloidal material isolated from natural soils and sediments. For the non-polar organic contaminants, colloid-facilitated transport was tested only with polystyrene latex microspheres [29,30], giving us little to assess the potential for natural colloids to facilitate the transport of such compounds. Most of these experiments present clear evidence that the presence of colloids accelerates the breakthrough of the contaminants (Figure 2).

Generally, the contaminants and colloids were selected to demonstrate that colloid-facilitated transport occurs and to assess the effect of other solution chemistry and porous media properties on colloid-facilitated transport. For example, high pH minimizes colloid deposition under most circumstances and increases the ability of colloids to facilitate contaminant transport [30,31]. High ionic strength reduced the colloid-facilitated transport of various transition metals by preventing the mobilization and transport of colloidal material from the soil porous medium [32]. In addition to its effect on colloid transport, high ionic strength also diminished the colloid-facilitated transport of cesium by kaolinite by competing for ion exchange sites on the kaolinite colloids [33]. Unfortunately, the laboratory experiments have not frequently been used to assess the extent of contaminant-colloid association needed to make colloid-facilitated transport significant. Only Sätmark et al. [31] included a non-sorbing compound (Na^+) as a control to demonstrate that strong association (such as that observed for Cs and iodate to goethite) is necessary for colloids to facilitate contaminant transport.

Table 1. A compilation of recent colloid-facilitated subsurface transport of contaminants in laboratory experiments. Travel times noted as either time or number of pore volumes (PV)

Contaminant	Colloids	Porous media	Travel time and distance	Comments	Reference
Co, Sr, Cs	clay minerals from porous media; neutron-activated for colloid tracing	course sand (90%) and clay or zeolite (10%)	56 h 15 cm	high sodium phosphate concentration from waste form	Torok et al. [131]
arsenate	hematite 100–300 nm 5–10 mg L^{-1}	alluvial sand and gravel; quartz, feldspar	up to 4 PV 0.8–3.4 m d^{-1} 2.5–5.1 cm	I 0.03 mol L^{-1}	Puls and Powell [174]
Cu, Pb, Ni, Cr	clay and organic matter	calcareous loam soil	~10 cm	0.1 mol L^{-1} NaCl 0.1 mol L^{-1} CaMg-acetate	Amrhein et al. [32]
phenanthrene, pyrene	polystyrene latex 76, 301 nm	glass 0.1, 0.2 mm quartz sand ~0.12 mm	250 PV 20 cm	0, 0.0005 mol L^{-1} KCl	Sojitra et al. [29]
Pb	clay minerals from soil; ~0.5–1 μm diameter	silt loam soil; non-calcareous	0.3–1 h 10–12 cm	0.02–0.050 mol L^{-1} NaCl 0.00015 mol L^{-1} CaCl$_2$	Grolimund et al. [175]
Cs	montmorillonite	quartz sand	30 PV	0.05–0.5 mol L^{-1} NaCl	Fauré et al. [80]
Cs	kaolinite 0–200 mg L^{-1} ~0.5 μm	quartz sand	100 PV 5.2 cm	I 0.002–0.1 mol L^{-1} pH 7.2	Saiers and Hornberger [33,134]
Cs, iodate	goethite 6–28 mg L^{-1} 0.12 μm	quartz sand	7.5–60 min 6.0 cm	I 0.001 mol L^{-1} pH 4.0, 10.5	Sätmark et al. [31]
Cs, Sr, Co	natural colloids from soil < 5 μm	sandy soil; quartz, feldspar	1–10 cm	pH 6.7	Tanaka and Ohnuki [133]
Ni	natural colloids from Lincoln sand 250 mg L^{-1}	Ottawa sand	10 cm	I 0.001 mol L^{-1} pH 8.3	Roy and Dzombak [30]
phenanthrene	latex microspheres 100 mg L^{-1}	Eustis sand	10 cm	I 0.001 mol L^{-1} pH 7.5, 9.8	Roy and Dzombak [30]
Cs	silica 100 nm 200 mg L^{-1}	glass beads 150–10 μm, 355–420 μm	10 PV 50 cm	pH 8.9 NaHCO$_3$/ Na$_2$CO$_3$	Noell et al. [135]
prochloraz (N-propyl-N-[2-(2,4,6-trichloro-phenoxy)ethyl] imidazole-1-carboxamide)	uncharacterized; mobilized from soil	undisturbed sandy loam soil	27 h 20 cm	2.5–13% in colloidal phase	de Jonge et al. [149]

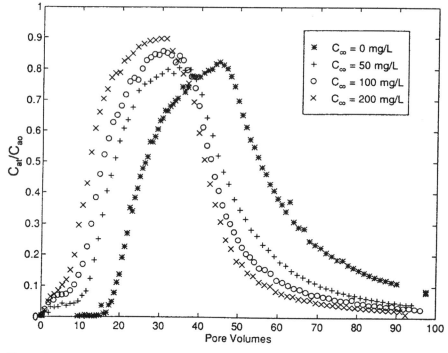

Figure 2. The presence of kaolinite colloids facilitates the transport of cesium in a quartz sand column. The normalized breakthrough of cesium-137 (a ratio of C_{at}, the total mobile ^{137}Cs concentration in the column effluent, and C_{ao}, the total concentration of ^{137}Cs in the column influent) through a quartz sand column as a function of the concentration of kaolinite colloids (C_{∞}). The number of pore volumes required for breakthrough decreases as the colloid concentration increases because a greater fraction of the cesium-137 is adsorbed to the colloids. The kaolinite colloids achieve breakthrough in the quartz sand column in one pore volume. From [134]. Reproduced by permission of the American Geophysical Union

1.3 COLLOID-FACILITATED TRANSPORT OF CONTAMINANTS IN FIELD EXPERIMENTS

Studies of colloid-facilitated transport in the field are dominated by actinides and transition metals (Table 2). Some of the field evidence for colloid-facilitated transport comes from assessments of actinide transport near underground deposits of uranium—so-called 'natural analog' studies [34–36]. These sites are of great interest for assessing the future performance of nuclear waste repositories. Detection of U, Th, and Ra associated with colloids in the groundwaters surrounding these deposits at distances from hundreds of meters to tens of kilometers seems to be strong evidence for colloid-facilitated transport; however, researchers must show that the colloids sampled at these distances

Table 2. A compilation of recent colloid-facilitated subsurface transport of contaminants in field experiments

Contaminant	Colloids	Porous media	Travel time and distance	Comments	Reference
U, Th	quartz, muscovite, lepidocrocite, 1:1 and 2:1 clay minerals; 10k MWCO	Alligator River U deposit, Australia; carbonate minerals, tuff; variable saturation	< 1–14 km	0.01–2.0% ^{238}U and 0.3–39% ^{230}Th in colloid phase	Airey [34]; Short et al. [35]
Mn, Co, Ce, Eu, Ru, Sb, Cs	quartz, feldspar, clay minerals	Nevada Test Site; fractured rhyolitic lava and tuff	~25 years 300 m		Buddemeier and Hunt [37]
Zn, Cu, Ag	30–60 mg L^{-1}	carbonate-rich sand, clay layers	20 years 30 m vertical transport	secondary sewage effluent irrigation	Magaritz et al. [40]
U, Th, Ra	illite, chlorite, kaolinite, iron oxyhydroxide, quartz, organic matter; 100–400 nm; 1–8 mg L^{-1}	Cigar Lake U deposit, Saskatchewan, Canada; fractured and geochemically altered sandstone	up to ~300 m	large (> 10^3 mL g^{-1}) and variable K_d values indicate 'irreversible' sorption	Vilks et al. [36]
Pu, Th	kaolinite, goethite, gibbsite; positively charged	sand, silt, clay; positively charged	6–11 days 1.02 km	pH 2.9–4.8; Pu and Th transported by colloids	Kaplan et al. [38]
PAHs	clay minerals, iron oxyhydroxide, iron sulfide, quartz > 10 nm, ~ 5 mg L^{-1}, 18% organic matter	marine sands; creosote contamination	80 m	5–35% in colloid phase; linear K_{oc}–K_{ow} relationship	Villholth [41]
Cs, Co, Eu, Pu	illite, smectite, zeolites 7 nm to > 1 μm	Nevada Test Site; fractured rhyolitic lava and tuff	1–80 m year^{-1} 1.3 km	> 90% in colloid phases	Kersting et al. [39]

from the deposits are not simply in equilibrium with the actinide content of the surrounding sediments. Likewise, detection of colloid-associated radionuclides near nuclear bomb test sites and nuclear waste sites [37–39] must show that the colloid-associated radionuclides are the only form to migrate over these distances. Only Magaritz *et al.* [40] were able to show that the metals associated with the colloid fraction recovered in their passive groundwater samplers were enriched relative to the metal content of the surrounding sediment.

In the only study to show colloid-facilitated transport of organic contaminants in the field, Villholth [41] showed that the extent of association of PAHs to colloids was linearly related to the octanol–water partition coefficient of the compounds at one field site and that colloid-facilitated transport was undetectable at another. The most significant difference between the sites was the amount of organic matter associated with the colloid phase.

The detection of colloid-associated actinides at the Cigar Lake site led Vilks *et al.* [36] to emphasize the need for essentially irreversible association between colloid and contaminant for colloid-facilitated transport to occur. For colloid-facilitated transport to be significant, the contaminant must remain bound to the colloid even when the colloid carries the contaminant away from the contaminant source and into regions the contaminant has not yet reached. If contaminant-colloid association were governed by equilibrium sorption, the contaminant would desorb from the colloids to re-equilibrate with the surrounding groundwater and subsurface materials. Some recent models of colloid-facilitated transport have shown very slow desorption kinetics are necessary for colloid-facilitated transport to be significant for both actinides [42] and nonpolar organic compounds [43].

Showing that colloid-facilitated transport of contaminants is occurring in the field is substantially more difficult than in the laboratory [44]. Usually, colloid-free control experiments cannot be conducted in the field to contrast contaminant transport in the absence and presence of the colloid. With this in mind, field studies must be carefully evaluated against the three criteria for colloid-facilitated transport. In addition, these three criteria should be used to design future field studies assessing colloid-facilitated transport.

The first criterion, that colloids be present, is ostensibly satisfied by all of the field studies—there certainly were colloids in the samples. To be thorough, though, we must evaluate the possibility that the colloids were mobilized during sampling and were not truly mobile in the groundwater. With the exception of Magaritz *et al.* [40], field studies have been conducted by pumping water samples from wells. A growing body of research indicates that sampling for colloids is representative only at low pumping rates—generally less than $1 \, L \, min^{-1}$ [45–49] or under passive sampling conditions [40,50]. At high pumping rates (greater than $1 \, L \, min^{-1}$), shear is expected to induce the mobilization of colloids not truly mobile in the groundwater. The depth of the wells at some sites limited researchers to high-capacity pumps and high pumping rates [37,39]. The high pumping rate problem can be overcome by pumping until the colloid concentration in the groundwater has stabilized, but such a procedure was not used in these field tests. Of course, high pumping rates could mobilize colloid-associated contaminants only if the contaminants had been transported to the vicinity of the sampling well, so the presence of contaminants, whether truly mobile with colloids or not, indicates some kind of anomalous transport. Preferential flow paths, such as those that might be found in the fractured tuff present at the Nevada Test Site, may also plausibly explain the rapid transport of the radionuclides.

One widely cited field study claiming colloid-facilitated transport has been disputed over proper identification of the contaminant flow path to the sampling locations. Penrose et al. [51] observed plutonium and americium associated with colloids about 3.4 km from the nearest waste source in an alluvial aquifer near Los Alamos National Laboratory. They contended that the colloids had facilitated the transport of these actinides through the alluvial aquifer. Recently, however, Marty et al. [52] used isotopic evidence to show that the plutonium could not have reached the sampling wells by subsurface transport. They contended that the plutonium was transported by surface flow and infiltrated into the aquifer at the sampling locations.

Some researchers have taken precautions to avoid assuming that colloid-facilitated transport has occurred. Notably, Gounaris et al. [53] detected metals and nonpolar organic compounds associated with colloidal phases in landfill leachate, but concluded that this association would not lead to colloid-facilitated transport. Assuming equilibrium between the aqueous, colloidal, and solid phases of the sediment, they calculated that the presence of colloids would result in decreases in retardation factors for strongly binding contaminants, but the resulting contaminant transport velocities were still negligible. It is clear from this type of calculation that the assumption of equilibrium will always indicate that colloid-facilitated transport will not be important because the contaminant will desorb from the colloids to the more abundant aquifer solids as the colloids move away from the contaminant source.

The second and third criteria, that the contaminant sorption to the colloids be strong and essentially irreversible and that the colloids and the associated contaminants were transported into contaminant-free regions of the subsurface environment, have been checked in only a piecemeal fashion. Vilks et al. [36] measured isotopic ratios of uranium and thorium to show that the actinides had been associated with colloids for as long as 8,000 years. Penrose et al. [51] measured no desorption of plutonium, but substantial desorption of americium, from colloids over equilibration periods of 2 days. A more direct check of both of these criteria would involve demonstrating that the subsurface material surrounding the sampling location is relatively free of contaminants. Only Magaritz et al. [40] showed that the colloids contained greater proportions of the contaminant metals than the surrounding sediments.

1.4 WHEN COLLOID-FACILITATED TRANSPORT DOES NOT OCCUR

Perhaps the most instructive assessment of colloid-facilitated transport is to examine the situations in which it did not occur (Table 3). Identifying the

Table 3. A compilation of recent laboratory and field experiments showing that colloid-facilitated transport did not significantly contribute to the subsurface transport of the contaminant. Travel times noted as either times or number of pore volumes (PV)

Contaminant	Colloids	Porous media	Travel time and distance	Comments	Reference
Ru, I, Te, Cs	clay minerals, 2–450 nm	River Glatt banks; glaciofluvial sand and gravel; quartz, feldspar, carbonates, illite, chlorite	2 days 5 m	<2.5%, in colloid phase; colloid filtration in riverbank	von Gunten et al. [56]
Cu, Cd, Zn	clay minerals, 2–450 nm	River Glatt banks (see above)	2 days 5 m	<2% in colloid phase; colloid filtration in riverbank	Waber et al. [57]
U	clay minerals, 2–450 nm	River Glatt banks (see above)	2 days 5 m	only 4% in colloid phase; colloid filtration in riverbank	Lienert et al. [58]
Pu, Am	uncharacterized; > 0.45 μm	Mortandad Canyon, Los Alamos National Laboratory; alluvial sand, silt, clay	~30 years 3.3 km	flow path disputed; isotopic evidence indicates surface flow	Penrose et al. [51]; Marty et al. [52]
Am, Cs, Cr, Ni, Cu, Cd, Pb, U	kaolinite, goethite, gibbsite; positively charged	sand, silt, clay; positively charged	6–11 days 1.02 km	pH 2.9–4.8; colloid concentrations too low; metals not adsorbed	Kaplan et al. [38,54]
Na	goethite 6–28 mg L^{-1} 0.12 μm	quartz sand	7.5–60 min 6.0 cm	weak Na–goethite association	Sätmark et al. [31]
atrazine (2-chloro-4-ethylamino-6-iso-propylamino-s-triazine	clay minerals mobilized from six soils	six undisturbed soils	6 PV 20 cm	2–18% transport enhancement not caused directly by association with colloids	Seta and Karathanasis [55,176]
PAHs	clay minerals, iron oxyhydroxide, iron sulfide, quartz > 10 nm ~ 5 mg L^{-1}, 10%, organic matter	glacial sands; creosote contamination	200 m	lower organic matter fraction	Villholth [41]

conditions that did not lead to colloid-facilitated transport will help delineate the transport criteria listed above. For example, how many colloids are enough to foster colloid-facilitated transport? How strong does the contaminant-colloid association have to be to allow colloid-facilitated transport to dominate? How resistant to deposition must the colloids be to observe colloid-facilitated transport?

Some of the laboratory studies described above explore changes in solution chemistry that diminish the importance of colloid-facilitated transport. Low pH and high ionic strength both inhibit colloid-facilitated transport by preventing the transport of colloids. The inclusion of sodium with cesium and iodate by Sätmark et al. [31] illustrated that colloid-facilitated transport does not occur for contaminants that do not bind strongly to colloids.

A greater number of field studies did not detect any significant colloid-facilitated transport. Kaplan et al. [38,54] investigated the transport of a set of actinides and transition metals in a contaminant plume characterized by low pH (2.9–4.8). Of the actinides and metals investigated, only plutonium showed significant colloid-facilitated transport. For the others, the colloid concentration was deemed to be too low. The low pH of the contaminant plume also inhibited extensive adsorption of the actinides and metals to the colloids. Other studies also cite weak contaminant–colloid interactions as the main factor in preventing colloid-facilitated transport. Seta and Karathanasis [55] observed only a minimal contribution of colloids to the transport of atrazine through saturated soils. Atrazine is a relatively soluble polar organic compound that does not sorb strongly to colloids. Villholth [41] detected colloid-facilitated transport of a suite of polycyclic aromatic hydrocarbons (PAHs) at only one of the two creosote-contaminated sites investigated. The main difference between the sites was the amount of organic matter associated with the colloidal phase. Because the extent of PAH–colloid association was linearly related to the compounds' octanol–water partition coefficient, the organic matter fraction of the colloids was the major factor controlling the extent of PAH sorption. In a series of riverbank filtration studies covering a wide range of radionuclides and transition metals, colloid-facilitated transport was limited by colloid deposition in the riverbank sediments [56–58]. In the adjacent river water, the radionuclides and metals were extensively associated with colloids, but neither the colloids nor the metals penetrated any significant distance into the riverbank sediments.

In the following sections, we will examine the processes of colloid generation, contaminant association with colloids, and colloid transport in detail to better delineate the conditions controlling colloid-facilitated transport.

2 THE SOURCES OF COLLOIDS IN SUBSURFACE ENVIRONMENTS

2.1 OVERVIEW OF COLLOID GENERATION IN SUBSURFACE ENVIRONMENTS

The first criterion for colloid-facilitated transport is that colloids must be present in sufficient amount to sorb a significant amount of the contaminant. To predict the potential for colloid-facilitated transport, we need to know the amount and nature of the colloids present in the subsurface environment. To know this, we need to know the sources of colloids—and the processes responsible for colloid generation—in the subsurface environment. McCarthy and Degueldre [59] presented a comprehensive review of colloids in the subsurface environment, concluding that colloid abundance is 'promoted by geohydrochemical perturbations, including those characteristic of waste disposal sites.' Chemical perturbations can lead to precipitation of supersaturated phases as colloids or mobilization of colloids from the subsurface material. Hydrodynamic perturbations can mobilize colloids from the subsurface material by shear. Geological perturbations (i.e., tectonic activity) are thought to be responsible for colloid mobilization in fractured media [60]. In the following sections, we will focus on the genesis of colloids by (1) precipitation of supersaturated phases, (2) release of existing colloids by chemical perturbations, and (3) release of existing colloids by hydrodynamic perturbations.

2.2 COLLOID PRECIPITATION

Colloids generated by precipitation of supersaturated phases fall into two categories: (1) colloids that are hazardous because they are composed of contaminants and (2) colloids that must associate with contaminants to be hazardous. In some circles, such colloids have been dubbed 'true colloids' and 'pseudo-colloids,' respectively [61]. The formation of colloids from supersaturated phases is governed by both equilibrium and kinetic considerations; supersaturation is necessary for colloids to precipitate, but precipitation may be kinetically limited by nucleation conditions. We must also keep in mind that the precipitation of colloidal phases may be followed by their dissolution if supersaturation is not maintained along the colloid flow path.

Researchers investigating high-level nuclear waste disposal have frequently noted the possibility of precipitation of hydrated oxides of actinides in the 'near-field environment,' the region around the waste that will be subjected to high actinide concentrations, high temperatures, and drastic geochemical gradients. Such conditions could be responsible for the reduction of a uranium(VI) solution to the less soluble uranium(IV), which would be followed by the formation of uranium(IV) oxyhydroxide colloids [62]. The likelihood of 'true colloid' generation may be overestimated because accurate characterization of the colloidal phase is not always performed. For example, Saltelli *et al.* [63] reported that 'americium colloids' leached from a vitreous waste form were effectively filtered by a glauconitic sand porous medium, but they did not determine whether the colloids were precipitated americium phases or americium adsorbed to colloidal phases generated by the glass degradation.

The precipitation of metal and metal sulfide colloid phases has been suggested as a possible means of metal accumulation in ore bodies. Laboratory studies have shown that both copper and mercury form colloidal sulfides in the 10–100 nm range [64,65]. Such small colloids would not be detected by typical groundwater sampling schemes; therefore, they might play a significant role in the bioavailability of these metals. Based on a surface water study, Benedetti and Boulègue [66] hypothesized that colloidal gold (Au(0)) may be the predominant gold species responsible for the supergene deposition of gold-bearing formations. Saunders [67] supported this concept by detecting gold textures in ore deposits that indicate colloidal transport. Gold complexes with humic substances have also been forwarded as the species contributing to supergene migration of gold [68].

The generation of so-called 'pseudo-colloids,' or colloids composed of non-hazardous compounds, has been reported under two different circumstances. The genesis of ferric oxyhydroxide colloids by oxidation of ferrous iron has been reported by Langmuir [69] and Liang *et al.* [70]. Langmuir [69] observed ferric oxyhydroxide particles in groundwater samples from coastal plain sediments in New Jersey where a geochemical gradient of increasing dissolved oxygen and one of decreasing ferrous iron concentrations were measured. Liang *et al.* [70] injected ferrous iron solutions into coastal plain sediments in South Carolina to measure the oxidation rate of ferrous iron. The product of the ferrous iron oxidation was ferric oxyhydroxide colloids. The colloidal stability of the ferric oxyhydroxide was enhanced by adsorption of natural organic matter. A geochemical gradient caused by a secondary sewage effluent plume was responsible for the formation of iron- and phosphorus-rich colloids in a Cape Cod, Massachusetts, aquifer [71]. The solubility product of vivianite ($Fe_3(PO_4)_2 \cdot 8H_2O$) was exceeded by a small amount in the groundwater. The size of the colloids, measured by dynamic light scattering and confirmed by scanning electron microscopy, was a relatively uniform 100 nm. It should be noted that sampling at the same locations in later years did not detect ferrous phosphate colloids, emphasizing that the conditions favoring colloid precipitation may be transitory.

2.3 RELEASE OF EXISTING COLLOIDS BY CHEMICAL PERTURBATIONS

Soils and sediments contain a wide distribution of particle sizes, including particles in the colloidal range from roughly 10 nm to 10 μm. These smallest particles are generally the most abundant in number [72,73]. These soil and sediment colloids are bound to larger grains by short-range chemical and physical interactions under geochemical conditions that prevailed during soil and sediment formation. When these geochemical conditions are changed, the forces binding these colloids to the grains can be reversed and colloid mobilization can be initiated. McCarthy and Degueldre [59] pointed out that waste disposal activities are the most common cause of such geochemical perturbations. Some natural processes, such as salt-water intrusion, can also change the forces binding colloids to grains. In addition, microbial interactions with soil particles can affect soil stability and colloid mobilization (Chapters 3 and 7, this book).

Decreases in ionic strength cause colloid mobilization (Figure 3), an effect that has been clearly seen in laboratory columns [74–84] and field-related studies ranging from secondary oil recovery [85,86] to the seawater–freshwater interface [87–89] to artificial recharge of aquifers [90,91]. Mobilization is caused by expansion of the electrostatic double layers around like-charged colloids and grains (e.g., negatively charged clay mineral colloids and quartz grains). The extent and rate of colloid release increases as ionic strength decreases [77,80].

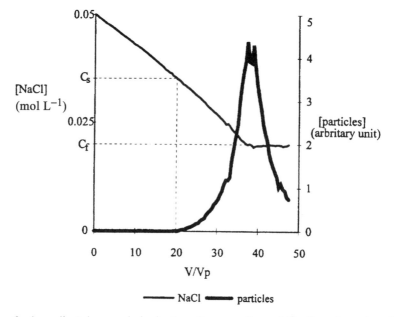

Figure 3. A gradient decrease in ionic strength causes the mobilization of montmorillonite colloids from a quartz sand (95%)/montmorillonite (5%) porous medium. Colloid release begins at $C_s = 0.035\,mol\,L^{-1}$ sodium chloride. The supply of mobilized colloids is exhausted in just a few pore volumes (V/V_p; the total volume V divided by a pore volume V_p) at the final sodium chloride concentration, $C_f = 0.019\,mol\,L^{-1}$. The colloid concentration was measured in the column effluent by absorption of ultraviolet light (280 nm wavelength). Reproduced from [80], Copyright (1996), with permission from Elsevier Science

This effect can in theory be predicted by DLVO theory [92,93]; however, in practice, surface heterogeneities, roughness, and poor characterization of surface properties make theoretical predictions unreliable. The rate of colloid release decreases over time—weakly bound colloids are mobilized first and strongly bound colloids are mobilized last, suggesting an exponential distribution of release rate coefficient activation energies [84]. If colloids and grains are oppositely charged, decreases in ionic strength have no effect on colloid mobilization because decreases in ionic strength strengthen the electrostatic attraction between oppositely charged surfaces [94,95].

Perturbations in solution composition can also affect colloid mobilization. At the same ionic strength, monovalent ions cause more release than bivalent ions [79,82,96,97]. For example, sodium chloride will mobilize more colloids than calcium chloride (Figure 4). The calcium ion is more likely to specifically complex with mineral surface sites, resulting in less electrostatic repulsion and release. In addition, the presence of calcium chloride during colloid deposition inhibits colloid release by low ionic strength sodium chloride solutions [82]. Anion composition can affect colloid mobilization, also—Seaman and Bertsch [78] observed that calcium chloride caused the release of positively charged colloids from an iron oxyhydroxide-coated sand, while calcium sulfate did not because sulfate directly reduces the surface charge of these colloids through adsorption.

Increases in solution pH generally cause colloid mobilization [74,79,94,95,98]. The increase in pH typically increases the negative charge on the colloid and grain surfaces, thus enhancing electrostatic repulsion. In sediments cemented by iron and aluminum oxides, this effect is dramatically apparent as the pH increases above the point of zero charge of the oxide coatings. When the surface charge of the oxide cement is reversed, colloid mobilization is rapid.

The addition of anions, surfactants, and reductants to sediments often results in colloid mobilization (Figure 5) and, in some cases, permeability reduction (Figure 6). Dodecanoic acid, ascorbic acid, and natural groundwater containing about $30\,\mathrm{mg\,L^{-1}}$ organic matter all promoted colloid mobilization in a iron oxyhydroxide-coated quartz sand [94]. Ascorbic acid appeared to mobilize the predominantly kaolinite colloids by dissolving the iron oxyhydroxide cement. The dodecanoic acid and organic matter-rich groundwater caused colloid release without iron oxyhydroxide dissolution by reversing the surface charge on the iron oxyhydroxide cement. Injection of phosphate, dodecyl sulfate, oxalic acid, and ascorbic acid into an iron oxyhydroxide- and silica-coated quartz sand all caused significant colloid release [95,98]. Phosphate and dodecyl sulfate mobilized colloids by reversing the surface charge on the iron oxyhydroxide coatings. Oxalic acid and ascorbic acid both reversed surface charge and dissolved (by ligand-promoted and reductive dissolution, respectively) the iron oxyhydroxide coatings. A mixture of phosphate and ascorbic acid was

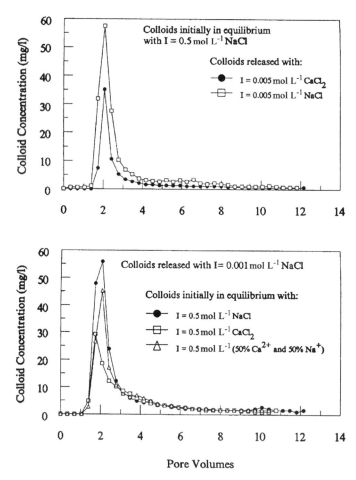

Figure 4. The effect of solution composition on the mobilization of colloids in porous media. In the upper graph, polystyrene latex microspheres deposited on glass in a $0.5 \, mol \, L^{-1}$ NaCl solution are mobilized to a greater extent by lower ionic strength solutions of NaCl than by CaCl$_2$. In the lower graph, the amount of microspheres mobilized decreases as the calcium concentration of the deposition solution increases. Reprinted from [83], Copyright (1997), with permission from Elsevier Science

most effective at mobilizing colloids in field 'push-pull' experiments. Surfactants have also been known to induce substantial decreases in formation permeability when used at concentrations typical of surfactant-enhanced remediation [99–101]. Such clogging is typically caused by excessive colloid mobilization followed by deposition in pore throats. The likelihood of substantial permeability reduction (up to two orders of magnitude) increases as the colloid-sized content of the soil increases. A bioremediation effort involving the

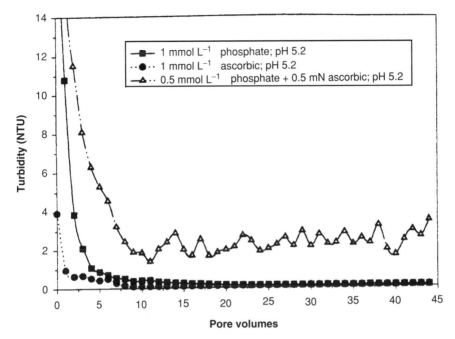

Figure 5. The combination of phosphate and ascorbic acid is much more effective at mobilizing colloids than either the phosphate or ascorbic acid alone. The colloids were mobilized from a southern Atlantic Coastal Plain sediment consisting mainly of ferric oxyhydroxide- and silica-coated quartz grains. Reduction by the ascorbic acid dissolved the ferric oxyhydroxides and adsorption of the phosphate imparted negative charge to the colloids; both of these mechanisms were necessary to mobilize the colloids. The colloid concentration is measured by turbidity as nephelometric turbidity units (NTU). Reprinted with permission from [95]. Copyright (1998) American Chemical Society

injection of a concentrated nutrient solution (phosphate, ammonium chloride, and hydrogen peroxide) also caused permeability reduction in a sand aquifer. Initially, the clogging was attributed to precipitation of iron oxides, but Weisner *et al.* [102] showed that the nutrient solutions caused extensive colloid mobilization that resulted in clogging.

Finally, colloids may be mobilized by the dissolution of cementing phases [94,95,98,103]. As discussed above, ascorbic acid dissolved the ferric oxyhydroxide cement in the Atlantic coastal plain sediments examined by Ryan and Gschwend [94] and Swartz and Gschwend [95,98]. Initially, Ryan and Gschwend [104,105] surmised that the same cement dissolution occurred naturally with natural organic matter acting as the electron donor for the reductive dissolution of the iron oxyhydroxides; however, laboratory experiments showed that the organic matter in the groundwater mobilized colloids by reversing surface charge of the iron oxyhydroxides rather than by dissolving them. Gschwend

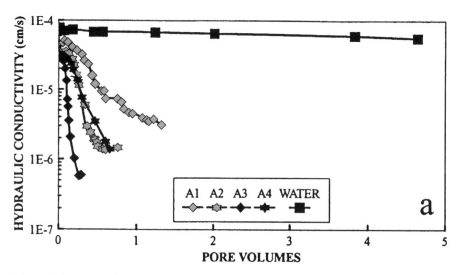

Figure 6. Decreases in the hydraulic conductivity of a loam soil (17% clay) caused by flushing with $0.01 \, mol \, kg^{-1}$ of anionic surfactants A1 (sodium dodecyl sulfate), A2 (sodium alpha olefin sulfonate), A3 (sodium dodecylbenzene sulfonate), and A4 (sodium laureth sulfate). Reprinted with permission from [100]. Copyright (1995) American Chemical Society

et al. [103] examined a fly ash disposal site in which low pH wastewater infiltrated into a calcium carbonate-cemented aquifer. The low pH groundwater dissolved the carbonate cement and released colloids.

2.4 RELEASE OF EXISTING COLLOIDS BY HYDRODYNAMIC PERTURBATIONS

In the saturated subsurface environment, hydrodynamic perturbations include (1) pumping to sample groundwater and (2) flow through fractures and other preferential flow paths. In both cases, the flow velocity in the subsurface environment is elevated above normal. Under these conditions, we expected colloid mobilization to be caused primarily by shear on the attached colloids. Extensive theoretical and experimental studies of the effect of shear on colloid release have been conducted in model systems of spherical particles attached to flat walls [106–111]. From these studies, we know that the rate and extent of colloid detachment increases as the flow velocity and particle size increase. Thus, pumping and fracture flow should increase colloid mobilization, with larger particles preferentially mobilized before smaller particles.

The rate of pumping is a major concern in the collection of representative colloid-containing groundwater samples [45–49]. Puls *et al.* [45] showed that increases in pumping rate from 0.6 to 92 L min^{-1} increased the amount and size of colloids in samples from a metal-contaminated alluvial aquifer. Under steady-state conditions, the sample turbidity was approximately the same over the range of pumping rates. Backhus *et al.* [46] showed that a temporary increase in the pumping rate to 1.0 L min^{-1} from 0.1 L min^{-1} caused a large increase in sample turbidity. Kearl *et al.* [47] used a down-hole video camera to detect an increase in particle size occurring when the pumping rate was increased from 0.1 to 1.45 L min^{-1}. In addition to increases in colloid concentration in the samples, increases in the associated contaminants were also observed [46,48,49]. The increase in contaminant levels provides great incentive to use low flow rate pumping for regulators concerned about samples that exceed limits and action levels. In addition, those who use low flow rate pumping can take satisfaction in collecting samples that are most representative of the truly mobile contaminant and colloid load.

Colloids are relatively abundant in samples taken from free-flowing fractures in very low permeability rock [60,112,113]. Because the flow through these fractures occurs at much greater velocity than the flow through the surrounding matrix, we can consider the fracture flow to be a hydrodynamic perturbation causing colloid mobilization. In the fractures, the primary minerals weather to secondary clay minerals and oxides, and these are the minerals that appear as colloids in the fracture flow samples. Characterization of the size distributions of these colloid samples shows that the distribution is weighted slightly toward the larger sizes [60], suggesting that erosion (shear) is controlling the size distribution.

2.5 NATURE AND ABUNDANCE OF COLLOIDS

McCarthy and Degueldre [59] assembled an extensive review of the types and abundances of colloids found in the subsurface environment. The composition of most of the colloids is similar to the composition of the subsurface environment, indicating that most colloids are mobilized from the subsurface materials by chemical or physical perturbations. In summary, we can expect the colloidal phase to be dominated by (1) clay minerals, which are naturally present in the colloid size range, (2) amorphous silica, (3) metal oxyhydroxides, including ferric iron, aluminum, and manganese, (4) carbonates, primarily calcium, (5) colloid-sized fragments of primary minerals, mainly quartz and feldspars, and (6) associated organic matter. The abundance of these colloids ranges from 0 to 100 mg L^{-1}, with the higher concentrations generally correlated with the greater chemical and hydrodynamic perturbations.

Since the review by McCarthy and Degueldre [59], most of the studies examining colloid nature and abundance in subsurface environments have shown similar results. Soils from the Savannah River Site in South Carolina generated mica or vermiculite, kaolinite, Ti-rich ferric oxyhydroxide, gibbsite, and quartz [38,114]. From coastal plain sediments at the Savannah River Site,

Seaman *et al.* [115] mobilized Al-rich goethite, kaolinite, and crandallite $(CaAl_3(PO_4)_2(OH)_2 \cdot H_2O)$. These Savannah River colloids are relatively unusual in having positively charged surfaces. Natural concentrations of these colloids were deemed too low to significantly contribute to colloid-facilitated transport [38]. The presence of organic matter as a coating is important for colloids of positive surface charge, such as ferric and aluminum oxyhydroxides

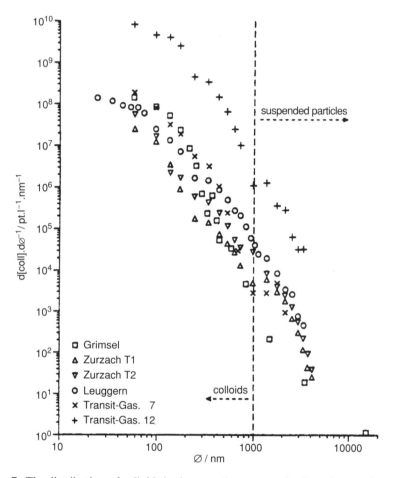

Figure 7. The distribution of colloid size in groundwater samples from fractured granite formations in Europe showing Pareto power law distributions ($N = \alpha d_p^{-\beta}$, where N is the number of colloids in a size range, d_p is the median colloid size in the range, and α and β are fitting parameters characteristic of the colloid size distribution). These colloid size distributions were measured by single particle counting with scanning electron microscopy. The number of colloids per size range is displayed on the y-axis and the median size of the range is displayed on the x-axis. Reprinted from [113], Copyright (1996), with permission from Elsevier Science

and the edges of clay minerals [25,104,114]. Atteia *et al.* [116] detected a mixture of mineral colloids including carbonates with coprecipitated metals in groundwater exiting a karst aquifer. The size distribution of these particles followed the Pareto power law; i.e., logarithmic decreases in colloid size corresponded to logarithmic increases in colloid number (Figure 7). Colloids recovered from fractures in granitic rocks have revealed similar size distributions [60,113]. Calcium carbonate colloids were also detected in limestone and calcic soils [117]. The transport of these colloids was implicated in the formation of calcic soil horizons.

Studies of deep crystalline rock formations being considered for nuclear waste disposal continue to produce new information about colloids in subsurface environments. In anoxic groundwater sampled from the Äpsö Hard Rock Laboratory in Sweden, Ledin *et al.* [112] found less than $0.1\,mg\,L^{-1}$ of a ferrous iron-rich colloid phase. Upon exposure to air, submicrometer ferric oxyhydroxide and calcium carbonate colloids formed in the groundwater samples. Degueldre *et al.* [113] extended an initial study of the colloid content of the Grimsel Test Site in Switzerland to two other sites in deep granitic rocks and found similar results: the colloid phases were dominated by clay minerals and amorphous silica in the submicrometer size range at concentrations of less than $100\,\mu g\,L^{-1}$. When calcium and sodium concentrations in the groundwater are low, however, colloid concentrations were as high as $10\,mg\,L^{-1}$. A granitic groundwater in Spain (El Berrocal) also contained mainly clay minerals, silica, and iron, titanium, and aluminum oxyhydroxides [118]. Depending on groundwater chemistry, a variety of metal carbonates and sulfides was also detected.

3 ASSOCIATION OF CONTAMINANTS WITH COLLOIDS

3.1 OVERVIEW OF CONTAMINANT ASSOCIATION WITH COLLOIDS

The second criterion for colloid-facilitated transport is the association of the contaminants with the colloids. We use 'association' to encompass the wide range of sorption mechanisms (e.g., complexation, ion exchange, hydrophobic partitioning) that may contribute to contaminant–colloid interactions. In this section, we will show that contaminant association with colloids must be (1) extensive, or 'strong,' and (2) essentially irreversible for colloid-facilitated transport to occur. Extensive contaminant–colloid association is typically characterized by a large distribution coefficient K_d. Irreversible colloid–contaminant association refers to slow desorption kinetics, where 'slow' must be considered relative to the velocity of flow through the subsurface environment.

The distribution coefficient is useful for qualitatively assessing the strength of the contaminant–colloid interaction, but we must remember that its use in describing the colloid-facilitated transport process is quite limited. First, distribution coefficients are inadequate for describing contaminant association with colloids when sorption is non-linear. Second, distribution coefficients represent equilibrium sorption under specific geochemical conditions; thus, they are unsuited for representing sorption when geochemical conditions are changing along a flow path [1]. Third, we must be keenly aware of the importance of desorption kinetics when considering the potential for colloid-facilitated transport. Vilks *et al.* [36] asserted that, 'if colloids are completely mobile in the subsurface, their impact on radionuclide migration will depend on colloid concentration and whether or not radionuclide attachment is reversible.' Colloid-facilitated transport will carry a contaminant from a zone in which the aqueous phase concentration of the contaminant is in equilibrium with the sorbed phase (both the mobile colloids and the immobile porous media grains) to a zone in which the contaminant is predominantly associated with the mobile colloids and not the aqueous phase. Desorption of the contaminants from the mobile colloids must be slow in this zone for colloid-facilitated transport to be important.

3.2 CONTAMINANT–COLLOID ASSOCIATION

In this section, we will examine the basic mechanisms of contaminant–colloid interactions for a variety of representative contaminants and colloids encountered in the subsurface environment. The contaminants examined will include cationic forms of metals (e.g., Cs^+, Cu^{2+}, PuO_2^+), anionic forms of metals (e.g., CrO_4^{2-}, AsO_4^{3-}), nonpolar organic compounds (e.g., polycyclic aromatic hydrocarbons, polychlorinated biphenyls, DDT) and polar organic compounds (e.g., atrazine). The colloids examined will include minerals with amphoteric surfaces (e.g., iron, aluminum, and manganese oxides, phyllosilicate edges, carbonates, and sulfides) and fixed charge surfaces (phyllosilicate faces), organic matter coatings on mineral phases, and bacteria. While the association of contaminants with organic matter coatings on colloids is addressed here, we do not address the association of contaminants with 'dissolved' organic matter (e.g., humic and fulvic acids) even though this association leads to facilitated transport of both metals and nonpolar organic compounds in the subsurface environment. The genesis and transport of dissolved organic matter is governed by mechanisms different from those addressed in this review for colloids.

The basic mechanisms of contaminant association with colloids include surface complexation, ion exchange, and hydrophobic partitioning. Each of these sorption mechanisms has been the subject of exhaustive research summarized in recent reviews [119–124].

3.2.1 Cationic Forms of Metals

Cationic forms of metals (cations) are the most frequently reported contaminants influenced by colloid-facilitated transport (Tables 1 and 2); therefore, we should expect that their association with colloids is strong and their desorption is slow. Surface complexation and ion exchange reactions dominate cation adsorption to mineral, organic, and biological surfaces. Some cations (e.g., Cu^{2+}, Pb^{2+}) form strong surface complexes with amphoteric oxides surfaces (e.g., iron, aluminum, manganese oxides, clay mineral edges). We classify the metal–ligand complexes that form as 'strong' if (1) the surface complex is formed even in the face of electrostatic repulsion between the cation and a positively charged oxide surface and (2) variations in ionic strength do not affect the extent of cation adsorption [119]. These conditions indicate the cation and surface ligand have formed a covalent bond. Adsorption of cations is limited by the number of reactive sites and the accumulation of positive charge on the oxide surface. The strength of a cation bond to a surface hydroxyl is related to the strength of the complex formed by the cation with hydroxyl ions in solution [120,125]; thus, metals such as Cu^{2+} and Pb^{2+} form stronger surface complexes than Co^{2+} and Cd^{2+}. A general order of the strength of surface complex formation is, from strongest to weakest [125],

$$Fe^{3+} > Th^{4+} > UO_2{}^{2+} > Al^{3+} > Cr^{3+} > Cu^{2+} > Ni^{2+} > Co^{2+} > Zn^{2+} > Cd^{2+}$$

For other cations, such as the alkali and alkaline earth metals, adsorption is relatively weak—electrostatic repulsion and increases in ionic strength decrease metal adsorption, indicating that covalent bonds are not formed. Cation adsorption to carbonate and sulfide surfaces is similar to their adsorption to oxide surfaces, but coprecipitation (incorporation of the cation into the carbonate or sulfide structure) must also be considered.

Cations will also adsorb to organic matter coatings on mineral surfaces. Major components of organic matter such as humic and fulvic acids readily adsorb to positively charged mineral surfaces (e.g., iron and aluminum oxides and clay edges at pH values below their pH_{pzc} values). Most natural waters contain sufficient organic matter to ensure near-complete coating of such mineral surfaces. In the aqueous phase, cations are readily bound by humic and fulvic acids by formation of complexes with carboxylic, phenolic, and sulfhydryl ligands [122]. For hard metals (e.g., $Fe^{3+}, Al^{3+}, UO^{2+}$), complexes with carboxylic and phenolic functional groups dominate metal binding, while sulfhydryl groups strongly bind soft metals like Hg^{2+}. In some cases, the presence of organic matter increases the adsorption of cations, suggesting that the cation complexes with the organic matter [126]. In other cases, the organic matter reduces cation adsorption, suggesting that it blocks sites for cation adsorption on the mineral surface.

Cation adsorption to clay mineral faces occurs by ion exchange. Ion exchange is typically defined as the replacement of one ion by another for ions residing in the diffuse double layer arising from the mineral surface. On the faces of clay minerals, the diffuse double layer is generated by negative surface charge arising from isomorphic substitution (e.g., replacement of Si^{4+} by Al^{3+}).

Clay minerals have high cation exchange capacities (10^{-5}–10^{-3} eq g^{-1}) because adsorption sites exist between clay layers composed of tetrahedral silica and octahedral alumina sheets. The extent of cation adsorption by ion exchange increases as the cation charge increases and the cation hydrated radius decreases:

$$Rb^+ > Cs^+ > K^+ > NH_4^+ > Na^+ > Li^+$$

$$Ba^{2+} > Sr^{2+} > Ca^{2+} > Mg^{2+}$$

The adsorption of cations to ion exchange sites is dominated by attractive electrostatic interactions. These electrostatic bonds are substantially weaker than the covalent bonds that bind cations to oxide surface ligands. The alkali and alkaline earth metals are most commonly the subjects of ion exchange investigations because their adsorption is dominated by ion exchange and not surface complexation.

The kinetics of cation surface complexation depend on the strength of the cation–aquo bond; the rate-determining step in adsorption is the removal of water molecules bound to the cation. Desorption is generally much slower than adsorption [127,128], suggesting that the breaking of the cation surface complex is the rate-determining step. If the strength of the cation surface complex controls desorption kinetics as the strength of the cation-aquo complex does for adsorption, then cations at the upper end of the Irving–Williams series will desorb most slowly. The desorption kinetics of cations bound to organic matter are expected to follow the same mechanism—desorption will be slowest for the cations that adsorb most strongly. For example, Penrose et al. [51] determined that plutonium desorption from natural colloids was unmeasurable over 2 days, while americium desorption was fast, a kinetics trend that follows the hydrolysis constants for these species. Similarly, Bunzl et al. [129] measured greater loss of americium than plutonium from grassland soils exposed to radionuclide fallout. Cerling and Turner [130] showed that strontium was desorbed more rapidly than cesium from iron and manganese oxide-coated gravels in a streambed. Recent work also indicates that desorption is initially fast and then much slower owing to the initial release of more weakly bound cations [127].

Ion exchange kinetics are dominated by the relative selectivity of cations and the rate of diffusion of the cation through clay minerals. As with surface complexation, hysteresis between adsorption and desorption is often observed for ion exchange reactions, but the hysteresis cannot be attributed to the strength of ion exchange bonds. Instead, the slow desorption of cations from clay minerals is the result of diffusion of the cations into the clay mineral interlayers following their initial adsorption. Similar sorption hysteresis in soil aggregates and other minerals containing microporosity can be attributed to diffusion of the cation.

Actual measurements of desorption rates in colloid-facilitated transport experiments show some trends that support the kinetic expectations. Torok et *al.* [131] estimated a half-life of 6.3 h for desorption of cesium from clay mineral colloids. Smith and Comans [132] surmised that the rapid adsorption of cesium to clay-rich lake sediments was followed by slow transfer to less-exchangeable sites (half-life of 50–125 days). Following this transfer, desorption was estimated as very slow (half-life on the order of 10 years). The corresponding adsorption half-life was only 1.4 min. Tanaka and Ohnuki [133] measured half-lives of 0.17, 0.35, 0.53 min for desorption of strontium, cesium, and cobalt from natural colloids, a trend that follows the relative selectivity and strength of binding exhibited by these cations. In contrast, Saiers and Hornberger [134] modeled the colloid-facilitated transport of cesium by kaolinite and obtained a model fit that resulted in cesium desorption that was faster than the adsorption (half-lives of 0.4 and 2.8 h, respectively). Noell *et al.* [135] used desorption half-lives of 3.5–4.2 min to model the release of cesium from amorphous silica colloids, while the adsorption half-lives that best fit the breakthrough data were sometimes longer (12 min) and sometimes shorter (2.3 min).

3.2.2 Anionic Forms of Metals

Oxyanions have also been identified as contaminants for which colloid-facilitated transport has occurred (Table 1). For oxyanions and other ligands, surface complexation dominates adsorption. Adsorption to organic matter is generally prevented by electrostatic repulsion because most of the ionized functional groups in organic matter are anionic. Similarly, adsorption of oxyanions by ion exchange is negligible because positively charged minerals in nature would generally be coated by organic matter.

Most transition metal oxyanions are capable of forming strong surface complexes with amphoteric oxides surfaces. As for transition metal cations, these oxyanion surface complexes are classified as 'strong' if the surface complex is formed when electrostatic repulsion prevails and variations in ionic strength do not affect the extent of adsorption. These conditions indicate the oxyanion and the surface metal have formed a covalent bond. Adsorption of oxyanions is limited by the number of reactive sites and the buildup of negative charge on the oxide surface. The strength of an oxyanion bond to a surface hydroxyl is related to the acidity of the oxyanion—weaker acids are more strongly bound to surface metal sites just as they more strongly bind protons. A general order of the strength of surface complex formation is, from strongest to weakest for adsorption to goethite [121],

$$AsO_4^{3-} > PO_4^{3-} > VO_4^{3-} > CrO_4^{2-} > SO_4^{2-} > SeO_4^{2-}$$

For anions such as sulfate and selenate, adsorption is relatively weak—electrostatic repulsion and increases in ionic strength decrease their adsorption, indicating that covalent bonds are not formed.

The kinetics of metal oxyanion adsorption and desorption follows the same general trends identified for cationic forms of metals. Desorption is usually much slower than adsorption owing the formation of strong surface complexes [136]. For arsenate and chromate, adsorption on goethite occurred via a fast ligand exchange resulting in monodentate surface complexes followed by a slow ligand exchange resulting in bidentate surface complexes. Following the order of surface complex strength, chromate desorption was faster than arsenate desorption.

3.2.3 Nonpolar Organic Compounds

Nonpolar organic compounds are driven from water by unfavorable interactions with the polar solvent. If organic matter is present, it will readily absorb nonpolar organic compounds as an organic solvent would readily partition these compounds from water [123,124,137]. To a limited extent, bare mineral surfaces also accept nonpolar organic compounds driven from the water, but organic matter must be essentially absent from the system for nonpolar organic compound partitioning to mineral surfaces to occur [138]. The extent of sorption into organic matter depends on the hydrophobicity of the nonpolar organic compound, a property described by the compound's octanol–water partition coefficient K_{ow}, and the amount of organic matter. Sorption of some nonpolar organic compounds is extensive because their interactions with water are so unfavorable, but this extensive sorption is not characterized as 'strong.' Unlike surface complexation, only relatively weak bonds (e.g., van der Waals attraction) are formed between the nonpolar organic compound and the sorbent, organic matter. Similarities between the nonpolar organic compound and organic matter structure favor enhanced partitioning; e.g., the extent of polycyclic aromatic hydrocarbon sorption increases with increasing aromatic content of the organic matter [139].

The organic matter responsible for the sorption of nonpolar organic compounds may be dissolved in the aqueous phase or present as coatings on inorganic colloids. The dissolved phase is often isolated from water samples by ultrafiltration with typical cutoffs of >1 nm or >500 Da molecular weight and <10 nm or $<100,000$ Da molecular weight. Both of these phases have been shown to be important in binding nonpolar organic compounds [53,140], but the transport of the dissolved organic matter and organic matter coatings is very different. Dissolved organic matter behaves as a ligand; its transport is limited by adsorption, mainly to positively charged minerals such as iron oxyhydroxides. Organic matter coatings are already adsorbed to mineral surfaces; thus, their transport follows that of the mineral colloids.

The kinetics of nonpolar organic compound sorption and desorption depend primarily on the rate of diffusion through the organic matter [124,141–145].

Significant hysteresis between sorption and desorption exists owing to diffusion of nonpolar organic compounds deeper into the organic matter after the initial absorption [146], a phenomenon that causes serious problems for pump-and-treat remediation of contaminated groundwater. In soils, desorption half-lives span the range of 0.3–140 days, with the shortest half-lives for smaller organic compounds (e.g., benzene, toluene, xylene) and the longest for larger organic compounds (e.g., polychlorinated biphenyls, chlorobenzenes). Desorption half-lives increase as the amount of organic matter increases, the mineral micro-porosity increases, and the molecular weight of the nonpolar organic compound increases. For a given contaminant, the amount of organic matter exerts the primary control on desorption rates [138,144]. Recent work indicates the presence of strongly and weakly bound contaminant fractions (or a distribution of binding site strengths) with correspondingly slower and faster desorption kinetics [142,143,145,147,148]. Extrapolating this slow desorption kinetics to colloids must be done with caution because the amount of organic matter associated with stationary soils is much greater than that associated with colloids. Because colloids will contain much less organic matter than soils and, more directly, the diffusion paths of contaminants out of the organic matter will be shorter, desorption from colloids should be significantly faster than desorption from soils.

3.2.4 Polar Organic Compounds

A variety of sorption mechanisms contribute to the sorption of polar organic compounds, especially those with ionizable functional groups, to colloidal phases. Depending on the structure of the polar organic compound, surface complexation, ion exchange, and hydrophobic partitioning may play a role in sorption [123,124]. For example, a polar organic compound such as atrazine contains amino functional groups that are capable of binding metals on the surfaces of oxide minerals. The same functional groups can protonate to form a cationic compound that can be adsorbed by ion exchange to clay minerals. Finally, the K_{ow} of atrazine is $10^{2.56}$, indicating that atrazine sorption may be aided by hydrophobic partitioning to organic matter. Predicting the sorption of a polar organic compound such as atrazine requires detailed knowledge of the composition of the colloidal phase.

The kinetics of polar organic compound sorption and desorption depend on which of the variety of sorption mechanisms dominate. As discussed above, the kinetics of desorption will be very slow if strong surface complexes are formed. If sorption is not strong, the desorption kinetics will be determined by the rate of the compound's diffusion from the interlayers and micropores of the minerals or organic matter.

3.3 CONTROL OF COLLOID-FACILITATED TRANSPORT BY CONTAMINANT-COLLOID ASSOCIATION

Our review of colloid-facilitated transport experiments in the field and laboratory reveals an interesting dichotomy—colloid-facilitated transport is a relatively frequent occurrence for metal cations and oxyanions, but a relatively rare occurrence for organic compounds, both nonpolar and polar. Only two studies showed colloid-facilitated transport of organic compounds by natural colloids, one in the laboratory [149] and one in the field [41]. Two other studies showing colloid-facilitated transport of organic compounds used polystyrene latex microspheres, which cannot be considered representative of natural colloids. In the laboratory study by de Jonge et al. [149], only 2.5–13% of the organic compound, prochloraz, was colloid associated, so true colloid-facilitated transport did not occur. In the field study by Villholth [41], the organic compound distribution in the sediments surrounding the down-gradient sampling wells was not measured, so we cannot be sure that true colloid-facilitated transport occurred. As noted in the introductory section, this is a shortcoming of virtually all of the colloid-facilitated transport field studies (the exception being the study by Magaritz et al. [40]).

The lack of clear evidence for significant colloid-facilitated transport of non-polar organic compounds points to the importance of desorption kinetics in determining the potential for colloid-facilitated transport (Figure 8). We know that metal cations and oxyanions can form strong covalent bonds with mineral colloids and associated organic matter, resulting in slow desorption kinetics [36]. The desorption of non-polar organic compounds from sediments is also quite slow, resulting in problems with remediation of organic compound-contaminated groundwater. Why, then, do we not find many examples of colloid-facilitated transport of non-polar organic compounds in the field? Desorption of non-polar organic compounds from colloids must be relatively fast. It appears that the manner in which organic matter is associated with mineral colloids must present short path lengths for diffusion of sorbed organic compounds out of the organic matter. Note that our exclusion of the many studies showing 'organic matter-facilitated transport' of organic compounds does not change this conclusion. While it is easy to facilitate the transport of organic compounds with dissolved organic matter in laboratory columns with short residence times [13,18,19], there is no field evidence that this process is important.

Recent efforts at modeling colloid-facilitated transport have emphasized the need for a kinetic approach to contaminant–colloid association [42,43,134, 150,151] and that irreversible sorption or very slow desorption will greatly increase colloid-facilitated transport distances (Figure 9). Roy and Dzombak [43] emphasized that slow desorption of the contaminant from the colloid is 'probably most important in enhancing transport distances of contaminants in natural systems with mobile colloids which usually occur at concentrations not

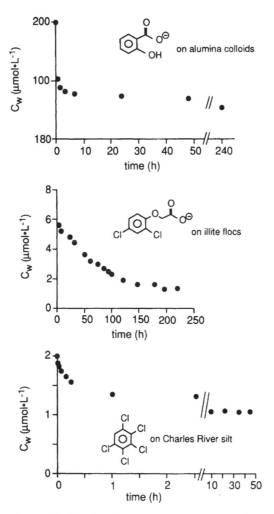

Figure 8. A comparison of the kinetics of contaminant desorption from suspended solids [124]. The vertical axes are the concentration of the compounds in water (C_w). The upper graph shows the desorption of salicylate from 20 nm aluminum oxide colloids. The center graph shows the desorption of (2,4-dichlorophenoxy)acetic acid from clay aggregates. The lower graph shows the desorption of pentachlorobenzene from silt-sized river sediment. The desorption half-life increases as the strength of the contaminant–colloid interaction increases, but the irreversibly sorbed residual varies from case to case. From [124]. Reprinted by permission of John Wiley & Sons, Inc.

exceeding a few mg L^{-1}.' Finally, it is important to remember that the rate of desorption must be considered relative to the rate of water flow. In the slow-moving groundwater of a deep aquifer, only the most strongly bound

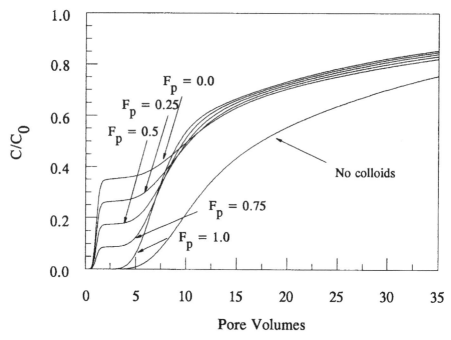

Figure 9. A model simulation of phenanthrene breakthrough as a function of the fraction of reversible sorption sites on colloids (F_p) showing that phenanthrene breakthrough is more rapid when phenanthrene sorption to the colloids is irreversible ($F_p = 0$). Sorption was set to be irreversible (the rate of phenanthrene exchange was set to $0\,h^{-1}$) at the sites not set as reversible. The phenanthrene inlet concentration was $C_0 = 1\,mg\,L^{-1}$. The colloid concentration was $100\,mg\,L^{-1}$. The equilibrium partition coefficient for the phenanthrene association with the colloids was $6{,}600\,mL\,g^{-1}$. The rates of colloid deposition and release were 0.1 and $0.3\,h^{-1}$. Reprinted from [43], Copyright (1998), with permission from Elsevier Science

contaminants will remain adsorbed to colloid long enough for colloid-facilitated transport to be considered significant. On the other hand, the transport of a wider range of contaminants could be facilitated by colloids in water infiltrating through a shallow unsaturated soil.

4 TRANSPORT OF COLLOIDS

4.1 OVERVIEW OF COLLOID TRANSPORT AND GOVERNING EQUATIONS

The transport of colloidal particles through granular porous media is governed by advection, dispersion, and the exchange of colloidal particles between the solid stationary matrix and bulk solution. This is commonly described by the advection–dispersion equation with appropriate terms for colloid deposition and release. For the simple case of one-dimensional transport (e.g., packed column) the governing equations are [12]:

$$\frac{\partial C}{\partial t} = D_p \frac{\partial^2 C}{\partial x^2} - v_p \frac{\partial C}{\partial x} - \frac{\rho_b}{\varepsilon} \frac{\partial S}{\partial t} \tag{1}$$

$$\frac{\rho_b \partial S}{\varepsilon \partial t} = k_d C - \frac{\rho_b}{\varepsilon} k_r S \tag{2}$$

This set of equations describes the colloidal particle concentration in suspension $C(x,t)$ and the amount of deposited colloidal particles per unit mass of the solid matrix $S(x,t)$ as a function of travel distance x and time t. Here, D_p is the hydrodynamic dispersion coefficient for colloidal particles, v_p the average interstitial velocity of colloidal particles, ρ_b the solid bulk density, ε the porosity, and k_d and k_r the colloid deposition and release rate coefficients, respectively. The initial colloid deposition and release rates are generally assumed to follow first-order kinetics. Such equations have been successfully utilized by several researchers to describe colloid or bacteria transport in laboratory scale columns [152,153].

For fixed chemical and physical conditions, the colloid release rate is often much lower than the colloid deposition rate [75,154]. Furthermore, in the presence of attractive double-layer forces between colloidal particles and matrix surfaces the deposition of colloidal particles is practically irreversible [155]. In such cases, the release term in equation (2) can be omitted and the transport equation simplifies to

$$\frac{\partial C}{\partial t} = D_p \frac{\partial^2 C}{\partial x^2} - v_p \frac{\partial C}{\partial x} - k_d C \tag{3}$$

The above equation cannot be applied when retained particles influence the rate of colloid deposition, as often observed in colloid breakthrough experiments utilizing large amounts of colloidal particles at the column inlet [156]. In this chapter, we will focus on deposition kinetics of colloidal particles at low surface coverage, which may be appropriate to colloidal transport in subsurface environments.

In the following subsections, we will discuss the role of two important processes that are integral components of the governing equations above, namely particle advection and particle deposition/filtration. Because subsurface environments are physically and geochemically heterogeneous, the effects of such heterogeneities on particle advection and deposition will be emphasized.

4.2 PARTICLE ADVECTION AND ROLE OF PHYSICAL HETEROGENEITY

Structural heterogeneities, such as macropores and fractures, are quite common in subsurface environments. Such physical heterogeneities may create preferential flow paths for the advective fluid (i.e., water) and, thus, markedly influence the transport of colloidal particles and solutes [12,157]. Preferential flow in continuous channels and cracks occurs in soils, particularly during periods of rapid water infiltration. Lenses or layers with different texture in sandy aquifers also result in significant physical heterogeneity.

The physical heterogeneity of subsurface porous media is reflected in the spatial variation of the hydraulic conductivity, an important parameter in modeling subsurface flow. The spatial variation of hydraulic conductivity results in a heterogeneous flow field that influences colloid transport and the resulting particle concentration distribution in the porous medium. For modeling purposes, two types of physical heterogeneity are commonly used: layered heterogeneity and random heterogeneity.

In a layered, physically heterogeneous subsurface porous medium, the porous medium is made up of several homogeneous layers. Thus, while each layer is homogeneous (i.e., with constant hydraulic conductivity), the entire system is heterogeneous. Porous media with fractures or lenses of different texture may be described as layered heterogeneous.

In random physical heterogeneity, the hydraulic conductivity is randomly distributed in the three-dimensional subsurface space. Substantial progress has been made in the past two decades to understand the random physical heterogeneity of groundwater aquifers. Evidence from field-scale hydraulic conductivity measurements indicates that the spatial distribution of hydraulic conductivity is lognormal [158,159]. It was also found that there exists a non-Gaussian behavior of the log-transformed hydraulic conductivity at relatively small scales, and that this non-Gaussian behavior shifts to Gaussian behavior as the length scale increases [160,161].

The influence of physical heterogeneity on colloidal transport is illustrated in the three examples below. These examples are taken from studies involving colloid transport in packed soil columns, a layered heterogeneous model aquifer, and a packed sand column. Figure 10 illustrates the transport of a conservative tracer and natural, *in situ* mobilized soil colloids in columns packed with heterogeneous soil aggregates [154]. The soil aggregates in the column comprised 17% clay, 66% silt, and 17% sand (by weight); the clay mineralogy was dominated by quartz, vermiculite, illite, and kaolinite. Inspection of the breakthrough curves reveals two distinct features. First, particle breakthrough occurs much earlier than the solute tracer; that is, colloidal particles in the soil column travel considerably faster than a conservative tracer. Second, only a fraction of the injected particles is recovered at the column effluent, as indicated by the reduced peak and area of the normalized particle breakthrough compared with that of a tracer. The enhanced transport of colloidal particles compared with that of a conservative tracer is attributed to the phenomenon of size exclusion [162]. Contrary to the solute tracer, a portion of the pore space of the soil matrix is not accessible to the mobile colloidal particles. In this physical mechanism of size exclusion, the pores through which particles cannot travel are basically those smaller than the colloidal particles. The second distinct feature of the particle breakthrough curves in Figure 10—the reduced peak and area of the normalized particle breakthrough compared with that of a solute tracer—is attributed to immobilization of colloidal particles on the surfaces of soil aggregates due to particle deposition. Particle deposition in porous media will be discussed in the next subsection.

The effect of layered-distributed physical (structural) heterogeneity on colloid transport is illustrated in Figure 11 [163]. The porous medium was divided into three horizontal layers parallel to the flow direction, with the hydraulic conductivity of the middle layer of the porous medium being twice as large as the hydraulic conductivity in the layers above and below. The colloid suspension is fed continuously (line injection) into the porous medium at the inlet boundary ($x = 0$), with 11 injection points set at 0.1 m intervals along the vertical z direction. Observations of the normalized concentration profiles over the entire two-dimensional porous medium domain are presented for $t = 0.75$ days. As shown, the fluid flows in the central layer faster than the other two layers, and most of the colloids migrate with the flow through the more permeable layer. This example points to the paramount importance of preferential flow paths in colloid transport.

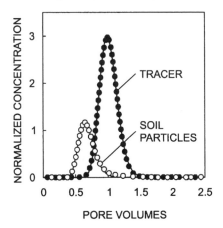

Figure 10. Transport behavior of natural colloidal particles (o) compared with a non-reactive tracer (•) in packed soil columns as a response to a pulse input. The results demonstrate the enhanced transport of the colloidal particles in the heterogeneous soil column due to the phenomenon of size exclusion. Reprinted with permission from [154]. Copyright (1998) American Chemical Society

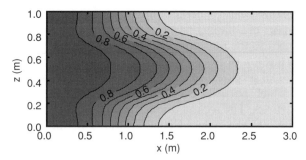

Figure 11. Effect of layered physical heterogeneity of porous media on colloid transport behavior. The contours describe the residual particle concentration (C/C_0) at an observation time of 0.75 days. The central layer (z between 0.3 and 0.7 m) has hydraulic conductivity $K = 100\,\text{m day}^{-1}$ and the layers above (z between 0.7 and 1.0 m) and below (z between 0 and 0.3 m) are with $K = 50\,\text{m day}^{-1}$. Colloids are continuously injected along the depth (z axis) on the left boundary of the porous medium ($x = 0$). Modified from [163]

The last example is from the work of Saiers *et al.* [153]. In this work, silica colloid transport experiments in a structured-heterogeneous porous medium were carried out. The column was packed with concentric layers of homogeneous sand of different grain size, aligned parallel to the flow direction, thus forming a preferential flow path through the coarse grained sand. The colloid and tracer breakthrough curves are depicted in Figure 12. The first breakthrough of colloidal particles occurred after 0.2 pore volumes, due to rapid transport along the preferential flow path. A second increase in colloid concentration was observed after about 2 pore volumes, due to slower transport in the fine grained matrix. Very similar breakthrough curves were also observed for the conservative solute tracer. As shown, the experimental data were described quite well by a transport model which considers two different hydraulic conductivities.

4.3 FILTRATION THEORY AND PARTICLE DEPOSITION KINETICS

4.3.1 Filtration Theory for a Model Porous Medium

Filtration theories describe the deposition (capture) of colloidal particles during the initial stage of filtration (the so-called 'clean-bed removal') based on fundamental concepts of mass transfer, hydrodynamics, and colloid and surface chemistry. The deposition of particles is represented by a single collector removal efficiency, usually denoted as η. The single collector efficiency is defined as the ratio of the overall particle deposition rate onto the collector to the convective transport of upstream particles toward the projected area of the collector. For an isolated spherical collector, the single collector removal efficiency is [164]

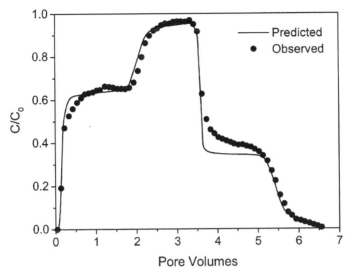

Figure 12. Model prediction and experimental breakthrough curve of colloidal silica particles through a structurally heterogeneous sand column with a distinct preferential flow path. The vertical axis is the normalized concentration (C, the concentration in a given pore volume, divided by C_0, the influent concentration). Reprinted from [153], Copyright (1994), with permission from Elsevier Science

$$\eta = \frac{I}{U C_0 \pi a_c^2} \tag{4}$$

where C_0 is the bulk concentration, U is the fluid approach velocity, and I is the actual deposition rate on collector of radius a_c. The single collector efficiency is then related to the removal efficiency of the entire granular filter medium through a simple mass balance. For a granular filter composed of uniform spheres, the result is [164,165]

$$\ln\left(\frac{C}{C_0}\right) = -\frac{3}{4}\frac{(1-\varepsilon)\eta L}{a_c} \tag{5}$$

Here ε is the porosity and L is the depth of the granular medium in the filter. This expression can also be viewed as the logarithmic attenuation in concentration of suspended particles traveling a distance L of porous medium.

The goal of the fundamental filtration theories is to predict η for a given suspension under known physical and chemical conditions. However, current theories fail to predict η when repulsive double layer interactions predominate [164,166]. As a result, it is necessary to combine an empirical factor in predicting η. In this approach, we multiply the single collector efficiency, η_0, determined from physical considerations, by an empirical collision (attachment) efficiency, α, which describes the fraction of collisions with filter grains that results in attachment. Thus, the overall single collector removal efficiency is

$$\eta = \alpha \eta_0 \tag{6}$$

where η_0 is the 'favorable' single collector removal efficiency, which is calculated without the inclusion of electric double layer interaction [164,165]. The collision efficiency, α, for a given colloidal suspension, solution chemistry, and filter medium can be determined from column experiments and is usually in the range of 10^{-3}–1 [7,164,166].

Based on numerical results obtained from the so-called 'trajectory analysis' for various physical conditions (in the absence of double layer interaction), Rajagopalan and Tien [167] proposed a correlation equation for η_0. This equation accounts for the effects of hydrodynamic (viscous) interaction and van der Waals attraction on colloid deposition rate. It is given by [164,167]

$$\eta_0 = 4.0 A_S^{1/3} \left(\frac{D_\infty}{d_c U} \right)^{1/3} + A_S N_{LO}^{1/8} N_R^{15/8} + 3.38 \times 10^{-3} A_S N_G^{1.2} N_R^{-0.4} \tag{7}$$

where A_S is a porosity-dependent parameter of the Happel sphere-in-cell model, and N_{LO}, N_R, and N_G are dimensionless parameters. The term N_{LO} characterizes the van der Waals attraction and is defined as

$$N_{LO} = \frac{4A}{9\pi \mu d_p^2 U} \tag{8}$$

where A is the Hamaker constant of the interacting media and d_p is the particle diameter. The parameter N_R is an aspect ratio given by

$$N_R = \frac{d_p}{d_c} \tag{9}$$

with d_c being the diameter of the collector. Lastly, N_G is a gravitational force number given by

$$N_G = \frac{(\rho_p - \rho)g d_p^2}{18 \mu U} \tag{10}$$

Here, g is the gravitational acceleration, μ is the fluid viscosity, and ρ_p and ρ are the density of particles and fluid, respectively.

The single collector efficiency η is related to the so-called filter coefficient λ_f through [164]

$$\lambda_f = -\frac{3}{2} \frac{1 - \varepsilon}{d_c} \eta \tag{11}$$

where ε is the effective porosity of the granular porous medium and d_c is the diameter of the collector grain. It can be shown that the particle deposition rate coefficient k_d used in equations (2) and (3) is related to the filter coefficient λ_f and the approach velocity U by the following simple relationship [12]:

$$k_d = \frac{\lambda_f U}{\varepsilon} = -\frac{U}{\varepsilon L}\ln(C/C_0) \tag{12}$$

An example illustrating the kinetics of particle deposition in a model granular porous medium is shown in Figure 13. In this figure, the deposition kinetics of two submicrometer-size latex particles in a column packed with uniform glass beads is presented. Experimental collision efficiencies of these suspensions at various ionic strengths were calculated from the measured particle breakthrough curves using equations (5) and (6). Results are presented as stability curves; that is, the logarithm of the collision efficiency α as a function of the logarithm of ionic strength (KCl). As seen, the deposition rate (or collision efficiency) increases as the salt concentration is increased until the critical deposition concentration is attained. The increase in the collision efficiency with salt concentration is attributed the reduced electrostatic double layer repulsion caused by double layer compression and charge screening. Above the critical deposition concentration the deposition rate is transport limited whereas below this concentration it is controlled by the repulsive electrostatic double layer interaction.

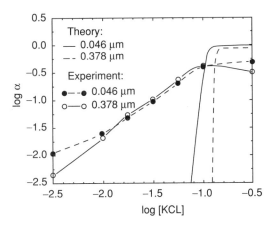

Figure 13. Experimental collision efficiencies (α) of two different suspensions of Brownian polystyrene latex colloids in flow through a column packed with uniform glass beads as a function of KCL concentration. The diameters of the particles are indicated in the figure. Theoretical collision efficiencies shown as lines without data points. Modified from [166]

4.3.2 Particle Deposition Kinetics in Natural Porous Media

Filtration theory can in principle be applied to describe the transport of mono-disperse colloidal particles through uniform natural granular media such as packed sand columns. Filtration theory has also been applied to explain the transport behavior of colloidal particles in sandy aquifers [168]. For most natural porous media, however, such calculations can lead only to qualitative agreement because of the physical and chemical heterogeneities of natural subsurface porous media. The example below illustrates that, despite the complexity of natural porous media, the deposition kinetics behavior is indeed not much different than the deposition behavior with model systems, as for the results shown earlier in Figure 13.

The deposition kinetics of soil colloids and synthetic latex particles in packed soil columns as a function of Na^+ and Ca^{2+} concentrations are depicted in Figure 14 [154]. The soil aggregates and natural colloids used are similar to those described in Figure 10. Results are presented as the logarithm of the deposition rate coefficient k_d, or the collision efficiency α, as a function of the logarithm of counter ion concentration. The results in Figure 14 demonstrate that calcium ions have a much more pronounced effect on the kinetics of particle deposition than sodium ions. Substantial particle deposition occurs at a much lower concentration of Ca^{2+} than Na^+. The critical deposition concentrations for Ca^{2+} (ca. 5 and 2 mmol L^{-1} for the natural and latex particles, respectively) are much lower than those with Na^+ (*ca.* 200 mmol L^{-1} for both the soil and latex particles). It is also rather remarkable that the colloid stability curves of the natural colloidal particles (top) are quite similar to those of the synthetic carboxyl latex particles (bottom). These curves are also similar to other stability curves displayed in previously published studies with model particles and collectors (as in Figure 13) [164], despite the marked heterogeneity of the natural colloidal particles in terms of chemical composition, particle size, and morphology. The deposition behavior with Na^+ and Ca^{2+} shown above is generally in qualitative agreement with the classical Derjaguin, Landau, Verwey, and Overbeek (DLVO) theory of colloidal stability and the well-known Schulze–Hardy rule [169], whereby divalent counterions reduce colloidal stability much more effectively than monovalent counterions.

4.3.3 Role of Geochemical Heterogeneity

Solid phases in terrestrial environments exhibit both physical and chemical heterogeneity [7]. Silicates and aluminosilicates are the dominant primary minerals in the stationary solid phase, especially quartz, feldspar, micas, and clays. In addition, carbonates and the oxides of iron, aluminum, and manganese represent an important group of accessory minerals that are often present as coatings on the primary minerals and as intergranular cement. Aquifer

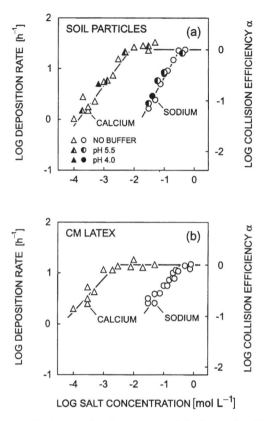

Figure 14. Influence of solution chemistry on particle deposition kinetics in natural porous media (natural soil columns). (a) Effect of electrolyte (NaCl or CaCl$_2$) concentration, counter ion valence (Na$^+$ and Ca^{2+}), and solution pH (pH 4.0 or 5.5, controlled by azide buffer) on deposition rate coefficients and experimental collision efficiencies for *in situ* mobilized soil particles. (b) Influence of electrolyte concentration and counterion valence on deposition rate coefficients and experimental collision efficiencies for model carboxyl latex particles. In both figures, solid lines serve to guide the eye. Reprinted with permission from [154]. Copyright (1998) American Chemical Society

sediments also contain a heterogeneous variety of complex organic molecules characterized by a wide range of molecular weights and compositions. These organic molecules are found attached to mineral surfaces as organic coatings or organic particulate matter occupying the interstitial regions of the stationary matrix.

Due to the presence of physical and chemical heterogeneities, most natural surfaces have an uneven, or heterogeneous charge distribution [170]. Surface charge heterogeneities can be classified in terms of scale as either macroscopic or microscopic. For deposition of colloidal particles onto heterogeneous surfaces, microscopic charge heterogeneity describes charge variations on a molecular level, whereas macroscopic charge heterogeneity refers to variations on the scale of colloidal particle dimensions or more. Microscopic charge heterogeneities arise from the regular arrangement of oppositely charged ions in the crystalline lattice, as well as from molecular-level structural defects such as kinks and screw dislocations. Macroscopic charge heterogeneity on natural surfaces results mainly from the presence of surface chemical impurities such as oxyhydroxide coatings [7,171].

Surface charge heterogeneity can be characterized by the distribution of local surface potential or charge on the mineral grains. Generally, it is not possible to assign exact values for variations in potential over the entire surface. However, a probability distribution of surface potentials may be assigned according to *a priori* knowledge or assumptions about surface characteristics. Song *et al.* [170] proposed two models to describe surface charge heterogeneity, namely the patchwise and random distribution models which are described below.

Patchwise heterogeneity implies that surface sites of equal potential are grouped together in macroscopic patches, each of which can be treated as a homogeneous surface. When considering large patches, it may be assumed that each patch behaves as a homogeneous, isolated surface in equilibrium with the bulk solution and that the interactions at patch boundaries can be neglected. In groundwater aquifers, macroscopic surface heterogeneities such as iron, aluminum, or manganese oxide patches on minerals are representative of large patchwise heterogeneities [7,170,172]. The simplest model of patchwise heterogeneity is the two-patch charge model, where the total surface area of the collector grains is divided into favorable and unfavorable fractions, as shown schematically in Figure 15.

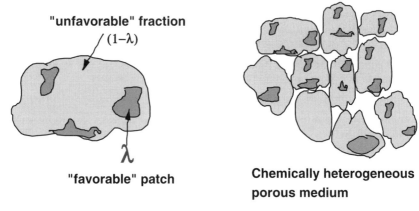

"unfavorable" fraction
(1–λ)

λ

"favorable" patch

Chemically heterogeneous porous medium

Figure 15. Schematic description of the two-patch model for patchwise geochemical heterogeneity. An isolated heterogeneous grain is shown on the left. A representative elementary volume of a geochemically heterogeneous porous medium is shown on the left

Random (continuous) heterogeneity indicates that sites of equipotential are randomly distributed over the entire surface. The random distribution model may be applied to collectors whose surfaces do not have an obvious patchwise arrangement of charge distribution, such as amorphous substances. Because there is often no information available on the distribution of surface potentials, a normal distribution is generally assumed.

The patchwise and random distribution models should be regarded as mathematical representations of surface charge heterogeneity because, at the present time, there are no available methods to characterize actual charge site distribution of natural grain surfaces. Despite their limited nature, however, these models are useful means of describing the characteristics of surface charge heterogeneity as applied to particle deposition and transport.

An example for the paramount effect of patchwise geochemical heterogeneity on colloid transport in porous media is shown in Figure 16. Colloid deposition experiments with colloidal silica particles flowing through columns packed with geochemically heterogeneous sand were carried out [173]. Patchwise geochemical heterogeneity was introduced to the granular porous medium by modifying the surface chemistry of a fraction of the quartz sand grains via iron oxyhydroxide coating. The initial ('clean bed') removal efficiency increases as the degree of geochemical heterogeneity (or the fraction of iron oxyhydroxide-coated sand) increases. At the initial stages of deposition, the colloid removal efficiency for the given conditions increases from little less than 2% for the clean quartz grains to 86% when 16% of the quartz grains are coated with iron oxyhydroxide. At the pH maintained during the column experiments (5.6–5.8), the iron oxyhydroxide-coated sand grains are positively charged and thus provide favorable surfaces for the deposition of the negatively charged silica colloids. The results suggest that the degree of geochemical variability and the availability of favorably charged surfaces may play a major role in determining the mobility of colloids in subsurface aquatic environments. In groundwater environments having a preponderance of iron oxyhydroxides as mineral coatings and low in organic carbon, colloid mobility is likely to be severely curtailed due to the favorable conditions for attachment of colloids to surfaces.

Calculated single collector efficiencies from the breakthrough curves in Figure 16 demonstrate that the following relationship holds for patchwise heterogeneous porous media surfaces [170]:

$$\eta = \lambda\eta_f + (1 - \lambda)\eta_u \tag{13}$$

where λ is the fraction of the grain surface which is favorable for deposition (as illustrated in Figure 15), and η_f and η_u are the single collector efficiencies for favorable (iron oxyhydroxide-coated sand) and unfavorable (clean sand) surfaces, respectively. Because $\eta_f \gg \eta_u$ for patchwise heterogeneous surfaces, equation (13) simplifies to

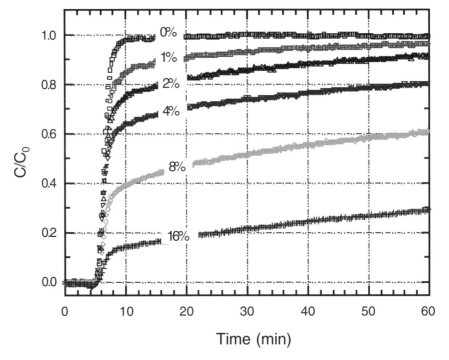

Figure 16. Silica particle transport in geochemically heterogeneous porous media. Experimental particle breakthrough curves correspond to columns packed with various fractions of iron oxyhydroxide-coated sand shown in the figure. The vertical axis is the normalized concentration (C, the concentration at a given time, divided by C_0, the influent concentration). Modified from [173]

$$\eta \approx \lambda \eta_f \tag{14}$$

Analysis of the breakthrough curve data in Figure 16 strongly supports this simplified expression [173].

5 SUMMARY AND FUTURE RESEARCH

In this review, we have examined colloid-facilitated transport of contaminants in the subsurface environment from the perspective of three criteria required for colloid-facilitated transport to occur. First, colloids must be present in sufficient quantity to sorb the contaminants. Colloids present in groundwaters include clay minerals, iron and aluminum oxides, calcium carbonate, fragments of quartz, feldspars, and other primary minerals, and the organic matter coatings bound to the surfaces of these colloids. Concentrations of these colloids reach $100\,mg\,L^{-1}$, but typical colloid concentrations are in the 0.01–$1\,mg\,L^{-1}$ range. Most of the colloids in the subsurface environment are generated by chemical and physical perturbations that mobilize colloid-sized soil and sediment particles. Some of the chemical perturbations causing colloid mobilization include the mixing of fresh water into more saline water (artificial recharge, seawater interfaces, secondary oil recovery), changes in pH, and the introduction of contaminants that can alter the surface charge of colloids and grains or dissolve the cement binding colloids to grains. Some colloids may be generated by precipitation of supersaturated phases, but their existence may be too transitory to be of great concern in colloid-facilitated transport. Further research is needed to relate the potential for colloid mobilization to the geologic setting. The effects of preferential flow in soil macropores and fractured media on colloid mobilization also merits more detailed attention in the future.

Second, the contaminants must bind strongly and essentially irreversibly to the colloids for colloid-facilitated transport to occur. The colloids must outcompete the surrounding mineral grains for contaminants in the contaminated area, carry these contaminants ahead of the contaminant plume, and resist desorbing these contaminants when the surrounding water and mineral grains are contaminant free. Contaminants that form surface complexes with colloids (e.g., transition metals, actinides) display the slowest desorption rates. In many cases, their adsorption can be considered irreversible with respect to the rate of groundwater flow. The rate of desorption of contaminants that sorb by weaker mechanisms (ion exchange, partition into organic matter) is limited mainly by diffusion. Their rate of desorption is still slow, but faster than that of surface-complexed contaminants. On the basis of desorption kinetics, metals and actinides are the most likely candidates for colloid-facilitated transport, a surmise confirmed by field evidence. In the future, some direct confirmation of the role of desorption kinetics in colloid-facilitated transport should be obtained, with proper consideration of the widely varying transport time scales in mind (e.g., short times during infiltration through the unsaturated zone, long times for release of actinides from a deep nuclear waste repository).

Third, the colloids carrying the contaminants must be mobile in the subsurface environment. Two factors can contribute to their mobility: minimizing collisions and avoiding attachment following collisions. Colloids of approximately 1 μm experience the fewest collisions in porous media at typical groundwater flow rates. Colloid collisions are also reduced in preferential flow paths (soil macropores, fractured media). Attachments are avoided by maintaining repulsive electrostatic interactions between colloids and grains. Generally, the chemical factors that encourage colloid generation also favor colloid transport—low ionic strength, high pH, and the presence of natural and anthropogenic species that increase surface charge or mask attachment sites. Heterogeneities in the subsurface environment lead to frequent encounters of colloids with oppositely charged mineral grains. Future research should focus on the roles of physical and chemical heterogeneity on colloid transport and better characterization of the surface properties of colloids and collector grains.

Laboratory studies suggest that colloid-facilitated transport should be a common occurrence; however, very few field studies have unequivocally demonstrated that it has occurred. To demonstrate colloid-facilitated transport in the field, future studies should focus on (1) characterizing the colloids present and developing plausible mechanisms for their generation, (2) assessing the reversibility of contaminant–colloid interaction and the presence of contaminant in the colloid, sediment, and aqueous phases beyond the contaminant plume, and (3) examining the mobility of the colloids in the subsurface environment.

LIST OF SYMBOLS

A	Hamaker's constant
A_s	porosity dependent parameter
a_c	collector radius
C	particle concentration
C_0	influent particle concentration
d_c	collector diameter
d_p	particle diameter
D_p	particle dispersion
D_∞	particle diffusion coefficient
g	gravitational acceleration constant
I	overall particle deposition rate onto a collector grain
k_d	particle deposition rate constant
k_r	particle release rate constant
L	filter bed depth
N_G	gravitational force dimensionless parameter
N_{LO}	van der Waals force dimensionless parameter
N_R	aspect ratio; $N_R = d_p/d_c$
S	amount of deposited colloidal particles per unit mass of the solid matrix
t	time
U	approach (superficial) velocity
v_p	average interstitial particle velocity
x	axial coordinate (distance)
α	collision (attachment) efficiency
ε	porosity
λ_f	filter coefficient
η_0	favorable single collector removal efficiency
η_f	single collector efficiency for favorable surface fraction

η_u single collector efficiency for unfavorable surface fraction
λ heterogeneity parameter for favorable fraction grain surface
λ_f filter coefficient
μ fluid viscosity
ρ density of the fluid
ρ_b solid bulk density
ρ_p density of suspended particles

REFERENCES

1. McCarthy, J. F. and Zachara, J. M. (1989). Subsurface transport of contaminants, *Environ. Sci. Technol.*, **23**, 496.
2. Champlin, J. B. F. and Eichholz, G. G. (1968). The movement of radioactive sodium and ruthenium through a simulated aquifer, *Wat. Resources Res.*, **4**, 147.
3. Champlin, J. B. F. (1971). The physics of fine-particle movement through permeable aquifers, *Soc. Petrol. Eng. J.*, **11**, 367.
4. Champlin, J. B. F. and Eichholz, G. G. (1976). Fixation and remobilization of trace contaminants in simulated subsurface aquifers, *Health Phys.*, **30**, 215.
5. Coles, D. G. and Ramspott, L. D. (1982). Migration of ruthenium-106 in a Nevada Test Site aquifer: discrepancy between field and laboratory results, *Science*, **215**, 1235.
6. Nyhan, J. W., Drennon, B. J., Abeele, W. V., Wheeler, M. L., Purtymun, W. D., Trujillo, G., Herrera, W. J., and Booth, J. W. (1985). Distribution of plutonium and americium beneath a 33-yr-old liquid waste disposal site, *J. Environ. Qual.*, **14**, 501.
7. Ryan, J. N. and Elimelech, M. (1996). Colloid mobilization and transport in groundwater, *Colloids Surf. A: Physicochem. Eng. Aspects*, **107**, 1.
8. Buffle, J. and van Leeuwen, H.P. (1992). *Environmental Particles*, Vol. 1, Lewis, Chelsea, MI.
9. Buffle, J. and van Leeuwen, H.P. (1993). *Environmental Particles*, Vol. 2, Lewis, Boca Raton, FL.
10. Swanton, S. W. (1995). Modelling colloid transport in groundwater; The prediction of colloid stability and retention behaviour, *Adv. Colloid Interface Sci.*, **54**, 129.
11. Ouyang, Y., Shinde, D., Mansell, R. S., and Harris, W. (1996). Colloid-enhanced transport of chemicals in subsurface environments: A review, *Crit. Rev. Environ. Sci. Technol.*, **26**, 189.
12. Kretzschmar, R., Borkovec, M., Grolimund, D., and Elimelech, M. (1999). Mobile subsurface colloids and their role in contaminant transport, *Adv. Agron.*, **66**, 121.
13. Dunnivant, F., Jardine, P. M., Taylor, D., and McCarthy, J. F. (1992). Cotransport of cadmium and hexachlorobiphenyl by dissolved organic matter through columns containing aquifer material, *Environ. Sci. Technol.*, **26**, 360.
14. Marley, N. A., Gaffney, J. S., Orlandini, K. A., and Cunningham, M. M. (1993). Evidence for radionuclide transport and mobilization in a shallow, sandy aquifer, *Environ. Sci. Technol.*, **27**, 2456.
15. Wang, E. X. and Benoit, G. (1996). Mechanisms controlling the mobility of lead in the spodosols of a northern hardwood forest ecosystem, *Environ. Sci. Technol.*, **30**, 2211.
16. McCarthy, J. F., Sanford, W. E., and Stafford, P. L. (1998). Lanthanide field tracers demonstrate enhanced transport of transuranic radionuclides by natural organic matter, *Environ. Sci. Technol.*, **32**, 3901.

17. Ballard, T. M. (1971). Role of humic carrier substances in DDT movement through forest soil, *Soil Sci. Soc. Am. Proc.*, **35**, 145.
18. Enfield, C. G. and Bengtsson, G. (1988). Macromolecular transport of hydrophobic contaminants in aqueous environments, *Ground Wat.*, **26**, 64.
19. Magee, B. R., Lion, L. W., and Lemley, A. T. (1991). Transport of dissolved organic macromolecules and their effect on the transport of phenanthrene in porous media, *Environ. Sci. Technol.*, **25**, 323.
20. Lindqvist, R. and Enfield, C. G. (1992). Biosorption of dichlorodiphenyltrichloroethane and hexachlorobenzene in groundwater and its implications for facilitated transport, *Appl. Environ. Microbiol*, **58**, 2211.
21. Kim, S. and Corapcioglu, M. Y. (1996). A kinetic approach to modeling mobile bacteria-facilitated groundwater contaminant transport, *Wat. Resources Res.*, **32**, 321.
22. Saiers, J. E. and Hornberger, G. M. (1996). Modeling bacteria-facilitated transport of DDT, *Wat. Resources Res.*, **32**, 1455.
23. Banaszak, J. E., Rittmann, B. E., and Reed, D. T. (1999). Subsurface interactions of actinide species and microorganisms: Implications for bioremediation of actinide–organic mixtures, *J. Radioanal. Nucl. Chem.*, **241**, 385.
24. Dohse, D. M. and Lion, L. W. (1994). Effect of microbial polymers on the sorption and transport of phenanthrene in a low-carbon sand, *Environ. Sci. Technol.*, **28**, 541.
25. Kaplan, D. I., Bertsch, P. M., Adriano, D. C., and Miller, W. P. (1993). Soil-borne mobile colloids as influenced by water flow and organic carbon, *Environ. Sci. Technol.*, **27**, 1193.
26. Jacobsen, O. H., Moldrup, P., Larsen, C., Konnerup, L., and Petersen, L. W. (1997). Particle transport in macropores of undisturbed soil columns, *J. Hydrol.*, **196**, 185.
27. Ryan, J. N., Illangasekare, T. H., Litaor, M. I., and Shannon, R. (1998). Particle and plutonium mobilization in macroporous soils during rainfall simulations, *Environ. Sci. Technol.*, **32**, 476.
28. Lægdsmand, M., Villholth, K.G., Ullum, M., and Jensen, K.H. (1999). Processes of colloid mobilization and transport in macroporous soil monoliths, *Geoderma*, **93**, 33.
29. Sojitra, I., Valsaraj, K. T., Reible, D. D., and Thibodeaux, L. J. (1995). Transport of hydrophobic organics by colloids through porous media 1. Experimental results, *Colloids Surf. A: Physicochem. Eng. Aspects*, **94**, 197.
30. Roy, S. B. and Dzombak, D. A. (1997). Chemical factors influencing colloid-facilitated transport of contaminants in porous media, *Environ. Sci. Technol.*, **31**, 656.
31. Sätmark, B., Albinsson, Y., and Liang, L. (1996). Chemical effects of goethite colloids on the transport of radionuclides through a quartz-packed column, *J. Contaminan Hydrol.*, **21**, 231.
32. Amrhein, C., Mosher, P. A., and Strong, J. E. (1993). Colloid-assisted transport of trace metals in roadside soils receiving deicing salts, *Soil Sci. Soc. Am. J.*, **57**, 1212.
33. Saiers, J. E. and Hornberger, G. M. (1999). The influence of ionic strength on the facilitated transport of cesium by kaolinite colloids, *Wat. Resources Res.*, **35**, 1713.
34. Airey, P. L. (1986). Radionuclide migration around uranium ore bodies in the Alligator Rivers region of the Northern Territory of Australia—Analogue of radioactive waste repositories—A review, *Chem. Geol.*, **55**, 255.
35. Short, S. A., Lowson, R. T., and Ellis, J. (1988). $^{234}U/^{238}U$ and $^{230}Th/^{234}U$ activity ratios in the colloidal phases of aquifers in lateritic weathered zones, *Geochim. Cosmochim. Acta*, **52**, 2555.

36. Vilks, P., Cramer, J. J., Bachinski, D. B., Doern, D. C., and Miller, H. G. (1993). Studies of colloids and suspended particles, Cigar Lake uranium deposit, Saskatchewan, Canada, *Appl. Geochem.*, **8**, 605.

37. Buddemeier, R. W. and Hunt, J. R. (1988). Transport of colloidal contaminants in groundwater: Radionuclide migration at the Nevada Test Site, *Appl. Geochem.*, **3**, 535.

38. Kaplan, D. I., Hunter, D. B., Bertsch, P. M., Bajt, S., and Adriano, D. C. (1994). Application of synchrotron x-ray fluorescence spectroscopy and energy dispersive x-ray analysis to identify contaminant metals on groundwater colloids, *Environ. Sci. Technol.*, **28**, 1186.

39. Kersting, A. B., Efurd, D. W., Finnegan, D. L., Rokop, D. J., Smith, D. K., and Thompson, J. L. (1999). Migration of plutonium in ground water at the Nevada Test Site, *Nature (London)*, **397**, 56.

40. Magaritz, M., Amiel, A. J., Ronen, D., and Wells, M. C. (1990). Distribution of metals in a polluted aquifer: A comparison of aquifer suspended material to fine sediments of the adjacent environment, *J. Contamin. Hydrol.*, **5**, 333.

41. Villholth, K. G. (1999). Colloid characterization and colloidal phase partitioning of polycyclic aromatic hydrocarbons in two creosote-contaminated aquifers in Denmark, *Environ. Sci. Technol.*, **33**, 691.

42. Smith, P. A. and Degueldre, C. (1993). Colloid-facilitated transport of radionuclides through fractures, *J. Contaminant Hydrol.*, **13**, 143.

43. Roy, S. B. and Dzombak, D. A. (1998). Sorption nonequilibrium effects on colloid-enhanced transport of hydrophobic organic compounds in porous media, *J. Contaminant Hydrol.*, **30**, 179.

44. Honeyman, B. D. (1999). Geochemistry—Colloidal culprits in contamination, *Nature (London)*, **397**, 23.

45. Puls, R. W., Clark, D.A., Bledsoe, B., Powell, R. M., and Paul, C. J. (1992). Metals in ground water: sampling artifacts and reproducibility, *Hazardous Waste Hazardous Mater.*, **9**, 149.

46. Backhus, D. A., Ryan, J. N., Groher, D. M., MacFarlane, J. K., and Gschwend, P. M. (1993). Sampling colloids and colloid-associated contaminants in ground water, *Ground Wat.*, **31**, 466.

47. Kearl, P. M., Korte, N. E., Stites, M., and Baker, J. (1994). Field comparison of micropurging vs. traditional ground water sampling, *Ground Wat. Monitoring Remediation*, 183.

48. Bangsund, W. J. and Kangas, K. W. (1998). Representative groundwater sampling using low-flow methods, *TAPPI J.*, **81(12)**, 129.

49. Creasey, C. L. and Flegal, A. R. (1999). Elemental analyses of groundwater: demonstrated advantage of low-flow sampling and trace-metal clean techniques over standard techniques, *Hydrogeol. J.*, **7**, 161.

50. Weisbrod, N., Ronen, D., and Nativ, R. (1996). New method for sampling groundwater colloids under natural gradient flow conditions, *Environ. Sci. Technol.*, **30**, 3094.

51. Penrose, W. R., Polzer, W. L., Essington, E. H., Nelson, D. M., and Orlandini, K. A. (1990). Mobility of plutonium and americium through a shallow aquifer in a semiarid region, *Environ. Sci. Technol.*, **24**, 228.

52. Marty, R. C., Bennett, D., and Thullen, P. (1997). Mechanism of plutonium transport in a shallow aquifer in Mortandad Canyon, Los Alamos National Laboratory, New Mexico, *Environ. Sci. Technol.*, **31**, 2020.

53. Gounaris, V., Anderson, P. R., and Holsen, T. M. (1993). Characteristics and environmental significance of colloids in landfill leachate, *Environ. Sci. Technol.*, **27**, 1381.

54. Kaplan, D. I., Bertsch, P. M., and Adriano, D. C. (1995). Facilitated transport of contaminant metals through an acidified aquifer, *Ground Wat.*, **33**, 708.
55. Seta, A. K. and Karathanasis, A. D. (1997). Atrazine adsorption by soil colloids and co-transport through subsurface environments, *Soil Sci. Soc. Am. J.*, **61**, 612.
56. von Gunten, H. R., Waber, U. E., and Krähenbühl, U. (1988). The reactor accident at Chernobyl: A possibility to test colloid-controlled transport of radionuclides in a shallow aquifer, *J. Contamin. Hydrol.*, **2**, 237.
57. Waber, U. E., Lienert, C., and von Gunten, H. R. (1990). Colloid-related infiltration of trace metals from a river to shallow groundwater, *J. Contamin. Hydrol.*, **6**, 251.
58. Lienert, C., Short, S. A., and von Gunten, H. R. (1994). Uranium infiltration from a river to shallow groundwater, *Geochim. Cosmochim. Acta*, **58**, 5455.
59. McCarthy, J. F. and Degueldre, C. (1993). Sampling and characterization of colloids and particles in groundwater for studying their role in contaminant transport. In *Environmental Particles*, Vol. 2, ed. Buffle, J. and van Leeuwen, H.P., Lewis, Boca Raton, FL, pp. 247–315.
60. Degueldre, C., Baeyens, B., Goerlich, W., Riga, J., Verbist, J., and Stadelmann, P. (1989). Colloids in water from a subsurface fracture in granitic rock, Grimsel Test Site, Switzerland, *Geochim. Cosmochim. Acta*, **53**, 603.
61. Kim, J. I. (1991). Actinide colloid generation in groundwater, *Radiochim. Acta*, **52/53**, 71.
62. Ho, C. H. and Miller, N. H. (1985). Effect of humic acid on uranium uptake by hematite particles, *J. Colloid Interface Sci.*, **106**, 281.
63. Saltelli, A., Avogadro, A., and Bidoglio, G. (1984). Americium filtration in glauconitic sand columns, *Nucl. Technol.*, **67**, 245.
64. Horzempa, L. M. and Helz, G. R. (1979). Controls on the stability of sulfide sols: colloidal covellite as an example, *Geochim. Cosmochim. Acta*, **43**, 1645.
65. Ravichandran, M., Aiken, G. R., Ryan, J. N., and Reddy, M. M. (1999). Inhibition of precipitation and aggregation of metacinnabar (mercuric sulfide) by dissolved organic matter isolated from the Florida Everglades, *Environ. Sci. Technol.*, **33**, 1418.
66. Benedetti, M. F. and Boulègue, J. (1991). Mechanism of gold transfer and deposition in a supergene environment, *Geochim. Cosmochim. Acta*, **55**, 1539.
67. Saunders, J. A. (1998). Silica and gold textures in bonanza ores of the Sleeper Deposit, Humboldt County, Nevada: Evidence for colloids and implications for epithermal ore-forming processes, *Econ. Geol.*, **89**, 628.
68. Bowell, R. J., Gize, A. P., and Foster, R. P. (1993). The role of fulvic acid in the supergene migration of gold in tropical rain forest soils, *Geochim. Cosmochim. Acta*, **57**, 4179.
69. Langmuir, D. (1969). Geochemistry of iron in a coastal-plain ground water of the Camden, New Jersey, area, *U.S. Geol. Survey Profess. Paper*, **650–C**, C224.
70. Liang, L., McCarthy, J. F., Jolley, L. W., McNabb, J. A., and Mehlhorn, T. L. (1993). Iron dynamics: Transformation of Fe(II)/Fe(III) during injection of natural organic matter in a sandy aquifer, *Geochim. Cosmochim. Acta*, **57**, 1987.
71. Gschwend, P. M. and Reynolds, M. D. (1987). Monodisperse ferrous phosphate colloids in an anoxic groundwater plume, *J. Contaminant Hydrol.*, **1**, 309.
72. Borkovec, M., Wu, Q., Degovics, G., Laggner, P., and Sticher, H. (1993). Surface area and size distributions of soil particles, *Colloids Surf. A: Physicochem. Eng. Aspects*, **73**, 65.
73. Grout, H., Wiesner, M. R., and Bottero, J. Y. (1999). Analysis of colloidal phases in urban stormwater runoff, *Environ. Sci. Technol.*, **33**, 831.
74. Kallay, N., Barouch, E., and Matijević, E. (1987). Diffusional detachment of colloidal particles from solid/solution interfaces, *Adv. Colloid Interface Sci.*, **27**, 1.

75. McDowell-Boyer, L. M. (1992). Chemical mobilization of micron-sized particles in saturated porous media under steady flow conditions, *Environ. Sci. Technol.*, **26**, 586.
76. Nocito-Gobel, J. and Tobiason, J. E. (1996). Effects of ionic strength on colloid deposition and release, *Colloids Surf. A: Physicochem. Eng. Aspects*, **107**, 223.
77. Ryan, J. N. and Gschwend, P. M. (1994). Effects of ionic strength and flow rate on colloid release: Relating kinetics to intersurface potential energy, *J. Colloid Interface Sci.*, **164**, 21.
78. Seaman, J. C., Bertsch, P. M., and Miller, W. P. (1995). Chemical controls on colloid generation and transport in a sandy aquifer, *Environ. Sci. Technol.*, **29**, 1808.
79. Kaplan, D. I., Sumner, M. E., Bertsch, P. M., and Adriano, D. C. (1996). Chemical conditions conducive to the release of mobile colloids from ultisol profiles, *Soil Sci. Soc. Am. J.*, **60**, 269.
80. Fauré, M.-H., Sardin, M., and Vitorge, P. (1996). Transport of clay particles and radioelements in a salinity gradient: Experiments and simulations, *J. Contamin. Hydrol.*, **21**, 255.
81. Fauré, M.-H., Sardin, M., and Vitorge, P. (1997). Release of clay particles from an unconsolidated clay-sand core: Experiments and modelling, *J. Contamin. Hydrol.*, **26**, 169.
82. Roy, S. B. and Dzombak, D. A. (1996). Colloid release and transport processes in natural and model porous media, *Colloids Surf. A: Physicochem. Eng. Aspects*, **107**, 245.
83. Roy, S. B. and Dzombak, D. A. (1998). Na^+–Ca^{2+} exchange effects in the detachment of latex colloids deposited in glass bead porous media, *Colloids Surf. A: Physicochem. Eng. Aspects*, **119**, 133.
84. Grolimund, D. and Borkovec, M. (1999). Long-term release kinetics of colloidal particles from natural porous media *Environ. Sci. Technol.*, **33**, 4054.
85. Khilar, K. C. and Fogler, H. S. (1984). The existence of a critical salt concentration for particle release, *J. Colloid Interface Sci.*, **101**, 214.
86. Sarkar, A. K. and Sharma, M. M. (1990). Fines migration in two-phase flow, *J. Petrol. Technol.*, 646.
87. Goldenberg, L. C., Magaritz, M., and Mandel, S. (1983). Experimental investigation on irreversible changes of hydraulic conductivity on the seawater–freshwater interface in coastal aquifers, *Wat. Resources Res.*, **19**, 77.
88. Goldenberg, L. C., Magaritz, M., Amiel, A. J., and Mandel, S. (1984). Changes in hydraulic conductivity of laboratory sand–clay mixtures caused by a seawater–freshwater interface, *J. Hydrol.*, **70**, 329.
89. Goldenberg, L. C. (1985). Decrease of hydraulic conductivity in sand at the interface between seawater and dilute clay suspensions, *J. Hydrol.*, **78**, 183.
90. Nightingale, H. I. and Bianchi, W. C. (1977). Ground-water turbidity resulting from artificial recharge, *Ground Wat.*, **15**, 146.
91. Brown, D. L. and Silvey, W. D. (1977). Artificial recharge to a freshwater-sensitive brackish-water sand aquifer, Norfolk, Virginia, *U.S. Geol. Survey Profess. Paper*, **939**.
92. Derjaguin, B. V. and Landau, L. (1941). Theory of the stability of strongly charged lyophobic sols and the adhesion of strongly charged particles in solutions of electrolytes, *Acta Physicochim. URSS*, **14**, 633.
93. Verwey, E. J. W. and Overbeek, J. Th. G. (1948). *Theory of the Stability of Lyophobic Colloids*, Elsevier, Amsterdam.
94. Ryan, J. N. and Gschwend, P. M. (1994). Effects of solution chemistry on clay colloid mobilization from an iron oxide-coated sand aquifer, *Environ. Sci. Technol.*, **28**, 1717.

95. Swartz, C. H. and Gschwend, P. M. (1998). Mechanisms controlling release of colloids to groundwater in a Southeastern Coastal Plain aquifer sand, *Environ. Sci. Technol.*, **32**, 1779.

96. Frenkel, H., Goertzen, J. O., and Rhoades, J. D. (1978). Effects of clay type and content, exchangeable sodium percentage, and electrolyte concentration on clay dispersion and soil hydraulic conductivity, *Soil Sci. Soc. Am. J.*, **42**, 32.

97. Shainberg, I., Rhoades, J. D., and Prather, R. J. (1980). Effect of low electrolyte concentration on clay dispersion and hydraulic conductivity of a sodic soil, *Soil Sci. Soc. Am. J.*, **45**, 273.

98. Swartz, C. H. and Gschwend, P. M. (1999). Field studies of in situ colloid mobilization in a Southeastern Coastal Plain aquifer, *Wat. Resources Res.*, **35**, 2213.

99. Allred, B. and Brown, G. O. (1994). Surfactant-induced reductions in soil hydraulic conductivity, *Ground Wat. Monitoring Remediation*, 174.

100. Allred, B. and Brown, G. O. (1995). *Surfactant-induced Reductions of Saturated Hydraulic Conductivity and Unsaturated Diffusivity, ACS Symp. Ser.*, **594**, American Chemical Society, Washington, DC, pp. 216–230.

101. Renshaw, C. E., Zynda, G. D., and Fountain, J. C. (1997). Permeability reductions induced by sorption of surfactant, *Wat. Resources Res.*, **33**, 371.

102. Wiesner, M. R., Grant, M. C., and Hutchins, S. R. (1996). Reduced permeability in groundwater remediation systems: Role of mobilized colloids and injected chemicals, *Environ. Sci. Technol.*, **30**, 3184.

103. Gschwend, P. M., Backhus, D. A., MacFarlane, J. K., and Page, A. L. (1990). Mobilization of colloids in groundwater due to infiltration of water at a coal ash disposal site, *J. Contamin. Hydrol.*, **6**, 307.

104. Ryan, J. N. and Gschwend, P. M. (1990). Colloid mobilization in two Atlantic Coastal Plain aquifers: Field studies, *Wat. Resources Res.*, **26**, 307.

105. Ryan, J. N. and Gschwend, P. M. (1992). Effect of iron diagenesis on the transport of colloidal clay in an unconfined sand aquifer, *Geochim. Cosmochim. Acta*, **56**, 1507.

106. O'Neill, M. E. (1968). A sphere in contact with a plane wall in a slow linear shear flow, *Chem. Eng. Sci.*, **23**, 1293.

107. Hubbe, M. A. (1984). Theory of detachment of colloidal particles from flat surfaces exposed to flow, *Colloids Surf.*, **12**, 151.

108. Sharma, M. M., Chamoun, H., Sita Rama Sarma, D. S. H., and Schechter, R. S. (1992). Factors controlling the hydrodynamic detachment of particles from surfaces, *J. Colloid Interface Sci.*, **149**, 121.

109. Amirtharajah, A. and Raveendran, P. (1993). Detachment of colloids from sediments and sand grains, *Colloids Surf. A: Physicochem. Eng. Aspects*, **73**, 211.

110. Das, S. K., Schechter, R. S., and Sharma, M. M. (1994). The role of surface roughness and contact deformation on the hydrodynamic detachment of particles from surfaces, *J. Colloid Interface Sci.*, **164**, 63.

111. Govindaraju, R. S., Reddi, L. N., and Kasavaraju, S. K. (1995). A physically based model for mobilization of kaolinite particles under hydraulic gradients, *J. Hydrol.*, **172**, 331.

112. Ledin, A., Karlsson, S., Düker, A., and Allard, B. (1994). Measurements in situ of concentration and size distribution of colloidal matter in deep groundwaters by photon correlation spectroscopy, *Wat. Research*, **28**, 1539.

113. Degueldre, C., Pfeiffer, H.-R., Alexander, W., Wernli, B., and Bruetsch, R. (1996). Colloid properties in granitic groundwater systems. I: Sampling and characterisation, *Appl. Geochem.*, **11**, 677.

114. Kaplan, D. I., Bertsch, P. M., and Adriano, D. C. (1997). Mineralogical and physicochemical differences between mobile and nonmobile colloidal phases in reconstructed pedons, *Soil Sci. Soc. Am. J.*, **61**, 641.
115. Seaman, J. C., Bertsch, P. M., and Strom, R. N. (1997). Characterization of colloids mobilized from southeastern coastal plain sediments, *Environ. Sci. Technol.*, **31**, 2782.
116. Atteia, O., Perret, D., Adatte, T., Kozel, R., and Rossi, P. (1998). Characterization of natural colloids from a river and spring in a karstic basin, *Environ. Geol.*, **34**, 257.
117. Baghernejad, M. and Dalrymple, J. B. (1993). Colloidal suspensions of calcium-carbonate in soils and their likely significance in the formation of calcic horizons, *Geoderma*, **58**, 17.
118. Turrero, M.-J., Gómez, P., Pérez del Villar, L., Moulin, V., Magonthier, M.-C., and Menager, M.-T. (1995). Relation between colloid composition and the environment of their formation: application to the El Berrocal site (Spain), *Appl. Geochem.*, **10**, 119.
119. Davis, J. A. and Kent, D. B. (1990). Surface complexation modeling in aqueous geochemistry. In *Mineral–Water Interface Geochemistry*, ed. Hochella, M. F. Jr. and White, A. F., Mineralogical Society of America, *Rev. Mineral.*, **23**, 177.
120. Dzombak, D. A. and Morel, F. M. M. (1990). *Surface Complexation Modeling: Hydrous Ferric Oxide*, John Wiley & Sons, New York.
121. Stumm, W. (1992). *Chemistry of the Solid-Water Interface*, Wiley-Interscience, New York.
122. Senesi, N. (1992). Metal-humic substance complexes in the environment. Molecular and mechanistic aspects by multiple spectroscopic approach. In *Biogeochemistry of Trace Metals*, ed. Adriano, D.C., Lewis, Boca Raton, FL, 429–496.
123. Senesi, N. (1993). Nature of interactions between organic chemicals and dissolved humic substances and the influence of environmental factors. In *Organic Substances in Soil and Water: Natural Constituents and Their Influences on Contaminant Behavior*, ed. Beck, A.J., Jones, K.C., Hayes, M.H.B., and Mingelgrin, U., Royal Society of Chemistry, Cambridge, 73–101.
124. Schwarzenbach, R. P., Gschwend, P. M., and Imboden, D. M. (1993). *Environmental Organic Chemistry*, John Wiley & Sons, New York.
125. Dugger, D. L., Stanton, J. H., Irby, B. N., McConnell, B. L., Cummings, W. W., and Maatman, R. W. (1964). The exchange of twenty metal ions with the weakly acidic silanol group of silica gel, *J. Phys. Chem.*, **68**, 757.
126. Ledin, A., Karlsson, S., Düker, A., and Allard, B. (1994). The adsorption of europium to colloidal iron oxyhydroxides and quartz—the impact of pH and an aquatic fulvic acid, *Radiochim. Acta*, **66/67**, 213.
127. Strawn, D. G., Scheidegger, A., and Sparks, D. L. (1998). Kinetics and mechanisms of Pb(II) sorption and desorption at the aluminum oxide-water interface, *Environ. Sci. Technol.*, **32**, 2596.
128. Lu, N., Cotter, C. R., Kitten, H. D., Bentley, J., and Triay, I. R. (1998). Reversibility of sorption of plutonium-239 onto hematite and goethite colloids, *Radiochim. Acta*, **83**, 167.
129. Bunzl, K., Flessa, H., Kracke, W., and Schimmack, W. (1995). Association of fallout $^{239+240}$Pu and ^{241}Am with various soil components in successive layers of a grassland soil, *Environ. Sci. Technol.*, **29**, 2513.
130. Cerling, T. E. and Turner, R. R. (1982). Formation of freshwater Fe–Mn coatings on gravel and the behavior of ^{60}Co, ^{90}Sr, and ^{137}Cs in a small watershed, *Geochim. Cosmochim. Acta*, **46**, 1333.

131. Torok, J., Buckley, L. P., and Woods, B. L. (1990). The separation of radionuclide migration by solution and particle transport in soil, *J. Contaminant Hydrol.*, **6**, 185.
132. Smith, J. T. and Comans, R. N. J. (1996). Modelling the diffusive transport and remobilisation of ^{137}Cs in sediments: The effects of sorption kinetics and reversibility, *Geochim. Cosmochim. Acta*, **60**, 995.
133. Tanaka, T. and Ohnuki, T. (1996). Colloidal migration behavior of radionuclides sorbed on mobile fine soil particles through a sand layer, *J. Nucl. Sci. Technol.*, **33**, 62.
134. Saiers, J. E. and Hornberger, G. M. (1996). The role of colloidal kaolinite in the transport of cesium through laboratory sand columns, *Wat. Resources Res.*, **32**, 33.
135. Noell, A. L., Thompson, J. L., Corapcioglu, M. Y., and Triay, I. R. (1998). The role of silica colloids on facilitated cesium transport through glass bead columns and modeling, *J. Contamin. Hydrol.*, **31**, 23.
136. Grossl, P. R., Eick, M. J., Sparks, D. L., Goldberg, S., and Ainsworth, C. C. (1997). Arsenate and chromate retention mechanisms on goethite. 2. Kinetic evaluation using a pressure-jump relaxation technique, *Environ. Sci. Technol.*, **31**, 321.
137. Karickhoff, S. W. (1981). Semi-empirical estimation of sorption of hydrophobic pollutants on natural sediments and soils, *Chemosphere*, **10**, 833.
138. Piatt, J. J., Backhus, D. A., Capel, P. D., and Eisenreich, S. J. (1996). Temperature-dependent sorption of naphthalene, phenanthrene, and pyrene to low organic carbon aquifer sediments, *Environ. Sci. Technol.*, **30**, 751.
139. Chin, Y.-P., Aiken, G. R., and Danielsen, K. M. (1997). Binding of pyrene to aquatic and commercial humic substances: The role of molecular weight and aromaticity, *Environ. Sci. Technol.*, **31**, 1630.
140. Backhus, D. A. and Gschwend, P. M. (1990). Fluorescent polycyclic aromatic hydrocarbons as probes of studying the impact of colloid on pollutant transport in groundwater, *Environ. Sci. Technol.*, **24**, 1214.
141. Mackay, A. A., Chin, Y.-P., MacFarlane, J. K., and Gschwend, P. M. (1996). Laboratory assessment of BTEX soil flushing, *Environ. Sci. Technol.*, **30**, 3223.
142. Werth, C. J. and Reinhard, M. (1997). Effects of temperature on trichloroethylene desorption from silica gel and natural sediments. 2. Kinetics, *Environ. Sci. Technol.*, **31**, 697.
143. Cornelissen, G., van Noort, P. C. M., Parsons, J. R., and Govers, H. A. J. (1997). Temperature dependence of slow adsorption and desorption kinetics of organic compounds in sediments, *Environ. Sci. Technol.*, **31**, 454.
144. Cornelissen, G., van Noort, P. C. M., and Govers, H. A. J. (1998). Mechanism of slow desorption of organic compounds from sediments: A study using model sorbents, *Environ. Sci. Technol.*, **32**, 3124.
145. Schlebaum, W., Schraa, G., and van Riemsdijk, W. (1999). Influence of nonlinear sorption kinetics on the slow-desorbing organic contaminant fraction in soil, *Environ. Sci. Technol.*, **33**, 1413.
146. Aochi, Y. O. and Farmer, W. J. (1997). Role of microstructural properties in the time-dependent sorption/desorption behavior of 1,2-dichloroethane on humic substances, *Environ. Sci. Technol.*, **31**, 2520.
147. Culver, T. B., Hallissey, S. P., Sahoo, D., Deitsch, J. J., and Smith, J. A. (1997). Modeling the desorption of organic contaminants from long-term contaminated soil using distributed mass transfer rates, *Environ. Sci. Technol.*, **31**, 1581.
148. Ten Hulscher, Th. E. M., Vrind, B. A., van den Heuvel, H., van der Velde, L. E., Beurskens, J. E. M., and Govers, H. A. J. (1999). Triphasic desorption of highly resistant chlorobenzenes, polychlorinated biphenyls, and polycyclic aromatic hydrocarbons in field contaminated soil, *Environ. Sci. Technol.*, **33**, 126.

149. de Jonge, H., Jacobsen, O. H., de Jonge, L. W., and Moldrup, P. (1998). Particle-facilitated transport of prochloraz in undisturbed sandy loam soil columns, *Soil Sci. Soc. Am. J.*, **27**, 1495.
150. van de Weerd, H. and Leijne, A. (1997). Assessment of the effect of kinetics on colloid facilitated transport in porous media, *J. Contamin. Hydrol.*, **26**, 245.
151. van de Weerd, H., Leijne, A., and van Riemsdijk, W. (1998). Transport of reactive colloids and contaminants in groundwater: Effect of nonlinear kinetic interactions, *J. Contamin. Hydrol.*, **32**, 313.
152. McCaulou, D. R., Bales, R. C., and McCarthy, J. F. (1994). Use of short-pulse experiments to study bacteria transport through porous media, *J. Contamin. Hydrol.*, **15**, 1.
153. Saiers, J. E., Hornberger, G. M., and Harvey, C. (1994). Colloidal silica transport through structured, heterogeneous porous media, *J. Hydrol.*, **163**, 271.
154. Grolimund, D., Elimelech, M., Borkovec, M., Barmettler, K., Kretzschmar, R., and Sticher, H. (1998). Transport of in situ mobilized colloidal particles in packed soil columns, *Environ. Sci. Technol.*, **32**, 3562.
155. Elimelech, M. (1991). Kinetics of capture of colloidal particles in packed beds under attractive double layer interactions, *J. Colloid Interface Sci.*, **146**, 337.
156. Song, L. and Elimelech, M. (1993). Dynamics of colloid deposition in porous media: Modeling the role of retained particles, *Colloids Surf. A: Physicochem. Eng. Aspects*, **73**, 49.
157. Toran, L. and Palumbo, A. V. (1992). Colloid transport through fractured and unfractured laboratory sand columns, *J. Contamin. Hydrol.*, **9**, 289.
158. Freeze, R. A. (1975). A stochastic-conceptual analysis of one-dimensional groundwater flow in nonuniform homogeneous media, *Wat. Resources Res.*, **11**, 725.
159. Sudicky, E. A. (1986) A natural gradient experiment on solute transport in a sand aquifer: Spatial variability of hydraulic conductivity and its role in the dispersion process, *Wat. Resources Res.*, **22**, 2069.
160. Painter, S. (1996). Evidence for non-Gaussian scaling behavior in heterogeneous sedimentary formations, *Wat. Resources Res.*, **32**, 1183.
161. Liu, H. H. and Molz, F. J. (1997). Comment on 'Evidence for non-Gaussian scaling behavior in heterogeneous sedimentary F. J. formations' by Scott Painter, *Wat. Resources Res.*, **33**, 907.
162. Kessler, J. H. and Hunt, J. R. (1994). Dissolved and colloidal contaminant transport in a partially clogged fracture, *Wat. Resources Res.*, **30**, 1195.
163. Sun, N., Elimelech, M., Sun, N. -Z., and Ryan, J. N. (2001). A two-dimensional model of colloid transport in physically and geochemically heterogeneous porous media, *J. Contamin. Hydrol.*, **49**, 173.
164. Elimelech, M., Gregory, J., Jia, X., and Williams, R. (1995). *Particle Deposition and Aggregation. Measurement, Modelling and Simulation*, Butterworth-Heinemann, Oxford.
165. Yao, K.-M., Habibian, M. T., and O'Melia, C. R. (1971). Water and wastewater treatment filtration: Concepts and applications, *Environ. Sci. Technol.*, **5**, 1105.
166. Elimelech, M. and O'Melia, C. R. (1990). Kinetics of deposition of colloidal particles in porous media, *Environ. Sci. Technol.*, **24**, 1528.
167. Rajagopalan, R. and Tien, C. (1976). Trajectory analysis of deep-bed filtration with the sphere-in-cell porous media model, *Am. Inst. Chem. Eng. J.*, **22**, 523.
168. Harvey, R. W. and Garabedian, S. P. (1991). Use of colloid filtration theory in modeling movement of bacteria through a contaminated sandy aquifer, *Environ. Sci. Technol.*, **25**, 178.

169. Hiemenz, P. C. (1986). *Principles of Colloid and Surface Chemistry*, Marcel Dekker, New York.

170. Song, L., Johnson, P. R., and Elimelech, M. (1994). Kinetics of colloid deposition onto heterogeneously charged surfaces in porous media, *Environ. Sci. Technol.*, **28**, 1164.

171. Coston, J. A., Fuller, C. C., and Davis, J. A. (1995). Pb^{2+} and Zn^{2+} adsorption by a natural aluminum- and iron-bearing surface coating on aquifer sand, *Geochim. Cosmochim. Acta*, **59**, 3535.

172. Ryan, J. N., Elimelech, M., Ard, R. A., Harvey, R. W., and Johnson, P. R. (1999). Bacteriophage PRD1 and silica colloid transport and recovery in an iron oxide-coated sand aquifer, *Environ. Sci. Technol.*, **33**, 63.

173. Johnson, P. R., Sun, N., and Elimelech, M. (1996). Colloid transport in geochemically heterogeneous porous media: Modeling and measurements, *Environ. Sci. Technol.*, **30**, 3284.

174. Puls, R. W. and Powell, R. M. (1992). Transport of inorganic colloids through natural aquifer material: Implications for contaminant transport, *Environ. Sci. Technol.*, **26**, 614.

175. Grolimund, D., Borkovec, M., Barmettler, K., and Sticher, H. (1996). Colloid-facilitated transport of strongly sorbing contaminants in a natural porous media: A laboratory column study, *Environ. Sci. Technol.*, **30**, 3118.

176. Seta, A. K. and Karathanasis, A. D. (1997). Stability and transportability of water-dispersible soil colloids, *Soil Sci. Soc. Am. J.*, **61**, 604.

Index

References to figures have the suffix 'f' and those to tables 't'.